BACTERIAL SPORE FORMERS
PROBIOTICS and EMERGING APPLICATIONS

Edited by:

Ezio Ricca, *Università Federico II, Naples, Italy*
Adriano O. Henriques, *Universidade Nova de Lisboa, Portugal*
and Simon M. Cutting, *Royal Holloway, University of London, UK*

🌐 *horizon bioscience*

Copyright © 2004
Horizon Bioscience
32 Hewitts Lane
Wymondham
Norfolk NR18 0JA
U.K.

www.horizonbioscience.com

British Library Cataloguing-in-Publication Data

A catalogue record for this book is available from the British Library

ISBN: 1-904933-02-5

Printed and bound in Great Britain by Cromwell Press, Trowbridge, Wiltshire.

Contents

Section I. General Aspects

Section II. Spores as Probiotics

Section III. Spores as Vaccine Vehicles and in Therapeutics

Appendices

Contributors

Jozef Anné
Laboratory of Bacteriology
Rega Institute for Medical Research
Leuven B-3000
Belgium

Rudolf R. Azizbekyan*
State Scientific Center
State Institute of Genetics and Selection of Industrial
Microorganisms
1-st. Dorozhny proezd b.1
Moscow 117545
Russia

Imrich Barák*
Institute of Molecular Biology
Slovak Academy of Sciences
Dúbravská cesta 21
845 51 Bratislava 45
Slovak Republic

Sofie Barbé
Laboratory of Bacteriology
Rega Institute for Medical Research
Leuven B-3000
Belgium

Teresa Barbosa
Instituto de Tecnologia Química e Biológia
Avenida da República
Apartado 127
2781-901 Oeiras Codex
Portugal

Giuseppina Cangiano
Dipartimento di Fisiologia Generale ed Ambientale
Università Federico II
via Mezzocannone 16
80134 Napoli
Italy

Stephen T. Cartman*
Department of Food and Environmental Safety
Veterinary Laboratories Agency (Weybridge)
Woodham Lane, Addlestone, New Haw, Surrey
KT15 3NB
UK

Marco Cassone
Laboratorio di Microbiologia Molecolare e Biotecnologia
(LAMMB)
Dipartimento di Biologia Molecolare
Università di Siena
53100 Siena
Italy

Annalisa Ciabattini
Laboratorio di Microbiologia Molecolare e Biotecnologia
(LAMMB)
Dipartimento di Biologia Molecolare
Università di Siena
53100 Siena
Italy

Teresa V. Costa
Instituto de Tecnologia Química e Biológica
Universidade Nova de Lisboa
Avenida da República
Apartado 127
2781-901 Oeiras Codex
Portugal

Simon M. Cutting*
School of Biological Sciences
Royal Holloway, University of London
Egham, Surrey, TW20 0EX
UK

Adam Driks*
Dept. of Microbiology and Immunology
Loyola University Medical Center
2160 S. First Ave.
Maywood, IL 60153.
USA

Le Hong Duc*
School of Biological Sciences
Royal Holloway, University of London
Egham
Surrey, TW20 0EX
UK

Maurilio De Felice
Dipartimento di Fisiologia Generale ed Ambientale
Università Federico II
via Mezzocannone 16
80134 Napoli
Italy

Andreas Felske*
GBF (German Research Centre for Biotechnology)
Division of Microbiology
Mascheroder Weg 1
D-38124 Braunschweig
Germany

Dagmar Fritze*
DSMZ-Deutsche Sammlung von Mikroorganismen und
Zellkulturen
Mascheroder Weg 1b
38124 Braunschweig
Germany

Alan C. Gange*
School of Biological Sciences
Royal Holloway
University of London
Egham, Surrey
TW20 0EX
UK

Karen J. Hagley
School of Biological Sciences
Royal Holloway
University of London
Egham, Surrey
TW20 0EX
UK

Adriano O. Henriques*
Instituto de Tecnologia Química e Biológica
Universidade Nova de Lisboa
Avenida da República
Apartado 127
2781-901 Oeiras Codex
Portugal

Huynh Anh Hong
School of Biological Sciences
Royal Holloway, University of London
Egham
Surrey, TW20 0EX
UK

Tomohiro Hosoi*
Tokyo Metropolitan Food Technology Research Center
1-9 Kanda Sakuma-cho
Chiyoda-ku
Tokyo 101-0025
Japan

Rachele Isticato
Dipartimento di Fisiologia Generale ed Ambientale
Università Federico II
via Mezzocannone 16
80134 Napoli
Italy

Kan Kiuchi
Department of Food Science and Nutrition
Faculty of Home Economics
Kyoritsu Women's University
2-2-1 Hitotsubashi
Chiyoda-ku
Tokyo 101-8433
Japan

Roberto M. La Ragione
Department of Food and Environmental Safety
Veterinary Laboratories Agency (Weybridge)
Woodham Lane, Addlestone, New Haw, Surrey
KT15 3NB
UK

Philippe Lambin
MAASTRO Lab
Maastricht Radiation Oncology
UNS 50/23
P.O. 616
6200 MD Maastricht
The Netherlands

Niall A. Logan*
Department of Biological and Biomedical Sciences
Glasgow Caledonian University
Cowcaddens Road
Glasgow G4 0BA
UK

Lígia O. Martins
Instituto de Tecnologia Química e Biológica
Universidade Nova de Lisboa
Avenida da República
Apartado 127
2781-901 Oeiras Codex
Portugal

Elinor McCartney*
Pen and Tec Consulting
Carrer Ordal 14
Mirasol
08195 Sant Cugat del Vallés
Barcelona
Spain

Lieve Van Mellaert
Laboratory of Bacteriology
Rega Institute for Medical Research
Leuven B-3000
Belgium

Nigel P. Minton*
Institute of Infection, Immunity and Inflammation
Queens Medical Centre
Nottingham
NG2 6DT
UK·

Peter Mullany*
Division of Infection and Immunity
Eastman Dental Institute for Oral Health Care Sciences
University College London (UCL)
University of London
256 Gray's Inn Road
London, WC1X 8LD
UK

Wayne L. Nicholson*
Department of Microbiology and Cell Science
University of Florida
Space Life Sciences Laboratory
Building M6-1025/SLSL
Kennedy Parkway & 5th Street
Kennedy Space Center, FL 32899
USA

Marco R. Oggioni*
Laboratorio di Microbiologia Molecolare e Biotecnologia
(LAMMB)
Dipartimento di Biologia Molecolare
Università di Siena
53100 Siena
Italy

Oliver J. Pennington
Institute of Infection, Immunity and Inflammation
Queens Medical Centre
Nottingham
NG2 6DT
UK

Irina Pinchuk
Laboratoire de Microbiologie
ENITA-University of Bordeaux
1, cours du Général de Gaulle
33175 Gradignan
France

Gianni Pozzi
Laboratorio di Microbiologia Molecolare e Biotecnologia
(LAMMB)
Dipartimento di Biologia Molecolare
Università di Siena
53100 Siena
Italy

Sirirat Rengpipat
Department of Microbiology
Faculty of Science
Chulalongkorn University
Bangkok 10330
Thailand

Ezio Ricca*
Dipartimento di Fisiologia Generale ed Ambientale
Università Federico II
via Mezzocannone 16
80134 Napoli
Italy

Adam P. Roberts
Division of Infection and Immunity
Eastman Dental Institute for Oral Health Care Sciences
University College London (UCL)
University of London
256 Gray's Inn Road
London, WC1X 8LD
UK

Karen Scott
Microbial Genetics
Gut Microbiology and Immunology Division
Rowett Research Institute
Greenburn Road
Bucksburn
Aberdeen
AB21 9SB
UK

Sonia Senesi*
Dipartimento di Patologia Sperimentale
Biotecnologie Mediche, Infettivologia ed Epidemiologia
Università di Pisa
Via San Zeno, 37-39
56126 Pisa
Italy

Cláudia R. Serra
Instituto de Tecnologia Química e Biológica
Universidade Nova de Lisboa
Avenida da República
Apartado 127
2781-901 Oeiras Codex
Portugal

Jan Theys
MAASTRO Lab
Maastricht Radiation Oncology
UNS 50/23
P.O. 616
6200 MD Maastricht
The Netherlands

Maria C. Urdaci*
Laboratoire de Microbiologie
ENITA-University of Bordeaux
1, cours du Général de Gaulle
33175 Gradignan
France

Rita Zilhão
Instituto de Tecnologia Química e Biológica
Universidade Nova de Lisboa
Avenida da República
Apartado 127
2781-901 Oeiras Codex
Portugal

* Corresponding author

Foreword

Those of us who study bacteria in the laboratory sometimes lose sight of how they behave in their natural settings and how they interact with other organisms. Microorganisms, such as the standard laboratory bacteria *Bacillus subtilis, Escherichia coli, Salmonella typhimurium,* and *Caulobacter crescentus,* are prized as powerful model systems for elucidating fundamental mechanisms of gene control, metabolism, cell division, motility, behavior and so forth. This fascinating and eclectic book reminds us of the artificial nature of laboratory studies and the full richness of the biology and practical applications of microorganisms when viewed in their natural context. The principal subject of **Bacterial Spore Formers: Probiotics and Emerging Applications** is members of the genus *Bacillus* and related genera, which are capable of producing a remarkable type of resting cell called the *endo*spore that can survive extremes of time and environment. Most laboratory investigations of endospore-forming bacteria are carried out on descendents of a single strain of *B. subtilis.* As aptly put by Wayne Nicholson in the opening chapter (1), "We [that is, those of us who study *B. subtilis* as a model system] have traded breadth for depth....many spore researchers are largely unaware of the staggering variety of environmental niches and life styles of which *Bacillus* spp. are capable..." Nicholson then proceeds to take us on an extraordinary journey from the soil to the guts of stingless bees to extravagant claims of spore longevity to space travel and theories of the origin of life. Next, valuable chapters on systematics and ecology by Dagmar Fritze (2) and Andreas Felske (3) introduce us to the full range of species (over 200 and growing) in the order Bacillales. The most notorious of all *Bacillus* species is *B. anthracis,* the causative agent of anthrax and the organism for which the ability of a bacterium to cause a disease was established. In chapter 4, Marco Oggioni and colleagues remind us of Koch's famous postulates and teach us about the diseases caused by *B. anthrax* and it relatives, such as the insect pathogen *B. thuringiensis.* Chapter 8 by Niall Logan extends this treatment of pathogenicity in a comprehensive review of the involvement of aerobic endospore-forming bacteria in human infections.

The most conspicuous feature of members of the Bacillales is, of course, their capacity to produce endospores, which their co-discoverer Robert Koch described as "pearls" (because of their striking oval, light-refracting appearance). And indeed chapters 5, 6, 7, and 9 are devoted to endospores. Imrich Barák (5) provides a comprehensive review of the molecular genetics of spore formation while Adriano Henriques (6) and colleagues and Adam Driks (7) provide a detailed account of the outer layer, the coat, and surface features of the spore, including the outer sack known as the exosporium that encases certain spores. We learn that the coat is more dynamic than generally recognized and that spores from different species are strikingly different in their outer structures, perhaps representing adaptations to diverse environmental niches. Fascinating chapters by Huynh Hong and Le Duc (9) and by Teresa Barbosa and colleagues (16) summarize evidence indicating that endospore-forming bacteria live in the gastrointestinal track of animals where they carry out repeated cycles of spore-formation, germination, growth, and re-sporulation and describe the properties of isolates of gut endospore formers. The theme of life in the gastrointestinal tract is extended by Peter Mullany and colleagues (chapter 10) who consider the various ways that DNA can be transferred between endospore-forming bacteria in their natural biological setting. **Bacterial Spore Formers: Probiotics and Emerging Applications** culminates with six chapters on practical applications of endospore-forming bacteria. Chapters by Sonia Senesi (11) and Tomohiro Hosoi and Kan Kiuchi (12) review the use of *Bacillus* spores as probiotics with possible therapeutic benefits for humans. Likewise, chapters by Stephen Cartman and Roberto La Ragione (13) and by Alan Gange and Karen Hagley (14) present the cases for the use of spores as alternatives to antibiotics in animal feed supplements and as plant protectants, respectively. One of the mechanisms by which probiotics are believed to exert their beneficial effects is through the production of antimicrobial agents, a topic that is reviewed in chapter 15 by Maria Urdaci and Irina Pinchuk. The extraordinary stability of spores makes them attractive as delivery agents for antigens and industrially important enzymes. Rachele Isticato and colleagues (17) bring us up to date on the use of fusions to coat proteins to display partner proteins on the spore surface and Simon Cutting (18) reports on successful strategies for using spores as oral vaccines. Most of the book is devoted to aerobic, spore-forming bacteria but the final chapter by Oliver Pennington and colleagues (19) reviews the remarkable use of clostridial spores in tumor therapy where the strict anaerobiosis of the bacterium restricts spore outgrowth to the hypoxic centers of solid tumors. Three valuable appendices provide a catalog of commercial probiotic products and of regulations governing their use.

I recommend **Bacterial Spore Formers: Probiotics and Emerging Applications** both to basic microbiologists, who will find it to be an eye-opener, and to industrial and medical microbiologists, who will find it to be a most valued resource.

Richard Losick
The Biological Laboratories
Harvard University

Books of Related Interest

Full details of all these books at: www.horizonbioscience.com

Chapter 1

Ubiquity, Longevity, and Ecological Roles of *Bacillus* Spores

Wayne L. Nicholson

SUMMARY

Our preceptions of bacterial spores have been undergoing recent dramatic changes, in part due to the upsurge of interest in the uses of spores of *Bacillus* spp. as probiotics in plant and animal health. In this chapter I will review the occurrence and diverse roles of *Bacillus* spp. and their endospores in the environment, with particular emphasis on the worldwide geographic distribution of spores, the ecology of spores in various niches, spore longevity, host-symbiont and host-parasite interactions among spores and higher organisms, and human exploitation of spores.

INTRODUCTION: A BRIEF HISTORY OF SPORE RESEARCH

Our perception of the importance and uses of bacterial spores has been undergoing a continuous and dramatic evolution since the independent discoveries of the spore by Cohn, Koch, and Tyndall in the latter part of the 19th century (Cohn, 1876; Koch, 1876; Tyndall, 1877). Historically, spores were best known as agents of human and animal disease (e.g., *Clostridium perfringens, C. tetani, Bacillus anthracis*) or food poisoning (e.g., *C. botulinum, B. cereus*). In keeping with the classical medical doctrine that "the only good bug is a dead bug", throughout the late 19th and early 20th century spores first became established as biodosimeters (e.g., *B. subtilis, B. stearothermophilus*) to assure the efficacy of autoclaves, disinfectants, and food preservation processes. Also during the first half of the 20th century, *B. anthracis* spores were developed and produced as biological weapons, an event the consequences of which we are dealing with to this day. Post World War II, spores and their associated products were first recognized and developed as natural insecticides (e.g., *B. popilliae, B. thuringiensis*) and several *Bacillus* spp. began to be exploited in the industrial production of postexponentially-expressed enzymes, antibiotics, and other secondary metabolites (e.g., *B. subtilis, B. polymyxa*).

In the late 1940's and throughout the 1950's, spurred by the dawning age of molecular biology, there began a period of increased interest in studying the mechanisms of cellular differentiation in spore-forming bacteria and the resistance properties of bacterial spores. These early studies concentrated on trying to discern common themes of differentiation using a variety of mesophilic and moderately thermophilic spore-formers, most notably *B. cereues, B. megaterium, B. licheniformis, B. stearothermophilus,* and *B. subtilis,* as well as various *Clostridium* spp. (for representative compilations of organisms studied towards the end of this time, see Halvorson,

1961), but progress in the field during this period was hampered by the lack of a *Bacillus* spp. with an easily-manipulated genetic system.

The multi-species approach to spore research was soon to become heavily skewed in favor of the *B. subtilis* model system as the result of a rapid succession of advances; Spizizen's discovery of natural DNA-mediated transformation in *B. subtilis* (Spizizen, 1958), Schaeffer's isolation and genetic analysis of *B. subtilis* sporulation mutants (Schaeffer et al., 1959), and Takahashi's discovery and demonstration of the utility of generalized *B. subtilis* transducing phage PBS-1 (Takahashi, 1961). In the following four decades, *B. subtilis* emerged as the second most extensively-studied bacterium (after *Escherichia coli*), and by far the best-understood Gram-positive microorganism.

Today, with a few notable exceptions, the vast majority of our knowledge about bacterial spores comes from laboratory studies of the descendents of a single microorganism, *B. subtilis* strain 168. We have traded breadth for depth, and as a result many spore researchers are largely unaware of the staggering variety of environmental niches and life styles of which *Bacillus* spp. are capable, and the new uses for spores which are only beginning to be exploited. One of the major goals of this book is to introduce the reader to some of these recent developments and trends in applied spore research. In this introductory chapter I will make an attempt to consider some of the known environmental roles and human uses of bacterial spores, concentrating mostly upon spores of *Bacillus* spp., the bacteria about which we possess the most information, and including other aerobic sporeformers as necessary. This review will build upon, revise, and update some of the themes which I have explored in previous reviews (Nicholson and Fajardo-Cavazos, 1997; Nicholson et al., 2000; Nicholson, 2002; Nicholson et al., 2002).

A BRIEF OVERVIEW OF SPORULATION, DORMANCY, AND GERMINATION

After half a century of intensive research in numerous laboratories, we have built a fairly detailed picture of the genetic, biochemical, and physiologic processes which occur during the differentiation of an actively-growing *Bacillus subtilis* cell into a dormant endospore (reviewed extensively in Sonenshein et al., 1993; Sonenshein et al., 2002). In the laboratory, *Bacillus* spores are usually prepared by cultivating the bacterium in batch culture at 37°C in a nutrient broth-based sporulation medium at a high growth rate and to high cell density until some essential nutrient, such as the carbon source, is exhausted from the medium (Schaeffer et al., 1965; Nicholson and Setlow,

1990). Nutrient exhaustion is sensed by the cell and triggers a cascade of coordinated genetic and biochemical events which culminate in the formation of an endospore which is highly resistant to harsh environmental insults such as heat, radiation, desiccation, chemical and enzymatic attack. These spores remain viable for many years when stored either in the dry state or in aqueous suspension. Although metabolically dormant, the spore is capable of constantly monitoring the nutritional status of its environment, and in the laboratory responds to rehydration and the appearance of exogenous nutrients by rapidly germinating and resuming vegetative growth. The details of these developmental processes are discussed extensively in Sonenshein et al. (2002).

Although very little is known about the growth, sporulation, or germination of *Bacillus* spp. in their natural habitats, it is not difficult to envision that the process probably differs significantly from the manner in which spores are prepared in the laboratory. Growth of sporeforming bacteria in their natural environments (e.g. soil, decaying organic matter, plant surfaces, insect and mammalian guts, etc.): (i) is almost certainly slower; (ii) probably takes place as microcolonies on and within solid substrates such as large soil particle aggregates (Drazkiewicz, 1994; Nicholson and Law, 1999); (iii) is subject to wide variations in temperature, humidity, nutrient and oxygen availability, and; (iv) probably occurs within multi-species consortia and/or in direct competition with other micro- and macroorganisms (Nicholson and Fajardo-Cavazos, 1997). How closely the laboratory models of *B. subtilis* differentiation and spore resistance properties approximate the biology of Bacilli in the environment is virtually unknown, but work in our lab has in recent years demonstrated that significant differences indeed exist between the laboratory model and the environmental situation, at least with regard to the spore UV resistance paradigm (Xue and Nicholson, 1996; Nicholson and Law, 1999; Riesenman and Nicholson, 2000; Slieman and Nicholson, 2000, 2001; reviewed in Nicholson and Fajardo-Cavazos, 1997; Nicholson et al., 2000, Nicholson, 2002; Nicholson et al., 2002).

UBIQUITY OF SPORES IN THE ENVIRONMENT

In considering the question "What is the role of *Bacillus* spores in the environment?", a simple and obvious answer immediately presents itself: to preserve and to propagate the genetic information contained within the bacterium. Based on the common observation that the most efficient method of inducing sporulation in the laboratory is by nutrient limitation, it is generally accepted that spore formation evolved as a mechanism for both spatial and temporal escape from local conditions unfavorable to growth (Priest, 1993; Slepecky and Leadbetter, 1994; Nicholson et al., 2000). As a device for preserving and dispersing genetic information in the environment, the spore is an incredible success. *Bacillus* spores can be found in environmental samples obtained from virtually anywhere on Earth, from the poles to the tropics, from oceans and freshwater environments, from the atmosphere, the terrestrial surface and deep subsurface (Priest, 1993; Nicholson et al., 2000; see Istock et al., 2001 for a detailed investigation of worldwide geographical distribution of *B. subtilis*). As

pointed out by Priest (1993), due to the resistance, longevity and widespread dispersal of spores it is at times uncertain whether spores found in environmental samples actually live in and contribute to the environment or have just accidentally accumulated. A sampling of some of the major spore reservoirs follows.

SPORES IN SOIL

It is generally accepted that the primary reservoir of sporeforming Bacilli is the soil, although we are almost completely ignorant of the physiology and population dynamics of germination, growth, and sporulation in the soil environment. We do know that, far from merely existing in soil as static dormant spores as originally proposed (Conn, 1916), a much more dynamic portrait of sporeforming *Bacillus* spp. in soils has been emerging over the past few decades. First noted were seasonal variations in the numbers, distribution, and predominance of various *Bacillus* spp. (Holding et al.,1965 and references therein), and later much evidence of interactions between soil Bacilli and the abiotic and biotic components of their environments has been uncovered and explored (see below). Evolution of and dynamic interactions between *Bacillus* spp. (Graham and Istock, 1981; Duncan et al., 1994) and their phages (Pantastica-Caldas et al., 1992) in soil have been documented, including intraspecies (Graham and Istock, 1979a; 1979b) and interspecies (Duncan et al., 1989) exchanges of genetic material in the soil environment. It has variously been proposed that transformation and natural competence of bacteria in the soil environment may have evolved from systems originally developed for the uptake of nucleic acid as a food source during times of starvation (Finkel and Kolter, 2001) or for purposes of DNA repair (Michod et al., 1988; Wojciechowski et al., 1989; Hoelzer and Michod, 1991; Michod and Wojciechowski, 1994), and further that these genetic exchange systems may have been precursors to sexual recombination and reproduction (Michod, 1989; Michod et al., 1990; Michod, 1993).

The distribution of *B. subtilis* cells in soil has been explored using probes consisting of fluorescent antibodies directed against either vegetative cells or spores of *B. subtilis* (Siala and Gray, 1974; Siala et al., 1974). These early studies indicated that while mainly spores were found in the mineral subsurface layers of soil, both vegetative cells and spores were found in the upper organic surface layers, associated mostly with particles of decaying organic matter. Siala and Gray (1974) further performed experiments to monitor the fate of *B. subtilis* cells or spores grown in laboratory media and applied to glass slides which were then placed into contact with soil. They showed that spore germination and growth of soil-inoculated *B. subtilis* was specifically associated with the growth and development of fungal hyphae in the soil. Although vegetative cells were observed growing in direct proximity to the fungal mycelia, it was uncertain whether the bacteria were simply using nutrients excreted by the living hyphae, or were causing death of the fungus and living off the products of hyphal lysis (Siala and Gray, 1974). In support of the latter notion, Podile and Prakash (1996) reported that *B. subtilis* strain AF-1 adhered to, colonized, and multiplied on *Aspergillus niger* mycelia; growth of strain AF-1 damaged and ultimately lysed the fungal cell wall, suggesting that some *Bacillus* spp. in soil could be predators of fungi.

The perception that Bacilli are merely free-living soil inhabitants is beginning to be revised, as numerous reports have documented the occurrence of *Bacillus* spp. in the rhizoplane and rhizosphere of various wild and cultivated plants (Seldin et al., 1984; Fradkin and Patrick, 1985; Kapoor and Kar, 1988; Li et al., 1992; Pandey and Palni, 1997). Indeed, we are beginning to appreciate that *Bacillus* spp. participate in a dynamic and complex soil ecology including interactions with resident plant roots, fungi, and nematodes. These studies have begun to offer the beginnings of a more complex and detailed portrait of the life of sporeforming *Bacillus* spp. in soil, and have led to the use of spore-forming bacteria in agriculture as agents for biological control of unwanted soil micro- and macroorganisms and growth-promoting agents in crop production (see below).

SPORES IN ROCKS
Extensive literature recently reported indicates that a wide variety of microbial life (called "endolithic" microbes) reside within rock extending from the surface regolith well into the deep subsurface rock itself (Fredrickson and Onstott, 1996; Nicholson et al., 2000; Nicholson, 2002). Using traditional culture-based isolation techniques, several *Bacillus* spp. have been obtained from the interiors of such locations as rock varnishes from Sonoran, Mohave, and Negev desert rocks (Palmer et al., 1986; Hungate et al., 1987; Nagy et al., 1991), deep subsurface boreholes (Boone et al., 1995; Balkwill, et al., 1997), near-subsurface Sonoran desert basalt (Benardini et al., 2003) and near-subsurface granite formations in the Santa Catalina mountain range of southern Arizona (W.L. Nicholson, unpublished). This list will doubtless grow as more sites are sampled and better isolation techniques devised. For example, using a non-culture (i.e., 16S rRNA)-based census technique, novel species of endolithic sporeforming bacteria have recently been identified within limestones from the supratidal-intertidal zone of the Atlantic coast off Florida (Andrews et al., 2002).

The question naturally arises: have spores found within rocks or ancient soils arisen from growth and sporulation of bacterial cells *in situ*, or have they simply become trapped by sedimentary deposition or groundwater percolation from overlying soil? Both scenarios are possible, and indeed not mutually exclusive. In favor of the "percolation" hypothesis, transport of bacterial spores through gravel aquifers (Pang et al., 1997; Sinton et al., 2000) and porous rock (Jang et al., 1983) has been demonstrated. In favor of the "*in situ* growth" hypothesis, the newly-described species *Bacillus infernus* was discovered growing vegetatively nearly 3 km below the Earth's surface in an aquifer which has likely been isolated from surface input for millenia (Boone et al., 1995).

Several *Bacillus* spp. (e.g., *B. brevis, B. licheniformis, B. mycoides, B. megaterium* and *B. subtilis*) have repeatedly been found, among other non-sporeforming bacteria, associated with the biodegradative activities which are slowly destroying old stone buildings and monuments in Europe (Tayler and May, 1991; Gomez-Alarcón et al., 1995; Flores et al., 1997; Stassi et

al., 1998; Turtura et al., 2000). At present it is unclear whether these *Bacillus* spp. are directly degradative to stone as pure cultures, or if they exert their activity optimally as complex consortia with other microorganisms (Turtura et al., 2000). Answering these questions will yield important insights into current attempts to halt the biodeterioration which is slowly erasing our ancient architectural and artistic heritage.

METAL OXIDATION: SPORES AS MINERS
Polymetallic nodules, commonly called "manganese nodules", are roughly spherical rock concretions found mostly on the sea bottom, but which have been recovered from freshwater sources as well. Manganese nodules vary in size from tiny particles visible only under a microscope to large pellets more than 20 centimeters across. They are formed around a nucleus (e.g., shell, bone fragment) in concentric layers consisting mostly of iron and manganese hydroxides. Nodules have excited the interest of mining concerns as they are enriched sources of elements such as Mn (27-30 %), Fe (6 %), Si (5%), Al (3%), Ni (1.25-1.5 %), Cu (1-1.4 %) and Co (0.2-0.25 %), with lesser amounts of Ca, Na, Mg, K, Ti and Ba (http://www.wikipedia.org/wiki/Manganese_nodule).

Metal deposition in nodules is recognized to occur by transformation of soluble metal ions in the water column, e.g., Mn(II), to insoluble precipitates consisting of Mn(III) and Mn(IV) oxides (reviewed in Brouwers et al., 2000). Although several processes are involved in the formation of nodules, the process which has sparked the most interest among microbiologists is the precipitation of metal hydroxides in nodules through the activity of microorganisms. Much research on manganese oxidation has centered on *Bacillus* strain SG-1, which was isolated from a near-shore manganese-rich sediment (Nealson and Ford, 1980). Dormant spores of *Bacillus* SG-1 catalyze the oxidation of Mn(II), the spores actually becoming encased in a shell consisting of Mn(III) and Mn(IV) oxides (Rosson and Nealson 1982). This arguably represents the first demonstration that metabolically inert spores are capable of altering their abiotic environment.

Bacillus SG-1 spores can increase the rate of Mn(II) oxidation by 4 to 5 orders of magnitude relative to abiotic rates (Hastings and Emerson, 1986). Biochemical and molecular genetic studies of Mn(II) oxidation by *Bacillus* SG-1 have revealed the involvement of a specific gene product, MnxG, which shares sequence similarity and biochemical properties with multicopper oxidases and is located in the outer spore layers (i.e., spore coat and exosporium) (Rosson and Nealson, 1982; de Vrind et al., 1986; van Waasbergen et al., 1996; Francis and Tebo 1999). The Mn(II)-oxidizing spore enzymes of these isolates are resistant to inactivation by heat (70 to 80°C), multiple freeze-thaw cycles, fixatives (UV and glutaraldehyde), SDS, lysozyme, and reductants (Rosson and Nealson, 1982; de Vrind et al., 1986; Francis and Tebo, 2002), suggesting that spores are stable enzymatic catalysts for the oxidative precipitation of metals in the environment (Francis and Tebo, 2002). Diverse marine *Bacillus* species, spores of which also display multicopper oxidases on their surfaces capable of catalyzing manganese oxidation, have recently been isolated from a number of marine sites in the San Diego, CA area (Francis and Tebo, 2002), demonstrating the widespread distribution of manganese-oxidizing spores in the environment.

The biological role(s) of manganese oxidation at the spore surface is currently a topic of debate. Some suggested functions have included: protection from damage by UV, adsorption and sequestration of toxic heavy metals, detoxification of reactive oxygen species, defense against predation or viral attack, and various oxidation-reduction schemes to obtain energy from organic and/or inorganic sources (reviewed in Brouwers et al., 2000). Due to this property of metal oxidation, it has been proposed that spores could be exploited as injectable nanoparticle biocatalysts to sequester heavy metals from contaminated sites. Unlike metabolically-active bacterial bioremediation agents, metal toxicity would not be an issue for dormant spores.

SPORES IN SPACE

From the subsurface we turn to space. Because of their notorious resistance and longevity, bacterial spores have been studied as possible candidates for natural transfer of life between the planets. This theory, variously dubbed "panspermia", "lithopanspermia", or "transpermia", postulates that viable microorganisms, residing within meteors ejected from a planet's surface by large impacts, could survive a trip through space to another planet (for extensive recent reviews, see Mileikowsky et al., 2000; Nicholson et al., 2000). For experimental convenience, the process has been divided into three phases; (i) impact-mediated ejection of crust material from the donor planet, (ii) travel through space, and (iii) entry and deposition onto the recipient planet (reviewed in Mileikowsky et al., 2000; Nicholson et al., 2000).

(i) Launch

Of the three phases listed above, it was historically considered that microbes could not survive ejection; however this view has changed based on new findings. First, a number of meteorites of lunar and martian origin have recently been described and characterized (Gladman et al., 1996; Thomas-Keprta et al., 2003) which have suffered relatively slight shock and heat stresses (Stöffler, 2000; Weiss et al., 2000) to which microbes might survive. Second, successive improvements in modeling the physics of impacts predicts the production of substantial numbers of high-speed, lightly-shocked ejecta from impact sites (Mileikowsky et al., 2000; Nicholson et al., 2000 and references therein; Head et al., 2002). Such high-speed ejecta arise close to the target surface in an area surrounding the impact site called the *spallation zone*. Within this zone, the energy of impact is translated mainly into upward acceleration of the target material rather than compression and heating (Melosh, 1985; reviewed in Nicholson et al., 2000). Therefore, endolithic microorganisms (such as bacterial spores) inhabiting rocks at or near the surface would be the most likely candidates for passengers in impact ejecta. Experimental evidence to date indicates that *B. subtilis* spores can survive the shock pressure (~32 GPa) (Stöffler, 2000; Horneck et al., 2001), heating (estimated to be only ~ 40-100°C) (Mileikowsky et al., 2000; Weiss et al., 2000; Horneck et al., 2001), acceleration (~ 3 x 10^6 m/sec^2)(Mastrapa

et al., 2001) and "jerk" (i.e., the rate of change of acceleration; ~ 6 x 10^9 m/sec^3)(Mastrapa et al., 2001) generated during an impact-mediated launch from Mars. However, it must be cautioned that these forces have been applied to spores singly, not in combination as would occur in nature. We are currently conducting ballistics experiments to test *B. subtilis* spore survival to all factors applied simultaneously, using the light 2-stage vertical gas gun at NASA Ames Research Center, Moffett Field, CA.

(ii) Transit through space

Once launched, spores in rock are subjected to the space environment itself, which places an entirely new set of physical factors limiting survival, including: space vacuum (~ 10^{-14} Pa) and the extreme desiccation which results; solar and galactic radiation (including UV, ionizing radiation, proton electron and heavy particle bombardment); and temperature extremes ranging from 4°K upwards, depending on distance, shielding, and orientation to the Sun (reviewed in Nicholson et al., 2000; Nicholson et al., 2002 and references therein). Spore survival to these conditions applied either singly or in combination have been studied either as laboratory simulations or by exposing *B. subtilis* spores directly to the space environment in low-Earth orbit (reviewed in Horneck, 1993; Nicholson et al., 2000). For long-term spore survival in space, shielding from solar UV radiation is essential, as spores exposed to full space as monolayers were inactivated within a few seconds by solar UV (Horneck, 1993). However, shielding can be quite minimal. *B. subtilis* spores were exposed to various components of the space environment for 2107 days (5.77 years) in near-Earth orbit on the Long Duration Exposure Facility (LDEF) (Horneck, 1993; Horneck et al., 1994). From initial populations of 10^8 spores, survival ranged from ~4 x 10^{-4} in spores exposed to full space in multilayers, to ~70% in spores artificially shielded only from solar UV but exposed to all other space conditions, a situation which may be analogous to the environment prevailing within impact ejecta (Horneck, 1993; Horneck et al., 1994, 2002; reviewed in Nicholson et al., 2000).

(iii) Entry and deposition

After a period in space, ejected rocks can be captured by the gravity field of another planet. It has been calculated that for a meteor of mass greater than ~1kg, its fall towards and impact into the surface of a recipient planet would pose no great limitation on spore survival, especially if the recipient planet possesses a substantial atmosphere. The passage of a meteor through Earth's atmosphere occupies less than a minute, far too little time for the heat generated to penetrate more than a few millimeters into the meteor. Further, atmospheric drag slows the meteor to terminal velocity, so landing is relatively gentle compared to launch (discussed in Passey and Melosh, 1980; Nicholson et al., 2000). However, this particular aspect of lithopanspermia theory has received the least amount of experimental attention to date. Using a sounding rocket at White Sands Missile Range, NM, we recently demonstrated that *B. subtilis* spores infused into granite carriers can survive atmospheric reentry at >2.2 km/sec from an altitude of ~250 km (L. Link and W.L. Nicholson, unpublished results).

One key factor limiting spore survival during interplanetary transfer, even if spores are well protected within a meteor, is time itself. Calculations based on orbital mechanics and launch velocities predict that a small fraction of the total ejecta from a Mars impact could be launched on trajectories intersecting Earth with passages of time of only a few months to a few years (Melosh, 1988; Moreno, 1988). However, from measurements of cosmic-ray exposure times, Gladman et al. (1996) calculated Mars-to-Earth transit times ranging from 0.7-15 million years for 10 meteorites found on Earth which originated from Mars. Could viable spores survive for such spans of time?

THE CONTROVERSY OF EXTREME SPORE LONGEVITY

Bacterial spores are probably the longest-lived cellular structures known; numerous reports document the isolation of spores ranging in age from several decades to hundreds of thousands of years, from environmental samples such as dried soil in herbarium collections, paleosols (ancient soils), ancient lake sediments, permafrost soils and ice cores (reviewed in Gest and Mandelstam, 1987; Kennedy et al., 1994; Potts, 1994 Nicholson et al., 2000; Nicholson, 2002).

Early attempts reporting the isolation of ancient microorganisms were poorly controlled and easily dismissed (Kennedy et al., 1994). Since the early days, experimental attempts to counter such criticisms have led to great improvements in both sampling technologies and the tightening of the quality controls of sampling and isolation protocols (reviewed in Kennedy et al., 1994; Vreeland and Rosenzweig, 2002). Given careful and controlled surface sterilization, sampling and culture protocols, it has been asserted in recent experiments that the ages of microorganisms or spores found match the age of the environmental sample from which they were obtained. However, the validity of this assumption must be rigorously tested in each case (as a caveat, see the above discussion of percolation of relatively recent spores through porous geologic strata). Still, there appears to be a direct correlation between the claimed age of ancient spores and the degree of skepticism with which the claim is met. For example, in recent years the scientific world has been treated to at least two reports of spore longevity spanning geologic time scales, both of which have been met with a high level of skepticism. In the mid-1990's there appeared a widely-publicized report that at least one viable *Bacillus* spp. spore had been isolated from the gut of a bee fossilized in Dominican amber for an astounding 25-40 million years (Cano et al., 1994; Cano and Borucki, 1995). Even this incredible age for a spore was dwarfed in the year 2000 by the report of the discovery of *Salibacillus* spp. isolate 2-9-3, which was recovered from a brine inclusion within a 250 million-year-old primary salt crystal obtained from the subterranean Salado Formation near Carlsbad, New Mexico (Vreeland et al., 2000). Several criticisms have been aimed at both these claims, generally centering around a few basic arguments which will be outlined below:

First, it is argued that the organisms recovered may represent recent contaminants arising either from ineffective surface sterilization of the sample, or from spores which have entered the ancient sample via microscopic fissures. While there is no way of completely eliminating this possibility, extensive documentation of rigorous sample selection criteria and sterilization control has been published for scientific scrutiny in the peer-reviewed literature (Cano and Borucki, 1995; Rosenzweig et al., 2000; reviewed in Vreeland and Rosenzweig, 2002).

Second, sporeformers isolated from ancient materials have been subjected to molecular phylogenetic analyses, either using 16S rRNA nucleotide sequences (Cano and Borucki, 1995; Vreeland et al., 2000) or amino acid sequences of protein-coding genes such as *recA* and *splB* (Maughan et al., 2002). Invariably, sequences from the putative ancient sporeformers have turned out to share considerable similarity to those of extant species (*B. sphaericus* in the case of the amber isolate; *Salibacillus marismortui* in the case of the brine isolate). This observation has been used to argue that, because the 16S rRNA sequences of the "ancient" sporeformers closely match those of "modern" bacteria, the isolates themselves are also likely to be "modern" (Graur and Pupko, 2001; Nickle et al., 2002). This second criticism highlights a fundamental problem with assigning absolute ages to microbes based on molecular sequence comparisons alone, and ignores several fundamental aspects of *Bacillus* biology. (i) In the absence of a microbial fossil record that can be linked to the geologic record, bacterial phylogenetic trees are constructed almost entirely from extant species. Because of this only relative, not absolute, ages for bacteria can be inferred. In an attempt to compensate for this deficit, various algorithms have been devised to transform percentages of nucleotide or amino acid similarity into "evolutionary rates" for bacteria. Such rates are valid only for the particular system under study, and are much more accurate when supported by a true fossil record. Very few systems exist which meet these criteria. For example, molecular evolutionary rates of DNA in the aphid endosymbionts *Buchnera* spp. are likely of rather high accuracy, because they are externally supported by, and calibrated to, the aphid fossil record (Ochman et al., 1999). But, as Ochman et al., (1999) themselves caution, they cannot be reliably used to calculate evolutionary rates in other bacteria. (ii) The number of generations per year, hence the evolutionary rate, experienced by enteric bacteria or obligate intracellular bacteria such as *Buchnera* spp. is rather constant and thus reliably calculated. Free-living, sporeforming *Bacillus* spp. on the other hand are subjected to a "feast or famine" existence, and in the spore state may not experience a single chromosomal replication event for years, decades, or millennia at a stretch, leading to a much jerkier evolutionary trajectory. Because of this fact, it is probably impossible to calculate a reliable molecular evolutionary rate for sporeformers. However, due to the very fact of dormancy, sporeformers likely exhibit a lower overall rate of evolution than do enteric bacteria (Maughan et al., 2002). (iii) The very concept of "modern" vs. "ancient" bacteria can itself be called into question. Due to cycles of weathering, sedimentary compression and uplift, soil is constantly being formed from eroded rock and ultimately being compressed back into rock. Thus, subsurface endolithic bacteria of indeterminate age are constantly being exposed in soil at the Earth's surface, being distributed over the surface on dust particles and in water, and ultimately re-entering the soil and bedrock by percolation and sedimentary deposition. Given geological cycling, is a newly-isolated bacterium "modern", thus a standard for comparison with a putative "ancient" bacterium, just because we humans obtained it from the environment yesterday?

A third critcism of the validity of claims of ancient spores relies on the argument that biological molecules, particularly DNA, cannot survive for such extended periods of time (Graur and Pupko, 2001; reviewed in Vreeland and Rosenzweig, 2002). Often quoted as substantiating evidence for this contention is a landmark review in which Lindahl (1993) discussed the various weaknesses in the chemical structure of DNA which render it susceptible to spontaneous degradation. This criticism overlooks several factors:

(i) The bulk of the studies cited in Lindahl (1993) were performed on naked DNA in aqueous solution, not on DNA within spores. In fact, this criticism conveniently overlooks a section contained in the very same review, entitled "Spores and Thermophiles", in which Lindahl (1993) states:

"There are biological strategies that overcome the intrinsic instability of DNA. Many microorganisms respond to hostile environmental conditions by sporulation, and DNA in some bacterial spores exhibit impressively increased resistance to heat- and radiation-induced damage." (Lindahl, 1993).

In support of this contention, substantial literature exists documenting the fact that DNA in bacterial spores is indeed in a much more protected and long-lived state than naked DNA. The high degree of bacterial spore resistance and longevity is explained in large part by detailed molecular mechanisms of DNA damage protection and repair (for recent reviews, see Setlow, 1995; Nicholson and Fajardo-Cavazos, 1997; Nicholson et al., 2000; Setlow, 2001).

(ii) It is currently unknown what effect the medium in which spores were trapped (amber or brine) would exert on either slowing or accelerating the rate of molecular degradation. Evidence exists that some molecular degradation processes are slowed by preservation in amber, most notably the racemization rate of amino acids (Poinar et al., 1996; Bada et al., 1999). Furthermore, studies on naked DNA embedded in plant resin (the precursor to amber) indicated that rapid drying after embedding was key to preserving DNA intact (Rogers et al., 2000). The effect of brine (i.e., saturated NaCl) on the degradation rate of biological molecules is currently unknown, but it has been suggested that one factor limiting the longevity of spores entrapped in brine inclusions is the naturally-occurring radioactive isotope potassium-40 (^{40}K) in brine, which may result in a radiation environment leading to the rapid fragmentation of DNA and accumulation of lethal "hits" in spores. To address this concern, the radiation environment prevailing inside subterranean brine inclusions was extensively modeled, and concluded not to be a significant impediment to 250 Ma survival of *Salibacillus* 2-9-3 spores (Nicastro et al., 2002). However, extrapolations using measurements of the ionizing radiation inactivation rate constants for several bacterial spores including *S. marismortui* led Kminek et al., (2003) to

the opposite conclusion, that the radiation environment limited survival of spores to less than 250 Ma.

I have recently taken an approach similar to that reported by Kminek et al. (2003) to experimentally address the question of extreme spore longevity (Nicholson, 2003). I reasoned that the longevity of a dormant spore is determined by the time elapsing between spore formation and its accumulation of a lethal "hit", that is, an amount of damage to (a) cellular target(s) sufficient to prevent the successful germination of the spore and subsequent growth of the bacterium. For the purposes of calculating spore longevity, the precise target or the exact cause of lethal damage is unimportant, the only parameter which matters is the elapsed time between spore formation and occurrence of the lethal "hit". Obviously we cannot wait around for millions of years, but the process of spore inactivation can be accelerated considerably by simply raising the temperature. Furthermore, the relationship between the velocity of spore inactivation and temperature is simple and well-established: plotting the log_{10} of the decimal reduction time (D-value) vs. temperature yields a straight line with negative slope (see Joslyn, 1983; Nicholson et al., 2000; Nicholson, 2003 for extensive discussions of spore inactivation kinetics). This inverse exponential relationship between the D-value and temperature implies the existence within spores of chemical reactions which lead ultimately to a lethal "hit", and that the velocity of these reactions are a strict function of temperature. Analyzing the published temperature inactivation data for spores from a number of *Bacillus* spp. (with the caveat that these studies did not use brine or amber as the spore environment) and extrapolation to ambient temperatures, I concluded that spore populations of mesophilic species stood a poor chance of survival for millions of years, whereas thermophilic spore populations might be predicted to survive such time scales with rather high probabilities (Nicholson, 2003). Present experiments in our lab directed towards experimentally testing a possible protective effect of brine on the thermal inactivation kinetics of spores are supported by the observations indicating that dry spores are inactivated more slowly than fully hydrated spores (Fox and Eder, 1969; Nicholson et al., 2000), and that solutes which lower the water activity of the medium in which spores are suspended (Murrell and Scott, 1966), including NaCl (Jagannath et al., 2003), have the net effect of increasing the survival time of spores at elevated temperatures.

SPOREFORMERS AS INSECT SYMBIONTS

An area which has been well-documented among entomologists, but virtually ignored by spore researchers is the symbiotic (literally, "living together") relationship which exists between sporeforming *Bacillus* spp. and a number of insects including: honeybees, solitary bees, and stingless bees (Gilliam et al., 1984; 1990) termites (Margulis et al., 1990; Sarkar, 1991); moths (Gilliam, 1985); leafrollers (McKillip et al., 1997); and cockroaches, millipedes, and sow bugs (Feinberg et al., 1999). Without a doubt this list will grow as new insects are tested.

What do these close insect-*Bacillus* associations imply? As is the case for obligate endosymbionts such as *Buchnera* spp. which colonize the guts of aphids (Moran and Baumann,

2000), a nutritional advantage to the insect has been postulated for the existence of *Bacillus* spp. in insect guts. For example, in cockroaches fed different diets, a positive correlation was observed between *B. cereus* population densities in the gut and the rate of insect weight gain (Feinberg et al., 1999). Other roles for insect-symbiotic *Bacillus* spp. have been postulated as well. A wide variety of *Bacillus* spp. have been found associated with brood provisions, pollen, larval feces, and in the alimentary canals of insects, including *B. licheniformis. B. cereus, B. subtilis, B. sphaericus, B. circulans, B. megaterium, B. alvei*, and *B. pumilus* (Gilliam et al., 1984; Gilliam, 1985; Gilliam et al., 1990a, 1990b) among others. Due to the metabolic capabilities and antibiotic production of these various *Bacillus* isolates, their role in the microecology of the insect nest has been postulated to be involved in conversion of various substrates to forms more nutritionally suitable for the insects, and prevention of overgrowth of pathogenic- or food-spoilage microorganisms (Gilliam et al., 1984; Gilliam, 1985; Gilliam et al., 1990). Thus is seems clear that the insect benefits from the association, but what about the *Bacillus* spp.? Unlike *Buchnera* spp. in the guts of aphids, the above-mentioned *Bacillus* species are obviously not obligate inhabitants of the insect gut, as they readily grow in laboratory culture, but their long-term residence in the insect gut likely implies some advantage to the bacterium, rather than passive accidental ingestion and passage through the alimentary tract. The issue of sporeformers as symbionts of higher animals is of crucial importance to the study of spores as probiotics, and will be discussed at length in subsequent chapters of this book.

SPORES AS DELIVERY VEHICLES FOR ANIMAL PATHOGENS

Several diseases of animals, including humans, are initiated by environmental contact of a susceptible host with spores of *Bacillus* spp. Below are summarized some of the best-studied. *B. thuringiensis* will be considered separately below as a biological control agent.

Paenibacillus larvae

P. larvae subsp. *larvae* is the causative agent of American foulbrood disease (AFB) of domestic honeybees. Because this bacterial disease has a major economic impact on honey, pollen, and wax production in apiculture, methods for detection and control of *P. larvae* have been the subjects of intensive study over the past decade. The spore, not the vegetative cell, is the infectious unit of *P. larvae* (Hornitzky, 1998a). Spores ingested by larvae germinate, grow to a high density, and sporulate within the larval host, killing it (Brodsgaard et al., 1998). During the process of cleaning infected chambers, nurse bees become contaminated and subsequently distribute spores throughout the entire colony. The disease is highly contagious and can quickly destroy an entire hive.

Detection of *P. larvae* in hives, larvae, and bee products has been an area of much research, and a plethora of molecular-based detection strategies have recently been described (for discussions, see Bakonyi et al., 2003; Genersch and Otten, 2003; Lauro et al., 2003). Currently, control of AFB relies on treating infected hives with antibiotics and maintaining good hygienic beekeeping practices. The only registered antibiotic for AFB treatment in the past several decades has been oxytetracycline (Evans, 2003), but widespread resistance in *P. larvae* has recently arisen (Miyagi et al., 2000). Substitute antibiotics are actively being sought and tested (Williams et al., 1998; Alippi et al., 1999; Elzen et al., 2002). Natural substances which inhibit the growth and/or spore germination of *P. larvae* have been identified in gut extracts of larvae and adult bees (Crailsheim and Riessberger-Galle, 2001; Riessberger-Galle et al., 2001), and in bee products (Hornitzky, 1998a; Wedenig et al., 2003). Two inhibitory proteins, one of which is the ~5 kDal defensin royalisin, have recently been identified and characterized from royal jelly (Bilikova et al., 2001; Bachanova et al., 2002).

Unlike our relatively good understanding of how AFB spreads within a hive, interhive (and worldwide) spread of AFB has been a more complex and difficult issue to grasp. The mere presence of *P. larvae* in a hive does not indicate active disease, and no strict correlation has been found between the concentration of *P. larvae* spores in honey and presence or absence of AFB in a hive (Hansen, et al., 1986). AFB does not appear to spread easily between closely-spaced hives by drifting bees (Hornitzky, 1998b). Mites which infect honeybee larvae were found to carry *P. larvae* spores, but did not efficiently spread AFB from hive to hive (Alippi et al., 1995), nor did the mites cause stress to larvae which made them more susceptible to AFB (Brodsgaard et al., 2000). However, DNA fingerprinting methods applied to worldwide isolates of *P. larvae* have demonstrated direct connections between *P. larvae* lineages in Europe, New Zealand, and South America (Alippi and Aguilar, 1998). It appears that hygienic practices by beekeepers themselves are key in AFB spread. AFB may be introduced into a hive along with live bees and brood, by feeding bees AFB-infected honey or pollen, by re-using infected honeycombs or hives, or by using contaminated beekeeping tools.

An interesting possibility for AFB control comes from the study of the hygienic behavior of bees themselves. Hives of the Asian honeybee, *Apis cerana*, show much more resistance to AFB than do those of the European honeybee, *Apis mellifera*, although *A. cerana* larvae are actually somewhat more susceptible to infection by *P. larvae*. Resistance appears to be due to the fact that *A. cerana* nurse bees are much more efficient at identifying and removing infected larvae from the hive, thus hindering spread of the disease (Chen et al., 2000). Increased hygienic behavior in *Apis mellifera* has also recently been selected for using a freeze-killed brood assay, and these hygiene-enhanced strains also demonstrated increased resistance to AFB (Spivak and Reuter, 2001).

Paenibacillus popilliae and *P. lentimorbus*

These two closely-related, former *Bacillus*, species are natural soil inhabitants which cause milky disease in the larvae of various Coleoptera (beetles). Both species are reported to produce a parasporal crystal similar to that to *B. thuringiensis* (Harrison et al., 2000). However, rather than killing the host outright, the parasporal crystal apparently aids in passage of the vegetative cell from the gut into the hemolymph of the larva, where the bacterium establishes parasitic vegetative growth (Zhang et al., 1997). Despite the fact that *P. popilliae* has been used as a biological control agent against Japanese beetles for the past 50 years (Stahly and Klein, 1992), the complex ecology of these microorganisms with their hosts in soil is just beginning to be unraveled. Doubtless, as in so many other systems, the advent of molecular fingerprinting techniques (Rippere et al., 1996; Tran et al., 1996; Harrison et al., 2000; Correa et al., 2001) will also accelerate the pace of ecological research on milky diesease. As with *P. larvae*, one factor limiting research on *P. popilliae* and *P. lentimorbus* is the inability to cultivate or produce spores of these organisms in the absence of the insect host, although some progress has been made in cultivating these species on cultured insect cell lines *in vitro* (Steinkraus et al., 1998). Issues associated with the use of *P. popilliae* and *P. lentimorbus* as biological control agents will be discussed further below.

Bacillus anthracis

The disease anthrax has been recognized in mammals for millenia, and studies on anthrax and its causative agent, *B. anthracis*, form an important part of the foundation both of spore research and of microbiology (for some original papers and excellent commentary, see Brock, 1998; 1999). Anthrax can be contracted by ingestion, inhalation, or cutaneous inoculation of *B. anthracis* spores. While the disease usually strikes grazing animals whose lifestyle places them in close contact with soil and dust, anthrax is also a zoonotic infection of humans; human infection generally occurs via accidental contact with hides, wool, or hair of infected animals. While natural anthrax is rare in industrialized countries, it is rather common in agricultural and developing societies. Most human cases are cutaneous, the hands and forearms being the most frequent inoculation site. Because of its rarity as a natural human pathogen, most anthrax cases are still easily treated with penicillin if diagnosis and treatment is timely (Oncul et al., 2002).

Systematic studies on the natural ecology of anthrax have mostly been limited to bison herds in northwestern Canada (Dragon et al., 1999) and grazing animals in Etosha National Park, Namibia (Lindque and Turnbull, 1994). Both studies have noted seasonal variations in anthrax incidence and differences in long-term survival of *B. anthracis* spores dependent upon soil type. Interestingly, in both studies it was found that anthrax preferentially struck adult male animals (Lindque and Turnbull, 1994; Dragon et al., 1999). In the case of Canadian bison it has been postulated that during dry periods following wet spring seasons *B. anthracis* spores become concentrated into low-lying wallows preferentially utilized by sexually-mature bulls (Dragon et al., 1999). In the Etosha study, presence of high concentrations of *B. anthracis* spores detected in scavenger feces collected from the vicinity of infected carcasses implicated scavengers as a route of anthrax spread during epizootics (Lindque and Turnbull, 1994). It has been postulated that such wild areas in which *B. anthracis* spore concentrations are high may serve as reservoirs of infection to surrounding domestic animals (Hugh-Jones and de Vos, 2002). Because of its rarity in developed countries,

interest in the ecology and epidemiology of anthrax had been waning considerably in recent years. However, recent anthrax bioterror attacks, in particular the attacks which occurred in October 2001 in the United States, have sparked an explosion of research into the biology of *B. anthracis* pathogenesis and the epidemiology of (primarily inhalational) anthrax. I will not reiterate the extensive recent literature on this topic, as it has been reviewed exhaustively elsewhere. For some recent reviews on anthrax as a bioterror weapon, see Kleitmann and Ruoff (2001), Atlas (2001; 2002), Inglesby et al. (2002); Noah and Crowder (2002), and Terriff et al. (2003).

Pasteuria penetrans and its relatives

Four species of *Pasteuria* are recognized: *P. ramosa, P. penetrans, P. thornei*, and *P. nishizawae*. *P. ramosa* was first described by Metchnikoff as a parasite of water fleas (*Daphnia* spp.), and the other three species are all known to be natural parasites of phytopathogenic nematodes in soil (reviewed by Sayre, 1993). The life cycle of *Pasteuria* is complex. The agent of infection in all cases is the spore, which attaches to the nematode cuticle upon contact via specific surface adhesion molecules. After the nematode has penetrated the plant root and begun to feed, the spores germinate, invade the nematode and establish a cycle of growth and sporulation, ultimately releasing up to ~2×10^6 infectious spores per nematode. Cycles of infection and spore release result in amplification of spores in soils to levels which become "suppressive" to the establishment of high nematode populations. Research on the biology of *Pasteuria* spp. has been hindered by the lack of an *in vitro* culture system, as the bacteria are obligate parasites of nematodes which themselves are obligate parasites of plants. Most research on their biology has concentrated on the development of these microorganisms as agents of biological control, which has gained a sense of urgency due to the phasing out for environmental reasons of the preferred nematode fumigant, methyl bromide.

SPORES AS TRACERS AND BIODOSIMETERS

The extreme stability of spores and ease of culture of most *Bacillus* spp. render their spores useful as environmental monitoring tools. Spores have been successfully used as flow tracers of ocean currents and tides (Upstill-Goddard et al., 2001), sewage discharges into rivers (Houston et al., 1989), and freshwater aquifers and soils (Hinojosa-Rebollar et al., 1995; Pang et al., 1998). Spores of non-pathogenic *Bacillus* spp. were recognized early as safe atmospheric tracers (Sattar et al., 1972) and they have been used extensively in modeling the aerosolization and atmospheric dispersal of biocontrol agents (Morris, 1977; Marthi et al., 1990). Spores of harmless *Bacillus* spp. are useful tracers of processes as diverse as tracking sources of contamination during food preparation (Leclerq et al., 1990; Eneroth et al., 1998) to mixing of activated sludge in sewage reactors (Horan et al., 1991). Spores have also been explored as tracers through biological systems in applications as diverse as determining ingestion rates in protzooans (Ben-Dov et al., 1994) to exploring the process of maternal-to-offspring transfer of normal cutaneous and intestinal microflora (Brunel and Gouet,

1989). Exploration of the new and exciting roles and uses of spores in higher organisms is of course the subject of many of the following chapters.

Spores have long been used as biodosimeters for verification of various disinfection and sterilization processes. A common example is the extensive use of *B. stearothermophilus* or *B. subtilis* spore strips in quality assurance protocols to test the efficiency of sterilization devices such as steam autoclaves, heating ovens, and ethylene oxide or hydrogen peroxide vapor reactors. In addition, *B. subtilis* spores have been used for more than 25 years as biological dosimeters for monitoring terrestrial solar UV flux, and offer a number of advantages over other biodosimetry systems, such as their: simplicity and ease of use and transport to and from environmental monitoring sites; stability upon prolonged storage, both before and after exposure; and the reproducibility of their UV inactivation responses. Since the last extensive review on the subject (Nicholson et al., 2000), additional worldwide solar UV and ozone column monitoring studies using spores have been performed in the North and South polar regions (Cockell et al., 2002; 2003), as well as in Europe and Asia (Munakata, 1999; 2000). More recently, personal spore dosimeters have been developed and commercialized for monitoring of individuals whose occupations expose them to high solar UV exposure, hence increased risk for skin cancers and cataracts (Quintern et al., 1997; Munakata et al., 1998; Moehrle et al. 2000a, 2000b; Moerhle, 2001; Moerhle et al., 2003).

Due to concerns regarding cost, chemical disinfection by-products, and the resistance of water-borne pathogens such as *Cryptosporidium parvum* oocysts to chlorine, UV radiation has increasingly become a viable option for municipal drinking water and wastewater treatment. Complex reactor and flow kinetics limit the usefulness of in-place photometric dosimeters in municipal applications, thus biodosimetry is the current "gold standard" for validating UV reactor efficacy. *B. subtilis* spores have recently become important biodosimeters for performance monitoring of UV disinfection reactors, leading to European UV disinfection standards for drinking water being based on *B. subtilis* spore biodosimetry (Hoyer 1998; 2000).

We recently showed that spores of wild-type laboratory strains of *B. subtilis* can serve as accurate and non-pathogenic surrogates for *B. anthracis* spores in UV treatment validation technologies (Nicholson and Galeano, 2003). However, different water-borne pathogens exhibit distinct UV inactivation kinetics; thus, accurate modeling of the efficacy of UV reactors on pathogens is best accomplished using non-pathogenic surrogates which mimic the inactivation kinetics of the corresponding pathogens as closely as possible. To address this issue, we recently demonstrated the utility of UV repair-defective mutant spores of *B. subtilis* as non-pathogenic surrogates whose UV inactivation kinetics closely mimic those of the water-borne pathogens *Cryptosporidium parvum* (Nicholson et al., 2001; Clancey et al., 2002) and three microsporidian (*Encephalitozoan*) species. (Marshall et al., 2003).

SPORES AS BIOLOGICAL CONTROL AGENTS

The use of spores of various *Bacillus* spp. as natural pesticides is not new, and "biopesticides" have long been touted as safe and environmentally-friendly alternatives to traditional pesticides. However, natural biopesticides also exert environmental impacts that often are not appreciated until long after their use has become institutionalized. For example, as mentioned above, *P. popilliae* has been in use for several decades as an agent to control Japanese beetle infestations, but widespread distribution of *P. popilliae* in the environment by humans has recently been brought into question, due to the fact that the bacterium encodes natural resistance to vancomycin, an important "last line of defense" antibiotic for human use (Rippere et al., 1998; Patel et al., 2000).

For control of lepidopteran larvae (caterpillars), *B. thuringiensis* (Bt) spores and parasporal crystal preparations have been a resounding commercial success: Bt accounts for greater than 90% of all marketed bioinsecticides, with a worldwide marker of over $100 million (Powell and Jutsum, 1993; Emmert and Handelsman, 1999). Despite the fact that Bt generally has performed better than traditional chemical pesticides in terms of target selectivity (Palmer, 1993; Hershey et al., 1995; Painter et al., 1996; Boyd and Boethel, 1998; Boisvert and Boisvert, 2000), concerns persist since Bt insecticides also kill potentially-beneficial nontarget predator insects. Such concerns have become especially acute coincident with the introduction and expression of the Bt crystal toxin directly within crop plants (Ferber, 2000; Saxena and Stotzky, 2000). Similarly, *B. sphaericus* spore preparations are widely used for control of dipteran species, particularly mosquito larvae, and studies to date also tend to indicate few noticeable ill effects on nontarget organisms (Aly and Mulla, 1987; Mulla et al., 1987; Vandenberg, 1990; Walton and Mulla, 1991).

The recent successes of *B. cereus* (Emmert and Handelsman, 1999) and *B. subtilis* (Turner and Backman, 1991; Backman, et al., 1994; Brannen and Kenney, 1997) strains as both natural fungicides and plant-growth-promoting bacteria when inoculated directly into soils as seed pretreatments has spawned a new generation of studies concerning the establishment and survival of sporeforming bacterial populations in soils (reviewed in van Veen et al., 1997). Van Elsas et al. (1986) first identified soil factors which limited the establishment of high spore titers of *B. subtilis* inocula introduced into soil as vegetative cells. Tokuda et al. (1995) demonstrated that these limitations could be overcome, and high-level populations of *B. subtilis* spores (10^7 spores/gm) could be established in soil by a combination of nutritional amendment of soil with a readily-metabolized carbon source such as glucose and raising the soil incubation temperature from 15°C to 25°C. Additional successes in enhancement of the establishment and survival of bacterial inoculant populations in soil have been achieved by encapsulation of microbes within a variety of polymeric materials, which can provide a source of both nutrition for the inoculant and protection of the inoculant from harmful soil factors (van Veen et al., 1997).

There is a synergistic increase in the efficacy of bioinsecticides which contain both spores and toxin crystals of *B. thuringiensis* (Miyasono et al., 1994; Johnson and McGaughey, 1996). However, when commercial preparations containing mixtures of *B. thuringiensis* spores and crystal inclusions are used for insect control, the spore component disappears rapidly from the field, attributable at least in part to solar UV-sensitivity of the spores (Griego and Spence, 1978; Dulmage and Aizawa, 1982; Benoit et al., 1990). Empirical attempts at circumventing this problem have concentrated on encapsulation of the biocontrol agent (Cokmus and Elcin, 1995; Skovmand and Bauduin, 1997; Morales-Ramos et al., 2000) or isolation of UV-resistant strains by UV irradiation of spores in the lab (Jones et al., 1991) or exposing spores to sunlight (Obeta, 1996).

Du and Nickerson (1996) presented evidence indicating that during *B. thuringiensis* sporulation, some of the insecticidal crystal toxin molecules produced actually insert into the spore coat where they are displayed for binding to specific receptors in the insect midgut, thereby facilitating spore attachment and germination in the insect host. This observation is reminiscent of the theme that spores can express surface proteins which interact with their environment (see the above discussion on manganese-oxidizing spores), and brings to mind the subject of using spores as nanoparticles. Du and Nickerson (1996) postulated that disruption of spore coat integrity by toxin insertion greatly enhances insect pathogenicity at the expense of a loss of some of the resistance properties of the spore. In support of this hypothesis, we recently demonstrated that mutational disruption of the spore coat layers in *B. subtilis* indeed resulted in decreased spore resistance to solar UV (Riesenman and Nicholson, 2000). The above sampling of studies underscores the importance of understanding the basic ecology of sporeformers in their natural habitats for the continued rational design of safe and effective biopesticides.

SPORES AS PROBIOTICS: CONCLUDING REMARKS

When I last reviewed the subject of human uses of spores (Nicholson, 2002), I felt it necessary to append a short section at the end of the article devoted to the newly-emerging field of "Spores as 'Probiotics' in Plant, Animal and Human Health". As is demonstrated amply in the following chapters of this book, there has been in the intervening 2 to 3 years a dramatic explosion of interest in the use of spores of various species of *Bacillus* as "probiotics", loosely defined as live microbes used as food or soil supplements which beneficially affect the growth or nutrition of the host, whether the host be plant or animal (Turner and Backman, 1991; Backman et al., 1994; Brannen and Kenney, 1997; Guillot, 1998; Emmert and Handelsman, 1999; Marta et al., 1999; Murphy et al., 2000; Verschuere et al., 2000).

Rather than offer a superficial overview of the subject of spores as probiotics, I will let the experts speak for themselves in the following chapters. In our extensive knowledge of the *B. subtilis* system we possess a solid foundation for future advances, and we are indeed embarking upon a fascinating new phase in applied spore research. I am looking forward eagerly to the discoveries of the next few years.

ACKNOWLEDGEMENTS
I thank Patricia Fajardo-Cavazos for critical reading of the manuscript. Aspects of work in our laboratory over the years has been generously supported by grants from the American Cancer Society, ASM, NASA, NIH, and USDA.

REFERENCES
Alippi, A.M. and Aguilar, O.M. (1998). Characterization of isolates of *Paenibacillus larvae* subsp. *larvae* from diverse geographical origin by the polymerase chain reaction and BOX primers. J. Invert. Pathol. *72*, 21-27.

Alippi, A., Albo, G.N., Leniz, D., Rivera, I., Zanelli, M.L., and Roca, A.E. (1999). Comparative study of tylosin, erythromycin and oxytetracycline to control American foulbrood of honey bees. J. Apicult. Res. *38*, 149-158.

Alippi, A.M., Albo, G.N., Marcangeli, J., Leniz, D., and Noriega, Alicia. (1995). The mite *Varroa jacobsoni* does not transmit American foulbrood from infected to healthy colonies. Exp. Appl. Acarol. *19*, 607-613.

Aly, C. and Mulla, M.S. (1987). Effect of two microbial insecticides on aquatic predators of mosquitoes. J. Appl. Entomol. *103*, 113-118.

Andrews, M., Sun, H.J. and Nealson, K.H. (2002). Novel bacterial 16S rRNA sequences from a marine endolithic community. Abstr. Gen. Meeting Amer. Soc. Microbiol. *102*, 333.

Atlas, R.M. (2001). Bioterrorism before and after September 11. Crit. Rev. Microbiol. *27*, 355-379.

Atlas, R.M. (2002). Responding to the threat of bioterrorism: a microbial ecology perspective: the case of anthrax. Internat. Microbiol. *5*, 161-167.

Bachanova, K., Klaudiny, J., Kopernicky, J., and Simuth, J. (2002). Identification of honeybee peptide active against *Paenibacillus larvae larvae* through bacterial growth-inhibition assay on polyacrylamide gel. Apidologie *33*, 259-269.

Backman, P.A., Brannen, P.M., and Mahaffee, W.F. (1994). Plant response and disease control following seed inoculation with *Bacillus subtilis*. In Proceedings of the third international workshop on plant growth-promoting rhizobacteria. M.H. Ryder, P.M. Stephens, and G.D. Bowen eds. (CSIRO, Australia), pp. 3-8.

Bada, J.L., Wang, X.-S., and Hamilton, H. (1999). Preservation of key biomolecules in the fossil record: current knowledge and future challenges, Phil. Trans. Royal Soc. London B-Biol. Sci. *354*, 77-87.

Bakonyi, T., Derakhshifar, I., Grabensteiner, E., and Nowotny, N. (2003). Development and evaluation of PCR assays for the detection of *Paenibacillus larvae* in honey samples: comparison with isolation and biochemical characterization. Appl. Environ. Microbiol. *69*, 1504-1510.

Balkwill, D.L., Reeves, R.H., Drake, G.R., Reeves, J.Y., Crocker, F.H., King, M.B. and Boone D.R. (1997). Phylogenetic characterization of bacteria in the subsurface microbial culture collection. FEMS Microbiol. Rev. *20*, 201-216.

Benardini, N., Sawyer, J., Venkateswaran, K.J. and Nicholson, W.L. (2003). Spore UV and acceleration resistance of endolithic *Bacillus pumilus* and *B. subtilis* isolates obtained from Sonoran desert basalt: implications for lithopanspermia. Astrobiology *3*, 709-717.

Ben-Dov, E., Zalkinder, V., Shagan, T., Barak, Z., and Zaritsky, A. (1994). Spores of *Bacillus thuringiensis* serovar *israelensis* as tracers for ingestion rates by *Tetrahymena pyriformis*. J. Invert. Pathol. *63*, 220-222.

Benoit, T.G., Wilson, G.R., Bull, D.L., and Aronson, A.I. (1990). Plasmid-associated sensitivity of *Bacillus thuringiensis* to UV light. Appl. Environ. Microbiol. *56*, 2282-2286.

Bilikova, K., Wu, G., and Simuth, J. (2001). Isolation of a peptide fraction from honeybee royal jelly as a potential antifoulbrood factor. Apidologie *32*, 275-283.

Boisvert, M. and Boisvert, J. (2000). Effects of *Bacillus thuringiensis* var. *israelensis* on target and nontarget organisms: a review of laboratory and field experiments. Biocontrol Sci. Technol. *10*, 517-561.

Boone, D.R., Liu, Y., Zhao, Z.-J., Balkwill, D.L., Drake, G.R., Stevens, T.O. and Aldrich, H.C. (1995). *Bacillus infernus* sp. nov., an Fe(III)- and Mn(IV)-reducing anaerobe from the deep terrestrial subsurface. Int. J. Systematic Bacteriol. *45*, 441-448.

Boyd, M.L. and Boethel, D.J. (1998). Susceptibility of predaceous hemipteran species to selected insecticides on soybean in Louisiana. J. Econ. Entomol. *91*, 401-409.

Brannen, P. M. and Kenney, D.S. (1997). Kodiak: A successful biological-control product for suppression of soil-borne plant pathogens of cotton. J. Indus. Microbiol. Biotechnol. *19*, 169-171.

Brock, T.D. (1998). Robert Koch: A Life in Medicine and Bacteriology (Washington, D.C.: ASM Press).

Brock, T.D. (1999). Milestones in microbiology (Washington, D.C.: ASM Press).

Brodsgaard, C.J., Ritter, W., and Hansen, H. (1998). Response of *in vitro* reared honey bee larvae to various doses of *Paenibacillus larvae larvae* spores. Apidologie *29*, 569-578.

Brodsgaard, C.J., Ritter, W., Hansen, H., and Brodsgaard, H.F. (2000). Interactions among *Varroa jacobsoni* mites, acute paralysis virus, and *Paenibacillus larvae larvae* and their influence on mortality of larval honeybees *in vitro*. Apidologie *31*, 543-554.

Brouwers, G.J., Vijgenboom, E., Corstjens, P. L. A. M., De Vrind, J. P. M., and De Vrind-De Jong, E. W. (2000). Bacterial Mn^{2+} oxidizing systems and multicopper oxidases: an overview of mechanisms and functions. Geomicrobiol. J. *17*, 1–24.

Brunel, A., and Gouet, P. (1989). Relationship between the digestive microflora of the newborn and maternal microflora in rodents as evidenced by a transit marker. Annales de Recherches Veterinaires. *20*, 461-472.

Cano, R.J. and Borucki, M.K. (1995). Revival and identification of bacterial spores in 25- to 40-million-year-old Domican amber. Science *268*, 1060-1064.

Cano, R J., Borucki, M.K., Higby-Schweitzer, M., Poinar, H.N., Poinar, G.O., Jr. and Pollard, K.J. (1994). *Bacillus* DNA in fossil bees: An ancient symbiosis? Appl. Environ. Microbiol. *60*, 2164-2167.

Chen, Y.W., Wang, C.H., An, J., and Ho, K.K. (2000). Susceptibility of the Asian honey bee, *Apis cerana*, to American foulbrood, *Paenibacillus larvae larvae*. J. Apicult. Res. *39*, 169-175.

Clancy, J.L., Hargy, T.M., Battigelli, D.A., Marshall, M.M., Korich, D.G., and Nicholson, W.L. (2002). Susceptibility of multiple strains of *Cryptosporidium parvum* to UV light (American Water Works Research Foundation Technical Report. Subject Area: Water Treatment).

Cockell, C., Rettberg, P., Horneck, G., Scherer, K., and Stokes, M.D. (2003). Measurements of microbial protection from ultraviolet radiation in polar terrestrial microhabitats. Polar Biol. *26*, 62-69.

Cockell, C.S., Rettberg, P., Horneck, G., Wynn-Williams, D.D., Scherer, K., and Gugg-Helminger, A. (2002). Influence of ice and snow covers on the UV exposure of terrestrial microbial communities: dosimetric studies. J. Photochem. Photobiol. B Biology *68*, 23-32.

Cohn, F. (1876). Untersuchungen über Bakterien. IV. Beiträge zur Biologie der Bacillen. Beiträge zur Biologie der Pflanzen. *2*, 249-276.

Cokmus, C., and Elcin, Y.M. (1995). Stability and controlled release properties of carboxymethylcellulose-encapsulated *Bacillus thuringiensis* var. *israelensis*. Pesticide Sci. *45*, 351-355.

Conn, H.J. (1916). Are sporeforming bacteria of any significance in soil under normal conditions? J. Bacteriol. *1*, 187-196.

Correa, M.M. and Yousten, A.A. (2001). Pulsed-field gel electrophoresis for the identification of bacteria causing milky disease in scarab larvae. J. Invert. Pathol. *78*, 278-279.

Crailsheim, K. and Riessberger-Galle, U. (2001). Honey bee age-dependent resistance against American foulbrood. Apidologie *32*, 91-103.

de Vrind, J. P. M., de Vrind-de Jong, E. W., de Voogt, J.-W. H., Westbroek, P., Boogerd, F. C. and Rosson, R.A. (1986). Manganese oxidation by spores and spore coats of a marine *Bacillus* species. Appl. Environ. Microbiol. *52*, 1096–1100.

Dragon, D.C., Elkin, B.T., Nishi, J.S., and Ellsworth, T.R. (1999). A review of anthrax in Canada and implications for research on the disease in northern bison. J. Appl. Microbiol. *87*, 208-213.

Drazkiewicz, M. (1994). Distribution of microorganisms in soil aggregates: effect of aggregate size. Folia Microbiol. *39*, 276-282.

Du, C., and Nickerson, K.W. (1996). *Bacillus thuringiensis* HD-73 spores have surface-localized Cry1Ac toxin: physiological and pathogenic consequences. Appl. Environ. Microbiol. *62*, 3722-3726.

Dulmage, H. T., and Aizawa, K. (1982). Distribution of *Bacillus thuringiensis* in nature. In Microbial and Viral Pesticides, E. Kurstak, ed. (New York: Marcel Dekker, Inc.), pp. 209-238.

Duncan, K.E., Ferguson, N., Kimura, K., Zhou, X., and Istock, C.A. (1994). Fine-scale genetic and phenotypic structure in natural populations of *Bacillus subtilis* and *Bacillus licheniformis*: implications for bacterial evolution and speciation. Evolution *48*, 2002-2025.

Duncan, K.E., Istock, C.A., Graham, J.B., and Ferguson, N. (1989). Genetic exchanges between *Bacillus subtilis* and *Bacillus licheniformis*: variable hybrid stability and the nature of bacterial species. Evolution *43*, 1585-1609.

Elzen, P., Westervelt, D., Causey, D., Rivera, R., Baxter, J., and Feldlaufer, M. (2002). Control of oxytetracycline-resistant American foulbrood with tylosin and its toxicity to honey bees (*Apis mellifera*). J. Apicult. Res. *41*, 97-100.

Emmert, E.A.B. and Handelsman, J. (1999). Biocontrol of plant disease: a (Gram-) positive perspective. FEMS Microbiol. Lett. *171*, 1-9.

Eneroth, A., Christiansson, A., Brendehaug, J., and Molin, G. (1998). Critical contamination sites in the production line of pasteurised milk, with reference to the psychrotrophic spoilage flora. Int. Dairy J. *8*, 829-834.

Evans, J.D. (2003). Diverse origins of tetracycline resistance in the honey bee bacterial pathogen *Paenibacillus larvae*. J. Invert. Pathol. *83*, 46-50.

Feinberg, L., Jorgensen, J., Haselton, A., Pitt, A., Rudner, R. and Margulis, L. (1999). *Arthromitus* (*Bacillus cereus*) symbionts in the cockroach *Blaberus giganteus*: dietary influences on bacterial development and population density. Symbiosis *27*, 109-123.

Ferber, D. (2000). New corn plant draws fire from GM food opponents. Science *287*, 1390.

Finkel, S.E. and Kolter, R. (2001). DNA as a nutrient: novel role for bacterial competence gene homologs. J. Bacteriol. *183*, 6288-6293.

Flores, M., Lorenzo, J., and Gomez-Alarcón, G. (1997). Algae and bacteria on historic monuments at Alcala de Henares, Spain. Int. Biodeterior. Biodegrad. *40*, 241-246.

Fox, K. and Eder, B.D. (1969). Comparison of survivor curves of *Bacillus subtilis* spores subjected to wet and dry heat. J. Food Sci. *34*, 518-521.

Fradkin, A. and Patrick, Z.A. (1985). Properties of bacteria isolated from surfaces of conidia of *Cochliobolus sativus* incubated in soil. Can. J. Microbiol. *31*, 411-416.

Francis, C. A., and Tebo, B.M. (1999). Marine *Bacillus* spores as catalysts for oxidative precipitation and sorption of metals. J. Mol. Microbiol. Biotechnol. *1*, 71–78.

Francis, C.A. and Tebo, B.M. (2002). Enzymatic manganese(II) oxidation by metabolically dormant spores of diverse *Bacillus* species. Appl. Environ. Microbiol. *68*, 874-880.

Fredrickson, J.K. and Onstott, T.C. (1996). Microbes deep inside the earth. Sci. Amer. *275*, 68-73.

Genersch, E. and Otten, C. (2003). The use of repetitive element PCR fingerprinting (rep-PCR) for genetic subtyping of German field isolates of *Paenibacillus larvae* subsp. *larvae*. Apidologie *34*, 195-206.

Gest, H. and Mandelstam, J. (1987). Longevity of microorganisms in natural environments. Microbiol. Sci. *4*, 69-71.

Gilliam, M. (1985). Microbes from apiarian sources: *Bacillus* spp. in frass of the greater wax moth. J. Invert. Pathol. *45*, 218-224.

Gilliam, M., Buchmann, S.L., and Lorenz, B.J. (1984). Microbial flora of the larval provisions of the solitary bees, *Centris pallida* and *Anthophora* sp. Apidologie *15*, 1-10.

Gilliam, M., Buchmann, S.L., Lorenz, B.J., and Schmalzel, R.J. (1990a). Bacteria belonging to the genus *Bacillus* associated with three species of solitary bees. Apidologie *21*, 99-106.

Gilliam, M., Roubik, D.W., and Lorenz, B.J. (1990b). Microorganisms associated with pollen, honey, and brood provisions in the nest of a stingless bee, *Melipona fasciata*. Apidologie *21*, 89-98.

Gladman, B.J., Burns, J.A., Duncan, M., Lee, P., and Levinson, H.F. (1996). The exchange of impact ejecta between the terrestrial planets. Science *271*, 1387-1392.

Gomez-Alarcón, G., Lorenzo, J., and Cilleros, Y.B. (1995). Weathering factors of granite in the building of the Royal Academy of Pharmacy. Anales de la Real Academia de Farmacia. *61*, 373-389.

Graham, J.B. and Istock, C.A. (1979a). Gene exchange and natural selection cause *Bacillus subtilis* to evolve in soil culture. Science *204*, 637-639.

Graham, J.B. and Istock, C.A. (1979b). Genetic exchange in *Bacillus subtilis* in soil. Mol. Gen. Genet. *166*, 287-290.

Graham, J.B. and Istock, C.A. (1981). Parasexuality and microevolution in experimental populations of *Bacillus subtilis*. Evolution *35*, 954-963.

Graur, D. and Pupko, T. (2001). The permian bacterium that isn't. Mol. Biol. Evol. *18*, 1143-1146.

Griego, V. M., and Spence, K.D. (1978). Inactivation of *Bacillus thuringiensis* spores by ultraviolet and visible light. Appl. Environ. Microbiol. *35*, 906-910.

Guillot, J.-F. (1998). Probiotics in animal nutrition. Cahiers Agricultures *7*, 49-54.

Halvorson, H.O. , ed. (1961). Spores II. (Minneapolis, MN, USA: Burgess Publishing Co.).

Hansen, H. and Rasmussen, B. (1986). The investigation of honey from bee colonies for *Bacillus larvae*. Tidssskrift for Planteavl. *90*, 81-86.

Harrison, H., Patel, R., and Yousten, A.A. (2000). *Paenibacillus* associated with milky disease in Central and South American scarabs. J. Invert. Pathol. *76*, 169-175.

Hastings, D., and Emerson, S. (1986). Oxidation of manganese by spores of a marine bacillus: kinetic and thermodynamic considerations. Geochim. Cosmochim. Acta *50*, 1819–1824.

Head, J.N., Melosh, H.J., and Ivanov, B.A. (2002). Martian meteorite launch: high-speed ejecta from small craters. Science *298*, 1752-1756.

Hershey, A.E., Shannon, L., Axler-Richard, R., Ernst, C. and Mickelson, P. (1995). Effects of methoprene and Bti (*Bacillus thuringiensis* var. *israelensis*) on non-target insects. Hydrobiologia *308*, 219-227.

Hinojosa-Rebollar, R.E., Hernandez-Delgadillo, R., Mesta-Howard, A.M., Tapia-Mendieta, M.P., and Ortigoza-Ferado, J. (1995). Biological flow tracers: growth and survival of *Bacillus subtilis* 65-8 under environmental stress. Revista Latinoamer. Microbiol. *37*, 43-53.

Hoelzer, M.A. and Michod, R.E. (1991). DNA repair and the evolution of transformation in *Bacillus subtilis*. III. Sex with damaged DNA. Genetics *128*, 215-224.

Holding, A.J., Franklin, D.A., and Watling, D.R. (1965). The microflora of peat-podzol transitions. J. Soil Sci. *16*, 45-59.

Horan, N.J., Parr, J. and Naylor, P.J. (1991). Evaluation of tracers for the determination of the mixing characteristics of activated sludge reactors. Environ. Technol. *12*, 603-608.

Horneck, G. (1993). Responses of *B. subtilis* spores to the space environment: results from experiments in space. Orig. Life Evol. Biosphere *23*, 37-52.

Horneck, G., Bücker, H., and Reitz, G. (1994). Long-term survival of bacterial spores in space. Adv. Space Res. *14*, 41-45.

Horneck, G., Rettberg, P., Reitz, G., Panitz, C., and Rabbow, E. (2002). Protection of bacterial spores in space, a contribution to the discussion on panspermia. Orig. Life Evol. Biosphere *32*, 542.

Horneck, G., Stöffler, D., Eschweiler, U., and Hornemann, U. (2001). Bacterial spores survive simulated meteorite impact. Icarus *149*, 285-290.

Hornitzky, M.A.Z. (1998a). The pathogenicity of *Paenibacillus larvae* subsp. *larvae* spores and vegetative cells to honey bee (*Apis mellifera*) colonies and their susceptibility to royal jelly. J. Apicultural Res. *37*, 267-271.

Hornitzky, M.A.Z. (1998b). The spread of *Paenibacillus larvae* subsp. *larvae* infections in an apiary. J. Apicultural Res. *37*, 261-265.

Houston, J., Learner, M.A. and Dancer, B.N. (1989). Selection of an antibiotic-resistant strain of *Bacillus subtilis* var. *niger* (*Bacillus globigii*) for use as a tracer in microbially rich waters. Water Res. *23*, 387-388.

Hoyer, O. (1998). Testing performance and monitoring of UV systems for drinking water disinfection. Water Supply *16*, 424-429.

Hoyer, O. (2000). The status of UV technology in Europe. IUVA News *2*, 22-27.

Hugh-Jones, M.E. and deVos, V. (2002). Anthrax and wildlife. Rev. Sci. Tech. Off. Int. Epiz. *21*, 359-383.

Hungate, B., Danin, A., Pellerin, N.B., Stemmler, J., Kjellander, P., Adams, J.B. and Staley, J.T. (1987). Characterization of manganese-oxidizing (manganese II to manganese IV) bacteria from Negev Desert (Israel) rock varnish: Implication in desert varnish formation. Can. J. Microbiol. *33*, 939-943.

Inglesby, T.V. O' Toole, T., Henderson, D.A., Bartlett, J.G., Ascher, M.S., Eitzen, E., Friedlander, A.M., Gerberding, J., Hauer, J., Hughes, J., McDade, J., Osterholm, M.T., Parker, G., Perl, T.M., Russell, P.K., and Tonat, K. (2002). Anthrax as a biological weapon, 2002: updated recommendations for management. J. Amer. Med. Assoc. *287*, 2236-2252.

Istock, C.A., Ferguson, N., Istock, N.L., and Duncan, K.E. (2001). Geographical diversity of genomic lineages in *Bacillus subtilis* (Ehrenberg) Cohn sensu lato. Organisms Diversity and Evolution *1*, 179-191.

Jagannath, A., Nakamura, I., and Tsuchido, T. (2003). Modelling the combined effects of pH, temperature and sodium chloride stresses on the thermal inactivation of *Bacillus subtilis* spores in a buffer system. J. Appl. Microbiol. *95*, 135-141.

Jang, L.K., Chang, P.W., Findley, J.E., and Yen, T.F. (1983). Selection of bacteria with favorable transport properties through porous rock for the application of microbial enhanced oil recovery. Appl. Environ. Microbiol. *46*, 1066-1072.

Johnson, D.E. and McGaughey, W.H. (1996). Contribution of *Bacillus thuringiensis* spores to toxicity of purified Cry proteins towards indianmeal moth larvae. Curr. Microbiol. *33*, 54-59.

Jones, D.R., Karunakaran, V., Burges, H.D., and Hacking, A.J. (1991). UV-resistant mutants of *Bacillus thuringiensis*. J. Appl. Bacteriol. *70*, 460-463.

Joslyn, L. (1983). Sterilization by heat. In Disinfection Sterilization and Preservation, S.S. Block, ed. (Philadelphia, USA: Lea and Febiger Publishers), pp. 3-46.

Kapoor, I.J. and Kar, B. (1988). Antagonistic effects of soil microbes *on Fusarium oxysporum* f. sp. *lycopersici*, causing tomato wilt. Int. J. Tropical Plant Dis. *6*, 257-262.

Kennedy, M.J., Reader, S.L., and Swierczynski, L.M. (1994). Preservation records of micro-organisms: evidence for the tenacity of life. Microbiology *140*, 2513-2529.

Klietmann, W.F. and Ruoff, K.L. (2001). Bioterrorism: Implications for the clinical microbiologist. Clin. Microbiol. Rev. *14*, 364-381.

Kminek, G., Bada, J.L., Pogliano, K., and Ward, J.F. (2003). Radiation-dependent limit for the viability of bacterial spores in halite fluid inclusions and on Mars. Radiat. Res. *159*, 722–729.

Koch, R. (1876). Untersuchungen über Bakterien. V. Die Aetiologie der Milzbrand Krankheit, begrandet auf Entwicklurgs geschichte des *Bacillus anthracis*. Beiträge zur Biologie der Pflanzen. *2*, 277-308.

Lauro, F.M., Favaretto, M., Covolo, L., Rassu, M., and Bertoloni, G. (2003). Rapid detection of *Paenibacillus larvae* from honey and hive samples with a novel nested PCR protocol. Int. J. Food Microbiol. *81*, 195-201.

Leclerq, P.M.N., Lalande, M., and Tissier, J.P. (1990). A method for assessment of the cleanability of equipment in the food processing industry. Sci. Aliments *10*, 17-41.

Li, C.-Y., Massicote, H.B. and Moore, L.V.H. (1992). Nitrogen-fixing *Bacillus* sp. associated with Douglas-fir tuberculate ectomycorrhizae. Plant and Soil *140*, 35-40.

Lindahl, T. (1993). Instability and decay of the primary structure of DNA. Nature *362*, 709-715.

Lindque, P.M. and Turnbull, P.C.B. (1994). Ecology and epidemiology of anthrax in the Etosha National Park, Namibia. Onderstepoort J. Vet. Res. *61*, 71-83.

Margulis, L., Olendzenski, L., and Afzelius, B.A. (1990). Endospore-forming filamentous bacteria symbiotic in termites: ultrastructure and growth in culture of *Arthromitus*. Symbiosis *8*, 95-116.

Marshall, M.M., Hayes, S., Moffett, J., Sterling, C.R., and Nicholson, W.L. (2003). Comparison of UV inactivation of spores of three *Encephalitozoon* species with that of spores of two DNA repair-deficient *Bacillus subtilis* biodosimetry strains. Appl. Environ. Microbiol. *69*, 683-685.

Marta, P., Brueckner, S., and Lueth, P. (1999). Plant growth promotion of different cultivated plants and biological control of soil-borne phytopathogenic fungi by *Bacillus subtilis* strain B2g. Zeitschrift fuer Pflanzenkrankheiten und Pflanzenschutz *106*, 74-81.

Marthi, B., Fieland, V.P., Walter, M., and Seidler, R.J. (1990). Survival of bacteria during aerosolization. Appl. Environ. Microbiol. *56*, 3463-3467.

Mastrapa, R.M.E., Glanzberg, H., Head, J.N., Melosh, H.J., and Nicholson, W.L. (2001). Survival of bacteria exposed to extreme acceleration: implications for panspermia. Earth Planet. Sci. Lett. *189*, 1-8.

Maughan, H., Birky, C.W., Jr., Nicholson, W.L., Rosenzweig, W.D., and Vreeland, R.H. (2002). The paradox of the "ancient" bacterium which contains "modern" protein-coding genes. Mol. Biol. Evol. *19*, 1637-1639.

McKillip, J.L., Small, C.L., Brown, J.L., Brunner, J.F., and Spence, K.D. (1997). Sporogenous midgut bacteria of the leafroller, *Pandemis pyrusana* (Lepidoptera: Tortricidae). Environ. Entomol. *26*, 1475-1481.

Melosh, H.J. (1985). Ejection of rock fragments from planetary bodies. Geology *13*, 144-148.

Melosh, H.J. (1988). The rocky road to panspermia Nature *332*, 687-688.

Michod, R.E. (1989). Origin and evolution of sex in prokaryotes. J. Cell. Biochem. Suppl *13C*, 87.

Michod, R.E. (1993). Genetic error, sex, and diploidy. J. Heredity *84*, 360-371.

Michod, R.E. and Wojciechowski, M.F. (1994). DNA repair and the evolution of transformation IV. DNA damage increases transformation. J. Evol. Biol.*7*, 147-175.

Michod, R.E., Wojciechowski, M.F., and Hoelzer, M.A. (1988). DNA repair and the evolution of transformation in the bacterium *Bacillus subtilis*. Genetics *118*, 31-40.

Michod, R.E., Wojciechowski, M.F., and Hoelzer, M.A. (1990). Evolution of sex in prokaryotes. In . UCLA Symposia on Molecular and Cellular Biology, New Series, vol. 122, M.T. Clegg and S.J. O'Brien, eds. (New York: Wiley-Liss Publishers), pp. 135-144.

Mileikowsky, C., Cucinotta, F.A., Wilson, J.W., Gladman, B., Horneck, G., Lindegren, L., Melosh, H.J., Rickman, H., Valtonen, M., and Zheng, J.Q. (2000). Natural transfer of viable microbes in space, part 1: From Mars to Earth and Earth to Mars. Icarus *145*, 391-427.

Miyagi, T., Peng, C.Y.S., Chuang, R.Y., Mussen, E.C., Spivak, M.S., and Doi, R.H. (2000). Verification of oxytetracycline-resistant American foulbrood pathogen *Paenibacillus larvae* in the United States. J. Invert. Pathol. *75*, 95-96.

Miyasono, M., Inagaki, S., Yamamoto, M., Ohba, K., Ishiguro, T., Takeda, R., and Hayashi, Y. (1994). Enhancement of delta-endotoxin activity by toxin-free spore of *Bacillus thuringiensis* against the Diamondback moth, *Plutella xylostella.*J. Invert. Pathol. *63*, 111-112.

Moehrle, M. (2001). Ultraviolet exposure in the Ironman triathlon. Med. Sci. Sports Exercise *33*, 1385-1386.

Moehrle, M., Dennenmoser, B., and Garbe, C. (2003). Continuous long-term monitoring of UV radiation in professional mountain guides reveals extremely high exposure. Int. J. Cancer *103*, 775-778.

Moehrle, M., Heinrich, L., Schmid, A., and Garbe, C. (2000a). Extreme UV exposure of professional cyclists. Dermatol. Basel *201*, 44-45.

Moehrle, M., Korn, M., and Garbe, C. (2000b). *Bacillus subtilis* spore film dosimeters in personal dosimetry for occupational solar ultraviolet exposure. Int. Arch. Occup. Environ. Health *73*, 575-580.

Morales-Ramos, L.H., McGuire, M.R., Galán-Wong, L.J. and Castro-Franco, R. (2000). Evaluation of pectin, gelatin and starch granular formulations of *Bacillus thuringiensis*. Southwestern Entomologist. *25*, 59-67.

Moran, N.A and Baumann, P. (2000). Bacterial endosymbionts in animals. Curr. Opinion Microbiol. *3*, 270-275.

Moreno, M.A. (1988). Microorganism transport from Earth to Mars. Nature *336*, 209.

Morris, O.N. (1977). Relationship between microbial numbers and droplet size in aerial spray applications. Canadian Entomologist. *109*, 1319-1324.

Mulla, M.S., Darwazeh, H.A., Davidson, E.W., Dulmage, H.T., and S. Singer. (1984). Larvicidal activity and field efficacy of *Bacillus sphaericus* strains against mosquito larvae and their safety to nontarget organisms. Mosquito News *44*, 336-342.

Munakata, N. (1999). Comparative measurements of solar UV radiation with spore dosimetry at three European and two Japanese sites. J. Photochem. Photobiol. B Biol. *53*, 7-11.

Munakata, N. (2000). Biomonitoring of solar-UV radiation at ten sites in Asia and Europe. J. Radiat. Res. *41*, 464.

Munakata, N., Ono, M., and Watanabe, S. (1998). Monitoring of solar-UV exposure among schoolchildren in five Japanese cities using spore dosimeter and UV-coloring labels. Japan. J. Cancer Res. *89*, 235-245.

Murphy, J.F., Zehnder, G.W., Schuster, D.J., Sikora, E.J., Polston, J.E., and Kloepper, J.W. (2000). Plant growth-promoting rhizobacterial mediated protection in tomato against tomato mottle virus. Plant Dis. *84*, 779-784.

Murrell, W.G. and Scott, W.J. (1966). The heat resistance of bacterial spores at various water activities. J. Gen. Microbiol. *43*, 411-425.

Nagy, B., Nagy, L.A., Rigaly, M.J., Jones, W.D., Krinsley, D.H. and Sinclair, N.A. (1991). Rock varnish in the Sonoran desert: microbiologically mediated accumulation of manganiferous sediments. Sedimentology *38*, 1153-1171.

Nealson K.H. and Ford, J. (1980). Surface enhancement of bacterial manganese oxidation: implications for aquatic environments. Geomicrobiol. J. *2*, 21–37.

Nicastro, A.J., Vreeland, R.H., and Rosenzweig, W.D. (2002). Limits imposed by ionizing radiation on the long-term survival of trapped bacterial spores: beta radiation. Int. J. Radiat. Biol. *78*, 891-901.

Nicholson, W.L. (2002). Roles of *Bacillus* spores in the environment. Cell. Mol. Life Sci. *59*, 410-416.

Nicholson, W.L. (2003). Using thermal inactivation kinetics to calculate the probability of extreme spore longevity: implications for paleomicrobiology and lithopanspermia. Orig. Life Evol. Biosphere *33*, 621-631.

Nicholson, W.L. and Fajardo-Cavazos, P. (1997). DNA repair and the UV resistance of bacterial spores: from the laboratory to the environment. Recent Res. Devel. Microbiol. *1*, 125-140.

Nicholson, W.L., Fajardo-Cavazos, P., Rebeil, R., Slieman, T.A., Riesenman, P.J., Law, J.F. and Xue, Y. (2002). Bacterial endospores and their significance in stress resistance. Antonie van Leeuwenhoek *81*, 27-32.

Nicholson, W.L. and Galeano, B. (2003). UV resistance of *Bacillus anthracis* spores revisited: validation of *Bacillus subtilis* spores as UV surrogates for spores of *B. anthracis* Sterne. Appl. Environ. Microbiol. *69*, 1327-1330.

Nicholson, W.L., Hargy, T.M., Marshall, M.M., and Durda, J.P. (2001). UV-sensitive spores of *Bacillus subtilis* strain provide a biodosimetry surrogate for *Cryptosporidium*. Proc. First Int. Cong. UV Technol. (CD-ROM).

Nicholson, W.L. and J.F. Law. (1999). Method for purification of bacterial endospores from soils: UV resistance of natural Sonoran desert soil populations of *Bacillus* spp. with reference to *B. subtilis* strain 168. J. Microbiol. Methods *35*, 13-21.

Nicholson, W.L., Munakata, N., Horneck, G., Melosh, H.J., and Setlow, P. (2000). Resistance of bacterial endospores to extreme terrestrial and extraterrestrial environments. Microbiol. Mol. Biol. Rev. *64*, 548-572.

Nicholson, W.L. and P. Setlow. (1990). Sporulation, germination, and outgrowth. In: Molecular Biological Methods for *Bacillus*, C.R. Harwood and S.M. Cutting, eds. (Sussex, England: John Wiley and Sons), pp. 391-450.

Nickle, D.C., Learn, G.H., Rain, M.W., Mullins, J.I., and Mittler. (2002). Curiously modern DNA for a "250-million-year-old" bacterium. J. Mol. Evol. *54*, 134-137.

Noah, D.L. and Crowder, H.R. (2002). Biological terrorism against animals and humans: a brief review and primer for action. J. Amer. Vet. Med. Assoc. *221*, 40-43.

Obeta, J.A.N. (1996). Effect of inactivation by sunlight on the larvicidal activities of mosquitocidal *Bacillus thuringiensis* H-14 isolates from Nigerian soils. J. Commun. Dis. *28*, 94-100.

Ochman, H., Elwyn, S., and Moran, N.A. (1999). Calibrating bacterial evolution. Proc. Natl. Acad. Sci. USA *96*, 12638-12643.

Oncul, O., Ozsoy, M.F., Gul, H.C., Kocak, N., Cavuslu, S., and Pahsa, A. (2002). Cutaneous anthrax in Turkey: a review of 32 cases. Scand. J. Infect. Dis. *34*, 413-416.

Painter, M.K., Tennessen, K.J., and Richardson, T.D. (1996). Effects of repeated applications of *Bacillus thuringiensis israelensis* on the mosquito predator *Erythemis simplicicollis* (Odonata: Libellulidae) from hatching to final instar. Environ. Entomol. *25*, 184-191.

Palmer, F.E., Staley, J.T., Murray, R.G.E., Counsell, T. and Adams, J.B. (1986). Identification of manganese-oxidizing bacteria from desert varnish. Geomicrobiol. J. *4*, 343-360.

Palmer, R.W. (1993). Short-term impacts of formulations of *Bacillus thuringiensis* var. *israelensis* de Barjac and the organophosphate temephos, used in blackfly (Diptera: Simuliidae) control, on rheophilic benthic macroinvertebrates in the middle Orange River, South Africa. Southern African J. Aquat. Sci. *19*, 14-33.

Pandey, A. and Palni, L.M.S. (1997). *Bacillus* species: the dominant bacteria of the rhizosphere of established tea bushes. Microbiol. Res. *152*, 359-365.

Pang, L., Close, M. and Noonan, M. (1998). Rhodamine WT and *Bacillus subtilis* transport through an alluvial gravel aquifer. Ground Water *36*, 112-122.

Pantastica-Caldas, M., Duncan, K.E., Istock, C.A., and Bell, J.A. (1992). Population dynamics of bacteriophage and *Bacillus subtilis* in soil. Ecology *73*, 1888-1902.

Passey, Q.R. and Melosh, H.J. (1980). The effect of atmospheric breakup on crater field formation. Icarus *42*, 211-233.

Patel, R., Piper, K., Cockerill, F.R. III, Steckelberg, J.M., and Yousten, A.A.. (2000). The biopesticide *Paenibacillus popilliae* has a vancomycin resistance gene cluster homologous to the enterococcal *vanA* vancomycin resistance gene cluster. Antimicrob. Agents Chemotherapy *44*, 705-709.

Podile, A.R. and Prakash, A.P. (1996). Lysis and biological control of *Aspergillus niger* by *Bacillus subtilis* AF 1. Can. J. Microbiol. *42*, 533-538.

Poinar, H.N., Hoess, M., Bada, J.L., and Paabo, S. (1996). Amino acid racemization and the preservation of ancient DNA. Science *272*, 864-866.

Potts, M. (1994). Desiccation tolerance of prokaryotes. Microbiol. Rev. *58*, 755-805.

Powell, K.A. and Jutsum, A.R. (1993). Technical and commercial aspects of biocontrol products. Pestic. Sci. *37*, 315-321.

Priest, F.G. (1993). Systematics and ecology of *Bacillus*. In *Bacillus subtilis* and Other Gram-Positive Bacteria: Biochemistry, Physiology and Molecular Genetics, A.L. Sonenshein, J.A. Hoch, and R. Losick, eds. (Washington, D.C.: ASM Press), pp. 3-16.

Quintern, L.E., Furusawa, Y., Fukutsu, K., and Holtschmidt, H. (1997). Characterization and application of UV detector spore films: the sensitivity curve of a new detector system provides good similarity to the action spectrum for UV-induced erythema in human skin. J. Photochem. Photobiol. B Biol. *37*, 158-166.

Riesenman, P.J. and Nicholson, W.L. (2000). Role of the spore coat layers in resistance of *Bacillus subtilis* spores to hydrogen peroxide, artificial UV-C, UV-B, and solar UV radiation Appl. Environ. Microbiol. *66*, 620-626.

Riessberger-Galle, U., von der Ohe, W., and Crailsheim, K. (2001). Adult honeybee's resistance against *Paenibacillus larvae larvae*, the causative agent of the American foulbrood. J. Invert. Pathol. *77*, 231-236.

Rippere, K., Johnson, J., Klein, M., and Yousten, A. (1996). DNA homologies among strains of milky disease bacteria. Abstr. Gen. Meeting Amer. Soc. Microbiol. *96*, 470.

Rippere, K., Patel, R., Uhl, J.R., Piper, K.E., Steckelberg, J.M., Kline, B.C., Cockerill, F.C. III, and Yousten, A.A. (1998). DNA sequence resembling *vanA* and *vanB* in the vancomycin-resistant biopesticide *Bacillus popilliae*. J. Infectious Dis. *178*, 584-588.

Rogers, S.O., Langenegger, K., and Holdenrieder, O. (2000). DNA changes in tissues entrapped in plant resins (the precursors of amber), Naturwissenschaften *87*, 70-75.

Rosenzweig, W.D., Woish, J., Peterson, J. and Vreeland, R. (2000). Development of a protocol to retrieve microorganisms from ancient salt crystals. Geomicrobiol. J. *17*, 185-192.

Rosson, R.A. and Nealson, K.H. (1982). Manganese binding and oxidation by spores of a marine *Bacillus*. J. Bacteriol *151*, 1027–1034.

Sarkar, A. (1991). Isolation and characterization of thermophilic, alkaliphilic, cellulose-degrading *Bacillus thermoalkalophilus* from termite (*Odontotermes obesus*) mound soil of a semiarid area. Geomicrobiol. J. *9*, 225-232.

Sattar, S.A., Synek, E.J., Westwood, J.C.N., and Neals, P. (1972). Hazard inherent in microbial tracers: reduction of risk by the use of *Bacillus stearothermophilus* spores in aerobiology. Appl. Microbiol. *23*, 1053-1059.

Saxena, D. and Stotzky, G. (2000). Insecticidal toxin from *Bacillus thuringiensis* is released from roots of transgenic Bt corn *in vitro* and *in situ*. FEMS Microbiol. Ecol. *33*, 35-39.

Sayre, R.M. (1993). *Pasteuria*, Metchnikoff, 1888. In *Bacillus subtilis* and Other Gram-Positive Bacteria: Biochemistry, Physiology and Molecular Genetics, A.L. Sonenshein, J.A. Hoch, and R. Losick, eds. (Washington, D.C.: ASM Press), pp. 101-111.

Schaeffer, P. Ionesco, H., and Jacob, F. (1959). Sur le déterminisme génétique de la sporulation bactérienne. C.R.H. Acad. Sci. Paris *249*, 577-578.

Schaeffer, P., J. Millet, and J.-P. Aubert. (1965). Catabolic repression of bacterial sporulation. Proc. Natl. Acad. Sci. USA *54*, 704-711.

Seldin, L., van Elsas, J.D. and Penido, E.C.G. (1984). *Bacillus azotofixans*, new species, a nitrogen-fixing species from Brazilian soils and grass roots. Int. J. Systematic Bacteriol. *34*, 451-456.

Setlow, P. (1995). Mechanisms for the prevention of damage to DNA in spores of *Bacillus* species. Ann. Rev. Microbiol. *49*, 29-54.

Setlow, P. (2001). Resistance of spores of *Bacillus* species to ultraviolet light. Environ. Mol. Mutagen. *38*, 97-104.

Siala, A. and Gray, T.R.G. (1974). Growth of *Bacillus subtilis* and spore germination in soil observed by a fluorescent antibody technique. J. Gen. Microbiol. *81*, 191-198.

Siala, A., Hill, I.R., and Gray, T.R.G. (1974). Populations of spore-forming bacteria in an acid forest soil, with special reference to *Bacillus subtilis*. J. Gen. Microbiol. *81*, 183-190.

Sinton, L.W., Noonan, M.J., Finlay, R.K., Pang, L., and Close, M.E. (2000). Transport and attenuation of bacteria and bacteriophages in an alluvial gravel aquifer. New Zealand J. Marine Freshwater Res. *34*, 175-186.

Skovmand, O. and Bauduin, S. (1997). Efficacy of a granular formulation of *Bacillus sphaericus* against *Culex quinquefasciatus* and *Anopheles gambiae* in West African countries. J. Vector Ecol. *22*, 43-51.

Slepecky, R.A. and Leadbetter, E.R. (1994). Ecology and relationships of endospore-forming bacteria: changing perspectives. In Regulation of Bacterial Differentiation, P.J. Piggot, C.P. Moran, Jr. and P. Youngman, eds. (Washington, D.C.: American Society for Microbiology), pp. 195-206.

Slieman, T.A. and Nicholson, W.L. (2000). DNA in dormant *Bacillus subtilis* spores exposed to artificial UV sources and to solar radiation accumulates single-strand breaks and cyclobutane pyrimidine dimers in addition to spore photoproduct. Appl. Environ. Microbiol. *66*, 199-205.

Slieman, T.A. and W.L. Nicholson. (2001). Role of dipicolinic acid in resistance of *Bacillus subtilis* spores exposed to artificial and solar UV radiation. Appl. Environ. Microbiol. *67*, 1274-1279.

Sonenshein, A.L., Hoch, J.A., and Losick, R., eds. (1993). *Bacillus subtilis* and Other Gram-Positive Bacteria: Biochemistry, Physiology and Molecular Genetics (Washington, D.C.: ASM Press).

Sonenshein, A.L., Hoch, J.A., and Losick, R., eds. (2002). *Bacillus subtilis* and Its Closest Relatives: From Genes to Cells (Washington, D.C.: ASM Press).

Spivak, M. and Reuter, G.S. (2001). Resistance to American foulbrood disease by honey bee colonies of *Apis mellifera* bred for hygienic behavior. Apidologie *32*, 555-565.

Spizizen, J. (1958). Transformation of biochemically deficient mutants of *Bacillus subtilis* by deoxyribonucleate. Proc. Natl. Acad. Sci. USA *44*, 1072-1078.

Stahly, D.P. and Klein, M.G. (1992). Problems with *in vitro* production of spores of *Bacillus popilliae* for use in biological control of the Japanese beetle. J. Invertebrate Pathol. *60*, 283-291.

Stassi, A., Zanardini, E., Cappitelli, F., Schiraldi, A., and Sorlini, C. (1998). Calorimetric investigations on the metabolism of *Bacillus* strains isolated from artistic stoneworks. Annal. Microbiol. Enzimol. *48*, 111-120.

Steinkraus, K.H., Granados, R.R., McKenna, K.A., Villani, M.G., Robbins, P.S., and Mortlock, R.P. (1998). Growth of *Paenibacillus larvae*, the causative agent of American foulbrood in honey bees and *Bacillus popilliae*, the causative agent of milky disease in beetle larvae in lepidopteran cell cultures. Acta Biotechnol. *18*, 123-133.

Stöffler, D. (2000). Maskelynite confirmed as diaplectic glass: indication for peak shock pressures of 45 GPa in all Martian meteorites. Proceedings of the 31st Lunar and Planetary Science Conference, Houston, TX.

Takahashi, I. (1961). Genetic transduction in *Bacillus subtilis*. Biochem. Biophys. Res. Commun. *5*, 171-175.

Tayler, S., and May, E. (1991). The seasonality of heterotrophic bacteria on sandstones of ancient monuments. Int, Biodeterior. *28*, 49-64.

Terriff, C.M., Schwartz, M.D., and Lomaestro, B.M. (2003). Bioterrorism: pivotal clinical issues: consensus review of the Society of Infectious Diseases Pharmacists. Pharmacotherapy *23*, 274-290.

Thomas-Keprta, K.L., Clemett, S.J., Bazylinski, D.A., Kirschvink, J.L., McKay, D.S., Wentworth, S.J., Vali, H., Gibson, E.K. Jr., and Romanek, C.S. (2003). Magnetofossils from ancient Mars: a robust biosignature in the martion meteorite ALH84001. Appl. Environ. Microbiol. *68*, 3663-3672.

Tokuda, Y., Takashi, A., and Shoda, M. (1995). Survival of *Bacillus subtilis* NB22 and its transformant in soil. Appl. Soil Ecol. *2*, 85-94.

Tran, L.M., Hilu, K. and Yousten, A.A. (1996). RAPD analysis of bacteria producing milky disease in scarab larvae. Abstr. Gen. Meeting Amer. Soc. Microbiol. *96*, 470.

Turner, J.T. and Backman, P.A. (1991). Factors relating to peanut yield increases after seed treatment with *Bacillus subtilis*. Plant Dis. *75*, 347-353.

Turtura, G.C., Perfetto, A., and Lorenzelli, P. (2000). Microbiological investigation on black crusts from open-air stone monuments of Bologna (Italy). Microbiol. Pavia. *23*, 207-228.

Tyndall, J. (1877). Further researches on the department and vital persistence of putrefactive and infective organisms from a physical point of view. Phil. Trans. Royal Soc. *167*, 149-206.

Upstill-Goddard, R.C., Suijlen, J.M., Malin, G. and Nightingale, P.D. (2001). The use of photolytic rhodamines WT and sulpho G as conservative tracers of dispersion in surface waters. Limnol. Oceanog. *46*, 927-934.

van Elsas, J.D., Dijkstra, A.F., Govaert, J.M. and van Veen, J.A. (1986). Survival of *Pseudomonas fluorescens* and *Bacillus subtilis* introduced into two soils of different texture in field microplots. FEMS Microbiol. Ecol. *38*, 151-160.

van Veen, J.A., van Overbeek, L.S. and van Elsas, J.D. (1997). Fate and activity of microorganisms introduced into soil. Microbiol. Mol. Biol. Rev. *61*, 121-135.

van Waasbergen, L. G., Hildebrand, M., and Tebo, B.M. (1996). Identification and characterization of a gene cluster involved in manganese oxidation by spores of the marine *Bacillus* sp. strain SG-1. J. Bacteriol. *178*, 3517–3530.

Vandenberg, J.D. (1990). Safety of four entomopathogens for caged adult honey bees (Hymenoptera: Apidae). J. Economic Entomol. *83*, 755-759.

Verschuere, L., Rombaut, G., Sorgeloos, P., and Verstraete, W. (2000). Probiotic bacteria as control agents in aquaculture. Microbiol. Mol. Biol. Rev. *64*, 655-671.

Vreeland, R.H. and Rosenzweig, W.D. (2002). The question of uniqueness of ancient bacteria. J. Indust. Microbiol. Biotechnol. *28*, 32-41.

Vreeland, R.H., Rosenzweig, W.D., and Powers, D.W. (2000). Isolation of a 250 million-year-old halotolerant bacterium from a primary salt crystal. Nature *407*, 897-900.

Walton, W.E. and Mulla, M.S. (1991). Integrated control of *Culex tarsalis* larvae using *Bacillus sphaericus* and *Gambusia affinis*: effects on mosquitoes and nontarget organisms in field mesocosms. Bull. Soc. Vector Ecol. *16*, 203-221.

Wedenig, M., Riessberger-Galle, U., and Crailsheim, K. (2003). A substance in honey bee larvae inhibits the growth of *Paenibacillus larvae larvae*. Apidologie *34*, 43-51.

Weiss, B.P., Kirschvink, J.L., Baudenbacher, F.J., Vali, H., Peters, N.T., MacDonald, F.A., and Wikswo, J.P. (2000). A low temperature transfer of ALH84001 from Mars to Earth. Science *290*, 791-795.

Williams, J.R., Peng, C.Y.S., Chuang, R.Y., Doi, R.H., and Mussen, E.C. (1998). The inhibitory effect of azadirachtin *on Bacillus subtilis, Escherichia coli*, and *Paenibacillus larvae*, the causative agent of American foulbrood in the honeybee, *Apis mellifera* L. J. Invert. Pathol. *72*, 252-257.

Wojciechowski, M.F., Hoelzer, M.A. and Michod, R.E. (1989). DNA repair and the evolution of transformation in *Bacillus subtilis*. II. Role of inducible repair. Genetics *121*, 411-422.

Xue, Y. and Nicholson, W.L. (1996). The two major DNA repair pathways, nucleotide excision repair and spore photoproduct lyase, are sufficient for the resistance of *Bacillus subtilis* spores to artificial UV-C and UV-B but not to solar radiation. Appl. Environ. Microbiol. *62*, 2221-2227.

Zhang, J., Hodgman, C.T., Krieger, L., Schnetter, W., and Schairer, H.U. (1997). Cloning and analysis of the first *cry* gene from *Bacillus popilliae*. J. Bacteriol. *179*, 4336-4341.

Chapter 2

Taxonomy and Systematics of the Aerobic Endospore Forming Bacteria: *Bacillus* and Related Genera

Dagmar Fritze

SUMMARY

The genus *Bacillus sensu stricto* presents itself today with about 90 species, and publications of new species appear with ever increasing frequency. Besides the well known *Bacillus* a small number of other Gram-positive, aerobic, endospore forming genera have been long-recognized and over the past eight years some twenty additional genera harbouring aerobic endospore-forming bacteria (AEFB) have been created. Together, these genera embrace more than 200 species. In terms of phylogenetic relationships, established on 16S rRNA sequence comparisons, these genera are classified in the order *Bacillales* together with a number of non-spore forming genera.

INTRODUCTION

Spores are particularly durable entities enabling an organism to survive unfavourable conditions, such as heat or drought, and have evolved in many groups of bacteria. Spore formation is found in anaerobic as well as aerobic organisms, and takes the form of both endo- or exospores. This review will focus on the aerobic endospore forming bacteria (AEFB) and thus will not discuss organisms from anaerobic endospore forming genera such as *Clostridium*, *Desulfotomaculum* or *Sporomusa* and also not the exospore forming genera including *Streptomyces* and related organisms.

However, a few exceptions will be made: e.g., the genus *Thermoactinomyces* forms true endospores and comparison of its rRNA sequences places it indeed within the radiation of the genus *Bacillus*, so it will, despite its conspicuous growth behaviour, be mentioned in this review. Additionally, a number of species will also be covered which have been described elsewhere as being anaerobic (e.g. *B. infernus, B. selenitireducens)*, or Gram-negative (e.g. *B. azotoformans, B. oleronius, B. horti)* or as being non-spore formers (e.g. *B. halodenitrificans, B. thermoamylovorans)*, but which, following the interpretation of their high rRNA sequence similarity, have been allocated to the genus *Bacillus*.

In 1876, Cohn and Koch independently demonstrated the presence of endospores in bacteria: Cohn described *B. subtilis* spores and demonstrated their heat resistance while Koch demonstrated the developmental life cycle of *B. anthracis*, from cell to resistant spore and from spore to cell. The ability to form spores offered itself as a defining feature of a taxon because it was such an unique characteristic. It was de Bary in 1884, who first incorporated sporulation as a characteristic of the genus *Bacillus* in his key for the classification of

bacteria. However, until the second decade of 1900, the use of this feature was not unequivocal and in a 1925 review of Buchanan (1925), where 35 bacterial classification keys of that time were compared, it could be seen that only in 19 of these was sporulation used to describe the genus *Bacillus*. Later classification schemes assigning all rod-shaped species to the genus *Bacillus,* resulted in a tremendous inflation of the genus and the subsequent problem of proper generic assignment of so many well known species. For example, since 1929 'Bacillus abortus' had been renamed *Brucella*, 'Bacillus acidophilus' as *Lactobacillus* and 'Bacillus aeruginosus' as *Pseudomonas*. These are just a few examples but the proper use of nomenclature has remained a problem for some time.

As an additional feature for classification of the genus *Bacillus,* aerobic growth had been introduced by Flugge as early as in 1907 in his identification keys and both characteristics, spore formation and aerobic growth, together with the statement 'usually Gram-positive' were more or less generally accepted from the late 1920's onwards.

The longevity of endospores is the predominant reason for the ubiquitous distribution of AEFB, which in turn results in their major impact on human activities. These organisms can be isolated from almost any natural habitat as well as from food, cosmetics, pharmaceuticals, etc. However, whether the individual site of isolation represents the real ecological habitat of a given organism or whether spores of that organism have been deposited there by wind or some other means can usually not be determined. Numerous points of potential entry into a given habitat must be taken into consideration (e.g. soil, dust, air, water, animals, food handling and processing equipment, transport vessels). On the other hand, virtually any habitat can be colonized by these organisms due to their remarkable diversity in physiological properties and thus in growth requirements (see Chapter 1).

Besides the concern of contamination and subsequent spoilage of materials by these organisms, certain taxa of AEFB have received considerable attention because of their outstanding applied importance. *Bacillus subtilis,* for example, is not only a renowned classical model organism for genetic research but is also widely used in traditional and industrial fermentation processes as well as in agriculture. Production of enzymes, antibiotics, antagonistic substances or surfactants or, on the other hand, degradation of xenobiotics are only some of the long list of applications of this and related organisms. Natto, a well known East Asian food, is based on the metabolic activity of 'Bacillus natto', (a strain of *B. subtilis)* on soy beans (see Chapter 12). Another example is *B. cereus,* of which certain non-toxin producing strains (e.g. *B. cereus* var. *'toyoi'*) have until recently been used as additives for animal feed (see Chapter 13). Interestingly, among the aerobic spore-forming bacteria *B. cereus* is the predominant causative agent for food poisoning and food spoilage (Chapter 4 and 8).

A number of opportunistic pathogens have been described in the AEFB, however, the only truly pathogenic organism (classified in risk group 3, the highest for bacteria), is *B. anthracis.*

While the term *'Bacillus'* is sometimes still thought of as embracing all Gram-positive, aerobic, endospore forming bacteria, it must be acknowledged that to date about 25 genera of AEFB comprising over 200 species have been validly published. To maintain such an enormous number of species in one single genus would be not acceptable in the long term. Accordingly, it is essential to establish additional taxa to account for the physiological and genetical breadth found in these organisms. An overview of the taxonomic development within the AEFB to the present day is outlined in Table 1 and will be discussed in more detail later.

WHY TAXONOMY? THE SPECIES CONCEPT IN BACTERIOLOGY

"Taxonomists are looked upon as providers of answers and not as sources of difficulties" (Pennington, 1994). However, name changes and reclassifications do need to occur as new information is gathered and sometimes these changes seemingly occur, for non-taxonomists, at an alarming frequency. Undoubtedly, a soundly based classification system depends on basic agreements and on a more universally accepted species definition that should not undergo permanent changes. While the ultimate goal of the modern phylogenetic approach is to assess the 'true', the 'natural' relationship of organisms, the species definitions in bacteriology are, to some extent, arbitrary and artificial, but also pragmatic. To introduce some control and clarity into systematics, bacterial taxonomists have given themselves operating rules for delimiting bacterial species. A major milestone in bacterial taxonomy was set by the introduction of the concept of species validation, and with the compilation and completion of the Approved Lists in 1980 (Skerman et al; updated in 1989) this year became the date for recognition of new names. That meant that names of species published before 1980 but not included in the Approved Lists

Table 1. Validly published genera and species of the aerobic endospore forming bacteria since 1980.

Appr. Lists 1980		Bergey's Manual 1986/89		up to 1996		up to Oct. 2003	
				Alicyclobacillus Wisotzkey et al. (1992)	3	*Alicyclobacillus*	8
						Ammoniphilus Zaitsev et al. (1998)	2
				Amphibacillus Niimura et al. (1990)	1	*Amphibacillus*	3
				Aneurinibacillus Shida et al. (1996)	2	*Aneurinibacillus*	3
						Anoxybacillus Pikuta et al. (2000)	3
Bacillus Cohn (1872)	31	*Bacillus*	34	*Bacillus*	79	*Bacillus*	88
				Brevibacillus Shida et al. (1996)	10	*Brevibacillus*	11
						Filobacillus Schlesner et al. (2001)	1
						Geobacillus Nazina et al. (2001)	10
						Gracilibacillus Wainö et al. (1999)	2
				Halobacillus Spring et al. (1996)	3	*Halobacillus*	5
						Jeotgalibacillus Yoon et al. (2001)	1
						Lentibacillus Yoon et al. (2002)	1
						Marinibacillus Yoon et al. (2001)	1
						Oceanobacillus Lu et al. (2001, 2002)	1
				Paenibacillus Ash et al. (1993, 1994)	12	*Paenibacillus*	45
						Salibacillus (1999)*	-
Sporolactobacillus Kitahara and Suzuki (1963)	1	*Sporolactobacillus*	1	*Sporolactobacillus*	1	*Sporolactobacillus*	5
Sporosarcina Kluyver and van Niel (1936)	1	*Sporosarcina*	2	*Sporosarcina*	1	*Sporosarcina*	6
				Sulfobacillus Golovacheva and Karavaiko (1978, 1991)	3	*Sulfobacillus*	3
						Thermaerobacter Takai (1999)	3
Thermoactinomyces Tsilinsky (1899)	5	*Thermoactinomyces*	7	*Thermoactinomyces*	6	*Thermoactinomyces*	6
						Thermobacillus Touzel et al. (2000)	1
						Ureibacillus Fortina et al. (2001)	2
						Virgibacillus Heyndrickx et al. (1998)	7
No of species	37		44		120		218
* the genus *Salibacillus* is no longer valid							

had no further standing in nomenclature and need not be taken into consideration when new species are described and named.

Through the adoption of the Bacteriological Code of Nomenclature (Lapage et al., 1975, 1992) and subsequent minor revisions, accepted rules for the description of a bacterial species exist. Laid down in these are established principles and recommendations to help to reduce nomenclatural difficulties. For example the original publication of the name of a new species should be made in the International Journal of Systematic and Evolutionary Microbiology (IJSEM, previously IJSB) or, if this publication has appeared elsewhere, through subsequent publication of the name of the species in the Validation List of the same journal. Another important and mandatory requisite recently added to the rules is, that the designated type strain of a new species must be deposited with two public service culture collections, in two different countries, preferably in two different regions of the world, from where this important reference material would be readily available to the scientific community worldwide. A list of type strains of aerobic endospore forming bacteria and their accession numbers in a selection of major microbial culture collections is provided in Table 2.

Recently an '*ad hoc* Committee for the Re-evaluation of the Species Definition in Bacteriology' of the International Committee for Systematic Bacteriology has reviewed and compiled the minimal descriptors for species delineation (Stackebrandt et al., 2002), updating the previous work of Wayne et al. (1987) by taking account of the introduction and availability of innovative methods for prokaryotic systematics. Besides a number of other recommendations, it was concluded that, for the time being, the parameters of DNA/DNA similarity and, whenever determinable, deltaT_m, remain the acknowledged standard for species delineation while it was agreed that new methods are needed to supplement and supplant this method. Species descriptions should include an almost complete 16S rDNA sequence determination and species should be identifiable by readily available methods (phenotypic and genotypic). It was also concluded that the phenotype continues to play a salient role in the decision about cut off points of genomic data for species delineation and that therefore it was necessary to base a species description on more than a single strain. The committee reinforced the earlier statement of Wayne et al. (1987) that new recommendations should be compatible with the current classification.

The availability of ever more sophisticated methods for the characterization of bacterial isolates (Figure 1) is responsible for much of the reclassification and delineation work, especially refinement of the genetic methodology that has had a major impact. However, a problem arises out of the methodological and taxonomical progress that service and research collections equally face. Newer taxonomic reclassification work is usually based on the type strain of a given species and, if at all, on an additional smaller set of strains of that species. This leaves, especially with the strain-rich traditional species, large numbers of strains in the various culture collections in the world unexamined.

In this context, an interesting taxonomic problem, that occurred and was resolved in the 1920's, should be mentioned since it concerns the origins of the original *B. subtilis* type strain. At that time, strains of two clearly different microorganisms were assigned the name of *B. subtilis* and held and used in many laboratories while originality was claimed for both of them. One of the strains could be traced

Table 2. List of species of aerobic endo-spore forming bacteria and collection numbers of their type strains (as of Dec. 2003).

Name of species	Collection Numbers of Type Strains					
	DSM	ATCC	CIP	JCM	LMG	NCIMB
Alicyclobacillus acidiphilus	DSM 14558					
Alicyclobacillus acidocaldarius						
subsp. *acidocaldarius*	DSM 446	ATCC 27009		JCM 5260	LMG 7119	NCIMB 11725
subsp. *rittmannii*	DSM 11297					
Alicyclobacillus acidoterrestris	DSM 3922	ATCC 49025			LMG 11152	NCIMB 13137
Alicyclobacillus cycloheptanicus	DSM 4006	ATCC 49028				
Alicyclobacillus herbarius	DSM 13609					
Alicyclobacillus hesperidum	DSM 12489					
Alicyclobacillus pomorum	DSM 14955					
Alicyclobacillus sendaiensis		ATCC BAA-609		JCM 11817		
Amphibacillus fermentum	DSM 13869					
Amphibacillus tropicus	DSM 13870					
Amphibacillus xylanus	DSM 6626	ATCC 51415		JCM 7361		
Ammoniphilus oxalaticus	DSM 11538	ATCC 700649				NCIMB 13535
Ammoniphilus oxalivorans	DSM 11537	ATCC 700648				NCIMB 13536
Aneurinibacillus aneurinilyticus	DSM 5562	ATCC 12856	CIP 104007	JCM 9024	LMG 15531	NCIMB 13690
Aneurinibacillus migulanus	DSM 2895	ATCC 9999	CIP 103841	JCM 8504		NCIMB 7096
Aneurinibacillus thermoaerophilus	DSM 10154				LMG 17165	NCIMB 13691
Anoxybacillus flavithermus	DSM 2641					
Anoxybacillus gonensis						NCIMB 13933
Anoxybacillus pushchinoensis	DSM 12423	ATCC 700785				
Bacillus agaradhaerens	DSM 8721	ATCC 700163				
Bacillus alcalophilus	DSM 485	ATCC 27647		JCM 5262	LMG 7120	NCIMB 10436
Bacillus amyloliquefaciens	DSM 7	ATCC 23350				NCIMB 12077

Bacillus anthracis		ATCC 14578				
Bacillus aquimaris	DSM 16205			JCM 11545		
Bacillus arseniciselenatis	DSM 15340	ATCC 700614				
Bacillus atrophaeus	DSM 7264	ATCC 49337		JCM 9070		NCIMB 12899
Bacillus azotoformans	DSM 1046	ATCC 29788	CIP 77.28			NCIMB 11859
Bacillus badius	DSM 23	ATCC 14574	CIP 58.52			NCIMB 9364
Bacillus barbaricus	DSM 14730					
Bacillus benzoevorans	DSM 5391	ATCC 49005				NCIMB 12555
Bacillus carboniphilus		ATCC 700100		JCM 9731		
Bacillus cereus	DSM 31	ATCC 14579	CIP 66.24		LMG 6923	
Bacillus chitinolyticus	DSM 11030					
Bacillus circulans	DSM 11	ATCC 4513			LMG 13261	
Bacillus clarkii	DSM 8720	ATCC 700162				
Bacillus clausii	DSM 8716	ATCC 700160	CIP 104718			NCIMB 10309
Bacillus coagulans	DSM 1	ATCC 7050	CIP 6025	JCM 2257	LMG 6326	NCIMB 9365
Bacillus cohnii	DSM 6307	ATCC 51227				
Bacillus decolorationis	DSM 14890				LMG 19507	
Bacillus edaphicus	DSM 12974					
Bacillus ehimensis	DSM 11029					
Bacillus endophyticus	DSM 13796		CIP 106778			NCIMB 13749
Bacillus fastidiosus	DSM 91	ATCC 29604				NCIMB 11326
Bacillus firmus	DSM 12	ATCC 14575	CIP 52.70	JCM 2512	LMG 7125	NCIMB 9366
Bacillus flexus	DSM 1320	ATCC 49095			LMG 11155	NCIMB 13366
Bacillus fumarioli					LMG 17489	NCIMB 13771
Bacillus funiculus	DSM 15141		CIP 107128	JCM 11201		
Bacillus fusiformis	DSM 2898	ATCC 7055				
Bacillus gibsonii	DSM 8722	ATCC 700164	CIP 104720			
Bacillus halmapalus	DSM 8723	ATCC 700165				
Bacillus haloalkaliphilus	DSM 5271	ATCC 700606				NCIMB 13457
Bacillus halodenitrificans	DSM 10037	ATCC 49067				
Bacillus halodurans	DSM 497	ATCC 27557				
Bacillus halophilus	DSM 4771	ATCC 49085				
Bacillus horikoshii	DSM 8719	ATCC 700161				
Bacillus horti	DSM 12751	ATCC 700778		JCM 9943		
Bacillus infernus	DSM 10277					
Bacillus insolitus	DSM 5	ATCC 23299	CIP 103268			NCIMB 11433
Bacillus jeotgali						
Bacillus krulwichiae				JCM 11691		NCIMB 13904
Bacillus laevolacticus	DSM 442	ATCC 23492	CIP 104421	JCM 2513	LMG 6329	NCIMB 10269
Bacillus lentus	DSM 9	ATCC 10840	CIP 52.74	JCM 2511		NCIMB 8773
Bacillus licheniformis	DSM 13	ATCC 14580	CIP 52.71	JCM 2505	LMG 6933	NCIMB 9375
Bacillus luciferensis					LMG 18422	
Bacillus megaterium	DSM 32	ATCC 14581	CIP 66.20	JCM 2506	LMG 7127	NCIMB 9376
Bacillus methanolicus		ATCC 51375				NCIMB 13113
Bacillus mojavensis	DSM 9205	ATCC 51516				NCIMB 13391
Bacillus mucilaginosus						
Bacillus mycoides	DSM 2048	ATCC 6462			LMG 7128	NCIMB 13305
Bacillus naganoensis	DSM 10191	ATCC 53909				
Bacillus nealsonii	DSM 15077	ATCC BAA-519				
Bacillus neidei	DSM 15031					
Bacillus niacini	DSM 2923					
Bacillus okuhidensis	DSM 13666			JCM 10945		
Bacillus oleronius	DSM 9356	ATCC 700005	CIP 104972			
Bacillus pallidus	DSM 3670	ATCC 51176				
Bacillus pseudalcaliphilus	DSM 8725	ATCC 700166				
Bacillus pseudofirmus	DSM 8715	ATCC 700159				NCIMB 10283
Bacillus pseudomycoides	DSM 12442					
Bacillus psychrodurans	DSM 11713					NCIMB 13837
Bacillus psychrosaccharolyticus	DSM 6	ATCC 23296				NCIMB 11729
Bacillus psychrotolerans	DSM 11706					NCIMB 13838
Bacillus pumilus	DSM 27	ATCC 7061	CIP 52.67	JCM 2508	LMG 7132	NCIMB 9369
Bacillus pycnus	DSM 15030					
Bacillus schlegelii	DSM 2000	ATCC 43741				NCIMB 13107
Bacillus selenitireducens	DSM 15326	ATCC 700615				
Bacillus silvestris	DSM 12223	ATCC BAA-269				
Bacillus simplex	DSM 1321	ATCC 49097				
Bacillus siralis	DSM 13140	CIP 106295				NCIMB 13601
Bacillus smithii	DSM 4216		CIP 103790	JCM 9076		
Bacillus sonorensis	DSM 13779					

Bacillus sphaericus	DSM 28	ATCC 14577	CIP 65.30	JCM 2502	LMG 7134	NCIMB 9370
Bacillus sporothermodurans	DSM 10599					NCIMB 13600
Bacillus subterraneus	DSM 13966	ATCC BAA-136				
Bacillus subtilis						
subsp. spizizenii	DSM 15029					
subsp. subtilis	DSM 10	ATCC 6051	CIP 52.65	JCM 1465	LMG 7135	NCIMB 3610
Bacillus thermantarcticus	DSM 9572					
Bacillus thermoamylovorans						
Bacillus thermocloacae	DSM 5250	ATCC 49805				NCIMB 13138
Bacillus thuringiensis	DSM 2046	ATCC 10792	CIP 53.137			NCIMB 9134
Bacillus tusciae	DSM 2912					
Bacillus vallismortis	DSM 11031					NCIMB 13505
Bacillus vedderi	DSM 9768	ATCC 700130				NCIMB 13458
Bacillus vulcani	DSM 13174		CIP 106305			
Bacillus weihenstephanensis	DSM 11821					
Brevibacillus agri	DSM 6348	ATCC 51663		JCM 9067		
Brevibacillus borstelensis	DSM 6347	ATCC 51668	CIP 104545	JCM 9022		NCIMB 13369
Brevibacillus brevis	DSM 30	ATCC 8246	CIP 52.86	JCM 2503	LMG 16703	NCIMB 9372
Brevibacillus centrosporus	DSM 8445	ATCC 51661		JCM 9071		
Brevibacillus choshinensis	DSM 8552	ATCC 51359	CIP 103838	JCM 8505		NCIMB 13345
Brevibacillus formosus	DSM 9885	ATCC 51669	CIP 104544	JCM 9169		NCIMB 13368
Brevibacillus invocatus		ATCC BAA-715	CIP 106911		LMG 18962	NCIMB 13772
Brevibacillus laterosporus	DSM 25	ATCC 64	CIP 52.83	JCM 2496	LMG 16000	NCIMB 9367
Brevibacillus parabrevis	DSM 8376	ATCC 10027	CIP 103840	JCM 8506		NCIMB 13346
Brevibacillus reuszeri	DSM 9887	ATCC 51665	CIP 104543	JCM 9170		NCIMB 13367
Brevibacillus thermoruber	DSM 7064					
Filobacillus milosensis	DSM 13259	ATCC 700960				
Geobacillus caldoxylosilyticus	DSM 12041	ATCC 700356				
Geobacillus kaustophilus	DSM 7263	ATCC 8005				NCIMB 8547
Geobacillus stearothermophilus	DSM 22	ATCC 12980	CIP 66.23	JCM 2501	LMG 6939	NCIMB 8923
Geobacillus subterraneus	DSM 13552					
Geobacillus thermocatenulatus	DSM 730					
Geobacillus thermodenitrificans	DSM 465	ATCC 29492				
Geobacillus thermoglucosidasius	DSM 2542	ATCC 43742				NCIMB 11955
Geobacillus thermoleovorans	DSM 5366	ATCC 43513				NCIMB 12698
Geobacillus toebii	DSM 14590					
Geobacillus uzenensis	DSM 13551					
Gracilibacillus dipsosauri	DSM 11125	ATCC 700347	CIP 105095			NCIMB 703027
Gracilibacillus halotolerans	DSM 11805	ATCC 700849				
Halobacillus halophilus	DSM 2266	ATCC 35676				NCIMB 2269
Halobacillus karajensis	DSM 14948				LMG 21515	
Halobacillus litoralis	DSM 10405	ATCC 700076				
Halobacillus salinus				JCM 11546		
Halobacillus trueperi	DSM 10404	ATCC 700077				
Jeotgalibacillus alimentarius				JCM 10872		
Lentibacillus salicampi		ATCC BAA-719	CIP 707807	JCM 11462		
Marinibacillus marinus	DSM 1297	ATCC 29841				NCIMB 2140
Oceanobacillus iheyensis	DSM 14371			JCM 11309		
Paenibacillus agarexedens	DSM 1327		CIP 107437			
Paenibacillus agaridevorans	DSM 1355		CIP 107436			
Paenibacillus alginolyticus	DSM 5050	ATCC 51185	CIP 103122	JCM 9068		NCIMB 12517
Paenibacillus alvei	DSM 29	ATCC 6344	CIP 66.18		LMG 6922	NCIMB 9371
Paenibacillus amylolyticus	DSM 11730	ATCC 9995		JCM 9906		NCIMB 13390
Paenibacillus apiarius	DSM 5581	ATCC 29575				NCIMB 13506
Paenibacillus azoreducens	DSM 13822		CIP 107224			NCIMB 13761
Paenibacillus azotofixans	DSM 5976	ATCC 35681			LMG 9815	NCIMB 12093
Paenibacillus borealis	DSM 13188					
Paenibacillus brasilensis	DSM 14914	ATCC BAA-413				
Paenibacillus campinasensis						
Paenibacillus chibensis	DSM 11731	ATCC 9966		JCM 9905		NCIMB 13574
Paenibacillus chinjuensis	DSM 15045			JCM 10939		
Paenibacillus chondroitinus	DSM 5051	ATCC 51184	CIP 103123	JCM 9072		NCIMB 12518
Paenibacillus curdlanolyticus	DSM 10247	ATCC 51898	CIP 104575			NCIMB 13444
Paenibacillus daejeonensis	DSM 15491			JCM 11237		
Paenibacillus dendritiformis						
Paenibacillus glucanolyticus	DSM 5162	ATCC 49278				NCIMB 12809
Paenibacillus glycanilyticus				JCM 11221		
Paenibacillus graminis	DSM 15220	ATCC BAA-95			LMG 19080	
Paenibacillus granivorans						

Paenibacillus illinoisensis	DSM 11733			JCM 9907		
Paenibacillus jamilae	DSM 13815					
Paenibacillus kobensis	DSM 10249					
Paenibacillus koleovorans				JCM 11186		
Paenibacillus koreensis						
Paenibacillus kribbensis				JCM 11465		
Paenibacillus larvae						
subsp. *larvae*	DSM 7030	ATCC 9545			LMG 9820	
subsp. *pulvifaciens*	DSM 3615	ATCC 49843	CIP 104764		LMG 6911	NCIMB 11201
Paenibacillus lautus	DSM 3035	ATCC 43898	CIP 52.74		LMG 11157	NCIMB 12780
Paenibacillus lentimorbus		ATCC 14707				NCIMB 11202
Paenibacillus macerans	DSM 24	ATCC 8244	CIP 66.19	JCM 2500	LMG 13281	NCIMB 9368
Paenibacillus macquariensis	DSM 2	ATCC 23464	CIP 103269		LMG 13289	NCIMB 9934
Paenibacillus naphthalenovorans	DSM 14203	ATCC BAA-206				
Paenibacillus nematophilus	DSM 13559					NCIMB 13845
Paenibacillus odorifer	DSM 15391	ATCC BAA-93			LMG 19079	
Paenibacillus pabuli	DSM 3036	ATCC 43899	CIP 103119	JCM 9074	LMG 11158	NCIMB 12781
Paenibacillus peoriae	DSM 8320	ATCC 51925	CIP 103812		LMG 14832	NCIMB 13389
Paenibacillus polymyxa	DSM 36	ATCC 842	CIP 66.22	JCM 2507	LMG 13294	NCIMB 8158
Paenibacillus popilliae		ATCC 14706				
Paenibacillus stellifer	DSM 14472					
Paenibacillus terrae				JCM 11466		
Paenibacillus thiaminolyticus	DSM 7262			JCM 8360		NCIMB 13158
Paenibacillus turicensis	DSM 14349					
Paenibacillus validus	DSM 3037	ATCC 43897	CIP 103120	JCM 9077	LMG 11161	NCIMB 12782
Sporolactobacillus inulinus	DSM 20348	ATCC 15538	CIP 103279	JCM 6014	LMG 11481	NCIMB 9743
Sporolactobacillus kofuensis				JCM 3419		
Sporolactobacillus lactosus				JCM 9690		
Sporolactobacillus nakayamae						
subsp. *nakayamae*	DSM 11696	ATCC 700379		JCM 3514		
subspec. *racemicus*	DSM 16324	ATCC 700381		JCM 3417	LMG 18785	
Sporolactobacillus terrae	DSM 11697	ATCC 700380		JCM 3516		
Sporosarcina aquimarina	DSM 14554			JCM 10887		
Sporosarcina globispora	DSM 4	ATCC 23301		JCM 2509		NCIMB 11434
Sporosarcina macmurdoensis	DSM 15428		CIP 107784			
Sporosarcina pasteurii	DSM 33	ATCC 11859				NCIMB 8841
Sporosarcina psychrophila	DSM 3	ATCC 23304				
Sporosarcina ureae	DSM 2281	ATCC 6473		JCM 2577		NCIMB 9251
Sulfobacillus acidophilus	DSM 10332	ATCC 700253				
Sulfobacillus disulfidooxidans	DSM 12064	ATCC 51911				
Sulfobacillus thermosulfidooxidans	DSM 9293					
Thermaerobacter marianensis	DSM 12885	ATCC 700841		JCM 10246		
Thermaerobacter nagasakiensis	DSM 14512			JCM 11223		
Thermaerobacter subterraneus	DSM 13965	ATCC BAA-137				
Thermoactinomyces dichotomicus	DSM 44778	ATCC 49854		JCM 9688		NCIMB 10211
Thermoactinomyces intermedius	DSM 43846	ATCC 33205		JCM 3312		
Thermoactinomyces peptonophilus	DSM 44666	ATCC 27302		JCM 10113		
Thermoactinomyces putidus	DSM 44608	ATCC 49853		JCM 8091		NCIMB 12324
Thermoactinomyces sacchari	DSM 43356	ATCC 27375		JCM 3137		NCIMB 10486
Thermoactinomyces vulgaris	DSM 43016	ATCC 43649		JCM 3162		NCIMB 11364
Thermobacillus xylanilyticus						
Ureibacillus terrenus	DSM 12654	ATCC BAA-384			LMG 19470	
Ureibacillus thermosphaericus	DSM 10633					
Virgibacillus carmonensis	DSM 14868				LMG 20964	
Virgibacillus marismortui	DSM 12325	ATCC 700626	CIP 105609			
Virgibacillus necropolis	DSM 14866				LMG 19488	
Virgibacillus pantothenticus	DSM 26	ATCC 14576	CIP 51.24		LMG 7129	NCIMB 8775
Virgibacillus picturae	DSM 14867				LMG 19492	
Virgibacillus proomii	DSM 13055				LMG 12370	NCIMB 13609
Virgibacillus salexigens	DSM 11483	ATCC 700290				

Collections are listed from which major proportions of the AEFB type strains are available. DSM = DSMZ - German collection of microorganisms and cell cultures, www.dsmz.de; ATCC: American type culture collection, www.atcc.org; CIP: Collecion of the Institute Pasteur, www.cip.fr; JCM: Japan Collection of Microorganisms, www.jcm.jp; LMG: Collection of the laboratory for microbiology in Ghent, www.lmg.be; NCIMB: National Collections of Industrial and Marine Bacteria, www.ncimb.uk; other collections from which considerable numbers of type strains of AEFB can be obtained are: CCM: www.ccm.cz; NRRL: www.nrrl.usa; CCUG: www.ccug.se; VKM: www.vkm.ru; NCTC: www.nctc.uk. More collections can be derived from WFCC-WDCM at www.wfcc.info .

Some of the above listed type strains are only available from specific collections: Paenibacillus campinasensis - KCTC 0364BP; *Paenibacillus dendritiformis* - BGSC 30A1; *Paenibacillus granivorans* - CBS 229.89; *Paenibacillus koreensis* - KCCM 40903 and KCTC 2393; *Thermobacillus xylanilyticus* - CNCM I-1017.

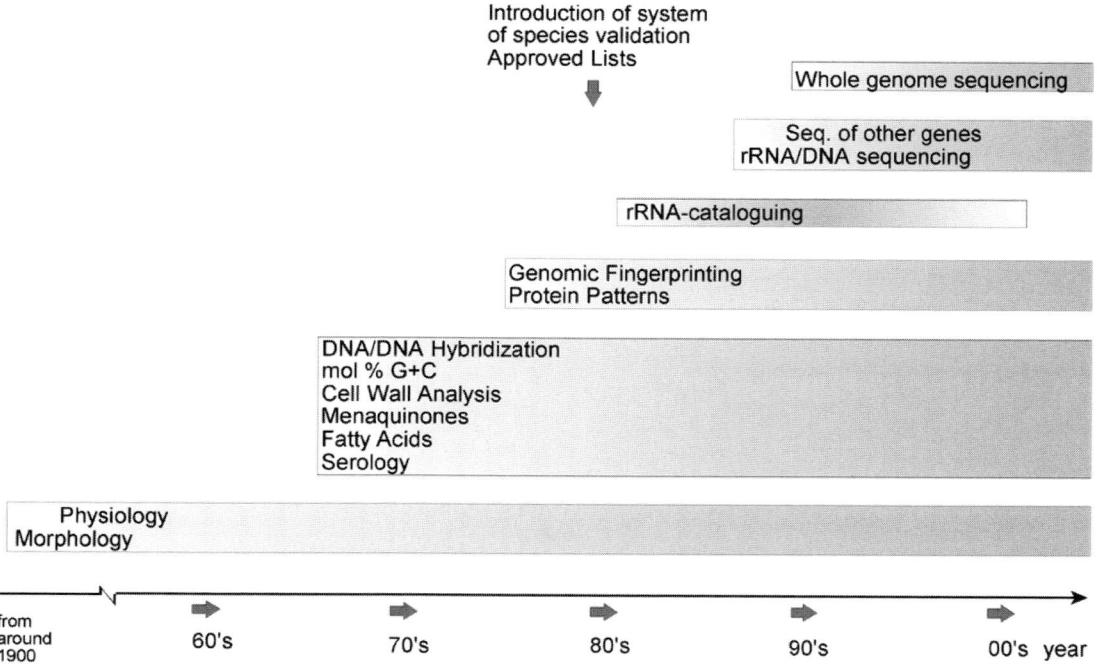

Figure 1. Development and Application of Taxonomic Methods over time.

back to the laboratory of Novy at the University of Michigan who had brought it from Koch's laboratory in 1888. The other could be traced back to a strain from the University of Marburg, said to have been isolated by Meyer and used in his 1899 work. Only by a meticulous comparative and detective work, especially by going back to the very early descriptions of that species, it could be resolved that the Marburg strain most closely represented the original culture of *B. subtilis* and it was this strain which was ultimately adopted by the Second International Congress for Microbiology in London 1936, as the neo-type of *B. subtilis*. The other strain, in all probability, was a member of the *B. cereus* group.

With a view to the culture collection problem mentioned above, it needs to be remembered that especially in smaller research collections, 'old' cultures of *B. subtilis* may be held, for which neither the classical nor the up-to-date assignment of the correct species name has been made and it is possible that cultures bearing the name of *B. subtilis* are in fact strains from the *B. cereus* group.

SPORE AND SPORANGIUM MORPHOLOGY AS A TAXONOMIC FEATURE IN THE GENERA OF AEFB

Being such an obvious characteristic, the shape of the spore, its position in and effect on the cell were used by Smith et al. (1952) and Gordon et al. (1973) as taxonomic characters and as a means to collate species into larger groups. In Table 3 an attempt is made to group all presently accepted species of AEFB into categories as defined by Gordon et al. (1973) plus some additional ones. The original category 1 is here divided into 1a (cells > 1 μm, spore ellipsoid, sporangium not swollen) and 1b (cells <1 μm, spore ellipsoid, sporangium not swollen),

while 2 (spore ellipsoid, sporangium swollen) and 3 (spore round, sporangium swollen) remained as before. Additional categories are established to include some of the new species: Category 4 includes those with round and ellipsoidal spores and swollen sporangium and Category 5 for those with round spores and non-swollen sporangia. Two others were established for groups that fall strongly out of the original scheme: one for organisms with round cells and round spores and one for organisms with mycelial growth and round spores while a further group embraces organisms in which spore formation has not been observed so far.

It is important to note that not only cell size and shape but also features such as the shape of spores and sporangia can be culture dependent and are thus features which should be determined using standardized culture conditions. An example from our own laboratory is the recently described species *B. psychrodurans* (Abd El-Rahman, 2002) that produces ellipsoid spores on Difco Marine Agar (Difco 2216) and round spores on CASO Agar (Merck 5458).

Even in the era of genetically based systematics, spore and sporangium morphology may still be of relevance. For example, in diagnostic identification, a straightforward microscopic examination by an experienced worker can rule out the presence of a '*B. cereus* like' organism.

RECOGNITION AND DEMONSTRATION OF SPORES

Several techniques are available for the recognition and demonstration of spores. One is the direct method of viewing stained or unstained mounts under the microscope while others are indirect methods such as pasteurisation or chemical determination of dipicolinic acid. Determination of

Table 3. Morphological groups of aerobic endospore forming bacteria, based on Gordon et al. (1973); amended and updated until 2003.

Group I. Cells rod shaped, Spores ellipsoid, Sporangia not swollen,

Subgroup 1: cell diameter > 1 μm; spore size:

B. anthracis	*B. benzoevorans*	*B. cereus*	*B. endophyticus*
B. fastidiosus	*B. funiculus*	*B. megaterium*	*B. mycoides*
B. nealsonii	*B. pseudomycoides*	*B. thuringiensis*	*B. weihenstephanensis*

Subgroup 2: cell diameter ≤ 1 μm; spore size:

Amm. oxalaticus (also II)	*Amm. oxalivorans* (also II)	*B. alcalophilus*	*B. amyloliquefaciens*
B. atrophaeus	*B. badius*	*B. clausii* (also II)	*B. coagulans* (also II)
B. edaphicus	*B. endophyticus*	*B. firmus*	*B. flexus*
B. fumarioli	*B. gibsonii*	*B. halmapalus*	*B. halophilus*
B. luciferensis (also II)	*B. krulwichiae*	*B. lentus*	*B. licheniformis*
B. mojavensis	*B. mucilaginosus*	*B. oleronius*	*B. pseudofirmus*
B. pumilus	*B. simplex*	*B. smithii* (also II)	*B. sonorensis*
B. sporothermodurans	*B. subtilis* (2 ssp.)	*B. vallismortis*	*B. vulcani* (also II)
Hal. litoralis (also II)	*Hal. trüperi* (also II)		

Group II. Cells rod shaped, Spores ellipsoid, Sporangia swollen; spore size:

Ali. acidiphilus	*Ali. acidocaldarius* (2 ssp.)	*Ali. acidoterrestris*	*Ali. cycloheptanicus*
Ali. herbarius	*Ali. hesperidum*	*Ali. pomorum*	*Amm. oxalaticus* (also I)
Amm. oxalivorans (also I)	*Amp. fermentum*	*Amp. tropicus*	*Amp. xylanus*
Ane. aneurinilyticus	*Ane. migulanus*	*Ane. thermoaerophilus*	*B. agaradhaerens*
B. aquimaris	*B. azotoformans*	*B. barbaricus*	*B. carboniphilus*
B. chitinolyticus	*B. circulans*	*B. clarkii*	*B. clausii* (also I)
B. coagulans (also I)	*B. cohnii*	*B. decolorationis*	*B. ehimensis*
B. fusiformis	*B. halodurans*	*B. horikoshii*	*B. horti*
B. jeotgali	*B. laevolacticus*	*B. luciferensis* (also I)	*B. marisflavi*
B. methanolicus	*B. naganoensis*	*B. niacini*	*B. okuhidensis*
B. pallidus	*B. pseudalcaliphilus*	*B. psychrosaccharolyticus*	*B. siralis*
B. smithii (also I)	*B. thermantarcticus*	*B. thermoamylovorans*	*B. thermocloacae*
B. tusciae	*B. vedderi*	*B. vulcani* (also I)	*Brb. agri*
Brb. borstelensis	*Brb. brevis*	*Brb. centrosporus*	*Brb. choshinensis*
Brb. formosus	*Brb. invocatus*	*Brb. laterosporus*	*Brb. parabrevis*
Brb. reuszeri	*Brb. thermoruber*	*Geo. caldoxilosilyticus*	*Geo. kaustophilus*
Geo. stearothermophilus	*Geo. subterraneus*	*Geo. thermocatenulatus*	*Geo. thermodenitrificans*
Geo. thermoglucosidasius	*Geo. thermoleovorans*	*Geo. toebii*	*Geo. uzenensis*
Gra. halotolerans	*Hal. litoralis* (also I)	*Hal. salinus*	*Hal. trüperi* (also I)
Oce. iheyensis	*P. agarexedens*	*P. agaridevorans*	*P. alginolyticus*
P. alvei	*P. amylolyticus*	*P. apiarius*	*P. azoreducens*
P. azotofixans	*P. borealis*	*P. brasilensis*	*P. campinasensis*
P. chibensis	*P. chinjuensis*	*P. chondroitinus*	*P. curdlanolyticus*
P. daejonensis	*P. dendritiformis*	*P. glucanolyticus*	*P. glycanilyticus*
P. graminis	*P. granivorans*	*P. illinoisensis*	*P. jamilae*
P. kobensis	*P. koleovorans*	*P. koreensis*	*P. kribbensis*
P. larvae (2 ssp.)	*P. lautus*	*P. lentimorbus*	*P. macerans*
P. macquariensis	*P. naphthalenovorans*	*P. nematophilus*	*P. odorifer*
P. pabuli	*P. peoriae*	*P. polymyxa*	*P. popilliae*
P. stellifer	*P. terrae*	*P. thiaminolyticus*	*P. turicensis*
P. validus	*Spl. inulinus*	*Spl. kofuensis*	*Spl. lactosus*
Spl. nakayamae (2 ssp.)	*Spl. terrae*	*Sul. acidophilus*	*Sul. disulfidooxidans*
Sul. thermosulfidooxidans	*Thb. xylanilyticus*	*Vir. carmonensis*	*Vir. marismortui*
Vir. necropolis	*Vir. picturae*	*Vir. salexigens*	

Group III. Cells rod shaped, Spores round, Sporangia swollen; spore size:

Ali. sendaiensis	*B. haloalkaliphilus*	*B. neidei*	*B. pycnus*
B. psychrodurans	*B. psychrotolerans*	*B. schlegelii*	*B. silvestris*
Fil. milosensis	*Gra. dipsosauri*	*Jeo. alimentarius*	*Sps. aquimarina*
Sps. globispora	*Sps. pasteurii*	*Sps. psychrophila*	*Tha. subterraneus*
Ure. terrenus			

Group IV. Cells rod shaped, Spores round and ellipsoid, Sporangia swollen; spore size:			
Ano. flavithermus	*B. sphaericus*	*Hal. karajensis*	*Len. salicampi*
Ure. thermosphaericus	*Vir. pantothenticus*	*Vir. proomii*	
Group V. Cells rod shaped, Spores round, Sporangia not swollen; spore size:			
Ano. pushchinoensis	*Ano. gonensis*	*B. insolitus*	*Mar. marinus*
Species Not Allocated to Above Morphological Groups:			
Round Cells, round Spores; spore size:			
Hal. halophilus	*Sps. ureae*		
Mycelial Growth, round Spores, swollen Sporangia; spore size:			
Tham. dichotomicus	*Tham. intermedius*	*Tham. peptonophilus*	*Tham. putidus*
Tham. sacchari	*Tham. vulgaris*		
No Spores demonstrated:			
B. arseniciselenatis	*B. halodenitrificans*	*B. infernus*	*B. selenitireducens*
B. subterraneus	*B. thermoamylovorans*	*Tha. marianensis*	*Tha. nagasakiensis*
Abbreviations of genera: *Ali.* = *Alicyclobacillus, Amm.* = *Ammoniphilus, Amp.* = *Amphibacillus, Ane.* = *Aneurinibacillus, Ano.* = Anoxybacillus, *B.* = Bacillus, *Brb.* = *Brevibacillus, Fil.* = *Filobacillus, Gra.* = *Gracilibacillus, Hal.* = *Halobacillus, Jeo.* = *Jeotgalibacillus, Len.* = *Lentibacillus, Mar.* = Marinibacillus, *Oce.* = *Oceanobacillus, P.* = *Paenibacillus, Sal.* = *Salibacillus, Spl.* = *Sporolactobacillus, Sps.* = *Sporosarcina, Sul.* = *Sulfobacillus, Tha.* = Thermaerobacter, *Tham.* = *Thermoactinomyces, Thb.* = *Thermobacillus, Ure.* = *Ureibacillus, Vir.* = *Virgibacillus,*			

dipicolinic acid, a compound typically found in endospores (Fahmy, 1985), may become necessary to definitively prove that an observed cell structure is, indeed, an endospore when no other method is possible. Pasteurisation is widely used in applied fields to prove the existence of spores but unless heat-treatment is extremely precise this method has its limitations. For example, heat treatment must be sufficient to expose every biological entity for enough time to kill all vegetative cells and phenomena such as clumping can enhance apparent survival. Another factor is that spores of different species can have different resistance properties, with some spores being unable to survive more than 55°C while others are able to survive near boiling temperatures! With those organisms with low spore heat resistance, selection for spores should be done by ethanol treatment (Koransky et al., 1978).

In older literature, staining procedures are described such as the malachite green stain for microscopic examination. If successfully stained, green spores are clearly apparent and compelling evidence for the presence of spores although, in practice, this method has proven unreliable. Another interesting method is the 'acid popping' of spores (Robinow, 1951; Warth, 1979) where a fixed microscopic mount is viewed while a strong acid is passed through the preparation. On contact with the acid, spores will rehydrate and lose their high refractility and turn immediately from bright to dark (Figure 2).

The easiest and quickest way to demonstrating endospores in a culture is by phase contrast microscopy (Figure 3) using an oil-immersion objective with x100 magnification. In simple native preparations highly refractile spores are easily seen as bright bodies within or outside dark contrasted cells. Lipid globules or air bubbles which might be present may have a somewhat similar appearance but are easily distinguished with some experience. For better viewing (and for micro-photography), it is sometimes useful to immobilize cells and to keep them in focus by using agar coated microscopic slides (Claus and Berkeley, 1986, p. 1109). This helps to better interpret shape and size of cells, sporangia and spores. This technique also avoids tumbling motions of the cell, which can often result in vertically positioned rod shaped cells which can then, when viewed 'end-on', resemble spores.

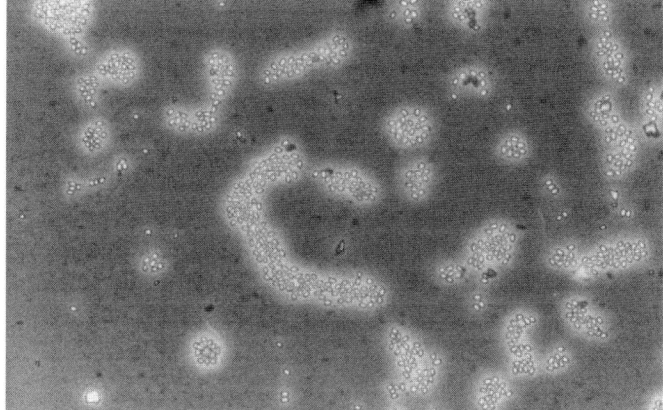
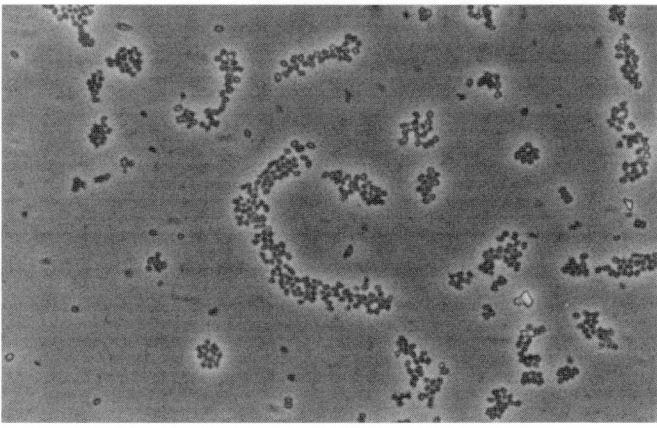

Figure 2. Phase contrast micrograph of a fixed spore preparation before and after "acid popping"; note loss of refractility.

PHYSIOLOGY AS A TAXONOMIC FEATURE IN THE GENERA OF AEFB

As mentioned before, virtually any habitat can be colonized by one of the many AEFB. Slepecky (1992) has compiled a comprehensive overview over the various sources of isolation of the species of *Bacillus* and their general growth requirements.

The physiological abilities of the genera and species hitherto described are reported to range from acidophilic (growth down to pH 1.5) to neutrophilic and alkaliphilic (growth up to pH 11), from psychrophilic (growth down to minus 5°C) to mesophilic and thermophilic (growth up to 78°C) as well as comprising heterotrophs and chemoautotrophs, N_2-fixing organisms, halophiles and facultative anaerobes. Figure 4 exemplifies the temperature and pH range covered by a selection of genera of AEFB. It can be seen that, despite the recent taxonomic efforts, the genus *Bacillus* still spans the widest physiological range. Table 4 specifies in more detail

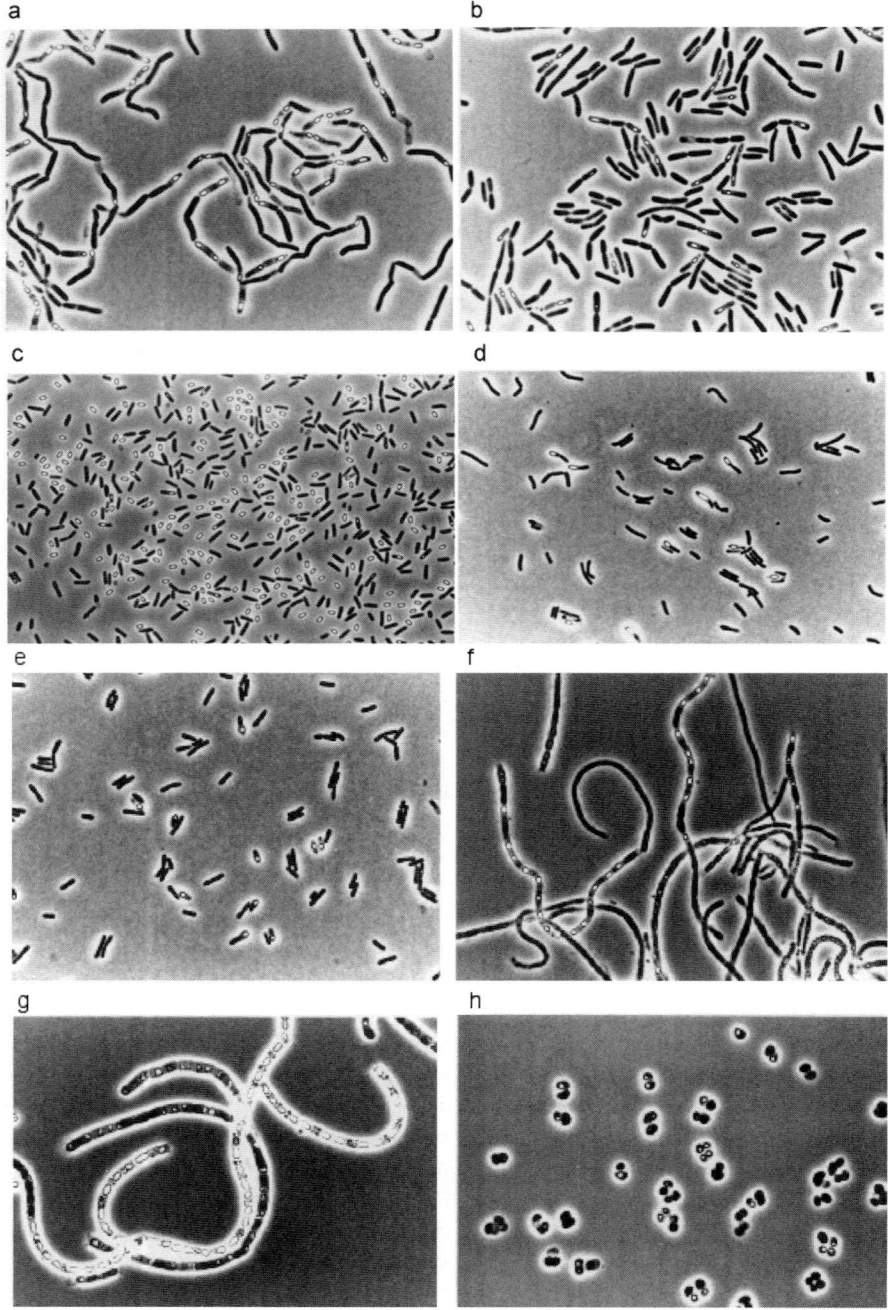

Figure 3. Phase contrast photomicrographs of type strains.
a. DSM 32^T *B. megaterium*, b. DSM 31^T *B. cereus*, c. DSM 10^T *B. subtilis* subsp. *subtilis*, d. DSM 22^T *Geo. stearothermophilus*, e. DSM 28^T *B. sphaericus*, f. DSM 5391^T *B. benzoevorans*, g. DSM 15141^T *B. funiculus*, h. DSM 2266^T *Hal.* (ex *Sporosarcina*) *halophilus*; all photographs at the same magnification.

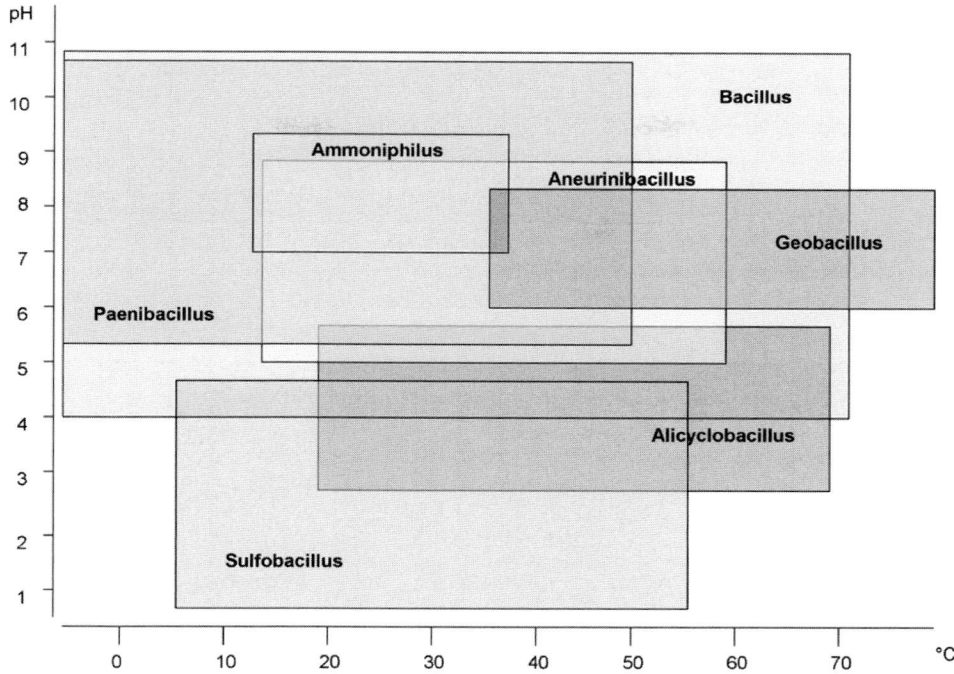

Figure 4. Physiological Diversity of Aerobic Endo-Spore Forming Bacteria.

the abilities of individual species to grow under certain culture conditions. Some species can be found in more than one group, taking into account their broader temperature and/or pH ranges. However, more species than indicated here may be able to grow under broader conditions, because the upper and lower limits of temperature or pH tolerances have not been determined for all species. This is especially true for the lower temperature limit and for the pH range. The cut off points of 5°C and 55°C as well as pH 4.5 and pH 9 have been selected with a view to the best discrimination of the species. As most research to date has been done in the mesophilic/neutral range, the group of organisms growing at 30°C/pH 7.0-7.2 is by far the largest and can therefore not be completely presented. More or less, all of the known species of aerobic endospore forming bacteria not included in the present table can be added to the mesophilic/neutral group in the central sector of Table 4.

Growth requirements of AEFB with respect to media composition may vary considerably. Numerous media have been formulated for the various kinds of aerobic spore forming bacteria (e.g. Atlas, 1995; Atlas 1997; DSMZ-Catalogue of Strains, www.dsmz.de). Some of these have been developed with the aim of simply optimising growth of a given organism, others with the aim of inducing or enhancing spore formation or to prevent loss of specific properties. Most aerobic sporeforming bacteria will grow on plain nutrient agar (0.8% peptone + 0.3% beef extract) with media containing soya peptone often giving the better results. Some species need the addition of glucose or other carbohydrates to nutrient agar while organisms like *B. sphaericus* need organic acids rather than carbohydrates. Other organisms have more specific

nutritional requirements, e.g *B. fastidiosus* cannot grow without uric acid or allantoin, *Sps. pasteurii* needs NH_4^+ and alkaline pH, and the species of *Ammoniphilus* are obligately ammonium dependent. Acidification or alkalisation of the basic medium may allow growth of most of the acidiphilic or alkaliphilic species. For some, however, more sophisticated media are needed. Figure 5 shows the maximum NaCl tolerance of a selection of species and their coinciding ability to grow over a certain pH range.

PHYSIOLOGICAL TESTS: DEFINITION AND STANDARDISATION

The use of physiological abilities as taxonomic markers in *Bacillus* has been evaluated mainly by Smith et al. (1952) and Gordon et al. (1973) and a set of about 30 classical tests compiled, which had been made more widely available through Claus and Berkeley (1986). With a view to the current size and diversity of the AEFB group, these tests are not sufficient to describe and discriminate a species, however, as applied for and tested with over 1000 strains by Gordon et al. (1973), these can provide a sound basis for comparisons. Furthermore, these tests have proven appropriate for this group of organisms to be studied. As far as specific physiological demands allow, these tests may be applied to other genera of AEFB and our own studies have shown that most of the tests are easily adapted to alkaline pH (Fritze et al., 1990). The same may be expected for acidic pH, however, care must be taken with solid media, where the agar must be autoclaved separately, since when autoclaved at a pH lower than approximately 5, the agar will not solidify on cooling.

Table 4. Upper and lower range of pH and temperature for growth of species of AEFB.

pH 9 / 5°C	pH 9 / 30°C	pH 9 / 55°C
B. agaradhaerens B. alcalophilus	B. agaradhaerens B. alcalophilus	
	B. clarkii	
	B. halodurans	B. halodurans
	B. haloalkaliphilus	
B. pseudalkaliphilus B. pseudofirmus	B. pseudalkaliphilus B. pseudofirmus	
		B. thermocloacae
	B. vedderi	
	B. coagulans B. halodurans B. licheniformis B. methanolicus B. smithii B. thermoamylovorans Brb. thermoruber Ure. thermosphaericus Thb. xylanilyticus	B. coagulans B. halodurans B. licheniformis B. methanolicus B. smithii B. thermoamylovorans Brb. thermoruber Ure. thermosphaericus Thb. xylanilyticus
B. insolitus B. weihenstephanensis B. psychrosaccharolyticus Mar. marinus Spo. globispora Spo. psychrophila	B. insolitus B. weihenstephanensis B. psychrosaccharolyticus Mar. marinus Spo. globispora Spo. psychrophila	
Pae. macquariensis	Ane. aneurinolyticus Ane. migulanus B. cereus, B. circulans, B. cohnii, B. clausii, B. gibsonii, B. halmapalus, B. horikoshii, B. horti, B. niacini, B. laevolacticus, B. lentus, B. subtilis, B. megaterium, B. mycoides, B. pumilus, B. thuringiensis Brb. brevis Pae. polymyxa	Ane. thermoaerophilus Geo. kaustophilus B. pallidus, B. schlegelii, B. tusciae Geo. stearothermophilus Geo. thermocatenulatus Geo. thermodenitrificans Geo. thermoglucosidasius Geo. thermoleovorans
pH 7 / 5°C	(pH 7 / 30°C only representative)	pH 7 / 55°C
not described to date	Ali. acidoterrestris Ali. hesperidum B. coagulans Sul. thermosulfidooxidans	Ali. acidoterrestris Ali. hesperidum B. coagulans Sul. thermosulfidooxidans
		Ali. acidocaldarius
	Ali. cycloheptanicus B. laevolacticus B. naganoensis Sul. acidophilus Sul. disulfidooxidans	
pH 4.5 / 5°C	pH 4.5 / 30°C	pH 4.5 / 55°C

When physiological testing is applied, the descriptions of Smith et al. (1952) and Gordon et al. (1973) should be precisely followed. For many physiological test media several recipes have been described in the literature and these will yield differing results. For example, problems often occur when testing for urease activity. This should be tested in a medium not containing peptone. Most *Bacillus* species are strongly proteolytic and so any rise in pH may arise from liberated ammonium. Similarly, tests for starch hydrolysis will yield different results depending on whether iodine or 95%

Growth up to % NaCl

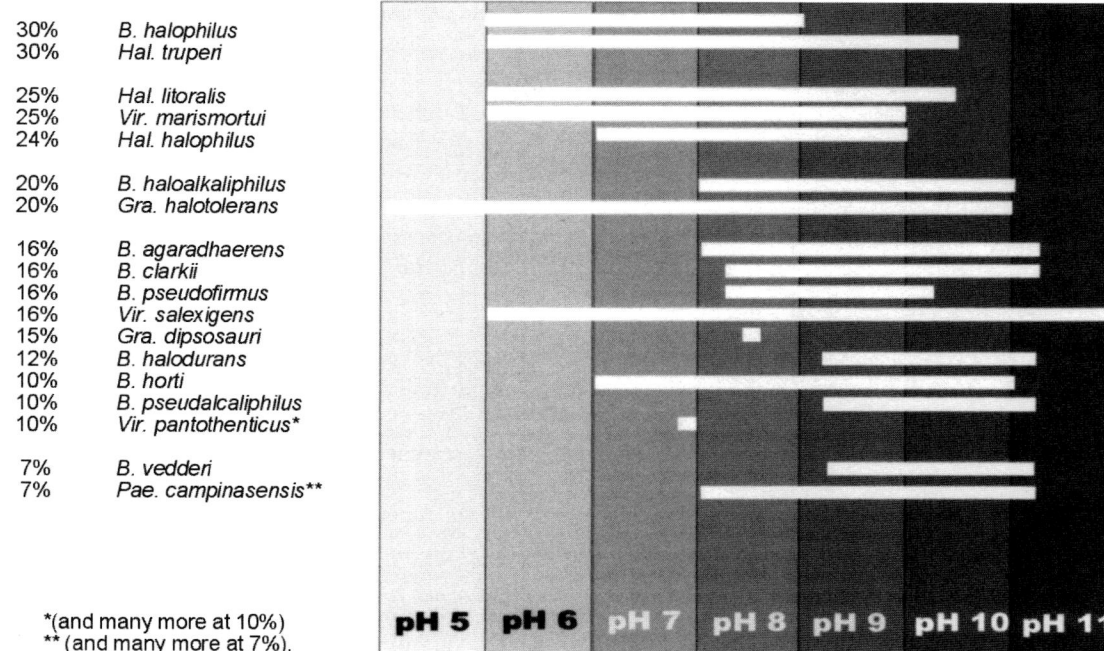

30%	*B. halophilus*
30%	*Hal. truperi*
25%	*Hal. litoralis*
25%	*Vir. marismortui*
24%	*Hal. halophilus*
20%	*B. haloalkaliphilus*
20%	*Gra. halotolerans*
16%	*B. agaradhaerens*
16%	*B. clarkii*
16%	*B. pseudofirmus*
16%	*Vir. salexigens*
15%	*Gra. dipsosauri*
12%	*B. halodurans*
10%	*B. horti*
10%	*B. pseudalcaliphilus*
10%	*Vir. pantothenticus**
7%	*B. vedderi*
7%	*Pae. campinasensis***

*(and many more at 10%)
** (and many more at 7%).

pH 5 pH 6 pH 7 pH 8 pH 9 pH 10 pH 11

Figure 5. Maximum NaCl tolerance and pH range of a selection of AEFB.

ethanol is used for revealing non-hydrolyzed starch. With the nitrate reduction test, a false negative result may be recorded if a combined reagent is used instead of the two separate solutions, as described by Gordon et al. (1973). Unfortunately, it is a fact that results of different authors are often difficult to compare because the methods applied were not thoroughly standardized. Therefore attempts to design dichotomous keys based on physiological characteristics for the identification of aerobic endo-spore forming bacteria (Slepecky, 1992) must fail because of the non-comparability of published properties of species.

Also based on physiological tests, a number of commercial identification kits have been developed (e.g. API, BIOLOG, VITEK, etc.) for the rapid and standardized identification of *Bacillus*. These systems are quite helpful, though only within their abilities. It should be kept in mind that results obtained with one system are only comparable with those obtained with the same system, and not with those obtained with other test kits or the conventional test tube/Petri dish tests. Another disadvantage of the commercial systems lies in the relatively low frequency of updating the underlying databases, especially for organisms of minor medical importance, such as environmental species. The result is, that the range of genera and species (e.g. of aerobic spore forming bacteria) included in these systems is usually relatively restricted, which makes the identification of new isolates difficult, if not false. As thoughtful interpretation of the results is needed, these systems cannot replace the experienced worker.

MODERN DEVELOPMENTS AND PRESENT STATUS IN TAXONOMY AND SYSTEMATICS OF THE AEFB

With growing numbers of species it has become necessary to develop additional means to prove identity or novelty of organisms. Methods for the determination of cellular compounds, referred to as chemotaxonomy, (e.g. cell wall composition, whole cell fatty acids or polar lipids), as well as the determination of genetic traits, such as mol% G+C content of the DNA and DNA-DNA hybridization have become available and proven useful (Figure 1). In an attempt to find a feature independent from growth conditions and present across all taxa, thus enabling direct comparison of all genera and species with the aim to establish a phylogenetically based system, analysis of 16S rRNA seemed to offer an ideal target. For the genus *Bacillus* the work of Ash et al. (1991) laid a first basis for revising the genus phylogenetically. Since then, determination of the 16S rRNA/DNA sequences has become the agreed basis for delineating taxa, however, it is becoming increasingly accepted that 16S rDNA methods are appropriate to determine approximate phylogenetic positions at the levels of genera and higher taxa, but they are not suitable to classify strains at the species level (Stackebrandt and Goebel, 1994). Here, phenotypic data need to be added. In Table 6, chemotaxonomical data are compiled which are available for the published genera, however, it must be kept in mind that data have been determined often for only a few strains and it can be seen that gaps with missing information preclude definitive interpretation and that the presently available type of information does not yet offer sufficient discriminatory power.

DISSECTION OF THE GENERA *BACILLUS* AND *SPOROSARCINA*

Four genera of AEFB are listed in the Approved Lists (1980): *Bacillus, Sporolactobacillus, Sporosarcina* and *Thermoactinomyces,* and all four are still valid today. *Sporolactobacillus* was formed from strains described as being intermediate to *Lactobacillus* and *Bacillus* and is differentiated from the latter genus mainly through their homolactic fermentation and their response to oxygen, since optimal growth occurs under microaerophilic conditions. *Thermoactinomyces* was originally created to harbour thermophilic *Actinomyces*-like organisms, where its mycelial-like growth places it easily within that group- but its true endospores and 16S rRNA/DNA similarities allocate it within the radiation of the genus *Bacillus* and related genera. However, the two remaining 'old' genera *Bacillus* and *Sporosarcina* have undergone considerable changes (see Table 5). To date, 10 novel genera have been created into which species have been placed that had been previously allocated to the genus *Bacillus*. The main argument for dissection for all of them was dissimilarities in their 16S rRNA/DNA sequences with the type species *B. subtilis*. For only some have additional salient characteristics been described. With *Alicyclobacillus,* embracing thermo-acidophilic organisms, an unusual type of fatty acid had been found: omega-alicyclic fatty acids, occurring as cyclohexan and cycloheptan fatty acids. Originally the genus *Anoxybacillus* had been described as being obligately anaerobic, however, recently Pikuta et al. (2003) revealed its aerotolerant, facultatively anaerobic nature and amended the species description. One common character of the species of *Geobacillus* is their ability to grow at high temperatures, however, others species such as *B. pallidus* (growth up to 70°C) or *Alicyclobacillus acidocaldarius* (growth up to 70°C) share this property. *Ureibacillus* is described as moderately thermophilic and as having an L-lys-D-asp peptidoglycan composition.

Most of the newly formed genera, such as *Aneurinibacillus, Brevibacillus* and *Paenibacillus,* differ from the parent genus and from each other in some phenotypic properties, but these are in most cases not exclusive. The genus *Gracilibacillus* has been described as being halotolerant, *Marinibacillus* and *Virgibacillus* are moderately halophilic. Two species of the genus *Bacillus (B. marismortui* and *B. salexigens)* had each been transferred to a new genus *Salibacillus* but as they were recently reclassified as *Virgibacillus,* the genus *Salibacillus* is no longer valid. It is interesting to see that the remaining core genus *Bacillus sensu stricto* still accommodates those 'traditional' species which are most widely known (often referred to as the '*subtilis*-group' or the '*cereus*-group') such as *B. subtilis* (the type species of the genus), *B. licheniformis,* and *B. pumilus* or *B. anthracis, B. cereus, B. mycoides,* and *B. thuringiensis,* as well as *B. megaterium, B. sphaericus* and many others, and is with more than 90 species by far the largest genus of the AEFB.

Originally designed for tetrads forming cocci, the genus *Sporosarcina* contained until 1995 two species: *Sporosarcina ureae* and *Sporosarcina halophila*. In 1996, one of the species, *Sporosarcina halophila,* was transferred (on the grounds of 16S rDNA sequences) to a newly formed genus, *Halobacillus,* together with two rod shaped species. It was thus renamed *Halobacillus halophilus* leaving *Sporosarcina ureae* the only species in that genus. On the other hand, quite recently, three species of the genus *Bacillus* namely *B. globisporus, B. psychrophilus* and *B. pasteurii* have been transferred to the genus *Sporosarcina,* as *Sporosarcina globispora, Sporosarcina psychrophila* and *Sporosarcina pasteurii.*

ADDITIONALLY FORMED GENERA OF AEFB

To date, the genera *Ammoniphilus, Amphibacillus, Filobacillus, Halobacillus, Jeotgalibacillus, Lentibacillus, Oceanobacillus, Sulfobacillus, Thermaerobacter* and *Thermobacillus* have been newly established to harbour aerobic endospore forming species containing new isolates (Table 1). Here, too, decisions to describe new genera were based predominantly on sequence analysis. Unfortunately, very often these genera contain only one species, which itself is sometimes based on one strain

Table 5. Species of the genera *Bacillus* and *Sporosarcina* (genera listed on the Approved Lists 1980) transferred to new genera (as of 2003).

Previous Species Name		Present Genus
B. acidocaldarius, B. acidoterrestris, B. cycloheptanicus	=>	*Alicyclobacillus*
B. aneurinilyticus, B. migulanus, B. thermoaerophilus	=>	*Aneurinibacillus*
'*B. flavothermus*'	=>	*Anoxybacillus*
B. agri, B. borstelensis, B. brevis, B. centrosporus, B. choshinensis, B. formosus, B. laterosporus, B. parabrevis, B. reuszeri, B. thermoruber	=>	*Brevibacillus*
B. kaustophilus, B. stearothermophilus, B. thermocatenulatus, B. thermodenitrificans, B. thermoglucosidasius, B. thermoleovorans	=>	*Geobacillus*
B. dipsosauri	=>	*Gracilibacillus*
B. marinus	=>	*Marinibacillus*
'*B. agar-exedens*' , *B. alginolyticus, B. alvei, B. amylolyticus, B. chondroitinus, B. curdlanolyticus, B. glucanolyticus, B. larvae, B. lautus, B. lentimorbus, B. macerans, B. macquariensis, B. pabuli, B. peoriae, B. polymyxa, B. popilliae, B. thiaminolyticus, B. validus*	=>	*Paenibacillus*
B. globisporus, B. pasteurii, B. psychrophilus	=>	*Sporosarcina*
B. thermosphaericus	=>	*Ureibacillus*
B. marismortui (>Salibacillus marismortui>), B. pantothenticus, B. salexigens (>Salibacillus salexigens>)	=>	*Virgibacillus*
Sporosarcina halophila	=>	*Halobacillus*
'...' indicates that the species had been described but not validly published; *B. marismortui* and *B. salexigens* had been reclassified as *Salibacillus* before being transferred to *Virgibacillus.* The genus *Salibacillus* is no longer valid.		

Table 6. Some chemotaxonomic data of genera of AEFB.

Genus	Cell wall composition	aerobic and/or anaerobic growth	Gram staining	Quinone	Major whole cell fatty acids	growth type
Alicyclobacillus	?	aerobic	Gram-+ Gram-v	MK7	omega-alicyclic fatty acids	thermo-acidophilic mod. thermophilic
Ammoniphilus	?	aerobic	Gram-v	MK7	?	obligate oxalotroph
Amphibacillus	meso-DAP	aerobic, fac. anaerobic	Gram-+	none	anteiso-C15 anteiso-C16 iso-C16, iso-C15	
Aneurinibacillus	meso-DAP	aerobic	Gram-+	MK7	iso-C15	
Anoxibacillus	?	aerobic fac. anaerobic	Gram-+	?	iso-C15 C16	mod. thermophilic alkaliphilic
Bacillus	mainly meso-DAP L-orn-D-glu, L-lys-D-asp, L-orn-D-asp,	aerobic, fac. and obl. anaerobic	Gram-+ Gram-v Gram--	MK7, v	v	v
Brevibacillus	meso-DAP	strictly aerobic fac. anaerobic	Gram-+ Gram-v	MK7	anteiso-C15 iso-C15	
Filobacillus	L-orn-D-glu	strictly aerobic	Gram-type + Gram-react. -	?	?	halophilic alkalitolerant
Geobacillus	meso-DAP	aerobic fac. anaerobic	Gram-+	MK7	iso-C15 iso-C16, iso-C17	thermophilic
Gracilibacillus	meso-DAP	aerobic	Gram-+	MK7	iso-C15 anteiso-C15 iso-C16 anteiso-C17	halotolerant
Halobacillus	L-orn-D-asp	aerobic	Gram-+	MK7	iso-C15 anteiso-C15 anteiso-C17	mod. halophilic
Jeotgalibacillus	L-lys direct	aerobic fac. anaerobic	Gram-v	MK7	iso-C15 anteiso-C15	mod. halophilic
Lentibacillus	meso-DAP	aerobic	Gram-v	MK7	anteiso-C15 iso-C16	halophilic
Marinibacillus	L-lys direct*	strictly aerobic	Gram-+	MK7	anteiso-C15 iso-C15	mod. halophilic
Oceanobacillus	?	aerobic	Gram-+	MK7	iso-C15 anteiso-C15 iso-C14	halotolerant alkalitolerant
Paenibacillus	meso-DAP	strictly aerobic fac. anaerobic	Gram-+ Gram-v Gram--	MK7	anteiso C15	
Sporolactobacillus	meso-DAP	aerobic fac. anaerobic	Gram-+	MK7	anteiso-C15 anteiso-C17 iso-C15, iso-C17	homolactic fermentation
Sporosarcina	L-lys-gly-D-glu L-lys-D-glu L-lys-ala-D-asp	strictly aerobic	Gram-+	MK7	?	
Sulfobacillus	?	aerobic	Gram-+	?	omega-alicyclic fatty acids	thermo-acidophilic, fac. autotroph
Thermaerobacter	?	strictly aerobic	Gram type + Gram react. v	?	iso-C17 anteiso-C17	thermophilic
Thermobacillus	?	aerobic	Gram--	MK7	?	mod. thermophilic
Thermoactinomyces	meso-DAP	aerobic	Gram--	MK7, MK8, MK9	?	mod. thermophilic
Ureibacillus	L-lys-D-asp	aerobic	Gram--	MK7, MK8 MK9	iso-C16	mod. thermophilic
Virgibacillus	meso-DAP	aerobic fac. anaerobic	Gram-+	MK7	iso-C15 anteiso-C15	mod. halophilic

+ = positive; - = negative; v = variable; * = alpha-carbonylgroup of glutamic acid is substituted by a glycine residue; polar lipids have been determined for some genera: *Lentibacillus*: phosphatidylglycerol (PG), diphosphatidylglycerol (DPG); *Ureibacillus*: PG, DPG; *Virgibacillus*: PG, DPG, phosphatidyletanolamin (PE).

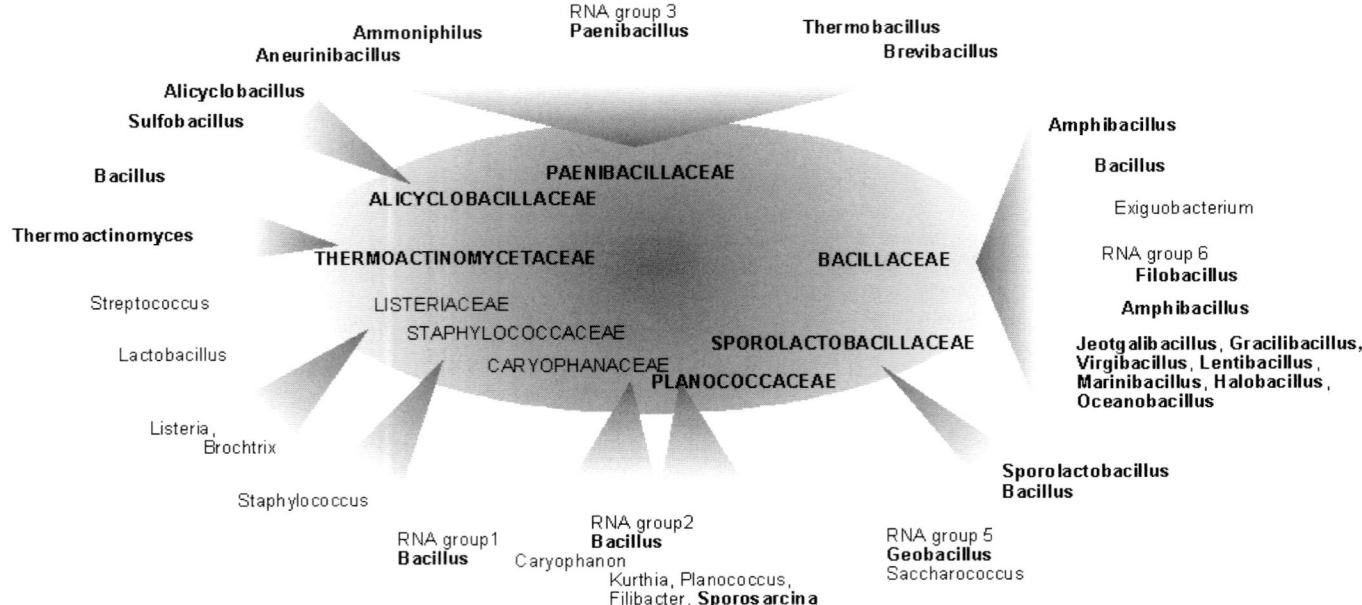

Figure 6. Systematic positions of the AEFB within the order Bacillales. Tentative harmonization of the taxonomic outline of Bergey's Manual (2003) and the schematic outline of the phylogenetic diversity of 16S rDNA of Stackebrandt and Swidersky (2002).

only. In these cases descriptions need to be interpreted carefully. As with the genera mentioned further above, only a few phenotypic characters have been determined which can help to distinguish between the genera.

Ammoniphilus has been described as a genus of ammonium-dependent and obligately oxalotrophic organisms. *Amphibacillus* has been described as being a link between the anaerobic genus *Clostridium* and the aerobic *Bacillus,* optimal growth occurs under microaerophilic conditions but spore formation occurs under aerobic as well as anaerobic conditions. In addition, the organisms of this genus lack catalase. The orn-D-glu type of the peptidoglycan characterizes the halophilic and alkalitolerant genus *Filobacillus.* For the moderately halophilic *Halobacillus* also an unusual peptidoglycan type, orn-D-asp, was established. Again a different cell wall composition, L-lys direct, was determined for *Jeotgalibacillus* which is also moderately halophilic. *Lentibacillus* is halophilic and *Oceanobacillus* is described as being halo- and alkalitolerant. The genus *Sulfobacillus* is another genus of thermoacidophilic organisms producing omega-alicyclic fatty acids (cyclohexan and cycloheptan fatty acids) - its species are facultative autotrophs and have a slightly lower pH optimum and a clearly lower pH minimum than those of the genus *Alicyclobacillus.* Both latter genera have closely related rRNA sequences and their nearest phylogenetic neighbours are the species *B. tusciae* and *B. schlegelii;* interestingly, both are described as being thermophilic and facultatively autotroph. In addition, *B. tusciae* is moderately acidophilic, while *B. schlegelii* is neutrophilic. *Thermobacillus* is a moderately thermophilic organism.

The genus *Thermaerobacter* is mentioned here because it is a Gram-positive, aerobically growing, endospore forming organism. Its 16S rDNA sequences, however, do not place the

three species of this genus within the radiation of *Bacillus* and related genera but with the anaerobic genera *Clostridium* and *Moorella.*

Whether these new definitions stand firm or are prone to renewed changes will be seen in future. With some of the new genera, differentiation from others may be difficult because of early decisions on cut off points on phylogenetic trees, when distances between branching points were still convincingly long. However, as these distances shrink, while more and more strains are sequenced and new species described, main points of branching become less obvious and may need to be interpreted in the light of additional phenotypic data, as was stated by the *ad hoc* committee (Stackebrandt et al., 2002). The individual decision of a taxonomist on where to cut off a genus may add to the difficulty in correlating phenotypic characters. Problems may also occur when sequence data are compared with those stored in public databases. It should to be taken into consideration that stored sequences may differ between strains of the same species, but it may happen that even for one and the same type strain several differing sequences can be retrieved, and it is often difficult to decide on the quality of these sequences. While it is clear that isolates should always be compared with the type strains of species, problems remain if several sequences are available. Unfortunately, inaccurate sequences cannot be removed from these databases and so selection and careful interpretation is necessary.

Consequent application of the 16S rRNA/DNA philosophy to higher taxa has resulted in grouping of the AEFB genera with well known non-spore forming genera. In the Taxonomic Outline of the new Bergey (URL dx.doi.org/ 10.1007/bergeysoutline) 10 families are placed within the order *Bacillales* of which the *Caryophanaceae, Listeriaceae* and *Staphylococcaceae* contain only non-spore-forming genera.

Within the families *Bacillaceae* and *Planococcaceae,* two (*Bacillaceae*) and three (*Planococcaceae*) non-spore-forming genera are placed side by side with the above described spore forming genera. In Figure 6 a tentative harmonisation of the taxonomic outline and the phylogenetic diversity of 16S rDNA as compiled by Stackebrandt and Swidersky (2002) is shown.

CONCLUSIONS AND OUTLOOK

Despite overwhelming developments, it seems as if the historical definitions of *Bacillus* may still be applied today for the whole group of AEFB. All members of the group are aerobic or facultatively anaerobic, though some anaerobic species are accepted following their allocation through sequence analysis, most of them are clearly Gram-positive at least in very young stages and most of them are, depending on culture conditions, able to form spores. Dissection of existing taxa and the description of new species and genera as well as phylogenetic reorganisations is an unfortunate fact of life and must continue. In all probability, only 0.1-1.0% of the existing bacterial diversity is currently known and we should expect the presently single-membered species and single-membered genera to steadily expand. Application of the available methods (Vandamme et al., 1996) across all taxa of the group will certainly help to clarify the picture while new and better methods need to be developed to find additional, more consistent discriminatory traits across the group of organisms.

ACKNOWLEDGEMENTS
Claudia Wahrenburg is thanked for taking microphotographs of type strains and for further valuable help.

REFERENCES
Abd EI-Rahman, H. A., Fritze, D., Sproeer, C. and Claus, D. 2002. Two novel psychrotolerant species: *Bacillus psychrotolerans* sp. novo and *Bacillus psychrodurans* sp. nov. containing ornithine in their cell walls. Int. J. Syst. Evol. Microbiol. *52,* 1-7.

Anon.: DSMZ catalogue of strains, www.dsmz.de

Anon.: Taxonomic Outline of the 2nd Ed. of Bergey's Manual of Systematic Bacteriology, URL dx.doi.org/10.1007/bergeysoutline

Ash, C., Farrow, J. A. E., Wallbanks, S. and Collins, M. D. 1991. Phylogenetic heterogeneity of the genus *Bacillus* revealed by comparative analysis of small-subunit-ribosomal RNA sequences. Lett. Appl. Microbiol. *13,* 202-206.

Ash, C., Priest, F. G. and Collins, M. D. 1993. Molecular identification of rRNA group 3 bacilli (Ash, Farrow, Wallbanks and Collins) using a PCR probe test. *Antonie van Leeuwenhoek 64,* 253-260.

Ash, C., Priest, F. G. and Collins, M. D. 1994. *Paenibacillus polymyxa* comb. nov. In: *Validation of the publication of new names and new combinations previously effectively published outside the IJSB.* List No. 51. Int. J. Syst. Bacteriol. *44,* 852.

Atlas, R.M. 1995. Handbook of media for environmental microbiology. CRC Press, Boca Raton. ISBN 0-8493-0603-5

Atlas, R.M. 1997. Handbook of microbiological media. CRC Press, Boca Raton. ISBN 0-8493-2638-9

Buchanan, R. E. 1925. General Systematic Bacteriology, Vol. 1, Baltimore, The Williams and Wilkins Co.

Claus, D. and Berkeley, R. C. W. 1986. Genus *Bacillus* Cohn 1872. Pages 1105-1139 in: Bergey's Manual of Systematic Bacteriology, vol. 2, eds. Sneath, P.H.A., Mair, N.S., Sharpe, M.E. and Holt, J.G., Bergey's Manual Trust, Williams and Wilkins, Baltimore.

Cohn, F. 1872. Untersuchungen ueber Bacterien. Beitr. Biol. Pflanz. 1, p. 127-224.

Cohn, F. 1876. Untersuchungen ueber Bacterien. IV. Beitr. Biol. Pflanz. 2, Heft 2, p. 249-277.

De Bary, A. 1884. Vergleichende Morphologie und Biologie der Pilze, Mycetozoen und Bacterien. Wilhelm Engelmann, Leipzig.

Fahmy, F., Mayer, F. and Claus, D. 1985. Endospores *of Sporosarcina halophila,* characteristics and ultrastructure. Arch. Microbiol. *140,* 338-342.

Fortina, M. G., Pukall, R., Schumann, P., Mora, D., Parini, C., Manachini, L. and Stackebrandt, E. 2001. *Ureibacillus* gen. nov., a new genus to accommodate *Bacillus thermosphaericus* (Andersson et al. 1995), emendation of *Ureibacillus thermosphaericus* and description of *Ureibacillus terrenus* sp. nov. Int. J. Syst. Evol. Microbiol. *51,* 447-455.

Fritze, D., Flossdorf, J. and Claus, D. 1990. Taxonomy of alkaliphilic *Bacillus* strains. Int. J. Syst. Bacteriol. *40,* 92-97.

Golovacheva, R. S. and Karavaiko, G. I. 1978. A new genus of thermophilic spore-forming bacteria, *Sulfobacillus.* Microbiology (Engl. Translation of Mikrobiologiya) *47,* 658-664.

Golovacheva, R. S. and Karavaiko, G. I. 1991. *Sulfobacillus* new genus, *Sulfobacillus thermosulfidooxidans* new species. In: Validation of the publication of new names and new combinations previously effectively published outside the IJSB. List No. 36. Int. J. Syst. Bacteriol. *41,* 179.

Gordon, R. E., Haynes W. C. and Pang, C. H. 1973. The genus *Bacillus.* U.S. Department of Agriculture, Handbook No. 427. Washington, D.C.

Heyndrickx, M., Lebbe, L., Kersters, K., De Vos, P., Forsyth, G. and Logan, N. A. 1998. *Virgibacillus,* a new genus to accommodate *Bacillus pantothenticus* (Proom and Knight 1950). Emended description of *Virgibacillus pantothenticus.* Int. J. Syst. Bacteriol. *48,* 99-106.

Kitahara, K. and Suzuki, J. 1963. *Sporolactobacillus* nov. subgen. J. Gen. Appl. Microbiol. *9,* 59-71.

Kluyver, A. J. and van Niel, C. B. 1936. Prospects for a natural classification of bacteria. Zentralbl. Bakteriol. Parasitenkd. Infektionskr. Hyg. Abt. II, *94,* 369-403.

Koch, R. 1876. Die Aetiologie der Milzbrandkrankheit. Beitr. Biol. Pflanz. 2, Heft 2, p. 277- 310.

Koransky, J. R., Allen, S. D. and Dowell, V. R. 1978. Use of ethanol for selective isolation of sporeforming microorganisms. Appl. Environ. Microbiol. *35,* 762-765.

Lapage, S. P., Sneath, P. H. A., Lessel, E. F., Skerman, V. B. D., Seeliger, H. P. R., and Clark, W. A. 1975. International Code of Nomenclature of Bacteria. Am. Soc. Microbiol., Washington D.C.

Lapage, S. P., Sneath, P. H. A., Lessel, E. F., Skerman, V. B. D., Seeliger, H. P. R., and Clark, W. A. 1992. International Code of Nomenclature of Bacteria. 1990 revision. ed. Sneath, P. H. A. American Society for Microbiology, Washington, DC, ISBN l-55581-039-X

Lu, J., Nogi, Y. and Takami, H. 2001. *Oceanobacillus iheyensis* gen. nov., sp. nov., a deep-sea extremely halotolerant and alkaliphilic species isolated from a depth of 1050 m on the Iheya Ridge. FEMS Microbiol. Lett. *205,* 291-297.

Lu, J., Nogi, Y. and Takami, H. 2002. *Oceanobacillus* new genus, *Oceanobacillus iheyensis* new species. In: Validation of the publication of new names and new combinations previously effectively published outside the IJSEM. List No. 85. Int. 1. Syst. Evol. Microbiol. *52,* 687.

Nazina, T. N., Tourova, T. P., Poltaraus, A. B., Novikova, E. V., Grigoryan, A. A., Ivanova, A. E., Lysenko, A. M., Petrunyaka, V. V., Osipov, G. A., Belyaev, S. S. and Ivanov, M. V. 2001. Taxonomic study of aerobic thermophilic bacilli: descriptions of *Geobacillus subterraneus* gen. nov., sp. nov. and *Geobacillus uzenensis* sp. nov. from petroleum reservoirs and transfer of *Bacillus stearothermophilus, Bacillus thermocatenulatus, Bacillus thermoleovorans, Bacillus kaustophilus, Bacillus thermoglucosidasius* and *Bacillus thermodenitrificans* to *Geobacillus* as the new combinations G. *stearothermophilus,* G. *thermocatenulatus,* G. *thermoleovorans,* G. *kaustophilus,* G. *thermoglucosidasius* and G. *thermodenitrificans.* Int. J. Syst. Evol. Microbiol. *51,* 433-446.

Niimura, Y., Koh, E., Yanagida, F., Suzuki, K.-I., Komagata, K. and Kozaki, M. 1990. *Amphibacillus xylanus* gen. nov., sp. nov., a facultatively anaerobic sporeforming xylan-digesting bacterium which lacks cytochrome, quinone, and catalase. Int. J. Syst. Bacteriol. *40,* 297-301.

Pennington, T. H. 1994. Molecular systematics and traditional medical microbiologists- problems and solutions. J. Med. Microbiol. *41,* 371-373.

Pikuta, E., Lysenko, A., Chuvilskaya, N., Mendrock, U., Hippe, H., Suzina, N., Nikitin, D., Osipov, G. and Laurinavichius, K. 2000. *Anoxybacillus pushchinensis* gen. nov., sp. nov., a novel anaerobic, alkaliphilic, moderately thermophilic bacterium from manure, and description of *Anoxybacillus flavithermus* comb. nov. Int. J. Syst. Evol. Microbiol. *50,* 2109-2117.

Pikuta, E., Cleland, D. and Tang, J. 2003. Aerobic growth *of Anoyxbacillus pushchinoensis* K1[T]: emended descriptions of *A. pushchinoensis* and the genus *Anoxybacillus*. Int. J. Syst. Evol. Microbiol. *53*, 1561-1562.

Robinow, C. F. 1951. Observations on the structure of *Bacillus* spores. J. Gen. Microbiol. *5*, 439-457.

Schlesner, H., Lawson, P. A., Collins, M. D., Weiss, N., Wehmeyer, U., Voelker, H. and Thomm, M. 2001. Int. J. Syst. Evol. Microbiol. *51*, 425-431.

Shida, O., Takagi, H., Kadowaki, K. and Komagata, K. (1996) Proposal for two new genera, *Brevibacillus* gen. nov. and *Aneurinibacillus* gen. nov. Int. J. Syst. Bacteriol. *46*, 939-946.

Slepecky, R. A. 1992. What is a *Bacillus?* Pages 1-21 in: Biology of Bacilli: Applications to Industry, eds. Doi, R. H. and McGloghlin, M. Butterworth-Heinemann, Boston, London, Oxford, Singapore, Sydney, Toronto, Wellington.

Skerman, V. B. D., McGowan, V. and Sneath, P. H. A. eds. 1980. Approved Lists of Bacterial Names. American Society for Microbiology, Washington, DC.

Skerman, V. B. D., McGowan, V. and Sneath, P. H. A. eds. 1989. Approved Lists of Bacterial Names. Amended Edition. American Society for Microbiology, Washington, DC.

Smith, N. R., Gordon, R. E. and Clark, F. E. 1952. Aerobic Sporeforming Bacteria U.S. Department of Agriculture, Monograph No. 16. Washington, D.C.

Spring, S., Ludwig, W., Marquez, M. C., Ventosa, A. and Schleifer, K.-H. 1996. *Halobacillus* gen. nov., with descriptions of *Halobacillus litoralis* sp. nov. and *Halobacillus trueperi* sp. nov., and transfer of *Sporosarcina halophila* to *Halobacillus halophilus* comb. nov. Int. J. Syst. Bacteriol. *46*, 492-496.

Stackebrandt, E. and Goebel, B.M. 1994. A place for DNA-DNA reassociation and 16S rRNA sequence analysis in the present species definition in bacteriology. Int. J. Syst. Bacteriol. *44*, 846-849.

Stackebrandt, E., Fredericksen, W., Garrity, G. M., Grimont, P. A. D., Kaempfer, P., Maiden, M. C. J., Nesme, X., Rossello-Mora, R., Swings, J. Triiper, H. G., Vauterin, L., Ward, A. C. and Whitman, W. B. 2002. Report of the Ad Hoc Committee for the Re-Evaluation of the Species Definition in Bacteriology. Int. J. Syst. Evol. Microbiol. *52*, 1034-1047.

Stackebrandt, E. and Swidersky, J. 2002. From Phylogeny to Systematics. Pages 8-22 in: Applications and Systematics of *Bacillus* and Relatives, eds: Berkeley, R., Heyndrickx, M., Logan, N. and De Vos, P. Blackwell Publishing. ISBN 0-632-05758-0.

Takai, K., Inoue, A. and Horikoshi, K. 1999. *Thermaerobacter marianensis* gen nov., sp. nov., an aerobic extremely thermophilic marine bacterium from the 11000 m deep Mariana Trench. Int. J. Syst. Bacteriol. *49*, 619-628.

Touzel, J. P., O'Donohue, M., Debeire, P., Samain, E. and Breton, C. 2000. *Thermobacillus xylanilyticus* gen. nov., sp. nov., a new aerobic thermophilic xylan-degrading bacterium isolated from farm soil. Int. J. Syst. Evol. Microbiol. *50*, 315-320.

Tsilinsky, P. 1899. On the thermophilic molds (in French). Annales de L'Institut Pasteur *13*, 500-505.

Vandamme, P., Pot, B., Gillis, M., De Vos, P., Kersters, K. and Swings, J. 1996. Polyphasic taxonomy, a consensus approach to bacterial systematics. Microbiol. Rev. *60*, 407-438.

Warth, A. D. 1979. Exploding spores. Spore Newsletter *6*, 4-6.

Wainoe, M., Tindall, B. J., Schumann, P. and Ingvorsen, K. 1999. *Gracilibacillus* gen. nov., with description of *Gracilibacillus halotolerans* gen. nov., sp. nov.; transfer of *Bacillus dipsosauri* to *Gracilibacillus dipsosauri* comb. nov., and *Bacillus salexigens* to the genus *Salibacillus* gen. nov., as *Salibacillus salexigens* comb. nov. Int. J. Syst. Bacteriol. *49*, 821-831.

Wayne, L. G., Brenner, D. J., Colwell, R. R., Grimont, P. A. D., Kandler, O., Krichevsky, M. I., Moore, L. H., Moore, W. E. C., Murray, R. G. E., Stackebrandt, E., Starr, M. P., and Trueper, H. G. 1987. Report of the *Ad Hoc* Committee on Reconciliation of Approaches to Bacterial Systematics. Int. J. Syst. Bacteriol. *37*, 463-464.

Wisotzkey, J. D., Jurtshuk, P., Jr., Fox, G. E., Deinhard, G. and Poralla, K. 1992. Comparative sequences analyses on the 16S rRNA (rDNA) of *Bacillus acidocaldarius, Bacillus acidoterrestris,* and *Bacillus cycloheptanicus* and proposal for creation of a new genus, *Alicyclobacillus* gen. nov. Int. J. Syst. Bacteriol. *42*, 263-269.

Yoon, J-H., Weiss, N., Lee, K-Ch., Lee, I-S., Kang, K. H. and Park, Y-H. 2001. *Jeotgalibacillus alimentarius* gen. nov., sp. nov., a novel bacterium isolated from jeotgal with L-Iysine in the cell wall, and reclassification of *Bacillus marinus* Rueger 1983 as *Marinibacillus marinus* gen. nov., comb. nov. Int. J. Syst. Evol. Microbiol. *51*, 2087-2093.

Yoon, J-H., Kang, K. H. and Park, Y-H. 2002. *Lentibacillus salicampi* gen. nov., sp. nov., a moderately halophilic bacterium isolated from a salt field in Korea. Int. J. Syst. Evol. Microbiol. *52*, 2043-2048.

Zaitsev, G., Tsitko, I. V., Rainey, F. A. et al. (1998) New aerobic ammonium-dependent obligately oxalotrophic bacteria: description of *Ammoniphilus oxaliticus* gen. nov., sp. nov. and *Ammoniphilus oxalivorans* gen. nov., sp. nov. Int. J. Syst. Bacteriol. *48:* 151-163.

Chapter 3

Ecology of *Bacillus* Species in Soil

Andreas D.M. Felske

SUMMARY

Are sporeforming bacteria of any significance in soil under normal conditions? This historical question is discussed here, in the light of recent culture-independent community analyses of soil DNA, that indicated hitherto unknown bacterial lineages as the predominant soil microbes. While early studies have suggested a prevalence of endospores, recent data indicate vegetative *Bacillus* cells among the predominant soil bacteria. Moreover, these abundant vegetative forms appeared to be hitherto unknown *Bacillus* spp., which were only recently accessible to cultivation. Here, a preliminary phylogenetic location of these new species is outlined and defined as the branch where almost half of all environmental 16S rDNA *Bacillus* clones from various studies accumulated. In contrast to the other predominant soil flora, this new soil *Bacillus* lineage appears to be at least, in part, cultivable. In future, this will allow seminal studies on specific physiological attributes of these new dominant soil *Bacillus* species. This review intends to draw more attention to this neglected group of abundant but at least, in part, cultivable soil bacteria.

INTRODUCTION

The genus *Bacillus* entered the stage from the very beginning of bacteriology. It was in 1905-that the Nobel Laureate Robert Koch demonstrated anthrax to be caused by *Bacillus anthracis*. It may well have been the seedling-like appearance of endospores and the observation of hatching cells, that allowed him to propose that these minute microorganisms were infectious agents. The endospore also aided the first steps of bacterial taxonomy since it provided one of the few distinct morphological features among bacteria when Ferdinand Cohn defined the endospore-forming bacteria as the genus *Bacillus*. Hence, it is perhaps not surprising that the endospore was and, is always, considered as the key factor in *Bacillus* ecology. Moreover, the endospore claimed an overwhelmingly dominant position, when specific culture approaches based on pasteurisation gave the impression, that almost all *Bacillus* cells in soil could be endospores. In 1916, this prompted H.J. Conn to ask "Are sporeforming bacteria of any significance in soil under normal conditions?", a question that is still relevant today.

Considering the apparent prevalence of endospores among *Bacillus* populations, one can formulate three possible scenarios: (i) the vegetative *Bacillus* is actually a rare soil bacterium, but endospore longevity mimics abundance. Hence, cultivation approaches recover populations which ceased to thrive a long time ago, and we gain a cumulative overview of *Bacillus* abundance for months, years or even decades. (ii) Under normal circumstances, *Bacillus* is not actually growing in soil, but can do so only at extraordinary occasions of nutrient input. Hence, *Bacillus* is not really a soil bacterium but more closely resembles a ubiquitous opportunist. Their abundance in soil is not so much due to adaptation to soil but rather with the fact that the ground is the prevalent place for death and decay. In this scenario endospores can acquire nutrients from natural processes of death and decay occurring at the soil surface, enabling germination and proliferation. (iii) *Bacillus* is not so much associated with the soil matrix but much more with soil animals. Here, Bacillus lives within the gastrointestinal tracts of soil-feeding animals and is excreted in the spore form on occasion in the faeces leading to a steady accumulation in the soil. Once excreted of course the spore could be ingested by other soil organisms providing a continual cycle of growth and sporulation.

There are some arguments and supportive data for each of the three scenarios. However, it would constitute a quite unusual finding for ecosystems, that opportunists or symbionts should be the predominant community members. The predominant community members are much more likely species at the bottom of the food chain, feeding on the most common carbon source and growing constantly and are widespread at the highest density among all organisms. An answer to this might have been provided by the culture-independent approach. This indicated, that *Bacillus* spp. might not be the predominant soil bacteria at all, but was more likely to be the largely unexplored bacterial divisions like Acidobacteria, Verrucomicrobia and not yet cultured lineages of α-Protobacteria. However, this point is far from being settled yet. Our common view on *Bacillus* in soil may still be challenged and could be altered significantly in the future. This chapter and the data presented herein is intended to prepare the ground for updating concepts on *Bacillus* ecology in soil

BASIC KNOWLEDGE OF *BACILLUS* ECOLOGY

The Endospores

The genus *Bacillus* was originally defined by the formation of endospores. This prominent structure provided spectacular features within the (once) otherwise plain bacterial morphology. Being the most heat-resistant living structures known, many studies concentrated on the endospore. Moreover, the extraordinary resistance of endospores directed the investigation of *Bacillus* not only intentionally but also practically. Heat treatments like pasteurisation were the most common techniques in selective isolation approaches for the recovery of *Bacillus* from soil and other environments. Sometimes, the potentially hazardous heat treatment could even increase the yield of colony forming units compared to the same experiment without heating. This phenomenon is called, 'heat activation' (Leuschner and Lillford, 1999) and reveals how misleading the cultivation approach can be. In this way, studies on *Bacillus* in the environment were usually restrained to and automatically focused on the presence of endospores. Since the spore is a dormant being of potentially extreme longevity, the counts of colony forming units on petri dishes were more or less estimations of spore accumulation in the given environment. The actual contribution of the bacterium to the metabolic turnover within this environment remained unclear, since this is provided by the vegetative form. Indeed, it is not that difficult to isolate obligate alkaliphilic *Bacillus* spp. from acid soils or thermophilic strains from cold environments. It has been found that alkaliphilic bacilli are widespread in soils, also acid ones, but are more abundant in soils of high pH with up to 10^6 cfu g^{-1} soil. This suggests transport and accumulation of endospores by wind and water also into unfavourable environments where they will remain inactive. However, extraordinary levels of viable endospores may still be a reasonable hint to recent growth. For instance, studies on a dried pond in Israel, containing dead mosquito larvae was the source of the insect pathogen *B. thuringiensis* (Margalit et al., 1985). Here, it would be reasonable to assume a true correlation between the circumstances of the sampling site and the isolated strain. Dissemination from dried ground surfaces by air currents is quite important for endospore dissemination, giving *Bacillus* spp. an efficient airborne transport system to overcome even huge distances between favourable environments. This is apparent in the case of strictly thermophilic bacteria from geographically isolated thermal regions. Where there is no connection between hot thermal locations, non-sporulating bacteria such as *Thermus* displayed signs of location-specific independent segregation (Williams, 1989), while thermophilic *Bacillus* spp. were more globally distributed with no detectable geographic specificity (White et al., 1989). Such findings suggest that the endospore is a key feature of *Bacillus* ecology, boosting the dissemination making the *Bacillus* life cycle quite different from other bacteria.

Endospores or Vegetative Cells? A Study on *B. subtilis*

Since the cultivation approach is biased in distinguishing endospores from vegetative cells, it would be favourable to have culture-independent methods. Indeed, it is possible to produce specific antisera against each of both forms (Norris and Wolf, 1961). Such antisera, fluorescently labelled, were used to detect *B. subtilis* in pine-forest soil (Siala et al., 1974). In principle, the use of antisera should be specific to absolute species, if not strain-specific, which may be convenient for strictly focused studies but limiting for a more general approach. Siala et al. (1974) collected samples from five soil profiles, stained them with fluorescent antibodies and analysed tens of thousands of soil particles for *B. subtilis* cells or spores. Only three profiles yielded vegetative cells, mostly on particles of organic matter

although mineral particles provided ca. 85 % of the available surface area. The appearance of *B. subtilis* cells was 2-3 orders of magnitude more on organic than on mineral particles. On organic matter, the cells were observed in microcolonies of an average of five cells, and the cells were also larger in size than on mineral particles, where microcolonies had the average size of two cells, i.e. mainly single cells occurred. On mineral particles, *B. subtilis* was virtually restricted to cryptocrystalline quartz (only 5-6 % of all mineral surface), while the more abundant quartz and aluminosilicate were almost devoid of *B. subtilis*. On organic matter particles *B. subtilis* could be found on all substrates of plant origin but not on the remains of fungi or soil animals. Only a few cells could be found on living roots but much more on dead roots. The highest scores yielded dead pine branches in the upper A_1 horizon with ca. 13 cells per mm^2, followed by dead pine roots with ca. 5 cells per mm^2, and other unidentified organic matter with ca. 3 cells per mm^2. The contribution of dead hyphae could not be resolved in this context. The living roots yielded only 0.02 cells per mm^2, living hyphae, sclerotia, exoskeletons and shells did not show any *B. subtilis* at all. Overall, the A_1 horizon appeared to be predominated by vegetative cells (67 %). The lower C horizon (at 8-10 cm depth) was apparently predominated by endospores (80 %), mainly found on dead roots (3 spores per mm^2). The relatively high abundance of vegetative cells detected here might be exceptional, but the authors nevertheless concluded that the idea of spore predominance in soil *Bacillus* populations had to be modified. Their study clearly suggested that *B. subtilis* is affiliated with decaying plant matter. Hence, it was the regular addition of organic matter by leaf fall or root growth and subsequent root death that might have provided the substrate for vegetative proliferation. With such experiments on spatial distribution the possible ecological niches of certain species may well be estimates only. However, it could be assumed that other *Bacillus* species would show quite different distribution patterns and ecological dependencies.

Longevity of Endospores: The case of *B. anthracis*

Anthrax is an infectious disease of herbivorous mammals and has caused epidemics in livestock and occasional infections in man. Interest in anthrax has increased recently due to its use in bioterrorism attacks. Since the infection is severe and often fatal, it is not surprising that the causative agent, *Bacillus anthracis*, is among the best-studied bacteria at all. Soil appears to be the reservoir for this organism, and naturally occurring anthrax infections tend to occur during dry periods following a wet period. The "incubator theory" suggests that outbreaks are linked to puddles covering and killing the grass cover, providing a suitable environment for the accumulation of *B. anthracis* populations at the surface, where they are exposed to livestock. When such puddles finally dry up and become dusty, the endospores can be disseminated and infect wounds of animals. Livestock in pastures without such areas of shallow flooding and subsequent drying up do not develop anthrax. Hence, dust appears to be is causative agent of infection and, of course, endospores may last indefinitely in such a dry environment. Since dust may easily be blown away and contaminate large areas, potentially the entire planet, the question is, why anthrax outbreaks were only of very limited regional impact, mostly in areas that have already a history of anthrax cases? The ease of endospore dissemination is in stark contrast to the apparently endemic anthrax occurrence (Van Ness, 1971). Interestingly, anthrax was linked with endemic soil environments long before the cause was identified. Livestock infections in hitherto unsuspected areas could always be traced to regions where the disease was already endemic in the soil. Anthrax outbreaks always resulted from contact with or ingestion of *B. anthracis* endospores in areas where anthrax is present or from animals or agricultural products derived from those infected areas. Hence, it is of utmost interest (i.e. for quarantine measures) to reveal where anthrax is already established in the soil and in which areas anthrax is likely to appear. In such soils, the longevity and high resistance of spores is an obvious problem for anthrax control. However, of special interest in the ecological context is the study of anthrax-resistant environments. Certain soil and water bodies which are not favourable for anthrax, either due to acidity, temperature or other factors, may rapidly eliminate vegetative cells and spores, even in cases of heavy contamination. The epidemiological evidence of anthrax indicates that compromised soils must maintain a continual *B. anthracis* lifecycle of endospores germinating to vegetative cells, their subsequent proliferation and then sporulation again. If this cycle is broken and endospores accumulate without further activation, the niches once conquered by *B. anthracis* will fall to competitors. Hence, *B. anthracis* may well disappear from a certain environment, even quite rapidly in unfavourable soils.

Nitrogen fixation

Some strains from the genus *Bacillus* (and genera derived from this) are capable of fixing nitrogen from the atmosphere. Recent studies suggest that nitrogen fixation is limited to a subset of *Paenibacillus* species (Achouak et al., 1999), but here the reader must be aware, that quite some name shifting was and is going on among the low-GC Gram positives. At the moment it looks as if all nitrogen-fixing *Bacillus* strains will fall under the genus *Paenibacillus*. Members of the species *Paenibacillus durus* (former *B. azotofixans*), *P. brasiliensis*, *P. macerans* and *P. polymyxa* were found to be associated with the rhizosphere and rhizoplane of plants (von der Weid et al., 2002). These bacteria may receive nutrients from the roots and in return constitute a nitrogen source for the plant. Studies on wheat showed that plants inoculated with *P. polymyxa* may have received up to 10 % of their nitrogen from bacteria (Kucey, 1988). Here, the observed enhancement of wheat growth might not only be due to improved nitrogen access but also to the bacterial production of plant growth promoters like gibberelin. For instance, some molecular hint on cell-to-cell interactions within nitrogen-fixing plant-bacterial associations in the rhizosphere provided a study of surface agglutinating proteins of *P. polymyxa* (Karpunina et al., 2001).

Association with animals

As proposed in the introduction, we should also consider the possibility that *Bacillus* spp. is more associated with soil animals than with the soil matrix itself. Members of the genus *Bacillus* can be isolated from the faeces of many animals, but this mostly represents transient presence within the intestine initiated by the ingestion of endospores with food. However, some species such as *B. licheniformis* and *B. laterosporus* may indeed be associated with the rumen of cattle and be involved in the degradation of hemicellulose (Williams and Withers, 1983). Nevertheless, the most important soil *Bacillus* spp. could never be linked to the cattle intestine. Therefore, this internal mode of proliferation and subsequent dissemination of spores within soil via faeces release could not be confirmed.

Looking within the soil body, the invertebrates may be much more important hosts. However, such insect-associated *Bacillus* spp. are obviously not obligate inhabitants of the insect, because they readily grow on standard media. Nevertheless, their continued presence in the intestines suggests that this is a favourable ecological niche for the bacteria and not only an accidental transient occurrence in the passage through the alimentary tract (see Chapter 9 for more discussion of the interaction of *Bacillus* species with the intestinal tracts of animals).

The insect pathogens *B. thuringiensis* and *B. sphaericus* grow within insect larvae in an invasive, i.e. pathogenic way, and thus indeed clearly proliferating within an animal, disseminating by insect movement and finally being released by the death and decay of the host. However, this pathogenic lifecycle could only be attributed to a minority population of, most likely, highly specialised bacteria, but never to predominant populations. Such organisms must exist in a careful equilibrium with their hosts, at a relatively low chance of infection. If infectious contact with the pathogen would be unavoidable for the host, the consequences would be a collapse of the population. This is obviously true, since *B. thuringiensis* and *B. sphaericus* could be effectively applied for pest control (Priest, 1992).

THE CULTURE-INDEPENDENT DNA APPROACH

With the introduction of DNA-based methods, the past decade has seen significant progress in bacterial ecology. The ever radiating application of nucleic acid techniques in microbial ecology has been concerned with phylogenetic relationships between bacteria as determined by 16S rDNA sequencing analysis. Although this molecular approach to bacterial communities could solve certain questions it also raised more. Like other environments, the soil in general has been shown to contain many more bacteria than have yet been isolated. It has been estimated that maybe less than 1% of all soil bacteria are accessible for standard cultivation techniques (Amann et al., 1995). Adding evidence, 16S rDNA-based surveys revealed hitherto unknown taxa as the main soil bacteria. Our lack of knowledge about soil bacteria communities is disturbing. There seems to be a huge diversity of inaccessible bacteria which are thriving under mesophilic conditions on a high diversity of nutrients, but only a 1% minority can be grown in culture. Analysis also covering this uncultured majority of microbial communities revealed *Acidobacteria* (Ludwig et al., 1997), novel α-Proteobacteria relatives (Liesack and Stackebrandt, 1992) or sometimes Verrucomicrobia (Lee et al., 1996) as the

Table 1. Culture-independent studies with appearance of *Bacillus*-like sequences in their clone libraries.				
	16S rDNA sequences			
Soil DNA-based study	Bacteria	*Bacillus*	New group	Reference
Clover-grass pasture, Wisconsin	180	31	25	Borneman et al., 1996
Forest and pasture soil, Amazonia	98	13	10	Borneman and Triplett, 1997
Grassland soils, The Netherlands	72	32	24	Felske et al., 1998
Contaminated landfill near Quebec, Canada	71	9	4	Lloyd-Jones and Lau, 1998
Four arid soils, Arizona	203	12	5	Dunbar et al., 1999
Rice paddy soil microcosms, Germany	57	13	10	Hengstmann et al., 1999
Brassica napus rhizosphere, UK	73	20	13	Macrae et al., 2000 and 2001
Chrysanthemum rhizosphere, The Netherlands	18*	4*	2*	Duineveld et al., 2001
* no clones but predominant excised DGGE bands; new group = *B. benzoevorans/niacini*-relatives				

main taxa of soil bacteria (likewise the other surveys listed in Table 1). Although representing the most frequently isolated soil bacteria, the genus *Bacillus* appears not that often in culture-independent studies on 16S rDNA directly extracted from soil. Hence, is it possible, that *Bacillus* spp. are in fact quite rare soil bacteria as proposed in the introduction? Indeed is it the endospore longevity coupled with the intrinsically biased cultivation approach that leads to the apparent high abundance of *Bacillus* spp. in soil?

Appearance of *Bacillus* in 16S rDNA Clone Libraries of Culture-independent Studies

In the past decade, several independent studies of microbial communities in soils have been performed by directly extracting DNA from soil samples, PCR-amplifying the 16S rDNA and subsequent cloning of the amplicons. This approach indicated that the most abundant soil bacteria were often remote relatives of known α-Proteobacteria and especially members of the division of *Acidobacteria* and also some Verrucomicrobia. The latter two are represented by only a handful of pure cultures each, and those isolates are again only remote relatives to the scores of soil 16S rDNA clones gained from these divisions. In general, *Acidobacteria* was found most often and in every

study. *Bacillus* however, was not always present in soil clone libraries. In total, numerous culture-independent studies revealed approximately 200 *Bacillus*–related 16S rDNA sequences from soil DNA. In contrast, the total number of bacterial 16S rDNA fragments directly retrieved from soil samples world-wide and posted within the EMBL database meanwhile approached approximately 10,000 (Summer 2003). Hence, the genus *Bacillus* constitutes only about 2% of all soil bacteria detected by this type of analysis. Among all the different culture-independent studies on soil bacteria, only one Dutch study (Felske et al., 1998) identified *Bacillus* as the most abundant taxon. Results from other sites around the world yielded only *Bacillus* in prominent numbers in some studies (Table 1).

Since detection of *Bacillus* 16S rDNA within the soil DNA fractions was very variable, it is worthwhile taking a closer look at the methods used for cell lysis. Unfortunately, all workers used different methods to release and purify nucleic acids from soil samples, however, to some extent this was justified and reflected the different kinds of soil. In general, different methods were tried and the most suitable then optimised making direct comparisons difficult. Nevertheless, looking at the primary lysis technique revealed an interesting pattern (Figure 1). Methods using detergents, enzymes or freezing and thawing to destroy

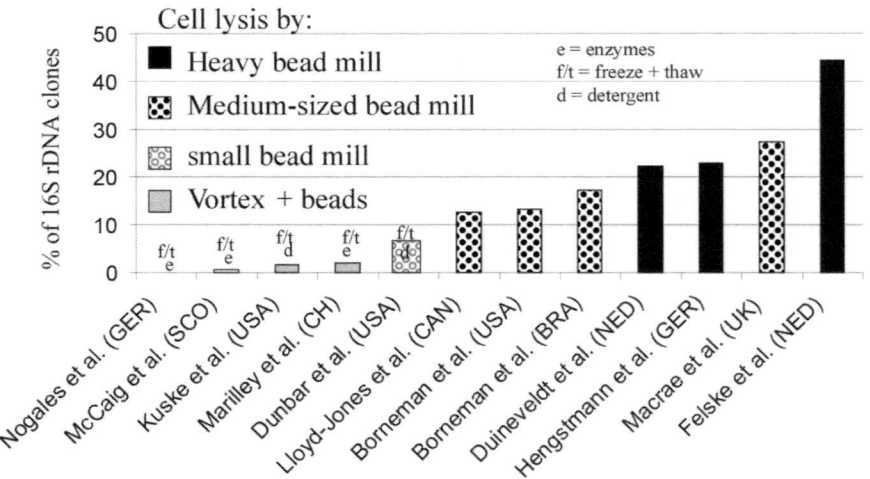

Figure 1. Comparison of relative *Bacillus* 16S rDNA contributions to various bacteria clone libraries made from directly extracted soil DNA from Europe and America. The percentage of *Bacillus* sequences among the bacterial 16S rDNAs varies from close to zero to up to more than 40%. A correlation with the applied cell lysis protocol might be suspected.

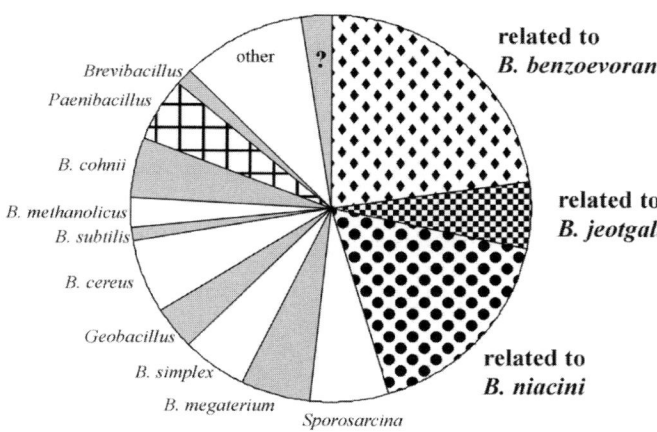

Figure 2. Phylogenetic affiliation of the approximately 200 16S rDNA clones related to *Bacillus* spp., representing predominant soil bacteria identified by culture-independent methods.

cells yielded little Bacillus DNA. Instead, those using a bead mill for cell lysis were highly effective. The heaviest instruments like the MSK cell homogeniser (from Braun-Melsungen, Germany), using the highest number and amplitude of eccentric rotations per minute, proved the most effective at releasing *Bacillus* DNA. Although no systematic comparitive has yet been made on methods for cell rupture we suspect that many DNA-based studies fail to identify *Bacillus* spp. Due to insufficient methods of cell lysis.

The modest appearance of *Bacillus*-like environmental 16S rDNA nevertheless revealed an interesting set of data. Approximately 200 cloned 16S rDNAs originating from *Bacillus* spp. were obtained from soil and nearly half of them were phylogenetically located in the grouping of *B. benzoevorans/B. niacini/B. jeotgali* (Figure 2). This is the first evidence that this phylogenetically quite coherent cluster of 16S rDNA sequences may well represent the most abundant soil *Bacillus* group worldwide. In the Dutch Drentse A grasslands,

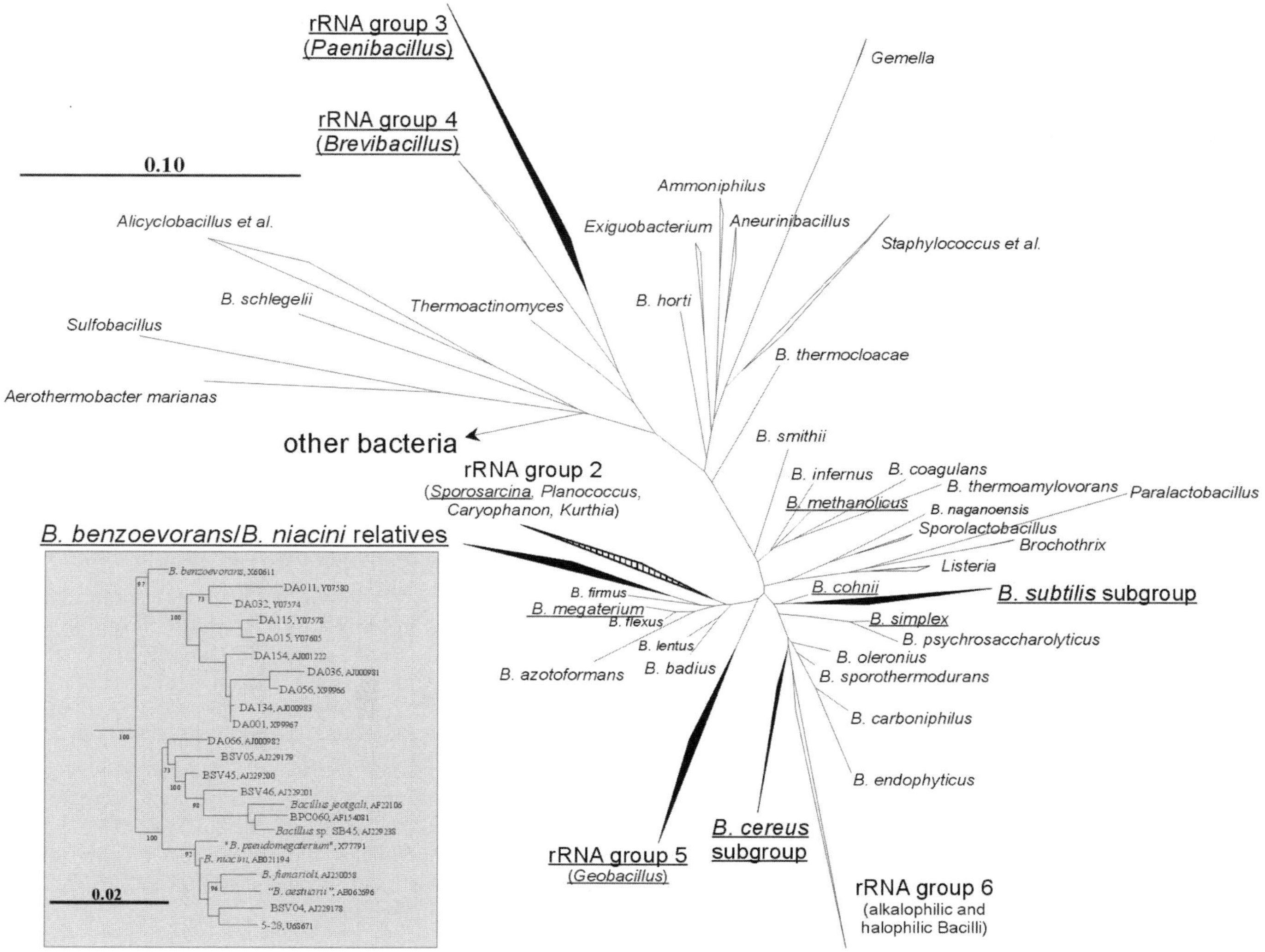

Figure 3. Phylogenetic tree of 16S rDNA sequences of *Bacillus* spp. and relative contributions of various taxa to the approximately 200 *Bacillus*–related sequences from clone libraries made from directly extracted soil DNA from Europe and America. Almost half of them (dotted) are not closely related to cultured species but form hitherto uncultured lineages related to each other.

the first case was found where *Bacillus* was the most abundant genus in a soil bacteria community (Felske, 1999). However, the predominant *Bacillus* group appeared to belong to this novel and uncharacterised *B. benzoevorans*-related lineage. Thus, even the genus *Bacillus*, seemingly a well studied soil bacteria carries new and unexplored lineages. A clone library of 16S rDNA was generated from directly extracted DNA from Drentse A grassland soils. Sequencing revealed that this collection of sequences was predominated by this novel lineage. In parallel, bacterial rRNA was isolated directly from extracted soil ribosomes. Subsequent RT-PCR using bacterial 16S rRNA specific primers allowed a community analysis using TGGE. The TGGE fingerprints showed a high similarity along a 1.5 km stretch of Drentse A grasslands. Five sequences of this novel cluster were readily detectable in these TGGE fingerprints. They were also quantified via multiple competitor RT-PCR analysis. 15% of bacterial ribosomes carried these five types. The most abundant 16S rRNA sequence, DA001, could be detected by fluorescent whole-cell hybridisation in soil bacteria suspensions.

Approximately half of all environmental 16S rDNA clones related to *Bacillus*, like the DA001 mentioned above, are falling in one separate branch of the *Bacillus* 16S rDNA phylogeny, here called the *B. benzoevorans*/*B. niacini* relatives (Figure 3). The studies listed in Table 1 detected such *Bacillus* relatives around the world. Similar studies not mentioned in this table yielded no or only few Gram-positives. The detailed view on this cluster (Figure 3) is based on only a few available sequences of high quality, since most environmental clone sequences are very short and sometimes of low quality. In the box of figure 3 we see the familiar situation that the next related described species occupied positions separated from most of the environmental clones (the same can be observed with *Acidobacteria* or Verrucomicrobia). Only strain SB45 may be very close to some environmental clones originating from the same soil. Further, this strain could be isolated with very high colony forming unit counts of 10^7-10^8 g^{-1} soil (Chin et al., 1999). Hence, this is an abundant isolate matching to abundant 16S rDNA clones which could be assumed to represent the predominant *Bacillus* lineage in this soil from rice microcosms.

Specific Detection of *Bacillus* with Cloning/DGGE
Recently, *Bacillus*-specific fingerprinting techniques have been developed (Garbeva et al., 2003). Such techniques allow a focus to be made on the *Bacillus* spp. fraction of bacterial communities. Garbeva et al. (2003) investigated another permanent grassland where the most abundant *Bacillus*-like clones (18% of them) were again showing sequence similarities to *B. benzoevorans*. Most of the other *Bacillus* clones analysed were related to the species *B. cereus*, *B. mycoides*, *B. niacini*, *B. pumilus*, *B. megaterium*, *B. thuringiensis*, *B. lentus*, and *B. coagulans*.

Some clones showed sequence affiliation with unidentified low-G+C% Gram-positive bacteria or uncultured *Bacillus* spp. This demonstrates that, with such *Bacillus*-specific molecular approaches, the detection of yet uncultured or unidentified *Bacillus* spp. from environmental samples is clearly improved.

Hence, the aim to reveal some major impact of agriculture on the soil *Bacillus* spp. could be achieved in a much more comprehensive way. Looking at agricultural soil under different regimes, i.e., permanent grassland, grassland recently turned to arable land, and arable land under agricultural rotation, differences in the *Bacillus*-related community structures between the treatments could be clearly detected. Higher diversity of the molecular profiles were consistently observed in the permanent grassland and the grassland turned into arable land, as compared to the arable land.

Molecular Detection of Endospores
Interestingly, the detection of endospores also succeeded with 16S rRNA-targeting fluorescent oligonucleotide probes (Fischer et al., 1995). However, it required spore permeabilization with sodium dodecyl sulphate, dithiothreitol and subsequent lysozyme incubation. This harsh treatment is necessary to allow relatively large molecules such as oligonucleotide probes to pass through the recalcitrant spore coat which consists of an often very thick layer of relatively insoluble proteins. Surprisingly, endospores often produce a more intese signal than vegetative cells. This may be due to (i) the fact, that ribosomes are dismantled and hence the remaining rRNA is freely accessible for interaction with the probe, (ii) the rRNA residues are at a higher concentration in the reduced endospore volume than in vegetative cells, (iii) the rRNA is not that easily washed out of the endospore during the harsh treatment in contrast to the remains of vegetative cells that are much more affected by rRNA loss, or (iv) the probes are more easily removed from the remains of vegetative cells during stringency washing steps than from the endospores. A recent experiment on Dutch grassland soils revealed dense globular clusters of tiny particles reacting with a *Bacillus*-specific probe (Figure 4). This observation may relate to another recently published study by Branda et al. (2001) who reported on a spatial organisation of sporulation in *B. subtilis*, which they called fruiting body formation. This interesting observation may be unique to one *B. subtilis* strain or could spore clustering be a more widespread phenomenon amongst soil *Bacillus* species?

NEW *BACILLUS*-ISOLATES FROM SOIL
Since classic cultivation approaches apparently neglected the most abundant and probably also ecologically most important bacteria, there is the need for major advances in the field of bacterial isolation. As shown in the preceding section, this even concerns the genus *Bacillus*, one of the best-studied bacterial taxa of all. The improvement of cultivation methods may partially overcome the inherent bias of representing natural bacterial communities by growing their members on artificial media in the laboratory. Since soil is such a complex environment, we often have no clue as to which factor of soil promotes growth that in turn requires an empirical approach to identification of culture media.

Described isolates related to this cluster
This *Bacillus* cluster predominated by environmental 16S rDNA clones, as defined by the accumulation of environmental 16S

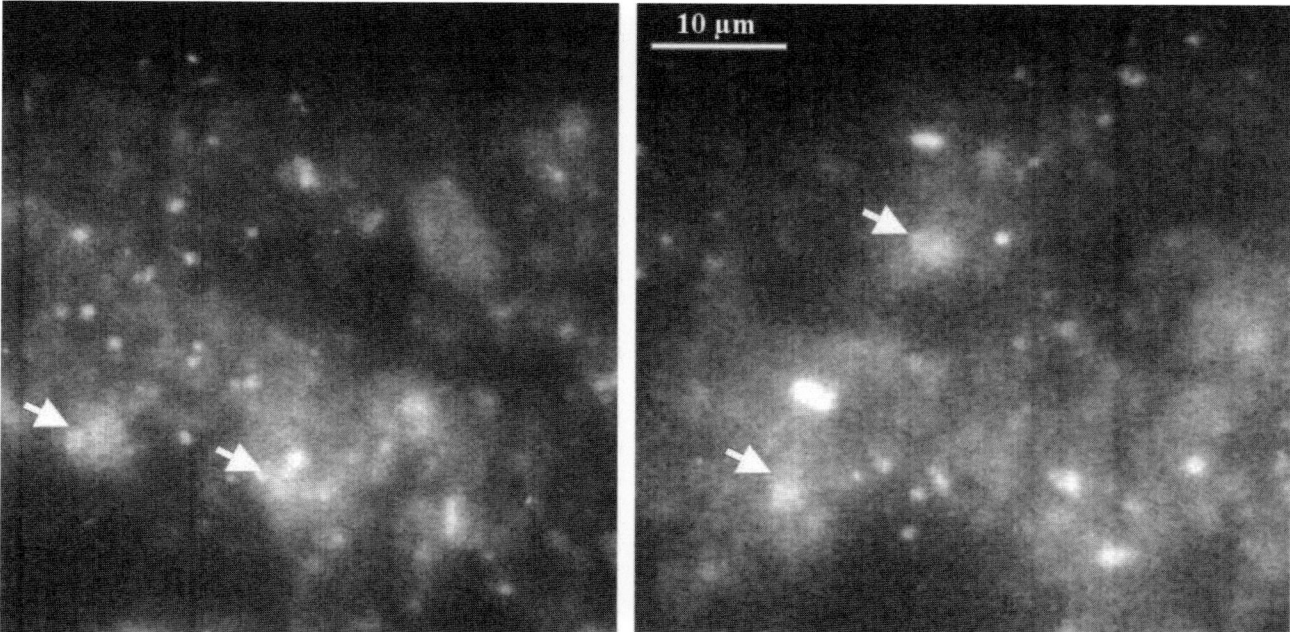

Figure 4. Detection of endospore clusters in soil by fluorescent *in situ* hybridisation. The blue signals originate from the universal DNA stain DAPI, the bright red signals from a Cy3-labelled *Bacillus*-specific probe.

rDNA sequences in a particular branch of the *Bacillus* rRNA subgroup 1, also includes some already described isolates. As in the case of Acidobacteria and Verrucomicrobia, the described isolates did not closely match the environmental 16S rDNA clones, but occupied separate branches (Figure 3). Firstly, there was *B. benzoevorans*, a species of only remote relationship to other *Bacillus* spp. (Pichinoty et al., 1984). More unusual properties have not been revealed and besides the initial description, there have been no follow-up studies. Another isolate, neither officially described as a new species nor studied in any further way, was called "*B. pseudomegaterium*" (Lafay et al., 1994). Both isolates were isolated using classic cultivation approaches without indications for particular abundance in soil. However, another study combining culture-independent analysis together with a cultivation approach identified this group as predominant in rice paddy soil microcosms (anaerobic) and also succeeded in identifying a matching isolate called SB45 (Hengstmann et al., 1999). Here, SB45 was isolated anaerobically on sugar substrate and grew aerobically but spores could never be observed. In the year 2000, the 16S rDNA sequence of *B. niacini* (Nagel and Andreesen, 1991) was made available and identified this unusual *Bacillus* as a very close relative to "*B. pseudomegaterium*". This has now become the best-studied relative, since utilisation of niacin has produced further studies. The aerobic *B. niacini* is a large, pleomorphic, endospore-forming rod and produces a blue pigment upon niacin degradation. Another species of more exotic origin, in Korean seafood, has recently been described as *Bacillus jeotgali* (Yoon et al., 2001). However, during the past decade none of these isolates have attracted much further attention but this will undoubtedly change in the near future.

The *B. niacini* relatives

The cluster of *B. niacini* relatives was firstly found to be important in soil environments by Hengstmann et al. (1999), and at that time named the "*Bacillus pseudomegaterium*" cluster (referring to the neither validly described nor further studied strain ATCC 49866). However, the recently sequenced 16S rDNA of the species *B. niacini* (Nagel and Andreesen, 1991) joined this cluster and represented therefore the first validly described species within, and consequently this species was used here to name this cluster. A few dozen 16S rDNA sequences directly retrieved from soil DNA (uncultured bacteria) from Europe and America also belong to this group. At the moment, this cluster accounts for approximately 15 % of all uncultured *Bacillus* 16S rDNA clones from directly extracted soil DNA. This apparent relative abundance of *B. niacini*-relatives in soil was also confirmed by a more *Bacillus*-specific culture-independent cloning and DGGE approach (Garbeva et al., 2003). Recently, some closely related isolates from various environments were published. Ando et al. (2001) isolated a hitherto unknown species from a Japanese mangrove swamp ("*B. aestuarii*"), and Logan et al. (2000) described the species *B. fumarioli* from geothermal soils in Antarctica. Janssen et al. (2002) cultured one member of this group (strain Ellin411 showed 96.5 % 16S rDNA sequence identity to *B. niacini*) from agricultural soil. The latter work is of extraordinary interest for bacterial soil ecology, since they succeeded in improving the efficacy of their cultivation approach and apparently getting much nearer to growing the most abundant soil bacteria.

A recent survey to isolate members of this group by recovering thousands of soil bacteria strains and subsequent high-throughput screening (Felske et al., 2003) tapped into the remarkable biodiversity within this group, multiplying the number of known strains. Some of these isolates could already be used to delineate and describe five hitherto unknown species (Heyrman et al., 2003): *B. novalis*, *B. vireti*, *B. soli*, *B. bataviensis* and *B. drentensis* with many more species from this group possibly awaiting discovery. Further, novel sequences of not yet published strains have recently appeared in the public databases. Garcia et al. (unpublished data) isolated strains of denitrifying *Bacillus* from tropical rice soils, including a few from the *B. niacini* neighbourhood. Venkateswaran et al. (unpublished data) investigated the microbial diversity of spacecraft assembly facilities (USA) and also isolated a *B. niacini*–relative. This raises the question of why it has been so difficult to culture before? According to Felske et al. (2003), the approach to isolate soil bacteria using classic dilute nutrient broth agar only yielded approximately 1% *B. niacini*–relatives among all picked colonies. Hence, it requires consideraable throughput of strains to find them among soil isolates. Efficient methods to do so may therefore have been the bottleneck previously, but modern microtiter format set-ups and fast and simple molecular methods such as PCR have allowed a high throughput in a much improved way.

Looking for the most suitable culture condition to enrich *B. niacini*-relatives on agar plates, it has turned out that mineral medium plus acetate at 37°C is best with up to 10% of all picked colonies being *B. niacini*-relatives. However, the isolates usually adapted quite rapidly to standard nutrient broth media and were often found to grow rapidly on this at room temperature. It remains unclear which ecological function the *B. niacini*-relatives may occupy in soil, but the high diversity among the isolates may indicate that there could be a very broad range of niches and functions being covered by these bacteria. Also the various sources and metabolic capabilities of recently studied strains closely related to this group hint to a remarkable functional diversity.

The *B. jeotgali* relatives

This separate cluster in the neighbourhood of *B. niacini* (Figure 3) is named after a recently described halophilic species from seafood (Yoon et al., 2001). Most uncultured members of this subcluster originated from anoxic paddy soil microcosms studied by Hengstmann et al. (1999). The true extent and diversity of this group is so far poorly understood. We recently isolated a couple of members apparently affiliated to this group but they are neither closely related to *B. jeotgali* nor to the sequences of yet uncultured soil bacteria. Hence, at the moment there is little conclusive information about the underlying phylogenetic relationships.

The *B. benzoevorans* relatives

The first case where a culture-independent approach identified the genus *Bacillus* as the most abundant taxon in a soil bacteria community was reported from the Dutch Drentse A grasslands (Felske, 1999). However, most of the *Bacillus*-related 16S rRNA sequences appeared to originate from a novel and uncharacterised lineage. A clone library of 16S rDNA was generated from directly extracted soil DNA. In parallel, bacterial rRNA was isolated from directly extracted soil ribosomes. Both fractions proved to be predominated by sequences closely related to each other but only remotely related to the next cultured and described species *B. benzoevorans*. Subsequent PCR or RT-PCR amplification with bacterial 16S rRNA specific primers allowed for community analysis via TGGE. The TGGE fingerprints showed a high similarity along the investigated 1.5 km stretch of Drentse A grasslands (Felske and Akkermans, 1998). Five sequences of this novel cluster were readily detectable in these TGGE fingerprints. The signals based on ribosomal RNA were also quantified via multiple competitor RT-PCR. Only these five types already contributed approximately 15% of all bacterial ribosomes. This predominance of the novel cluster in the rRNA fraction extracted from intact soil ribosomes suggested their activity. The most abundant 16S rRNA sequence, DA001, could be detected by fluorescent whole-cell hybridisation in soil bacteria suspensions. Apparently vegetative 2 μm-rods of at least 10^8 cells g^{-1} soil were detected with a multiple staining approach. Looking at the phylogenetic position of this cluster, it represents only a tiny section within the highly diverse *Bacillus* phylogeny (Figure 3). This *Bacillus* cluster predominated by environmental 16S rDNA clones, as defined by the accumulation of environmental 16S rDNA sequences in a particular branch of the *Bacillus* rRNA subgroup 1, is remotely related to *B. benzoevorans*, a large *Bacillus* growing rapidly in sheathed filaments and capable of the utilisation of various aromatic compounds (Pichinoty et al., 1984). More unusual properties of this species have not been revealed and besides the description, no follow-up studies have been reported. More selective media and/or high-throughput screening of large numbers of strains will be required to isolate more strains from this branch of *Bacillus*. Recently, a TGGE-guided attempt to identify such strains among hundreds of isolates from Dutch grassland soil failed to match with the ones detected in a clone library from the same source (Felske et al., 1999). At this time, the extent, meaning and definition of a particular cluster was unclear and mainly based on a number of sometimes very short environmental 16S rDNA sequences of variable quality. Nevertheless, a handful of strains closely related to the clones (>98% 16S rDNA sequence similarity) emerged in this study. A variety of media was used in this cultivation campaign and different strains of this cluster could be isolated by using mineral media amended with acetate. This isolation strategy for recalcitrant bacteria, i.e. the use of passively cell wall-penetrating carbon sources like acetate, was recently described by Button et al. (1998). The use of media with aromatic compounds selective for *B. benzoevorans* (Pichinoty et al., 1984) did not yield new *Bacillus* strains, even *B. benzoevorans*. However, this picture changed somewhat with our recent resumption of isolation. This time, benzoate was used very successfully to isolate different novel strains along with *B. benzoevorans*. This cultivation strategy is now applied to generate a comprehensive culture collection of several thousand strains which will be rapidly screened for the presence of this cluster by molecular methods. Among the numerous isolates members of this cluster will be detected by specific oligonucleotide probes targeting 16S rDNA. This screening is required because the isolation approach is still not selective. The number of novel *Bacillus* colonies on the plates appeared to be well below 5% of the total colony count. The low specificity of the described cultivation approach was more directed by exclusion of other bacteria than promoting the new cluster. Hence, the number of colony forming units was in the range of 10^6 g^{-1} soil, just similar or even lower than the yields obtained with common standard media. The new *Bacillus* lineage accounted for 10^4 g^{-1} soil. A rather specific isolation approach should not only select for but also promote the new lineage and boost the number of colonies to levels more comparable to the culture-independent measurements (approaching 10^8-10^9 colonies g^{-1} soil). Nevertheless, a start has been made and the first described species of this cluster should be published soon.

OUTLOOK

What is true for bacteria in general also holds true for the genus *Bacillus* in particular. We already know many species but we have only just started to recognise the truly abundant and probably important members of this taxon. Since the yet unexplored relatives of *B. benzoevorans* and *B. niacini* may contribute a significant fraction to all bacterial ribosomes in soil, they should also be responsible for major bacterial metabolic activities. Further, these relatives of *B. benzoevorans* and *B. niacini* may well be the most abundant *Bacillus* in soils. Due to the fact that members of the Gram positive genus *Bacillus* are notoriously recalcitrant to cell lysis, they have probably been neglected by many culture-independent DNA-based studies on soil bacteria. Since *Bacillus* DNA is best isolated using harsh extraction methods, the integrity of the extracted DNA will suffer. Hence, the increasingly popular metagenomic approach, that requires especially long DNA fragments from environmental samples is unlikely to identify the genus *Bacillus*.

As demonstrated in recent studies, this hitherto neglected fraction of *Bacillus* soil diversity is at least in part cultivable. Therefore, *Bacillus* (at least relatives of *B. benzoevorans* and *B. niacini*) may appear abundant as vegetative cells. Looking at their probably significant metabolic contribution to microbial activity in soil, they should be considered among the "housekeeping" soil bacteria. If this holds true, the focus on *Bacillus* ecology may well move away from the endospore, and their metabolism may be the true key.

REFERENCES

Achouak, W., Normand, P., and Heulin, T. (1999). Comparative phylogeny of *rrs* and *nifH* genes in the Bacillaceae. Int. J. Syst. Bacteriol. *49*, 961-967.

Amann, R.I., Ludwig, W., and Schleifer, K.H. (1995). Phylogenetic identification and *in situ* detection of individual microbial cells without cultivation. Microb. Rev. *59*, 143-169.

Ando, Y., Mitsugi, N., Yano, K., and Karube, I. (2001). Isolation of a bacterium from mangrove soil for degradation of sea sludge. Appl. Biochem. Biotechnol. *95*, 175-182.

Borneman, J., Skroch, P.W., O'Sullivan, K.M., Palus, J.A., Rumjanek, N.G., Jansen, J.L., Nienhuis, J., and Triplett, E.W. (1996). Molecular microbial diversity of an agricultural soil in Wisconsin. Appl. Environ. Microbiol. *62*, 1935-1943.

Borneman, J., and Triplett, E.W. (1997). Molecular microbial diversity in soils from Eastern Amazonia: evidence for unusual microorganisms and microbial population shifts associated with deforestation. Appl. Environ. Microbiol. *63*, 2647-2653.

Branda, S.S., Gonzalez-Pastor, J.E., Ben-Yehuda, S., Losick, R., and Kolter, R. (2001). Fruiting body formation by *Bacillus subtilis*. Proc. Natl. Acad. Sci. U S A *98*, 11621-11626.

Button, D.K., Robertson, B.R., Lepp, P.W., and Schmidt, T.M. (1998). A small, dilute-cytoplasm, high-affinity, novel bacterium isolated by extinction culture and having kinetic constants compatible with growth at ambient concentrations of dissolved nutrients in seawater. Appl. Environ. Microbiol. *64*, 4467-4476.

Conn, H.J. (1916). Are spore-forming bacteria of any significance in soil under normal conditions? Technical Bulletin of the New York State Agricultural Experiment Station *51*, 187-195.

Duineveld, B.M., Kowalchuk, G.A., Keijzer, A., van Elsas, J.D., and van Veen, J.A. (2001). Analysis of bacterial communities in the rhizosphere of chrysanthemum via denaturing gradient gel electrophoresis of PCR-amplified 16S rRNA as well as DNA fragments coding for 16S rRNA. Appl. Environ. Microbiol. *67*, 172-178.

Dunbar, J., Takala, S., Barns, S.M., Davis, J.A., and Kuske, C.R. (1999). Levels of bacterial community diversity in four arid soils compared with cultivation and 16S rRNA gene cloning. Appl. Environ. Microbiol. *65*, 1662-1669.

Felske, A., and Akkermans, A.D.L. (1998). Spatial homogeneity of the most abundant bacterial 16S rRNA molecules in grassland soils. Microb. Ecol. *36*, 31-36.

Felske, A., Wolterink, A., van Lis, R., and Akkermans, A.D.L. (1998). Phylogeny of the main bacterial 16S rRNA sequences in Drentse A grassland soils (The Netherlands). Appl. Environ. Microbiol. *64*, 871-879.

Felske, A. (1999). Reviewing the DA001-files: A 16S rRNA chase on suspect #X99967, a *Bacillus* and Dutch underground activist. J. Microbiol. Meth. *36*, 77-93.

Felske, A., Heyrman, J., Balcaen, A., and De Vos, P. (2003). Isolation and biodiversity of hitherto not described soil bacteria related to *Bacillus niacini*. Microb. Ecol., in press.

Garbeva, P., van Veen, J.A., van Elsas, J.D. (2003). Predominant *Bacillus* spp. in agricultural soil under different management regimes detected via PCR-DGGE. Microb. Ecol. *45*, 302-316.

Hengstmann,U., Chin,K.J., Janssen,P.H., and Liesack,W. (1999). Comparative phylogenetic assignment of environmental sequences of genes encoding 16S rRNA and numerically abundant cultivable bacteria from an anoxic rice paddy soil. Appl. Environ. Microbiol. *65*, 5050-5058.

Heyrman, J., Vanparys, B., Logan, N.A., Balcaen, A., Rodriguez-Diaz, M., Felske, A., and De Vos, P. (2003). *Bacillus novalis* sp. nov., *Bacillus vireti* sp. nov., *Bacillus soli* sp. nov., *Bacillus bataviensis* sp. nov. and *Bacillus drentensis* sp. nov., five new species isolated from the Drentse A grasslands. Int. J. Syst. Evol. Microbiol., in press.

Janssen, P.H., Yates, P.S., Grinton, B.E., Taylor, P.M., and Sait, M. (2002). Improved culturability of soil bacteria and isolation in pure culture of novel members of the divisions *Acidobacteria*, *Actinobacteria*, *Proteobacteria*, and *Verrucomicrobia*. Appl. Environ. Microbiol. *68*, 2391-2396.

Karpunina, L.V., Mel'nikova,U.Iu., Konnova,S.A., and Abros'kina,O.M. (2001). Role of the agglutinating proteins of bacilli and rhizobia in bacterial interaction. Mikrobiologiia, *70*, 519-524.

Kucey, R.H.N. (1988). Alteration of wheat root systems and nitrogen fixation by associative nitrogen-fixing bacteria measured under field conditions. Can. J. Microbiol. *34*, 735-739.

Lee, S.-Y., Bollinger, J., Bezdicek, D., and Ogram, A. (1996). Estimation of the abundance of an uncultured soil bacterial strain by a competitive quantitative PCR method. Appl. Environ. Microbiol. *62*, 3787-3793.

Leuschner, R.G., and Lillford, P.J. (1999). Effects of temperature and heat activation on germination of individual spores of *Bacillus subtilis*. Lett. Appl. Microbiol. *29*, 228-232.

Liesack, W., and Stackebrandt, E. (1992). Occurrence of novel groups of the domain *Bacteria* as revealed by analysis of genetic material isolated from an Australian terrestrial environment. J. Bacteriol. *174*, 5072-5078.

Lloyd-Jones, G., and Lau, P.C. (1998). A molecular view of microbial diversity in a dynamic landfill in Quebec. FEMS Microbiol. Lett. *162*, 219-226.

Logan, N.A., Lebbe, L., Hoste, B., Goris, J., Forsyth, G., Heyndrickx, M., Murray, B.L., Syme, N., Wynn-Williams, D.D., and De Vos, P. (2000). Aerobic endospore-forming bacteria from geothermal environments in northern Victoria Land, Antarctica, and Candlemas Island, South Sandwich archipelago, with the proposal of *Bacillus fumarioli* sp. nov. Int. J. Syst. Evol. Microbiol. *50*, 1741-1753.

Ludwig, W., Bauer, S.H., Bauer, M., Held, I., Kirchhof, G., Schulze, R., Huber, I., Spring, S., Hartmann, A., and Schleifer, K.H. (1997). Detection and *in situ* identification of representatives of a widely distributed new bacterial phylum. FEMS Microbiol. Lett. *153*, 181-190.

Macrae, A., Rimmer, D.L., and O'Donnell, A.G. (2000). Novel bacterial diversity recovered from the rhizosphere of oilseed rape (*Brassica napus*) determined by the analysis of 16S ribosomal DNA. Antonie Van Leeuwenhoek *78*, 13-21.

Macrae, A., Lucon, C.M.M., Rimmer, D.L., and O'Donnell, A.G. (2001). Sampling DNA from the rhizosphere of *Brassica napus* to investigate rhizobacterial community structure. Plant and Soil *233*, 223-230.

Margalit, J. and Dean, D. (1985). The story of *Bacillus thuringiensis* var. *israelensis* (B.t.i.). J. Am. Mosq. Control Assoc. *1*, 1-7.

Nagel, M., and Andreesen, J.R. (1991). *Bacillus niacini* sp. nov., a nicotinate-metabolizing mesophile isolated from soil. Int. J. Syst. Bacteriol. *41*, 134-139.

Norris, J.R., and Wolf, J. (1961). A study of antigens of the aerobic spore-forming bacteria. J. Appl. Bacteriol. *24*, 42-56.

Pichinoty, F., Asselineau, J., and Mandel, M. (1984). Biochemical characterization of *Bacillus benzoevorans* sp. nov., a new filamentous, sheated mesophilic species which degrades various aromatic acids and phenols. Ann. Microbiol. *135*, 209-217.

Priest, F.G. (1992). Biological control of mosquitoes and other biting flies by *Bacillus sphaericus* and *Bacillus thuringiensis*. J. Appl. Bacteriol. *72*, 357-369.

Siala, A., Hill, I.R., and Gray, T.R.G. (1974). Populations of spore-forming bacteria in an acid forest soil, with special reference to *B. subtilis*. J. Gen. Microbiol. *81*, 183-190.

Van Ness, G.B. (1971). Ecology of anthrax. Science *172*, 1303-1307.

von der Weid, I., Duarte, G.F., van Elsas, J.D., and Seldin, L. (2002). *Paenibacillus brasilensis* sp. nov., a novel nitrogen-fixing species isolated from the maize rhizosphere in Brazil. Int. J. Syst. Evol. Microbiol. *52*, 2147-2153.

White, D., Priest, F.G., and Sharp, R.J. (1989). Preliminary taxonomic studies on 1000 isolates of thermophilic bacilli. In: Microbiology of extreme environments and ist potential for biotechnology. M.S. da Costa, J.C. Duarte and R.A.D. Williams, ed. Elsevier Applied Science, New York. p. 387.

Williams, A.G., and Withers, S.E. (1983). *Bacillus* spp. in the rumen ecosystem. Hemicellulose depolymerases and glycoside hydrolases of *Bacillus* spp. and rumen isolates grown under anaerobic conditions. J Appl Bacteriol, *55*, 283-292.

Williams, R.A.D. (1989). Biochemical taxonomy of the genus *Thermus*. In: Microbiology of extreme environments and ist potential for biotechnology. M.S. da Costa, J.C. Duarte and R.A.D. Williams, ed. Elsevier Applied Science, New York. pp. 82-97.

Yoon, J.-H., Kang, S.-S., Lee, K.-C., Kho, Y.H., Choi, S.H., Kang, K.H., and Park, Y.-H. (2001). *Bacillus jeotgali* sp. nov., isolated from jeotgal, Korean traditional fermented seafood. Int J Syst Bacteriol, *51*, 1087-1092.

Chapter 4

Pathogenic Bacilli: *Bacillus anthracis* and Close Relatives

Marco R. Oggioni, Annalisa Ciabattini, Marco Cassone, and Gianni Pozzi

SUMMARY

Species of the genus *Bacillus* are ubiquitous, and with a few exceptions, non-pathogenic bacteria. The direct involvement in human and animal disease of *Bacillus anthracis* is the prime example on which Koch formulated his postulates. The contribution of this species and that of the other closely related species to human disease are reviewed in this article. This review confirms the extremely low pathogenic potential for most of these species with the obvious exception of *B. anthracis* and *B. cereus*. Altogether the other species of the genus *Bacillus*, being non-pathogenic and not harbouring drug resistance genes, appear to meet the requirements of the classification systems used for food and feed safety, including both the European QPS system (Qualified Presumption of Safety) and the American GRAS system (Generally Recognised As Safe).

INTRODUCTION

Species of the genus *Bacillus* are ubiquitous endospore forming Gram positive bacteria. The primary habitat of these metabolic diverse microorganisms is believed to be the soil where prolonged survival is guaranteed by the heat and exsiccation resistant spores. Nutrient rich environments permitting germination and outgrowth are the rhizosphere (Foldes et al., 2000) and possibly the intestine of invertebrates (Jensen et al., 2003). Due to their ubiquity in soil and extreme heat resistance, spores commonly contaminate a variety of foods. For example, they are ingested in the diet as contaminants (Rosenkvist and Hansen, 1995; Viljoen and von Holy, 1997), used as probiotics in humans and livestock (Green et al., 1999; Jadamus et al., 2002; Oggioni et al., 1998; Senesi et al., 2001) or consumption in selected fermented foods (e.g., natto) (Wang and Fung, 1996). Despite this continued exposure human *Bacillus* infections are extremely rare.

DEFINITION OF THE CAUSAL RELATIONSHIP OF MICROORGANISM TO DISEASE

When bacteria were first discovered, their ability to cause disease was not immediately obvious and had to be established by the application of a strict set of rules. Robert Koch (1843-1910) formulated his four postulates (Figure 1), derived from his work on *Bacillus anthracis*, to prove the microbial causation of disease (Koch, 1998a). While Koch's postulates are still fundamental rules, their fulfilment is difficult in a variety of cases, including infections due to opportunistic pathogens and infections of immuncompromised host (no achievement of the 1st postulate), infections due to non cultivable species (no achievement of 2nd postulate), and infections in which there is no animal model (no achievement of the 3rd postulate). When applying Koch's postulates to infections of the species of the genus *Bacillus*, with the exception of anthrax for which they were developed, a number of problems are commonly encountered. These typically are, (i) the ubiquity of *Bacillus* spores in the environment, (ii) the belief that species of this genus are non pathogenic (iii) recognition of *Bacillus* species by clinical laboratories as 'contaminants' and, (iv) the difficulty of unambiguous species identification using standard carbohydrate metabolism tests. Taken together these facts lead to a probable underestimation of the incidence of *Bacillus* infections, and an insufficient clinical and microbiological description of Bacillus species and disease.

DISEASES CAUSED BY *BACILLUS ANTHRACIS*

Members of the *Bacillus cereus/Bacillus thuringiensis* phylogenetic group, which includes *B. anthracis*, are virtually indistinguishable by 16S rRNA sequence analysis, with as much variability present among the multiple rDNA operons of an individual species as among different isolates (Pannucci et al., 2002). Multilocus enzyme electrophoresis and analysis of nucleotide polymorphisms suggest they may represent a single species (Cherif et al., 2002; Helgason et al., 2000). Most isolates of the species *B. cereus, B. thuringiensis*, and *B. anthracis* carry plasmids that vary in number and size, with some plasmids exceeding 500 kb (Pannucci et al., 2002). In fact, the virulence factors of *B. anthracis* and *B. thuringiensis* are encoded by plasmid-carried genes (Berry et al., 2002; Mock and Fouet, 2001; Okinaka et al., 1999).

Anthrax is an acute infectious disease caused by *B. anthracis* (Koch, 1998b). Anthrax most commonly occurs in wild and domestic animals (cattle, sheep, goats and other herbivores), but it can also occur in humans when they are exposed to infected animals or tissue from infected animals (cutaneous infection in farmers; respiratory infection in mill workers). The use of

1.	The specific organism should be shown to be present in all cases of animals suffering from a specific disease but should not be found in healthy animals.
2.	The specific microorganism should be isolated from the diseased animal and grown in pure culture on artificial laboratory media.
3.	This freshly isolated microorganism, when inoculated into a healthy laboratory animal, should cause the same disease seen in the original animal.
4.	The microorganism should be reisolated in pure culture from the experimental infection.

Figure 1. Koch's postulates (1884).

the attenuated Sterne veterinary vaccine in the 1930's lead to a worldwide reduction in livestock cases (Turnbull, 2000). This decrease in animal cases was accompanied by a decrease in human cases. Unfortunately there are still regions were anthrax is hyper-endemic or endemic and this generally parallels the quality of public health and veterinary services in the respective regions (Hugh-Jones, 1999) (http://www.vetmed.lsu.edu/whocc/mp_world.htm).

Human anthrax can be classified into three major clinical forms depending on the route of infection, these being, cutaneous, inhalation, and gastrointestinal diseas. By far the most frequent form of anthrax is Cutaneous anthrax that occurs after contact with anthrax spores with any region of exposed skin. Spores are taken up by macrophages and germinate intracellularly both at the site of infection and in the regional lymph nodes, ultimately leading to toxin production. Clinically the lesion forms 1-10 days after exposition as a pruritic painless papule that enlarges and erodes within a day or two leaving a necrotic ulcer which may leave a black eschar. Regional lymphadenopathy is also evident (Bortolussi and Kennedy, 1999). If untreated fatality rates are approximately 10-20%, but antibiotic treatment drastically reduces the development of severe sequelae. Gastrointestinal disease results from ingestion of poorly cooked contaminated meat (Kanafami et al., 2003). This form of anthrax is reported to be responsible for only about 1% of human anthrax cases, but with a case fatality rate of up to 75% if untreated. Inhalation anthrax is an extremely severe, but fortunately very rare disease (before the Sverdlovsk outbreak in 1979 only 30 cases had been described) (Bortolussi and Kennedy, 1999; Meselson et al., 1994).

The clinical view of inhalation and gastrointestinal anthrax show some striking similarities. Both cause regional hemorrhagic lymphadenitis (respectively mediastinal and mesenteric) with an important toxemic and septicemic component. In both cases bacteria are cultured from the lymph nodes and rods (not spores) are visible upon pathological examination of diseased lymph nodes (Bortolussi and Kennedy, 1999; Kanafami et al., 2003). In the case of inhalational disease clinical features include a incubation period of up to 20 days, a prodromic flu-like phase and a second fulminant phase consisting of severe dyspnea and shock with a mean survival of 1 day. Disease after spore inhalation develops when fine spore suspensions reach the alveoli where they are taken up by macrophages. At this stage the spores are transported by the macrophages through the pulmonary lymphatics to hilar and mediastinal lymph nodes. Germination of the organism has been demonstrated within macrophages (Guidi-Rontani et al., 1999) and linked to toxin production. Cytotoxicity is a result of the activity of the lethal factor (LF) which is synthesized within the macrophage upon germination and before its release from the lysing cell (Guidi-Rontani et al., 2001). The resulting septicemic phase of disease correlates to multiplication in the bloodstream of the vegetative cells, which are protected from phagocytosis by the poly-D-glutamate capsule. Inhalation anthrax is the disease targeted when preparing anthrax spores for biological warfare or bioterrorism. Following the intentional release of anthrax spores in 2001 in the USA 22 cases of anthrax (11 inhalational,

11 cutaneous) were identified; of which 5 of the inhalational cases were fatal (Jernigan et al., 2002).

Clinical anthrax in animals has distinct forms depending on the host. In cattle and sheep, typically the symptomatic phase of the disease may evolve in just one to two hours. The very first indication of disease may be sudden death of the animal! Clinical signs include fever, muscle tremors, respiratory distress, and convulsions, but often go unnoticed. Sudden death in an animal without prior signs is one of the most important signs leading to suspicion of an anthrax infection. After death, there may be hemorrhagic nasal, buccal, or anal exudates, lack of rigor mortis, and the blood does not clot (www.aphis.usda.gov/vs/). In all countries national guidelines exist to report and control anthrax in the veterinary field.

Protection against anthrax, conferred by spores of *B. anthracis*, was described 120 years ago (by Greenfield in 1880 and by Pasteur in 1881). Presently the most widely used vaccine for the prevention of anthrax in animals is the Sterne-strain vaccine. This vaccine is based on a toxigenic live attenuated variant of *B. anthracis* developed by Sterne in 1937 (strain34F2). This non-encapsulated strain lacks plasmid pOX2 which codes for capsule formation. In Russia and China live spore vaccines equivalent to the Sterne strain are in use (Strain 55), while in Italy animals are vaccinated with a strain (Carbosap) which still carries both virulence plasmids, being nevertheless attenuated in virulence (Adone et al., 2002; Fasanella et al., 2001). The residual virulence of all these live vaccine strains discouraged their use in humans (Turnbull, 2000). To overcome this problem, acellular vaccines have been developed as response to threats of biological warfare or bioterrorism originating soon after the second world war. In the USA a cell-free culture filtrate adsorbed onto aluminium hydroxide from non-encapsulated non-proteolytic strain V770-NP-R is used for human vaccination. The UK vaccine is an alum precipitate of cell-free culture supernatant of the non-encapsulated toxigenic strain 34F2 (Sterne). None of these two vaccines is available for the general public. Before the 2001 attacks these acellular vaccines were criticised for their complicated vaccination schedule, insufficient vaccine memory conferred, lot-to-lot variability and the potential for localized reactions (Oggioni et al., 2003; Turnbull, 1991; Turnbull, 2000). More recent reports underline the safety and efficacy of these vaccines (Lange et al., 2003) and in the USA millions of doses of anthrax vaccine were committed by NIAID to two biotechnology companies (Cohen, 2003). However, there is still the need for a next generation anthrax vaccine.

DISEASES CAUSED BY *BACILLUS CEREUS*

Bacillus cereus is an ubiquitous soil bacterium associated with gastrointestinal disease and some rare but more serious non-gastrointestinal infections (Callegan et al., 2002; Drobniewski, 1993). While detection of *B. anthracis* in a clinical sample is indicative of its causal correlation to the disease the "carriage of *B. cereus* in stools is transient and its presence at any one time reflects solely its intake with foods" (Turnbull and Kramer, 1985). Criteria for the confirmation of *B. cereus* mediated diarrhoea should therefore include detection of at least 10^5 CFU per gram of faecal sample, identification of the same *B. cereus* isolate in food and most importantly detection of enterotoxin in food or faeces (Tan et al., 1997). The epidemiology of gastrointestinal disease changes significantly for different geographic areas, with prevalence varying from 1.3% (USA) to 17.8% (Finland) of all gastrointestinal toxic-infections. Still, the number of reported cases in different countries is not comparable because of the diversity in reporting procedures (Kotiranta et al., 2000).

There are two distinct forms of gastrointestinal disease caused by *B. cereus,* which are emetic and diarrhoeal syndrome (Drobniewski, 1993; Kotiranta et al., 2000). Both types of food poisoning are due to preformed toxin in the food and are not associated to production of toxin in the human intestinal tract (Drobniewski, 1993). The rapid-onset emetic syndrome is characterized by nausea and vomiting. The emetic syndrome is self limiting with recovery within 24 hours. One case with hepatic failure and death was described in 1997 (Mahler et al., 1997). Boiled rice that is held for prolonged periods at ambient temperature and then quick-fried before serving is a frequent cause, although dairy products or other foods may also be responsible. The diarrhoeal syndrome has a slower onset and diarrhoea and abdominal pain occurs 8 to 16 hours after consumption of contaminated food. Symptoms resolve autonomously within 12 hours and antibiotic therapy is not needed (Drobniewski, 1993). Diarrhoeal syndrome is associated with a variety of foods in which heat-resistant endospores survive cooking or pasteurization and then germinate and multiply when the food is inadequately refrigerated.

B. cereus emetic syndrome is caused by emetic toxin named cereulide. The toxin consists of a ring structure of three repeats of four amino and/or oxy acids: [D-O-Leu-D-Ala-L-O-Val-L-Val]$_3$. This ring structure (dodecadepsipeptide) has a molecular mass of 1.2 kDa, and is chemically closely related to the potassium ionophore valinomycin (Agata et al., 1994). The emetic toxin is resistant to heat (up to 121°C), tolerates extreme pH values between 2 and 11, and is stable to pepsin and trypsin treatment (Shinagawa et al., 1996). The toxin molecule is thought to be produced via enzymatic modification of media components rather than being ribosomally synthesized (Agata et al., 1995; Granum and Lund, 1997). The production of the toxin occurs during sporulation of the culture, and an *in vitro* assay demonstrated that the toxin produces a vacuolar response in Hep-2 tissue culture cells (Agata et al., 1994). Cerulide induces vomiting by binding to the 5-HT$_3$ subtype of serotonin receptors of the afferent vagus nerve (Agata et al., 1995).

Enterotoxin-mediated diarrhoeal food poisoning infections seems to be mediated by many proteins with different characteristics. A three component haemolysin named HBL (hemolysin with binding and lytic components) with enterotoxic activity has been purified and characterised (Beecher et al., 1995; Beecher and Macmillan, 1991; Granum and Lund, 1997). This enterotoxin complex is composed of a binding component B of 35 kDa, and two lytic components L$_1$ and L$_2$ (36 and 45 kDa) and all three components are required for maximal haemolytic-, cytotoxic- and dermonecrotic-activity, vascular permeability and fluid accumulation in ligated rabbit ileal loops (Beecher et al., 1995). Studies of interactions of HBL with erythrocytes,

suggested that the B protein is the component that binds HBL to the target cells, and that L$_1$ and L$_2$ have lytic functions (Beecher and Macmillan, 1991). Another model suggests that the three components of HBL bind to target cells independently and then constitute a membrane-attacking complex responsible for the osmotic lysis of cells (Beecher and Wong, 1997). All three proteins of the HBL are transcribed from one operon (*hbl*) (Ryan et al., 1997). *hblC* (transcribing L$_2$) and *hblD* (transcribing L$_1$) encode proteins of 447 aa and 384 aa. L$_2$ has a signal peptide of 32 aa and L$_1$ a signal peptide of 30 aa. The B protein, transcribed from *hblA*, consists of 375 aa, with a signal peptide of 31 aa (Heinrichs et al., 1993).

Another three component enterotoxin complex associated with diarrhoeal food poisoning infections is the non-hemolytic NHE protein (non hemolytic enterotoxin) (Lund and Granum, 1996). There is a high degree of identity between the N-terminal amino acids sequence of L$_1$ of hemolysin BL and the 39 kDa subunit of NHE, and also between parts of L$_2$ and the 45 kDa protein (Granum and Lund, 1997) suggesting similar roles in different *B. cereus* strains (McKillip, 2000). The third part of the complex is a 105 kDa protein, that may be the binding component of the multi-component toxin in studies of interactions between NHE and Vero cells (Lund and Granum, 1997). The two other components are probably not able to bind to these cells alone.

A third toxin, recently reported, is the single component T enteroxin, which seems to possess similar biological activity to HBL and NHE (Agata et al., 1995). However the three-component toxins are actually the ones referred to as etiological causes of diarrhoeal food poisoning infections due to *B. cereus* (Lund and Granum, 1997). Some strains produce both HBL and NHE, while other strains contain genes for only one of them (Granum et al., 1996). The enterotoxin T gene has recently been shown to be absent in 57 of 95 strains of *B. cereus*, and in five out of seven strains involved in food poisoning (Granum et al., 1996). A series of other toxins, including Cereolysin O, phosphatidylinositol-specific phospholipase C, sphingomyelinase, and a hemolysin provisionally designated hemolysin IV, have been identified and differently related to virulence (Kotiranta et al., 2000 and Chapter 8).

Systemic infections caused by *B. cereus* are not common and generally linked to predisposing host risk factors which result in immune-compromission (leukaemia, drug abuse etc.). Olszewski and colleagues (Olsewski et al., 1999) reported a frequent finding of *B. cereus* and other *Bacillus* species in the lymph nodes of patients with filarial lymphedema. The percentage of recovery of *Bacillus* increased in lymph nodes obtained from patients with active recurrent inflammatory episodes, and was associated in some cases to positive blood cultures. No positive lymph node was found in healthy controls. This points to the hypothesis of intramacrophage germination, with survival of bacilli and consequent disease only in cases of impaired immune responses.

Ocular infections can occur in immunocompetent patients. In 1993, Drobniewski (Drobniewski, 1993) listed 35 cases of ocular infections and 36 systemic non-gastrointestinal infections reported in this century. Systemic disease (endocarditis, pneumonia and meningitis) (Drobniewski, 1993; Gaur et al., 2001) and ocular infections (endophthalmitis and panophtalmitis) are determined by traumatic or surgical introduction of spores or bacteria. Despite aggressive therapeutic and surgical intervention, frequently *B. cereus* endophthalmitis results in loss of functional vision or blindness, within 1 to 2 days (Callegan et al., 2003; Kotiranta et al., 2000). The severity of systemic infections is strictly linked to the underlying predisposing factors and timing of therapeutic intervention.

In France, until recently, *B. cereus* strain IP5832 was licensed (as the probiotic Bactisubtil) for oral treatment of diarrhoea for human use (Hoa et al., 2000; Spinosa et al., 2000a; Spinosa et al., 2000b). and was withdrawn 15/2/2001 although a similarly named product is marketed in Portugal by Aventis Pasteur (see Appendix 1). In 1988 a Belgian study reported that IP5832 was responsible for 1% of septic episodes in cancer patients, all of whom had received this probiotic as empiric therapy for tube-feeding related diarrhoea (Richard et al., 1988). The IP5832 strain was used as a probiotic for livestock until its recent withdrawl in the EU based upon its potential virulence traits (see Chapter 13 and Appendix II).

UNCONFIRMED SUSPECTS ON TOXICITY AND ALLERGENICITY OF *BACILLUS THURINGIENSIS*

Bacillus thuringiensis is a ubiquitous soil microorganism that has gained recent popularity for its ability to control certain insect pests in a natural, environmentally friendly manner. This feature makes *B. thuringiensis* the principal component of some commercial pesticides. Unlike typical nerve-poison insecticides, *B. thuringiensis* acts by producing large crystal proteins inclusions (crystal toxins) during sporulation that react with the cells of the gut lining of susceptible insects. These *B. thuringiensis* proteins paralyse the digestive system, and the infected insect stops feeding within hours. *B. thuringiensis*-affected insects generally die from starvation, which can take several days. Unlike most insecticides, *B. thuringiensis* pesticides do not have a broad spectrum of activity, so they do not kill beneficial insects, such as honeybees. The most commonly used strain of *B. thuringiensis* in pesticides is *kurstaki* strain, which kills only leaf- and needle-feeding caterpillars. In the past decade, *B. thuringiensis* strains have been developed that control certain types of fly larvae (i.e. *israelensis* strain); these are widely used against larvae of mosquitoes, black flies and fungus gnats.

In recent years, the close relationship between *B. thuringiensis* and the bacterium *B. cereus*, frequently associated with food-borne outbreaks of gastrointestinal disease, has been confirmed. Because of these similarities, *B. thuringiensis*-based products have been subjected to extensive evaluation both prior to and subsequent to becoming commercially available (Siegel, 2001b). Epidemiological studies have been conducted on several occasions after aerial applications of *B. thuringiensis* subsp. *kurstaki* over populated areas. The first epidemiological study that specifically addressed the possibility of increased incidence of infection and food-poisoning associated with large-scale spraying of *B. thuringiensis* subsp. *kurstaki* was performed in Oregon 1985-1986 years (Green et al., 1990). At the beginning

some *B. thuringiensis*-positive specimens were identified, but upon further examination, 52 of the 55 isolates were assessed to be probable contaminants. Of the three remaining isolates, *B. thuringiensis* subsp. *kurstaki* could neither be ruled in nor be ruled out as a cause of the patient's disease (Green et al., 1990).

The second study concerned the short-term health effects following the aerial applications of Foray 48B pesticide to control European gypsy moth populations and was carried out in Victoria, British Columbia (Capital Regional District, 1999; Valadares de Amorim et al., 2001). The study included a survey of the general health of the population, monitoring admission to hospital emergency departments and clinical surveillance of patients, and examining the records of many telephone calls. Among clinical isolates, 325 were identified as positive for *B. thuringiensis* of which 43 were pure cultures. 13 of the pure isolates came from genital-urinary culture, 10 from culture of eyes, and 9 from skin cultures. *B. thuringiensis* subsp. *kurstaki* bacteria were also isolated from nasal swabs taken from patients living inside and outside of the spray zone prior to the application of Foray 48B. Nevertheless, the incidence of *B. thuringiensis* subsp. *kurstaki* in the nares of the human population increased significantly within the spray zone after the first aerial application of Foray 48B and by the time the spray program was completed, a significant increase above prespray levels was detected both inside and outside of the spray zone (Valadares de Amorim et al., 2001). Despite this exposure, however, the human health surveillance program failed to detect any correlation between the aerial application of *B. thuringiensis* subsp. *kurstaki* and short-term health effects in the general adult population, in emergency room visits, or in aggravation of asthma symptoms in children. Recent studies on the effects of aerial spraying with *B. thuringiensis* subsp. *kurstaki* on asthma of children, again demonstrate that there is no difference in symptoms between subjects and controls, neither before nor after the spray and no evidence of adverse effects from the use of the biological pesticides was found (Pearce et al., 2002). Taken together, the epidemiological studies, laboratory research (Siegel, 2001b) and the prolonged use of *B. thuringiensis*-based pesticides in operational control programs, confirm that the isolates of *B. thuringiensis* used in commercial products are safe.

DISEASES CAUSED BY OTHER *BACILLUS* SPECIES

The main problem when searching the literature for *Bacillus* infections is the difficulty of unequivocal species identification. In most cases identification is based on biochemical characteristics assayed using the API 50 CH (Biomerieux) system, which is based on the performance of carbohydrate metabolism tests (Logan and Berkeley, 1984). Unfortunately results of this assay are not always very easy to interpret, leading frequently to misidentifications or unsuccessful identification (reporting of *Bacillus* spp.). Even with the availability of 16S rRNA sequencing the species identification within the genus *Bacillus* is not always unequivocal. Considering only case reports of the last 4 years (1999-2003) infections due to "non-

Table 1. Disease due to "non pathogenic" Bacilli in industrial use.

Species	Product	Clinical cases	Comment	Reference
B. cereus	Probiotic "Bactisubtil"	Four cases of sepsis in cancer patients	Over the counter drug licensed for human use in France, withdrawn from the market in 2001, but still licensed for veterinary use by EC	Richard et al., 1988
B. clausii	Probiotic "Enterogermina"	One case of recurrent septicaemia in end stage cancer patient and possibly one episode of cholangitis in transplant recipient	Over the counter drug licensed for human use in Italy for over 30 years	Oggioni et al., 1998; Spinosa et al., 2000a; Spinosa et al., 2000b
B. licheniformis, B. pumilus	Drain cleaner	Bacteraemia due to intentional injection in drug addict and recurrent bacteraemia due to intentional injection in patient with history of psychiatric illness	Biological/organic drain cleaner	Galanos et al., 2003; Hannah and Ender, 1999
B. thuringiensis	Insecticide	One possible case of corneal ulcer in farmer	Hundreds of products marketed worldwide	Siegel, 2001a

pathogenic" *Bacillus* species are very few. These include one endophthalmitis (Tandon et al., 2001) and one endocarditis (Krause et al., 1999) due to *B. circulans*, one endocarditis due to *B. popilliae* (Wu et al., 1999), two cases of bacteraemia due to *B. licheniformis* (Galanos et al., 2003; Hannah and Ender, 1999), one report of recurrent septicaemia due to *B. clausii* (Oggioni et al., 1998) and the re-examination of one case of cholangitis due to *B. clausii* (Spinosa et al., 2000b). The most important consideration on these limited number of cases is that all patients had severe underlying diseases compromising the immune system and thus facilitating infection. When considering previous reports and reviews on "non-pathogenic" *Bacillus* infections the state of immune compromisation of patients was always reported (Banerjee et al., 1988; Blue et al., 1995; Cotton et al., 1987; Woo et al., 2001). In this recently growing population of immune compromised patients the true prevalence of infections due to non pathogenic *Bacillus* species may be greatly underestimated, due to the frequent habit to discard any samples yielding *Bacillus* spp. as 'contaminants'. Studies on the clinical incidence of *Bacillus* associated bacteraemia agree that between 1 and 5 percent of positive blood cultures from cancer or transplant patients are positive for species of the genus *Bacillus* (Banerjee et al., 1988; Blue et al., 1995; Cotton et al., 1987; Woo et al., 2001). Interestingly no blood cultures yielding *Bacillus* are reported in non immune suppressed patients. Most authors tend to link *Bacillus* bacteraemia to indwelling central venous catheters (Matsumoto et al., 2003), while two reports link the bacteraemia to the use of oral probiotics (Oggioni et al., 1998; Richard et al., 1988; Spinosa et al., 2000b). Moreover, bacilli can be found in mesenteric lymph nodes in healthy dogs (Dahlinger et al., 1997). The germination and outgrowth of these non-pathogenic *Bacillus* in lymphoid organs in severely immune suppressed individuals has been compared to some features of anthrax pathogenesis in the immune competent (Spinosa et al., 2000a). A further issue was pointed by Woo and colleagues who studied culture-negative febrile episodes. They demonstrated that a significant proportion of septicemic episodes are due to difficult-to-culture cell-wall-deficient bacteria, of which *Bacillus* species account for 50 percent (Woo et al., 2001).

ANTIMICROBIAL DRUG RESISTANCE

The alarming increase in antimicrobial drug resistance is principally due to the spread of mobile genes forced by selection. One of the prime requirements for rapid spread of any genetic trait, carried by mobile genetic elements, is that the microbes "live" in an environment exposed to the respective selective pressure. In the case of antimicrobial drug resistance the environment where microbes are exposed to selective pressure is the body of humans, pets and farm animals. To envisage a selective pressure towards spread of antimicrobial drug resistance in any species of the genus *Bacillus* vegetative multiplication within humans or animals has to be demonstrated. Fortunately this is highly improbable for most species of the genus *Bacillus* since no species is part of the normal human flora despite daily uptake in food or feed (e.g., contaminants or starter cultures of alkaline fermentation, bread, natto) (Foldes et al., 2000; Wang and Fung, 1996) and widespread use as oral probiotics (Oggioni et al., 2003). The absence of selective pressure and thus of changes in the natural drug resistance profile of species of the genus *Bacillus* have been recognised in the past and even proposed as taxonomic criteria (Reva et al., 1995). This work very efficiently confirms that resistance in natural isolates is a stable property reflecting adaptation of strains to conditions in ecological niches occupied by certain species (Reva et al., 1995). The stability of the epidemiologic distribution of mobile genetic elements is even more remarkable when considering that plasmids of *Staphylococcus aureus*, a prime reservoir of drug resistance determinants, efficiently replicate in *Bacillus* spp. (Gryczan et al., 1978). Even if no significant spread of drug resistance has been observed for species of the genus *Bacillus* a wide variety of drug resistance determinants have been characterised in these species. Table 2 reports a list of the some of the most important determinants sequenced so far.

Systemic *Bacillus* infections are best treated with vancomycin and or ciprofloxacin (Das et al., 2001; Weber et al., 1988) to which essentially all *Bacillus* species are sensitive. Since for most species of this genus the major providers of automatised drug susceptibility testing systems do not report reference values, most laboratories are unable to provide reliable reports on drug susceptibility profiles or report drug susceptibility profiles by use of NCCLS breakpoints for staphylococci (Mohammed et al., 2002). Clinically, one of the most important resistance phenotypes is the natural resistance of *B. cereus* isolates to penicillin and other β-lactam-containing compounds. In contrast to *B. cereus* susceptibility to penicillin

Table 2. Drug resistance determinants of the genus *Bacillus*.

GenBanK accession	Resistance determinant
Macrolide resistance	
L08389	*B. anthracis ermJ* gene
AF480456	*B. cereus ermB* gene
M29832	*B. licheniformis ermD* gene
M77505	*B. licheniformis ermK* gene
M15332	*B. sphaericus ermG* gene
Aminoglycoside resistance	
X03364	*B. circulans aph* gene for amino-glycoside phosphotransferase
AJ494863	*B. circulans btrW, btrV, btrU, btrR* and *btrN* genes
AF539790	*B. clausii* aminoglycoside nucleotidyltransferase (*aadD2*) gene
AJ506108	*B. clausii ant(4')-Ib* gene for aminoglycoside 4'-O-nucleotidyltransferase
NC_002116	*B. sp.* plasmid pRBH1 kanamycin nucleotidyltransferase
M26879	*B. subtilis* aminoglycoside 6-adenylyltransferase (*aadK*) gene
Tetracycline resistance	
AY212949	*B. amyloliquefaciens* strain *tetB* gene
X51366	*B. cereus* plasmid pBC16 tetracycline resistance gene
AF491293	*B. sp.* R89 Tn916-like transposon *tetM* gene
X08034	*B. subtilis* GSY908 *tet* gene for tetracycline-resistance
X58999	*B. subtilis* IS1K transposable element DNA tetBS908
D00006	*B. subtilis* plasmid pNS1981 tetracycline resistance gene (*tet*)
D12567	*B. subtilis tetBSR* gene
Chloramphenicol resistance	
K01810	*B. pumilus cat-66* gene
K01812	*B. pumilus cat-86* gene
M17117	*B. pumilus* chloramphenicol acetyltransferase gene
Y12398	*B. subtilis* plasmid pGR71 DNA, chloramphenicol acetyltransferase
M12657	*B. subtilis* Tn9-derived *cat* gene
Penicillin resistance	
AF367983	*B. anthracis* beta-lactamase I (*bla1*) gene
AF367984	*B. anthracis* beta-lactamase II (*bla2*) gene
X06599	*B. cereus* beta-lactamase I gene
M11189	*B. cereus* beta-lactamase II gene
M15195	*B. cereus* beta-lactamase III gene
X05798	*B. licheniformis blaI* gene
X62244	*B. mycoides blacI* gene for beta-lactamase I
Y10006	*B. sp. bla* gene
M15350	*B. sp.* strain 170 beta-lactamase gene
X83424	*B. thuringiensis bla* gene for beta-lactamase

is a common trait of *B. anthracis*. Still the two *B. anthracis bla* genes (Table 1) confer resistance to ampicillin to both *B. subtilis* and *E. coli* (Chen et al., 2003). Due to this fact, even if ß-lactam agents, particularly penicillin, have been used worldwide to treat anthrax in humans, these agents should be discouraged for the prophylaxis and treatment of anthrax. The prophylaxis recommended for adults and children to prevent inhalational anthrax consists in therapy with ciprofloxacin or doxycycline for 60 days (Jefferds et al., 2002; Williams et al., 2002). Other agents like rifampicin, vancomycin, imipenem, chloramphenicol, penicillin and ampicillin, clindamycin, and clarithromycin may also be suggested (Brook, 2002).

THE SAFETY OF *BACILLUS*

Annually there are world wide a few thousands of cases of cutaneous anthrax (not fatal if treated), 20 to 100 times as many cases of anthrax in livestock, few severe non-gastrointestinal cases due to *B. cereus,* a difficult to quantify number of cases of food poisonings, also due to *B. cereus,* and an incalculable number of fatal intoxications of caterpillars and insects due to intentional release of *B. thuringiensis* spores. These three pathogens belong to nearly indistinguishable species (*B. cereus* group) within the genus *Bacillus,* which accounts for at least other 130 recognised species (NCBI taxonomy web site). For all the non-*cereus* species, even if ubiquitous in soil, present as food constituents, as food or feed contaminants, or as human probiotics, no significant epidemiological association to any disease has ever been drawn. Except for bacteraemia in severely immunocompromised patients no clinical cases have been reported. This suggests then that non-*cereus Bacillus* are safe and can be used for industrial preparations for agricultural, veterinary and human use,.

In summary, the non-*cereus* species of the genus *Bacillus,* being non-pathogenic and not harbouring drug resistance genes, meets all requirements of the classification systems for food and feed safety, both the European QPS system (Qualified Presumption of Safety) (http://europa.eu.int/comm/food/fs/sc/scan/index_en.html) and the American GRAS system (Generally Recognized As Safe) (http://www.cfsan.fda.gov/~dms/opanoti.html).

REFERENCES

Adone, R., Pasquali, P., La Rosa, G., Marianelli, C., Muscillo, M., Fasanella, A., Francia, M., and Ciuchini, F. 2002. Sequence analysis of the genes encoding for the major virulence factors of *Bacillus anthracis* vaccine strain "Carbosap". J Appl Microbiol. *93,* 117-121.

Agata, N., Mori, M., Ohta, M., Suwan, S., Ohtani, I., and Isobe, M. 1994. A novel dodecapsipeptides, cerulide, isolated from *Bacillus cereus* causes vacuole formation in Hep-2 cells. FEMS Microbiol. Lett. *121,* 31-34.

Agata, N., Ohta, M., Mori, M., and Isobe, M. 1995. A novel dodecapeptide, cereulide, is an emetic toxin of *Bacillus cereus*. FEMS Microbiol. Lett. *129,* 17-20.

Banerjee, C., Bustamante, C.I., Wharton, R., Talley, E., and Wade, J.C. 1988. *Bacillus* infections in patients with cancer. Arch Intern Med. *148,* 1769-74.

Beecher, D.J., and Macmillan, J.D. 1991. Characterization of the components of hemolysin BL from *Bacillus cereus*. Infect. Immun. *59,* 1778-1784.

Beecher, D.J., Schoeni, J.L., and Wong, A.C.L. 1995. Enterotoxin activity of hemolysin BL from *Bacillus cereus*. Infect. Immun. *63,* 4423-4428.

Beecher, D.J., and Wong, A.C.L. 1997. Tripartite hemolysin BL from *Bacillus cereus*. Hemolytic analysis of component interaction and model for its characteristics paradoxical zone phenomenon. J. Biol. Chem. *272,* 233-239.

Berry, C., O'Neil, S., Ben-Dov, E., Jones, A.F., Murphy, L., Quail, M.A., Holden, M.T., Harris, D., Zaritsky, A., and Parkhill, J. 2002. Complete sequence and organization of pBtoxis, the toxin-coding plasmid of *Bacillus thuringiensis* subsp. israelensis. Appl. Environ. Microbiol. *68,* 5082-5095

Blue, S.R., Singh, V.R., and Saubolle, M.A. 1995. *Bacillus licheniformis* bacteraemia: five cases associated with indwelling central venous catheters. Clin. Infect. Dis. *20,* 629-633.

Bortolussi, R., and Kennedy, W. 1999. Aerobic Gram-positive bacilli. In: *Infectious Diseases,* Armstrong D, Cohen J, eds. London: Mosby, pp 15.1-15.20

Brook, I. 2002. The prophylaxis and treatment of anthrax. Int. J. Antimicrob. Agents. *20,* 320-325

Callegan, M.C., Kane, S.T., Cochran, D.C., and Gilmore, M.S. 2002. Molecular mechanisms of Bacillus endophtalmitis pathogenesis. DNA Cell. Biol. *21*, 367-373

Callegan, M.C., Kane, S.T., Cochran, D.C., Gilmore, M.S., Gominet, M., and Lereclus, D. 2003. Relationship of plcR-regulated factors to *Bacillus* endophtalmitis virulence. Infect. Immun. *7*, 3116-3124.

Capital Regional District (1999). Human health surveillance during the aerial spraying for control of North American gypsy moth on southern Vancouver Island, British Columbia. Report to the administrator, Pesticides Control Act, Ministry of Environment, Lands, and Parks, Province of British Columbia. Office of the Medical Health Officer and Director of Research, Capital Health Region, Victoria, British Columbia, Canada.

Chen, Y., Succi, J., Tenover, F.C., and Koehler, T.M. 2003. Beta-lactamase genes of the penicillin-susceptible *Bacillus anthracis* Sterne strain. J. Bacteriol. *185*, 823-830

Cherif, A., Borin, S., Rizzi, A., Ouzari, H., Boudabous, A., and Daffonchio, D. 2002. Characterization of a repetitive element polymorphismpolymerase chain reaction chromosomal marker that discriminates *Bacillus anthracis* from related species. J. Appl. Microbiol. *93*, 456-462

Cohen, J. 2003. Anthrax vaccine. NIAID's $233 million problem put on hold. Science *300*, 1065

Cotton, D.J., Gill, V.J., Marshall, D.J., Gress, J., Thaler, M., and Pizzo, P.A. 1987. Clinical features and therapeutic interventions in 17 cases of Bacillus bacteraemia in immunosuppressed patient population. J. Gen. Microbiol. *25*, 672-674

Dahlinger, J., Marks, S.L., and Hirsch, D.C. 1997. Prevalence and identity of translocating bacteria in healthy dogs. J Vet Intern Med *11*, 319-322.

Das, T., Choudhury, K., Sharma, S., Jalali, S., and Nuthethi, R. 2001. Clinical profile and outcome in Bacillus endophtalmitis. Ophtalmology *108*, 1819-1825.

Drobniewski, F.A. 1993. *Bacillus cereus* and related species. Clin. Microbiol. Rev. *6*, 324-338

Fasanella, A., Losito, S., Trotta, T., Adone, R., Massa, S., Ciuchini, F., and Chiocco, D. 2001. Detection of anthrax vaccine virulence factors by polymerase chain reaction. Vaccine *19*, 4214-4218.

Foldes, T., Banhegyi, I., Herpai, Z., Varga, L., and Szigeti, J. 2000. Isolation of *Bacillus* strains from the rhyzosphere of cereals and in vitro screening for antagonism against phytopathogenetic, food-borne pathogenic and spoilage micro-organisms. J. Appl. Microbiol. *89*, 840-846.

Galanos, J., Perera, S., Smith, H., O'Neal, D., Sheorey, H., and Waters, M.J. 2003. Bacteremia due to three *Bacillus* species in a case of Munchausen's syndrome. J. Clin. *41*, 2247-2248

Gaur, A.H., Patrick, C.C., McCullers, J.A., Flynn, P.M., Pearson, T.A., Razzouk, B.L., Thompson, S.J., and Shenep, J.L. 2001. *Bacillus cereus* bacteremia and meningitis in immunocompromised children. Clin. Infect. Dis. *32*, 1456-1462

Granum, P.E., Andersson, A., Gayther, C., de Giffel, M.C., Larsen, H., Lund, T., and O'Sullivan, K. 1996. Evidence for a further enterotoxin complex produced by *Bacillus cereus*. FEMS Microbiol. Lett. *141*, 145-149

Granum, P.E., and Lund, T. 1997. *Bacillus cereus* and its food poisoning toxins. FEMS Microbiol. Lett. *157*, 223-228

Green, D.H., Wakely, P.R., Page, A., Barnes, A., Baccigalupi, L., Ricca, E., and Cutting, S. 1999. Characterization of two Bacillus probiotics. Appl. Environ. Microbiol. *65*, 4288-4291.

Green, M., Heumann, M., Sokolow, R., Foster, L.R., Bryant, R., and Skeels, M. 1990. Public health implications of the microbial pesticide *Bacillus thuringiensis*: an epidemiological study, Oregon, 1985-86. Amer. J. Public Health *80*, 848-852.

Gryczan, T.J., Contente, S., and Dubnau, D. 1978. Characterization of *Staphylococcus aureus* plasmids introduced by transformation in *Bacillus subtilis*. J. Bacteriol. *134*, 318-329.

Guidi-Rontani, C., Levy, M., Ohayon, H., and Mock, M. 2001. Fate of germinated *Bacillus anthracis* spores in primary murine macrophages. Mol. Microbiol. *42*, 931-938.

Guidi-Rontani, C., Weber-Levy, M., Labruyère, E., and Mock, M. 1999. Germination of *Bacillus anthracis* spores within alveolar macrophages. Mol. Microbiol. *31*, 9-17.

Hannah, W.N., and Ender, P.T. 1999. Persistent *Bacillus licheniformis* bacteremia associated with an international injection of organic drain cleaner. Clin. Infect. Dis. *29*, 659-661.

Heinrichs, J.H., Beecher, D.J., Macmillan, J.M., Zilinskas, B.A. 1993. Molecular cloning and characterization of the *hblA* gene encoding the B component of hemolysin BL from *Bacillus cereus*. J. Bacteriol. *175*, 6760-6766.

Helgason, E., Okstad, O.A., Caugant, D.A., Johansen, H.A., Fouet, A., Mock, M., Hegna, I., and Kolsto, A.B. 2000. *Bacillus anthracis, Bacillus cereus, and Bacillus thuringiensis*: one species on the basis of genetic evidence. Appl. Environ. Microbiol. *66*, 2627-2630.

Hoa, N.T., Baccigalupi, L., Huxham, A., Smertenko, A., Van, P.H., Ammendola, S., Ricca, E., and Cutting, S.M. 2000. Characterization of *Bacillus* species used for oral bacteriotherapy and bacterioprophylaxis of gastrointestinal disorders. Appl. Environ. Microbiol. *60*, 5241-5247

Hugh-Jones, M. 1999. 1996-1997 Global Anthrax Report. J. Appl. Microbiol. *87*, 191

Jadamus, A., Vahjen, W., Schafer, K., and Simon, O. 2002. Influence of the probiotic strain *Bacillus cereus* var. *toyoi* on the development of enterobacterial growth and on selected parameters of bacterial metabolism in digesta samples of piglets. J. Anim. Physiol. Anim. Nutr. (Berl) *86*, 42-54.

Jefferds, M.D., Laserson, K., Fry, A.M., Roy, S., Hayslett, J., Grummer-Strawn, L., Kettel-Khan, L., and Schuchat, A. 2002. Adherence to antimicrobial anthrax prophylaxis among postal workers. Emerg. Infect. Dis. *8*, 1138-1144.

Jensen, G.B., Hansen, B.M., Eilenberg, J., and Mahillon, J. 2003. The hidden lifestyles of *Bacillus cereus* and relatives. Environ. Microbiol. *5*, 631-640.

Jernigan, D.B., Raghunatan, P.L., Bell, B.P., Brechner, R., Bresnitz, E.A., Butler, J.C., et al. 2002. Investigation of bioterrorism-related anthrax, United States, *2001*, epidemiologic findings. Emerg. Infect. Dis. *8*, 1019-1028.

Kanafami, Z.A., Ghossain, A., Sharara, A.I., Hatem, J.M., and Kanij, S.S. 2003. Endemic gastrointestinal anthrax in 1960s Lebanon: clinical manifestations and surgical findings. Emerg. Infect. Dis. *9*, 520-525

Koch, R. 1998a. Die Aetiologie der Tuberculose. In: Milestones in Microbiology: 1556 to 1940., Brock TD, ed. ASM Press, pp 116

Koch, R. 1998b. Untersuchungen uber BaKterien V. Die Aetiologie der Milzbrand-Krankheit, begruendet auf die Entwicklungsgeschichte des *Bacillus anthracis*. In: Milestones in Microbiology: 1556 to 1940, Brock TD, ed. ASM Press, pp 89.

Kotiranta, A., Lounatmaa, K., and Haapasalo, M. 2000. Epidemiology and pathogenesis of *Bacillus cereus* infections. Microbes Infect *2*, 189-198.

Krause, A., Gould, F.K., and Forty, J. 1999. Prosthetic heart valve endocarditis caused by *Bacillus circulans*. J. Infect. *39*, 160-162.

Lange, J.L., Lesikar, S.E., Rubertone, M.V., Brundage, J.F. 2003. Comprehensive systematic surveillance for adverse effects of Anthrax Vaccine Adsorbed, US Armed Forces, 1998-2000. Vaccine *21*, 1620-1628.

Logan, N.A., and Berkeley, R.C. 1984. Identification of *Bacillus* strains using the API system. J. Gen. Microbiol. *130*, 1871-1882.

Lund, T., and Granum, P.E. 1996 Characterization of a non-haemolytic enterotoxin complex from *Bacillus cereus* isolated after a foodborne outbreak. FEMS Microbiol. Lett. *141*, 151-156.

Lund, T., and Granum, P.E. 1997. Comparison of biological effect of the two different enterotoxin complexes isolated from three different strains of *Bacillus cereus*. Microbiology *143*, 3329-3339.

Mahler, H., Pasi, A., Kramer, J.M., Shulte, P., Senging, A.C., Bar, W., and Krahenbuhl, S. 1997. Fulminant liver failure in association with the emetic toxin of *Bacillus cereus*. New Engl. J. Med. *336*, 1142-1148.

Matsumoto, S., Suenaga, H., Naito, K., Sawazaki, M., Hiramatsu, T., and Agata, N. 2003. Management of suspected nosocomial infections: an audit of 19 hospitalized patients with septicemia caused by *Bacillus* species. Jpn. J. Infect. Dis. *53*, 196-202

McKillip, J.L. 2000. Prevalence and expression of enterotoxins in *Bacillus cereus* and other *Bacillus* spp., a literature review. Antonie van Leeuwenhoek *77*, 393-399

Meselson, M., Guillemin, J., Hugh-Jones, M., Langmuir, A., Popova, I., Shelokov, A., and Yampolskaya, O. 1994. The Sverdlovsk anthrax outbreak of 1979. Science *266*, 1202-1208.

Mock, M., and Fouet, A. 2001. Anthrax. Ann Rev Microbiol *55*, 647-671.

Mohammed, M.J., Marston, C.K., Popovic, T., Weyant, R.S., and Tenover, F.C. 2002. Antimicrobial susceptibility testing of *Bacillus anthracis*: comparison of results obtained by using the National Committee of Clinical Laboratory Standards broth microdilution reference and Etest agar gradient diffusion methods. J. Clin. Microbiol. *40*, 1902-1907.

Oggioni, M.R., Ciabattini, A., Cuppone, A.M., and Pozzi, G. 2003. *Bacillus* spores for vaccine delivery. Vaccine *21*, S96-S101.

Oggioni, M.R., Galieni, P., Bigazzi, C., Pozzi, G., and Valensin, P.E. 1998. Recurrent septicaemia in in immunocompromised patient due to probiotic strains of *Bacillus subtilis*. J. Clin. Microbiol. *36*, 325-326.

Okinaka, R.T., Cloud, K., Hampton, O., Hoffmaster, A.R., Hill, K.K., Keim, P., Koehler, T.M., Lamke, G., Kumano, S., Mahillon, J., Manter, D., Martinez, Y., Ricke, D., Svensson, R., and Jackson, P.J. 1999. Sequence and organization of pXO1, the large *Bacillus anthracis* plasmid harboring the anthrax toxin genes. J. Bacteriol. *181,* 6509-6518.

Olsewski, W.L., Jamal, S., Manokaran, G., Pani, S., Kumaraswami, V., Kubicka, U., Lukomska, B., Tripathi, F.M., Swoboda, E., Meisel-Mikolajczyk, F., Stelmach, E., and Zaleska, M. 1999. Bacteriological studies of blood, tissue fluid, lymph and lymph nodes in patients with acute dermatolymphangioade nitis (DLA) in course of "filarial" lymphedema. Acta Tropic *73,* 217-224.

Pannucci, J., Okinaka, R.T., Sabin, R., and Kuske, C.R. 2002. *Bacillus anthracis* pXO1 plasmid sequence conservation among closely related bacterial species. J. Bacteriol. *184,* 134-141.

Pearce, M., Habbick, B., Williams, J., Eastman, M., and Newman, M. 2002. The effects of aerial sprying with Bacillus thuringiensis Kurstaki on children with asthma. Can. J. Public Health *93,* 21-25.

Reva, O.N., Vyunitskaya, V.A., Reznik, S.R., Kozachko, L.A., and Smirnov, V.V. 1995. Antibiotic susceptibility as a taxonomic characteristic of the genus *Bacillus*. Int J Syst Bacteriol *45,* 409-411

Richard, V., Van der Auwera, P., Snoeck, R., Daneau, D., and Meunier, F. 1988. Nosocomial bacteremia caused by *Bacillus* species. Eur J. Clin. Microbiol. Infect. Dis. *7,* 783-785.

Rosenkvist, H., and Hansen, A. 1995. Contamination profiles and characterisation of *Bacillus* species in wheat bread and raw materials for bread production. Int. J. Food Microbiol. *26,* 353-363.

Ryan, P.A., Macmillan. J.M., and Zilinskas, B.A. 1997. Molecular cloning and characterizations of the genes encoding the L_1 and L_2 components of hemolysin BL from *Bacillus cereus*. J. Bacteriol. *179,* 2551-2556.

Senesi, S., Celandroni, F., Tavanti, A., and Ghelardi, E. 2001. Molecular characterization and identification of *Bacillus clausii* strains marketed for use in oral bacteriotherapy. Appl. Environ. Microbiol. *67,* 834-839.

Shinagawa, K., Ueno, Y., Hu, D., Ueda, S., and Sugii, S. 1996. Mouse lethal activity of a HEp-2 vacuolation factor, cerulide, produced by *Bacillus cereus* isolated from vomiting-type food poisoning. J Vet Med Sci *58,* 1027-1029.

Siegel, J.P. 2001a. The mammalian safety of *Bacillus thuringiensis*-based insecticides. J. Invertebr. Pathol. *77,* 13-21.

Siegel, J.P. 2001b. The mammalian safety of *Bacillus thuringiensis*-based insecticides. J. Invertebr. Pathol. *77,* 13-21.

Spinosa, M.R., Braccini, T., Ricca, E., De Felice, M., Morelli, L., Pozzi, G., and Oggioni, M.R. 2000a. On the fate of ingested *Bacillus* spores. Res. Microbiol. *151,* 361-368.

Spinosa, M.R., Wallet, F., and Oggioni, M.R. 2000b. The trouble in tracing opportunistic pathogens: cholangitis due to *Bacillus* in a French hospital caused by a strain related to an Italian probiotic? Microb. Ecol. Health Dis. *12,* 99-101.

Tan, A., Heaton, S., Farr, L., and Bates, J. 1997. The use of Bacillus diarrhoeal enterotoxin (BDE) detection using an ELISA technique in the confirmation of the aetiology of *Bacillus*-mediated diarrhoea. J. Appl. Microbiol. *82,* 677-682.

Tandon, A., Tay-Kearney, M.L., Metcalf, C., and McAllister, L. 2001. *Bacillus circulans* endophtalmitis. Clin. Experiment. Ophtalmol. *29,* 92-93.

Turnbull, P.C. 1991. Anthrax vaccines: past, present and future. Vaccine *9,* 533-539.

Turnbull, P.C., and Kramer, J.M. 1985. Intestinal carriage of *Bacillus cereus*. J. Hyg. (London) *95,* 629-638.

Turnbull, P.C. 2000. Current status of immunization against anthrax: old vaccines may be here to stay for a while. Curr. Opin. Infect. Dis. *13,* 113-120.

Valadares de Amorim, G., Whittome, B., Shore, B., and Levin, D.B. 2001. Identification of *Bacillus thuringiensis* subsp.*kurstaki* strain HD1-like bacteria from environmental and human samples after aerial spraying of Victoria, British Columbia, Canada, with Foray 48B. Appl. Environ. Microbiol. *67,* 1035-1043.

Viljoen, C.R., and von Holy, A. 1997. Microbial populations associated with commercial bread production. J. Basic Microbiol. *37,* 439-444.

Wang, J., and Fung, D.Y. 1996. Alkaline-fermented foods: a review with emphasis on pidan fermentation. Crit. Rev. Microbiol. *22,* 101-138.

Weber, D.J., Saviteer, S.M., Rutala, W.A., and Thomann, C.A. 1988. In vitro susceptibility of *Bacillus* spp. to selected antimicrobial agents. Antimicrob. Agents Chemother. *32,* 642-645.

Williams, J.L., Noviello, S.S., Griffith, K.S., Wurtzel, H., Hamborsky, J., Perz, J.F., et.al. (2002), Anthrax postexposure prophylaxis in postal workers, Connecticut, 2001. Emerg. Infect. Dis. *8:* 1133-1137.

Woo, P.C., Wong, S.S., Lum, P.N., Hui, W.T., and Yuen, K.Y. 2001. Cell-wall deficient bacteria and culture-negative febrile episodes in bone-marrow-transplant recipients. Lancet *357:* 675-679.

Wu, J.Y., Hong, T.C., Hou, C.J., Chou, Y.S., Tsai, C.H., and Yang, D.I. 1999. *Bacillus popiliae* endocarditis with prolonged complete heart block. Am. J. Med. Sci. *317:* 263-265.

Chapter 5

Sporulation in *Bacillus subtilis* and Other Bacteria

Imrich Barák

SUMMARY

Bacterial endospores and the organisms that form them have been the focus of intense research interest for many decades. Of all spore-forming bacteria only *Bacillus subtilis* became a model to study the remarkable process of spore formation and the factors controlling this simple example of unicellular differentiation. This review summarizes important advances in our understanding of gene expression during spore formation with a particular emphasis on the developmental stages that lead to the initiation of sporulation and a survey of what is now known of this process in other *Bacilli* and *Clostridia* species.

INTRODUCTION

Some bacteria possess complex developmental programmes that drive environmental adaptation and cell differentiation. A few well-studied examples of differentiation processes are known in bacteria. For example, *Caulobacter crescentus*, a Gram-negative bacterium, can form two distinctly different cell types and its life–cycle is characterized by changing from one type to the other. One form, a stalked cell, can produce a swarm cell but it cannot itself become a swarm cell. This newly formed cell type is motile and can lose a flagellum and develop a stalk. The first morphological feature of swarm cell formation is an asymmetric cell division and this cell division cycle is governed by unknown signals. They seem to act via a pathway resembling a two-component regulatory system or a phosphorelay (Shapiro and Losick, 1997). Differentiation from non-motile to motile cells, and vice versa, also occurs in species of *Hyphomicrobium* and *Rhodomicrobium*.

In some bacteria differentiation can result in the formation of a resting cell – either a spore or a cyst. Resting cells may help to spread the organism and/or allow survival in a hostile environment. Under suitable conditions a spore or cyst can germinate and form a new vegetative cell. There are two basic types of spores – endospores and exospores. Endospores are formed by species of *Bacillus, Clostridium, Desulfotomaculum, Thermactinomyces, Sporolactobacillus,* and *Sporosarcina* genera. Interestingly, endospore formation is not restricted to Gram-positive organisms as listed above, but also some Gram-negative microorganisms as *Sporomusa acidovorans, Sporomusa malonica, Sporomusa ovata, Sporomusa termitida, Acetonema longum* and *Coxiella burnetii* can produce them.

An endospore is formed within a mother cell in a compartmentalized structure referred to as a prespore or forespore. It can exist in a state of dormancy for a very long time and is highly resistant to many hostile factors as extreme temperature, pH, various chemical agents, radiation and desiccation (see Chapter 1). On the other hand exospores are produced by septation and fragmentation of hyphae by several species of the Actinomycetes genus. These spores lack specialized structures such as the cortex and spore coat, typical of endospores and thus are less resistant than endospores to hostile environmental conditions. Although they are less metabolically active than vegetative hyphae they are not completely dormant.

The process of endospore formation in *B. subtilis* is the most studied example of differentiation in bacteria. The morphological events during sporulation have been divided, according to electron-microscopic studies, into basic stages with the first one defined as stage 0 and the last one, consisting in the release of resistant spores, as stage VII, with the whole process taking about eight hours to complete in laboratory conditions. More than 200 gene products have been identified as being involved in this process. Generally, these genes are named *spo* followed by a number according to the morphological stage at which mutations in them arrest spore development. For example, mutations in the *spoIIIA* gene halts spore development at stage III and the process does not proceed to stage IV.

This chapter will present a brief overview of the sporulation process in *Bacillus subtilis* and examines what is known of this process in other *Bacilli* and *Clostridia* species. Recent reviews (Errington, 1993; Stragier and Losick, 1996; Errington, 2003) and the recently published book "*Bacillus subtilis*: From Cells to Genes and from Genes to Cells" (Sonenshein et al., 2002) have extensively addressed the various aspects of the sporulation process in *B. subtilis*.

BACILLUS SUBTILIS CELL CYCLE

The life cycle of *B. subtilis* comprises three different processes: vegetative growth, sporulation and germination (the spore outgrowth) (Figure 1). Vegetative growth occurs when nutrients are available and is characterized by cells growing logaritmically by symmetric division. When nutrients become limiting, *B. subtilis* might initiate the sporulation process. However, sporulation is not the only developmental option for the vegetative cell that can also follow diverse routes, expressing genes needed for general adaptation to adverse conditions, or genes needed to scavenge alternative nutrients and to increase competitiveness against other species which are using the same scarce energy resources. In such conditions various proteases and degradative enzymes (amylases, xylanases, cellulases etc.), alternate chemotaxis pathways, different antibiotics and transport functions are expressed and/or activated. Some cells become competent for uptake of exogenous DNA, which is

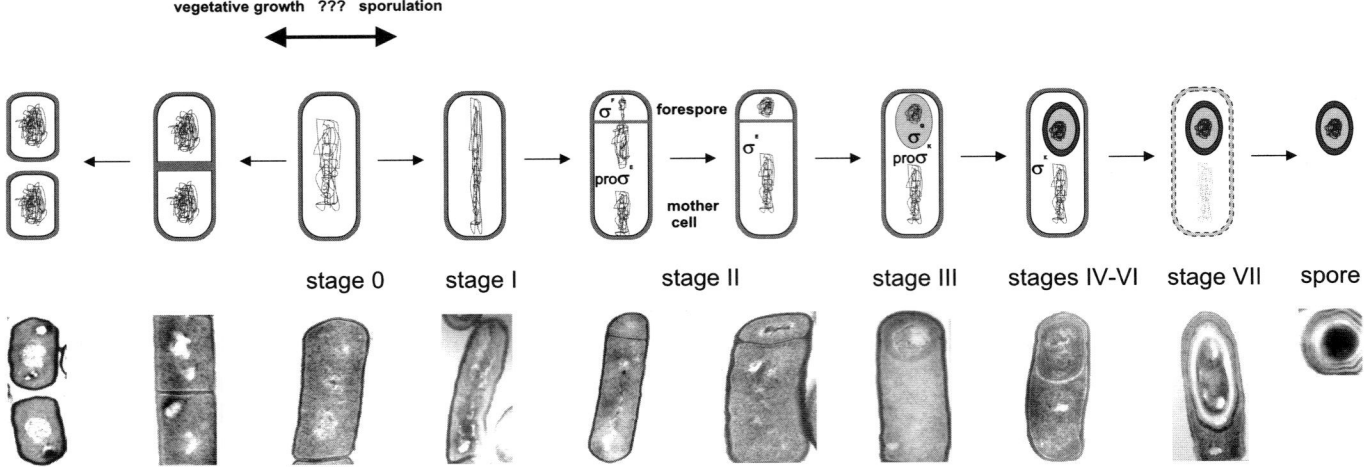

Figure 1. The life cycle of *Bacillus subtilis*. Electron microscopic representation of thin sectioned cells at different stage of development are at the bottom. More detailed information is provided in the text.

introduced into the recipient's chromosome and may confer selective advantages to future generations living under adverse conditions (Dubnau, 1999).

Initiation of sporulation can then be viewed as the last opportunity for cell survival where surrounding conditions prevent normal cell growth. Sporulation though, is a relatively high energy-cost process and before entering into an irreversible program of cell differentiation the cell must sense not only the decreased levels of available nutrients but also that sufficient resources are still available to produce a spore. Sporulation as a last resort for survival is supported by the recent discovery of cannibalism in *B. subtilis* (Gonzalez-Pastor et al., 2003). Here, the 'early' sporulating cells produce and export a killing factor which kills the non-sporulating sibling cells releasing a final pool of nutrients.

The sporulation process is activated by complex regulatory circuits and begins when the threshold concentration of phosphorylated Spo0A is reached. Upon entry into sporulation, the replication origin regions of two daughter chromosomes that arise from the last round of vegetative DNA replication migrate to the extreme opposite poles of the sporangium – stage I of sporulation (Glaser et al., 1997; Lin et al., 1997). Electron microscopy of thin section of the cell in this stage shows an axial filament formation of the chromatin (Figure 1). The first clear morphological step in sporulation is the formation of an asymmetric septum at stage II. This septum divides the cell into a large cell, called the sporangium or mother cell, and a much

smaller cell, called the prespore. Early in development, the prespore is engulfed by the mother cell in a process resembling eukaryotic phagocytosis and creates a compartmentalized chamber referred to as the forespore at stage III. The small forespore next develops into a highly resistant spore and the mother cell lyses – stage IV to VII (Figure 1). Finally, the spore can germinate and enter into vegetative growth when conditions improve.

A great deal is known about the regulation of gene expression during sporulation in *B. subtilis*. The initiation of sporulation is controlled by the key transcription regulator, the Spo0A protein. After the decision to initiate sporulation has been made, phosphorylated Spo0A, Spo0A~P, leads to transcriptional repression of many vegetative genes. This allows RNA-polymerase containing either the housekeeping sigma factor σ^A, or the alternative sigma factor σ^H, to initiate the transcription of early sporulation-specific genes. Spo0A therefore produces profound changes in the global pattern of gene expression in the cell (Fawcett et al., 2000; Molle et al., 2003).

Gene expression during sporulation is orchestrated through the activity of five sporulation specific sigma factors (σ^H, σ^F, σ^E, σ^G and σ^K), the last four of which are compartment specific, activated either in the forespore or in the mother cell (Figure 2). All these sigma factors as subunits of RNA polymerase recognize specific sets of promoter sequences and allow transcription of different genes.

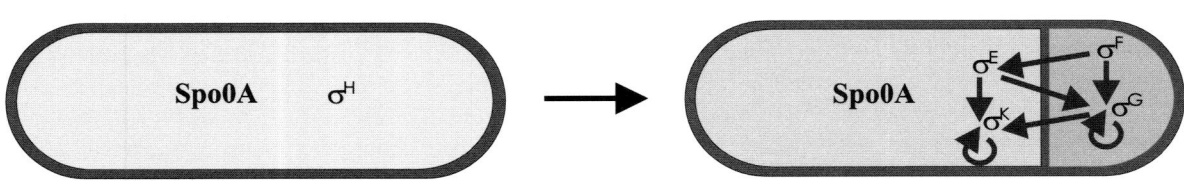

Figure 2. Crisscross regulation of cell-type-specific gene expression during sporulation.

Morphological changes result from a carefully regulated program of gene expression and genes must be expressed at the proper time and level, and in the proper compartment. A cascade of five σ factors can account for much of the temporal and spatial regulation of gene expression during sporulation (Losick and Stragier, 1992). Figure 2 summarizes the order of action and the location of Spo0A and the five sporulation specific σ factors. The three arrows that cross the sporulation septum and prespore membrane, respectively, represent three instances of intercompartmental communication. Each of these communication events is a signal transduction pathway that begins with an active σ factor in one compartment and leads to activation of the next σ factor in the cascade. Each of the four compartment-specific σ factors (σ^F, σ^E, σ^G and σ^K) is inactive upon initial synthesis and requires a subsequent activation step. In the case of σ^F this σ factor is produced in the pre-divisional cell and becomes active after the formation of the polar septum and specifically only in the forespore.

INITIATION OF THE SPORULATION

The initiation of sporulation is a complex process where at least three types of input are integrated by the cell before the decision to sporulate is made. These signals include availability of nutrients, population density and cell cycle signals. Sophisticated and sensitive signal transduction pathways are essential to monitor accurately all these signals. The main monitoring signal is based on the so-called two-component system, where signal acquisition involves autophosphorylation of a sensor histidine kinase and transduction takes place when the kinase phosphorylates its cognate response regulator protein.

The key molecule in transducing the signals indicative of a deteriorating cellular environment into the activation of genes required for sporulation is the response regulator, Spo0A. Spo0A phosphorylation and thus the entire initiation of sporulation is under the control of an expanded two-component sensory signalling system termed a phosphorelay (Figure 3) (Hoch, 1998). In this system, environmental signals trigger the ATP-dependent autophosphorylation of one of at least five sensor kinases, KinA, KinB, KinC, KinD and KinE, on a specific histidine residue (Stephenson and Hoch, 2002). The phosphoryl group then migrates via Spo0F and Spo0B to an aspartic acid residue in the response regulator, Spo0A. The complexity of this regulatory cascade is specified also by the existence of three known Spo0F~P specific phosphatases (Rap proteins) (Perego, 1998) and a Spo0A~P phosphatase (Spo0E) (Ohlsen et al., 1994). These phosphatases disrupt the flow of phosphate to Spo0A and the concentration of Spo0A~P reaches a sporulation-inducing level only when the Rap proteins are antagonized by specific pentapeptides (Perego, 1998). How the activity of Spo0E is neutralized is not yet known though. Clearly, the existence of multiple phosphatases could provide the potential to respond to several different signals.

The master control element in the decision to sporulate is the Spo0A protein. If a threshold concentration of Spo0A~P is attained, sporulation commences. Phosphorylation of Spo0A activates its latent transcription activation and transcription repression properties by stimulating its binding to seven base pair DNA consensus sequences 5' TGTCGAA 3', also known as 0A-boxes, which are present in multiple copies upstream of Spo0A-activated genes (Baldus et al., 1994; Strauch et al.,

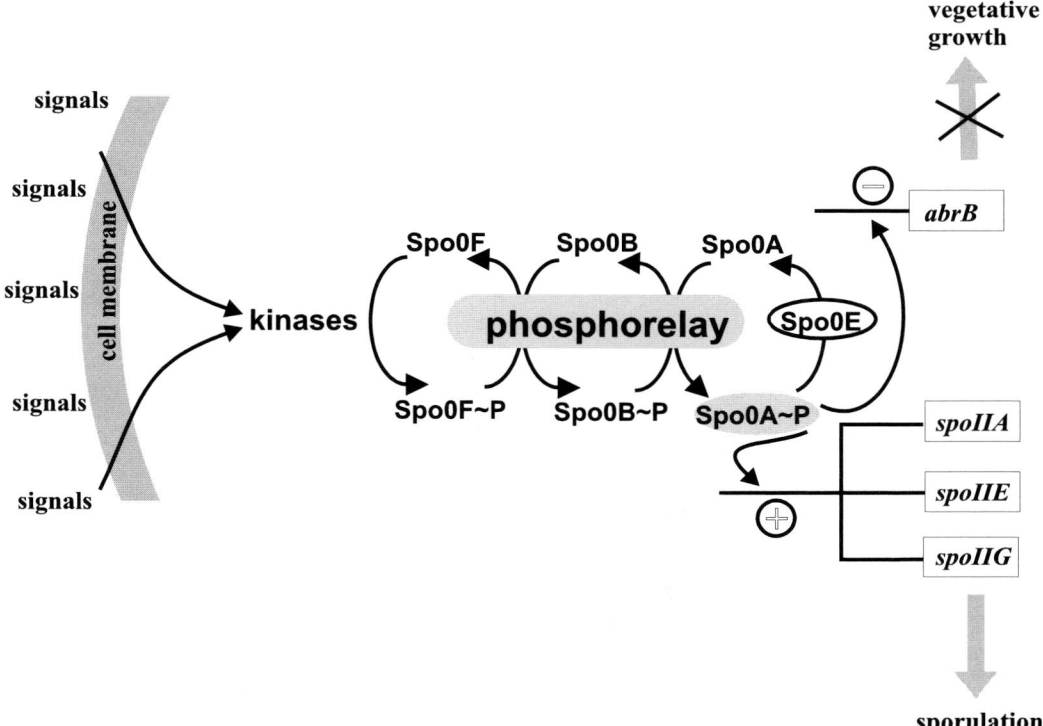

Figure 3. Initiation of the sporulation process: the phosphorelay. More detailed information is provided in the text.

1990). Spo0A~P activates transcription from the promoters of the *spoIIA* (Trach et al., 1991; Wu et al., 1991), *spoIIG* (Baldus et al., 1994; Satola et al., 1991) and *spoIIE* (York et al., 1992) operons which encode the forespore (σ^F) and mother cell (σ^E)-specific RNA polymerase sigma factors and a protein phosphatase SpoIIE which localizes to the sporulation septum, respectively. It also activates transcription of genes encoding the phosphorelay components Spo0F (Strauch et al., 1993) and Spo0A itself (Chibazakura et al., 1991; Strauch et al., 1992). Transcription of *spoIIA* depends on RNA polymerase containing the σ^H subunit (E-σ^H) whereas *spoIIG* transcription is directed by E-σ^A (Kenney et al., 1988; Wu et al., 1992). σ^A is the primary vegetative phase sigma factor of *Bacillus subtilis*, while σ^H is the first sporulation specific sigma factor. Spo0A is unusual in activating transcription from both σ^A- and σ^H-dependent promoters. Spo0A~P is also a negative regulator of transcription of *abrB* (Strauch et al., 1990), *kinA* (Hoch, 1993) and at higher concentrations, of *spo0F* and *spo0A* (Strauch et al., 1992). AbrB is itself a repressor of transcription of *spo0H* which encodes σ^H (Weir et al., 1991). The importance of Spo0A as an activator and a repressor of transcription of many stationary-phase as well as sporulation-specific genes is emphasized by the finding that more than 500 transcripts in *B. subtilis* are at least 3-fold directly or indirectly dependent on Spo0A (Fawcett et al., 2000).

Spo0A from *B. subtilis* consists of a single polypeptide chain of 267 residues that form two domains of similar size, an N-terminal regulatory domain and a C-terminal trans-activation domain. The regulatory domain contains all the signature residues characteristic of the large family of response regulators. In contrast, the sequence of the C-terminal domain is conserved only among Spo0A homologues from endospore forming bacteria. It contains a helix-turn-helix DNA-binding motif (Brown et al., 1994; Zhao et al., 2002). A proteolytic fragment, encompassing the N-domain (N-Spo0A), remains a substrate for the phosphorelay while the complementary C-terminal fragment (C-Spo0A) is able to bind to DNA and activate transcription. This implies that the N-domain is an inhibitor of the function of the C-domain and that this inhibition is overcome by phosphorylation, at Asp56. In addition to genetic studies Spo0A has more recently become the subject

of intensive biophysical studies. It has been established that Spo0A forms dimers upon phosphorylation and that dimer formation is mediated principally by the N-terminal receiver domains (Asayama et al., 1995; Lewis et al., 2002). The crystal structures of the isolated receiver (N-Spo0A) and effector (C-Spo0A) domains of Spo0A from *B. stearothermophilus* have been solved (Lewis et al., 2000a; Lewis et al., 2000b) the former in the phosphorylated form (Lewis et al., 1999). Recently, further insight as to the mode of DNA-binding by Spo0A has been gleaned by the determination of the crystal structure of the effector domain of Spo0A in complex with 0A boxes (Zhao et al., 2002). Whilst the isolated effector domain is monomeric in solution, in the structure of the DNA complex, C-Spo0A forms a dimer. It is also known that the isolated effector domains, created by genetic deletion or proteolysis, are capable of binding to DNA and altering the levels of transcription of Spo0A-controlled genes (Bird et al., 1993). Truly unphosphorylated Spo0A is incapable of binding 0A boxes *in vitro* (Ladds et al., 2003), and thus the unphosphorylated protein cannot regulate transcription. As in the case of other response regulators, it would appear that in Spo0A the function of the receiver domain is to inhibit the effector domain and that phosphorylation serves to overcome this inhibition. What is not known at present is (i) the nature of the inter-subunit interactions in Spo0A dimers, (ii) the nature of the inter-domain interactions responsible for inhibition in unphosphorylated Spo0A and (iii) whether a common surface on the receiver domain is responsible for both sets of interactions.

Activated Spo0A protein is required, at the morphological level, for the switch from symmetric to asymmetric septation at the onset of sporulation. Through regulation of expression of specific set of genes, Spo0A also regulates chromosome partitioning, the initiation of differential gene expression in forespore and, as recently shown, at later stages during sporulation it becomes a compartment specific transcription factor which further regulates expression of genes in the mother cell (Fujita et al., 2003). Taken together, Spo0A appears to be an extraordinarily complicated transcriptional regulator, reflecting its indispensable role in driving a developmental decision to initiate and proceed with the sporulation process.

CHROMOSOME PARTITIONING DURING SPORULATION

In vegetatively growing *B. subtilis* cell division and segregation of the newly duplicated chromosomes are coordinated and occur through a complex, highly regulated mechanism (see below). At the beginning of sporulation, with the appearance of the asymmetric septum the two chromosomes instead of splitting form an elongated structure known as an axial filament. Localization experiments of *oriC* by using the Lac repressor (targeted to a copy of *lacO* placed near *oriC*) and the chromosome segregation protein Spo0J by means of GFP fusions, revealed that axial filament formation is accompanied by migration of the *oriC* regions towards opposite poles of the cell (Glaser et al., 1997; Lin et al., 1997). Thus, the chromosomes in this predivisional sporangium are oriented with their replication origin regions to opposite poles of the

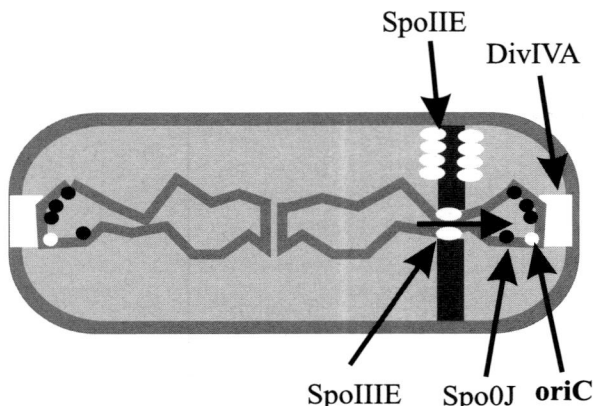

Figure 4. Outline of chromosome segregation during *Bacillus subtilis* sporulation.

cell. Interestingly, the anchoring sites of chromosomes are in region lying about 150-300 kbp away from *oriC*. One protein candidate involved in recruiting the chromosomes to this site is the DivIVA protein (Thomaides et al., 2001), which forms an oligomer-like structures (Muchova et al., 2002). The movement of the two chromosomes is under the control of the phosphorelay system (Graumann et al., 2001). There are three known proteins involved in chromosome segregation during sporulation – Spo0J, RacA and the Soj DNA-binding proteins. Soj has an ability to undergo co-operative relocation from nucleoid to nucleoid (Marston et al., 1999a) or pole to pole (Quisel et al., 1999) and this movement requires the Spo0J protein, which binds to condensation regions of the chromosome near the *oriC*. Also, RacA binds to the chromosome and is part of the mechanism that attaches the two chromosomes to the poles, most likely by contacting the DivIVA protein already localized at the cell pole (Figure 4) (Ben Yehuda and Losick, 2003; Wu and Errington, 2003).

The sporulation septum bisects the axial filament leaving only one third of one chromosome in the forespore, creating the transient genetic asymmetry (described below) (Wu and Errington, 1994). The remaining two-thirds of the chromosome is only transferred 10-20 minutes later from the mother cell into the forespore via a conjugation-like mechanism directed by the SpoIIIE partitioning protein (Wu and Errington, 1997). The SpoIIIE protein is targeted to the septum (Wu and Errington, 1997) and has ATP-dependent DNA-tracking activity with a direct role in DNA transfer (Bath et al., 2000). The SpoIIIE hydrophobic amino-terminal domain is involved in targeting of the protein to the sporulation septum (Bath et al., 2000).

ASYMMETRIC CELL DIVISION

Cell division, often called septation, consists of invagination of cytoplasmic membrane and concurrent peptidoglycan synthesis. Most of the molecules involved in this process are now known but mechanisms controlling where, how and when cells form the division septum are still controversial. The earliest event in the cell division cycle is the formation of the FtsZ ring at the future septum site. FtsZ is a highly conserved GTPase with a high degree of similarity with the tubulins (eukaryotic cytoskeletal proteins). During vegetative growth the FtsZ ring forms at the midcell and the cell is then divided at this site. At least two distinct mechanisms are involved in accurate placement of the division machinery: the Min system and nucleoid occlusion. The Min system functions mainly to prevent the possibility of division at the cell poles. In *E. coli* the Min system consists of MinC, MinD and MinE proteins. MinD activates MinC inhibition function and MinE is a topological factor allowing relief of division inhibition in the central region of the cell (deBoer et al., 1992).

Recent localization experiments in living cells revealed that MinD, which is a membrane-associated ATPase (deBoer et al., 1991), oscillates from pole to pole in dividing cells (Raskin deBoer, 1999a; Raskin deBoer, 1999b). This amazing oscillation repeats about every 10-20 seconds. The division inhibitor MinC can not oscillate on its own, but it binds to MinD and co-oscillates with the same pattern as MinD (Hu and Lutkenhaus,

1999; Raskin and deBoer, 1999b). The MinE was originally shown to form the ring-like structure at the mid-cell site (Raskin deBoer, 1997). However, recent studies have shown that in living cells the localization of MinE protein also undergoes a rapid oscillation what is coupled to MinD oscillation (Fu et al., 2001). This oscillating movement of MinD and MinE is co-dependent because lack of MinE causes uniform distribution of MinD around the cytoplasmic membrane (Rowland et al., 2000).

Taken together, all these results raise the question of how the oscillation works? Recently, molecular structures of MinC, MinD and MinE have been solved (Cordell and Lowe, 2001; King et al., 2000). Although, there are few models explaining this protein oscillation phenomenon (Hu and Lutkenhaus, 2001; Margolin, 2001), further genetic, localization and structural studies will be required for a more complete understanding of this process.

B. subtilis, like *E. coli* and a wide variety of other prokaryotes, has MinC and MinD homologues, and these are important for the prevention of asymmetric septation during vegetative growth. However, *B. subtilis* lacks MinE and the MinC - MinD oscillation has not been observed (Marston et al., 1998; Marston et al., 1999b). In *B. subtilis*, the DivIVA protein serves as the topological factor of mid-cell division (Cha and Stewart, 1997; Edwards and Errington, 1997). However, DivIVA is not a MinE homolog, and in comparison to MinE's ability to form dimers, forms higher oligomers (Zhang et al., 1998; Muchova et al., 2002). Moreover, the DivIVA mechanism of action is totally different with respect to MinE. DivIVA requires FtsZ and other cell division proteins for its localization to the division sites late in their maturation (Edwards and Errington, 1997; Marston et al., 1998). Unlike most of the other division proteins, it is then retained at the completed cell poles. The MinCD complex is also recruited to division sites, partly independently of DivIVA, but DivIVA is needed to bind the complex at the poles to block the asymmetric division in newly formed daughter cells (Marston et al., 1998; Marston and Errington, 1999a; Marston and Errington, 1999b). The DivIVA/MinCD division system appears to have no direct role in initiation of FtsZ ring formation at the mid-cell site but rather it inhibits the division at the polar sites. DivIVA seems to have a second function during the sporulation process, specifically at the stage of prespore chromosome segregation (Thomaides et al., 2001). Possibly, the most important factor of mid-cell division site selection in *B. subtilis* is the position of the nucleoid. Recent studies of cell division in outgrowing spores indicated the crucial role of nucleoid position for FtsZ ring formation (Harry et al., 1999). The DivIVA/MinCD system seems to be involved in inhibition of division at the polar nucleoid-free region sites and it is not crucial for mid-cell site selection (Harry, 2001). The proposed nucleoid occlusion mechanism (Woldringh et al., 1991) is very attractive but is still a poorly defined model. This model proposes that the nucleoid has a negative effect on division wherever it occupies space in the cell. Thus, the mid-cell site appears and disappears cyclically during vegetative growth with rounds of chromosome replication (Errington et al., 2003).

As mentioned above, the first clear morphological feature of sporulation is the formation of an asymmetric septum that bisects the bacterial cell into two unequally sized compartments, the larger mother cell and the smaller forespore. The proper positioning of this sporulation septum is dependent on Spo0A activity, which mediates the assembly of division proteins near the cell pole instead of at mid-cell. Such a positional switch could be partially triggered through the activity of the sporulation specific protein SpoIIE, also a component of the sporulation septum, that is expressed as a result of Spo0A activity (Barak and Youngman, 1996; Feucht et al., 1996; Khvorova et al., 1998). The central domain of SpoIIE is involved in oligomerization of the protein and is responsible for interaction with FtsZ, the protein essential for cell division (Lucet et al., 2000; Prepiak et al., 2001). In *B. subtilis*, the division septum is a relatively thick structure containing a substantial amount of peptidoglycan (PG) that separates the two daughter cells at cytokinesis. In contrast, the asymmetric septum formed during sporulation is much thinner and most of the PG separating the two lipid bilayers that comprise the septal structure is removed soon after septation is completed. The pliable septal structure then migrates toward one pole of the cell, eventually engulfing the forespore compartment, which will mature into a dormant endospore.

Western blot analysis of fractionated cell extracts from mother cell and forespore, using polyclonal antibody against the globular part of SpoIIE, showed the possibility to target this protein specifically into asymmetric septa (Barak et al., 1996). These results were surprising, considering the fact that *spoIIE* expression starts before the asymmetric sporulation septum begins to form and SpoIIE should be detected in the membrane fraction of the forespore and mother cell. Further localization experiments, with SpoIIE-GFP fusions in *B. subtilis* (Arigoni et al., 1995) and *B. megaterium* (Barak et al., 1996) using fluorescent microscopy, clearly showed the localization of SpoIIE in the asymmetric sporulation septum. There are at least two possibilities to explain these results. SpoIIE protein can specifically recognize the septation site or it can be built specifically into already formed septa. The second event is not very likely in view of electron-microscopy studies of SpoIIE mutant cells (Barak and Youngman, 1996; Illing and Errington, 1991). It seems, that presence of SpoIIE is necessary at the beginning of the formation of the sporulation septa. Some results support the proposal that asymmetric septation is a modified form of vegetative septation and uses the same basic machinery, including proteins as FtsZ, FtsA and others.

What is known about the activation of asymmetric division during sporulation? Spo0A and σ^H proteins are clearly involved in the shift of cell division to polar sites (Levin and Losick, 1996; Ben Yehuda and Losick, 2002). This shift is mediated by specific contact of two proteins FtsZ and SpoIIE (Khvorova et al., 1998; Lucet et al., 2000; Ben Yehuda and Losick, 2002). It was previously suggested that the switch in FtsZ ring localization is effected by a mechanism which blocks formation of such ring at the mid-cell and releases the polar sites inhibition (Levin and Losick, 1996; Barak et al., 1998). The asymmetric division occurs at about the time when normal medial division would occur. Therefore, in the cell with two complete chromosomes

where sporulation was initiated, medial division has to be blocked. Mutant cells affected in Spo0A, divide at mid-cell site without FtsZ ring relocalization effect (Levin and Losick, 1996). However, recently it was shown that the switch from medial to polar Z rings is accomplished by a spiral like structure of FtsZ that grows from mid-cell outward toward the cell poles, where it is converted into bipolar rings (Ben Yehuda and Losick, 2002). Interestingly, the process is reversible and both FtsA and EzrA were shown to co-localize with the FtsZ. SpoIIE plays a crucial role in this process, possibly by activating the bipolar Z rings formation and by stabilizing such structures.

The next event, after the FtsZ and SpoIIE rings formation, involves repositioning of the entire division machinery to the cell poles. An interesting aspect of this process is that potential division sites are actually at both cell poles and they are used for septa formation in the disporic mutants, as observed in *spoIIE*, *spoIIAA*, *spoIIAC* (coding σ^F), *spoIIGA* and *spoIIGB* (coding σ^E) mutants. It is crucial, that in wild type cells the division occurs at only one of the poles. It is known that σ^E–dependent genes block the maturation of the second polar division site (Eichenberger et al., 2001).

Although, the asymmetric cell division during sporulation resembles the vegetative division, we do not know the mechanisms that are involved in accurate placement of the division machinery. Firstly, it appears that it does not involve the Min system because mutations in *minC* and *minD* have little effect on the sporulation frequency (Cha and Stewart, 1997; Edwards and Errington, 1997; Barak et al., 1998). However, it is not possible to exclude a partial role of Min system during sporulation in light of the fact that a sporulation-like septum is misplaced from its normal polar site in a small proportion of *minD* mutant cells (Barak et al., 1998; Thomaides et al., 2001). Secondly, the polar septum during sporulation overcomes nucleoid occlusion and constricts around the nucleoid. The nature of the effector that overcomes or eliminates the spatial veto exerted by the nucleoid is not known.

Although, the spatial regulation of vegetative and sporulation cell division significantly differ, both processes use essentially the same protein machinery, except SpoIIE protein that is a specific component of the sporulation septum. *B. subtilis* has homologues of most *E. coli* division proteins, including FtsZ, FtsA, FtsL, FtsQ (DivIB in *B. subtilis*), FtsW (YlaO in *B. subtilis*) and PBP3 (PBP2B in *B. subtilis*) (Errington, 2001). The hierarchy of assembly of mid-cell and sporulation division appears to be similar for both processes (again except SpoIIE).

ACTIVATION OF THE FIRST COMPARTMENT SPECIFIC SIGMA FACTORS σ^F AND σ^E

The σ^F and σ^E factors are both synthesized prior to septum formation and the mere presence of these factors does not account for compartmentalized gene expression. The activation of σ^F is triggered by a complex mechanism (Figure 5). Immediately after σ^F is synthesized, it is held inactive by a stable interaction with a dimer of its anti-σ-factor SpoIIAB that is encoded by a gene that belongs to the same operon (Duncan and Losick, 1993; Min et al., 1993; Campbell and Darst, 2000; Campbell et al., 2002). In

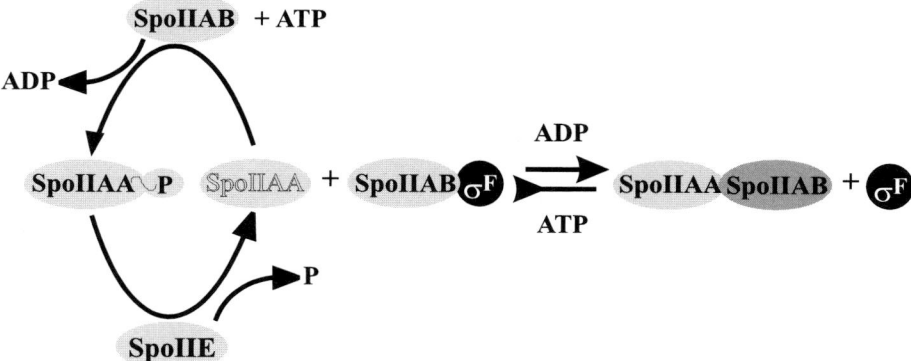

Figure 5. Activation of σ^F factor.

the presence of unphosphorylated molecules of SpoIIAA, which is an anti-anti-σ-factor, SpoIIAB binds to SpoIIAA and active σ^F is released (Duncan et al., 1996). However, SpoIIAB as a protein-kinase phosphorylates SpoIIAA, which is subsequently not able to bind SpoIIAB. An anti-σ factor is then free again to block σ^F (Min et al., 1993). After SpoIIAB has phosphorylated SpoIIAA, it remains in a complex with ADP. It is maintained in a special conformational state, in which it is able to bind another dephosphorylated SpoIIAA, forming a ternary complex that decays very slowly. This mechanism could be important from the point of retention of an anti-σ-factor in relatively stable complexes (Lee et al., 2000).

There is an enormous amount of genetic, biochemical and molecular biology data regarding the activation of σ^F and recently the crystal structure of key protein components have been solved. The crystal structure of phosphorylated and unphosphorylated form of SpoIIAA was solved from *Bacillus sphaericus* (Seavers et al., 2001). This single domain globular protein consists of a central sheet of five β strands, and four α helices. On one face, the β-pleated sheet is embedded with a pair of α-helices, while the other face is exposed to the solvent. The phosphorylation site, dephosphorylation of which is important for activation of σ^F, is located at the N-terminus of the helix α2 so that the phosphoryl group points into the solvent. Comparisons between unphosphorylated and phosphorylated SpoIIAA had shown, that structural changes accompanying phosphorylation are only slight. Extended hydrophobic surfaces together with flexibility in helix α3 and the following loop indicate that this protein often undergoes a disorder to order transition and therefore has a disposition for forming stable protein-protein interactions (Seavers et al., 2001). Also the structure of the anti-σ factor SpoIIAB in the complex with σ^F of *B. stearothermophilus* was resolved (Campbell et al., 2002). The SpoIIAB - σ^F complex consists of a SpoIIAB dimer bound to one σ^F molecule so that σ^F binding directly to the dimer interface makes contact with both SpoIIAB monomers. Dimerization interactions occur among β1 and α1 of each SpoIIAB monomers. The ATP – binding site is represented by Asn50, which plays a key role in chelating Mg^{2+}, critical for ATPase or kinase activity of the GHKL superfamily members (Campbell et al., 2002). The catalytic site is formed by Glu46, which functions as a catalytic base in the kinase reaction.

An additional crucial component of the σ^F activation process is the 92 kDa SpoIIE protein, a Ser/Thr sporulation specific phosphatase, which allows dephosphorylation of the SpoIIAA-phosphate. SpoIIE is also an important constituent of the sporulation septum and the functions of SpoIIE belong to the most important key events that enable proper activation of σ^F (Duncan et al., 1995). The *spoIIE* locus was defined genetically as the site of mutations which blocked development at the stage of asymmetric septation (Piggot et al., 1973; Barak and Youngman, 1996; Feucht et al., 1996). Electron microscopy studies clearly showed that SpoIIE protein plays an important role in septal morhogenesis (Barak and Youngman, 1996; Feucht et al., 1996). A σ^F-independent involvement of SpoIIE in septum assembly is also suggested by the fact that null mutations in *spoIIE* cause a thick-septum phenotype, and this is consistent with a direct or indirect role for SpoIIE in peptidoglycan deposition, since the thick-septum phenotype is apparently caused by the incorporation of excess peptidoglycan into the space between the septum membranes. SpoIIE is an integral membrane protein and has a multidomain structure (Arigoni et al., 1999; Barak et al., 1996). Its N-terminal domain (domain I) is hydrophobic and contains 10 membrane-spanning segments, which target the protein to the membrane. Interestingly, SpoIIE is localized to the site of asymmetric septation in the form of a SpoIIE ring (E-ring) and later on, stays sequestered into the sporulation septum membrane (Arigoni et al., 1995; Barak et al.,1996). Domain I is followed by a large central domain (domain II), which is involved in oligomerization of SpoIIE and in the interaction with the cell division protein FtsZ (Lucet et al., 2000) and thus domain II is likely involved in the Spo0A-dependent switch in the location of FtsZ rings (Khvorova et al., 1998). Finally, the C-terminal domain (domain III) possesses a serine phosphatase activity capable of dephosphorylating the regulatory protein SpoIIAA, which in turn controls the activity of σ^F (Duncan et al., 1995). Domain III is structurally related to eukaryotic PP2C protein phosphatases, which are involved in regulating stress response pathways in both prokaryotes and eukaryotes (Adler et al., 1997). SpoIIE protein plays a crucial role in the sporulation septum assembly and in coupling σ^F activation to asymmetric division and thus confining σ^F activity to the forespore.

Although, most of the proteins responsible for this process are now known, the mechanism that delays σ^F activation until the sporulation septum is completed and that confines its activity only into prespore is not yet fully understood. It is important to mention that all the proteins involved in σ^F activation are made before asymmetric septation and thus this process involves mechanisms that are not easy to study and has been the subject of considerable speculation. To address the question of the proper activation of σ^F, many experiments have been focused on this topic and many models have been suggested. There are a few basic models, which try to explain σ^F compartment-specific activation and thus establishing the first step in the cell-type specific gene expression.

The first model claims that the ratio between septum-bound SpoIIE-phosphatase and cytoplasmic SpoIIAB-kinase is higher in the smaller prespore, compared to the larger mother cell. This might result in relatively higher concentrations of dephosphorylated SpoIIAA in the prespore compartment. This would bind to SpoIIAB and cause the release of an active σ^F (Duncan et al., 1995). It was also argued that this difference is possibly reinforced by a difference in the ratio of ATP/ADP in the two compartments, and might be sufficient to account for compartment-specific activation of σ^F (Duncan et al., 1995). However, bacterial species with globular cell shapes, such as *Sporosarcina ureae*, use probably a very similar mechanism of sporulation but sporulation septa in this case are positioned at the mid-cell site (Chary et al., 2000). In addition, it was shown that the rate-determining step in partner switching of SpoIIAB is the dissociation of a SpoIIAB-ADP complex and this is not affected by changes in the concentration of ATP (Najafi et al., 1997).

A second model proposes that the SpoIIE-phosphatase is sequestrated only to the prespore side of the sporulation septum and SpoIIAA can be dephosphorylated by SpoIIE preferentially in the forespore compartment. This was suggested on the basis of fluorescence microscopy localization studies of SpoIIE (Wu et al., 1998). However, the dephosphorylation of SpoIIAA does not require septation (Feucht et al., 1999; King et al., 1999). This model also raises other questions – how such SpoIIE localisation in growing lipid bilayers can be achieved? In addition, other experiments have shown that cells with truncated forms of SpoIIE, that lack the transmembrane domains and do not localize into the sporulation septum still can sporulate very efficiently (about 30% of the wild type strain) (Arigoni et al., 1999).

A third model suggests the importance of the transient genetic asymmetry that occurs during chromosome segregation immediately after creation of the sporulation septum (Figure 6A). This process seems to be responsible for the beginning and also for the timing of spatial differentiation of gene expression (Frandsen et al., 1999). During sporulation, the chromosomes have an extended, rod-like structure, the axial filament, and the polar septum forms over one of the two daughter chromosomes, a phenomenon in marked contrast with normal vegetative septation. The transient genetic asymmetry is based on the fact that there is a short interval when the two chromosomes are not equally distributed between two cell compartments. The sporulation septum actually bisects one of the chromosomes, so that during this interval only 30% of the genetic material is present in the prespore and this includes the part of chromosome close to the replication origin. The remaining part of the chromosome stays for 15-20 minutes imprisoned in the mother cell (Wu and Errington, 1998). The importance of such transient genetic asymmetry for σ^F activation specifically in the forespore, came from studies of Frandsen et al. (1999). They constructed a set of *B. subtilis* strains lacking both the genes essential for σ^F activation, *spoIIE* and *spoIIAA*. Interestingly, they were able to force such cells to sporulate by moving a copy of *spoIIAC* (which encodes σ^F) from its original position to the origin-proximal positions that is trapped in the forespore. This experiment clearly demonstrated that transient genetic asymmetry could be used to activate σ^F specifically in the forespore without the key activators, SpoIIE and SpoIIAA. The remaining question is how this transient genetic asymmetry is used in the wild type cell. A possible explanation would be that an hypothetical inhibitor of SpoIIE is encoded by a gene located far from the replication origin (Figure 6A) therefore trapped in the mother cell for a short time. This could cause a transient absence of the inhibitor molecules in the prespore (Frandsen et al., 1999). However, the existence of such an inhibitor has not yet been confirmed (Feucht et al., 1999; King et al., 1999) Another possible explanation involves the transient *spoIIAB* genetic asymmetry (Figure 6B). This gene enters the prespore late and its protein product SpoIIAB (subject to proteolysis) is temporarily not replenished (Dworkin and Losick, 2001; Pan et al., 2001). The position effect of *spoIIAB* was only observed in a mutant background where SpoIIE was no longer localized to the polar septum. Thus, it is likely that at least two different mechanisms act in parallel to activate σ^F specifically in the

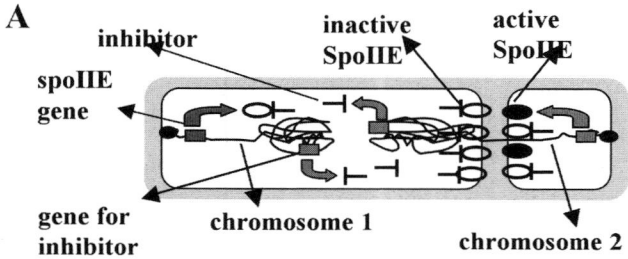
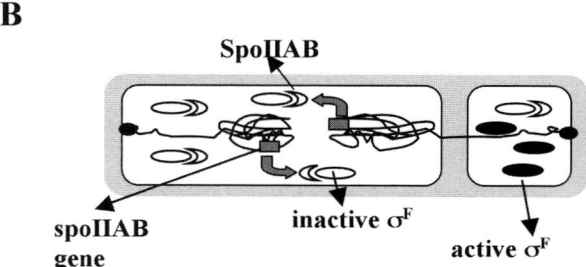

Figure 6. Recent models of σ^F activation specifically in the forespore.

forespore. The first mechanism involves localization of SpoIIE to the sporulation septum which increase the relative effective concentration of its phosphatase activity in the forespore. The second mechanism involves the transient genetic asymmetry of *spoIIAB*, the product of which is, in addition, proteolysed in the forespore thus reducing the level of the anti- σ^F factor relative to σ^F.

A similar transient genetic asymmetry was shown to be important for activation of the next sigma factor, σ^E, which is synthesized before septation as an inactive pro-σ^E. The N-terminal pro sequence is removed by the membrane-associated protease, SpoIIGA (Stragier et al., 1988). The activity of SpoIIGA is regulated by the SpoIIR protein which is under the control of σ^F (Hofmeister et al., 1995). The *spoIIR* gene is located at the chromosome near to the origin and is transcribed as early as possible following σ^F activation. Relocation of *spoIIR* to the origin-distal positions resulted in a delay and a reduction in σ^E activation (Khvorova et al., 2000; Zupancic et al., 2001). Thus, transient genetic asymmetry of *spoIIR* appears to regulate the proper temporal activation of σ^E. Although, σ^F-SpoIIGA-SpoIIR pathway explains the timing of σ^E activation, it does not explain why σ^E is specifically activated only in the mother cell. There are two possible mechanisms explaining this phenomenon. First, it was proposed that pro-σ^E is sequestered on the mother cell face of the sporulation septum and thus active σ^E is produced only in the mother cell (Ju et al., 1997). Second, it was hypothesised that σ^E and/or pro-σ^E are proteolysed specifically in the forespore (Ju et al., 1998).

MORPHOLOGY AND GENE EXPRESSION CHANGES DURING σ^G AND σ^K ACTIVATION

σ^G is synthesized in the forespore when the engulfment process is completed and the forespore appears as a free protoplast inside of the mother cell. Clearly, activation of σ^G is under the control of this morphological checkpoint. Formation of the forespore protoplast is under the control of σ^E dependent gene products SpoIID, SpoIIM and SpoIIP (Rong et al., 1986; Smith and Youngman, 1992; Frandsen and Stragier, 1995). Compartmentilazation of σ^G is allowed by σ^F dependent transcription of the *spoIIIG* gene (coding σ^G) only

in the forespore. However, an important, unresolved question remains – how σ^G activation is delayed until the engulfment process completes. σ^G activation is dependent on SpoIIAB, the anti-sigma factor of σ^F. The formation of the SpoIIAA-SpoIIAB complex releases σ^F activation but it is not sufficient to activate σ^G. However, it was proposed that gene products of the eight-cistron *spoIIIA* operon, transcribed by σ^E, activate σ^G in the forespore through SpoIIAB by an unknown mechanism (Kellner et al., 1996).

The last compartment-specific sigma factor which is activated during sporulation is σ^K. This sigma factor is activated specifically in the mother cell compartment and is dependent on at least three levels of regulation. Firstly, its formation is dependent on mother cell chromosome rearrangement. In vegetatively growing cells, *sigK* gene, coding σ^K, is interrupted by a 40 kbp DNA *skin* element (Kunkel et al., 1990). Secondly, this *skin* element is excised from the mother cell chromosome by the SpoIVCA recombinase whose transcription is under σ^E control. Thirdly, the primary product of σ^K, like its mother cell-predecessor σ^E, is subject to a post-translational modification during which the precursor protein pro-σ^K is cleaved by a membrane-bound protease, SpoIVFB (Cutting et al., 1991). The protease is held inactive by two other mother cell-specific, proteins SpoIVFA and BofA (Cutting et al., 1991; Ricca et al., 1992). The activation signal for pro-σ^K comes from the forespore via SpoIVB. SpoIVB, whose gene is transcribed by σ^G in the forespore, is also a protease, which is self-cleaved into several proteolytically active fragments (Hoa et al., 2002; Wakeley et al., 2000). A recent model for pro-σ^K cleavage into active σ^K is that SpoIVB cleaves BofA and/or SpoIVFA and thus triggers release of the protease activity of SpoIVFB in the mother cell.

COMPARISON OF SPORULATION PROCESS IN DIVERSE ENDOSPORE-FORMING BACTERIA

B. subtilis has proven to be a rewarding model to study the process of spore formation and to understand the general concepts behind unicellular differentiation. A challenging question is now to understand if mechanisms controlling sporulation in *B. subtilis* are applicable to other spore-formers.

Table 1. Similarities and differences in sporulation process between *B. subtilis* and of different *Bacilli*, *Clostridia* and *Sporosarcina* strains[1].

Stages	Other *Bacilli*		*Clostridia*		*Sporosarcina*	
	Differences	Similarities	Differences	Similarities	Differences	Similarities
0	*spo0E* *abrB* is not present in all *Bacilli*	Phosphorelay, *spo0A, J, B, F, rapA, kinA*	No phosphorelay, *spo0E, B, F, rapA, kinA, abrB* is not present in all *Clostridia*	*spo0A, spo0J*	not known (nk)	nk
1		*spo0H, divIVA*		*spo0H, divIVA*	nk	nk
2		*spoIIA(A-C), E, G(A,B), B, D, M, P, Q*		*spoIIA(A-C), E, G(A, B), B, D, M, P, Q*	Symmetrically positioned septum	nk
3		*spoIIIA(A-H), E, G, J*		*spoIIIA(A-H), E, G, J*	nk	*spoIIIE*
4		*spoIVA, B*		*spoIVA, B*	nk	nk
5		*spoVA(A-F), B, D, E, G, K, M, R, S, T,*	*spoVK*	*spoVA(CDE), B, D, E, G, M, R, S, T,*	nk	nk
6-7	*cotC,T,V*	*cotA, D, E, JC*	*cotC, T, V*	*cotJC*	nk	nk

[1]All the shown characteristics are mostly based only on the genomic sequences of *B. subtilis, B. anthracis, B. stearotermophilus, B. holodurans, Clostridium acetobutylicum* and *C. difficile*. Note, that not all *B. subtilis* sporulation genes are listed in the table. Not present are those which are present only in some *Bacilli* and/or some *Clostridia* strains. Nk, not known.

Many endospore-forming bacteria are important in agriculture, industry and medicine and this makes it important to study the sporulation process in biotechnologically relevant spore formers other than *B. subtilis*.

A high degree of similarity among *B. subtilis* and other Gram-positive endospore-formers so far studied has been observed at the morphological level and with respect to gene expression. However, there are significant differences when we examine bacteria such as *Sporosarcina urea*, which is spherical and where there is no clear morphological asymmetry at the level of septum formation. The link between asymmetry in morphology and gene expression regulation is obviously not applicable in this symmetrically sporulating bacterium. It is likely that the study of significant variations in the sporulation processes in different bacteria will help us to understand the evolution of this remarkable cell differentiation process.

The long-recognized diversity among the endospore-forming bacteria is emphasized by nutritional diversity and by different habitats in which these bacteria live. Both these two aspects seem to have a great impact on differences in the initiation of sporulation. The completion of several genome sequences of endospore-forming bacteria (*B. subtilis, B. anthracis, B. stearotermophilus, B. holodurans, Clostridium acetobutylicum* and *C. difficile*) has provided powerful tools to compare the sporulation processes among all these spore-formers. The first detailed genomic comparison of sporulation processes among these bacteria was recently made in an excellent review by Stragier (2002). Some of the main similarities and differences among the sporulation processes in these bacteria are presented in Table 1. The initiation of endospore formation seems to have one universal control mechanism that is mediated through the Spo0A protein. However, genes coding proteins of the phosphorelay are missing from the genome of *Clostridia* species and the concentration of active Spo0A~P must be controlled by a different mechanism. Perhaps, this reflects the distinct habitats and metabolic abilities of *Clostridia* and *Bacilli* species. The second important variation is at the stage of spore coat formation, where only a few sporulation specific genes are conserved among analyzed endospore-forming bacteria (Table 1).

While the general mechanisms leading to endospore formation seem to have evolved similarly in Gram-positive bacteria, little information is available about these processes in Gram-negative bacteria. In *Coxiella burnetii,* a Gram-negative, obligate intracellular pathogen, spore formation was analyzed by electron-microscopic studies (McCaul, 1991). The recent completion of *C. burnetii* genome sequence did not reveal homologues of sporulation genes (Seshadri et al., 2003), with the only exception of the previously described homologue of SpoIIIE (Oswald and Thiele, 1993). The possible explanation of this lack of similarity between Gram-positive and Gram-negative sporeformers could be that *C. burnetii* as well as the other Gram-negative sporeformers listed above, have diverged early during evolution.

ACKNOWLEDGEMENTS

Author would like to thank to members of his laboratory, Katarina Muchova, Zuzana Chromikova, Patrik Florek for help with manuscript preparation. The work in the author's laboratory is supported by grant 2/1004/21 from the Slovak Academy of Sciences and The Wellcome Trust Grant 066732/Z/01/Z.

REFERENCES

Adler, E., Donelladeana, A., Arigoni, F., Pinna, L.A., and Stragier, P. 1997. Structural relationship between a bacterial developmental protein and eukaryotic PP2C protein phosphatases. Mol. Microbiol. *23*, 57-62.

Arigoni, F., Guerout-Fleury, A.M., Barak, I., and Stragier, P. 1999. The SpoIIE phosphatase, the sporulation septum and the establishment of forespore-specific transcription in *Bacillus subtilis*: a reassessment. Mol. Microbiol. *31*, 1407-1415.

Arigoni, F., Pogliano, K., Webb, C.D., Stragier, P., and Losick, R. 1995. Localization of protein implicated in establishment of cell-type to sites of asymmetric division. Science *270*, 637-640.

Asayama, M., Yamamoto, A., and Kobayashi, Y. 1995. Dimer form of phosphorylated Spo0A, a transcriptional regulator, stimulates the Spo0F transcription at the initiation of sporulation in *Bacillus subtilis*. J. Mol. Biol. *250*, 11-23.

Baldus, J.M., Green, B.D., Youngman, P., and Moran, C.P. 1994. Phosphorylation of *Bacillus subtilis* transcription factor Spo0A stimulates transcription from the *spoIIG* promoter by enhancing binding to weak OA boxes. J. Bacteriol. *176*, 296-306.

Barak, I., Behari, J., Olmedo, G., Guzman, P., Brown, D.P., Castro, E., Walker, D., Westpheling, J., and Youngman, P. 1996. Structure and function of the *Bacillus* SpoIIE protein and its localization to sites of sporulation septum assembly. Mol. Microbiol. *19*, 1047-1060.

Barak, I., Prepiak, P., and Schmeisser, F. 1998. MinCD proteins control the septation process during sporulation of *Bacillus subtilis*. J. Bacteriol. *180*, 5327-5333.

Barak, I., and Youngman, P. 1996. SpoIIE mutants of *Bacillus subtilis* comprise two distinct phenotypic classes consistent with a dual functional role for the SpoIIE protein. J. Bacteriol. *178*, 4984-4989.

Bath, J., Wu, L.J., Errington, J., and Wang, J.C. 2000. Role of *Bacillus subtilis* SpoIIIE in DNA transport across the mother cell-prespore division septum. Science *290*, 995-997.

Ben Yehuda, S., and Losick, R. 2002. Asymmetric cell division in *Bacillus subtilis* involves a spiral-like intermediate of the cytokinetic protein FtsZ. Cell *109*, 257-266.

Ben Yehuda, S., Rudner, D.Z., and Losick, R. 2003. RacA, a bacterial protein that anchors chromosomes to the cell poles. Science *299*, 532-536.

Bird, T.H., Grimsley, J.K., Hoch, J.A., and Spiegelman, G.B. 1993. Phosphorylation of Spo0A activates its stimulation of *in vitro* transcription from the *Bacillus subtilis spoIIG* operon. Mol. Microbiol. *9*, 741-749.

Brown, D.P., Ganova-Raeva, L., Green, B.D., Wilkinson, S.R., Young, M., and Youngman, P. 1994. Characterization of Spo0A homologs in diverse *Bacillus* and *Clostridium* species identifies a probable DNA-binding domain. Mol. Microbiol. *14*, 411-426.

Campbell, E.A., and Darst, S.A. 2000. The anti-sigma factor SpoIIAB forms a 2, 1 complex with sigma(F), contacting multiple conserved regions of the sigma factor J. Mol. Biol. *300*, 17-28.

Campbell, E.A., Masuda, S., Sun, J.L., Muzzin, O., Olson, C.A., Wang, S., and Darst, S.A. 2002. Crystal structure of the *Bacillus stearothermophilus* anti-sigma factor SpoIIAB with the sporulation sigma factor, σ^F. Cell *108*, 795-807.

Cha, J.H., and Stewart, G.C. 1997. The *divIVA* minicell locus of *Bacillus subtilis*. J. Bacteriol. *179*, 1671-1683.

Chary, V.K., Hilbert, D.W., Higgins, M.L., and Piggot, P.J. 2000. The putative DNA translocase SpoIIIE is required for sporulation of the symmetrically dividing coccal species *Sporosarcina ureae*. Mol. Microbiol. *35*, 612-622.

Chibazakura, T., Kawamura, F., and Takahashi, H. 1991. Differential regulation of Spo0A transcription in *Bacillus subtilis*: glucose represses promoter switching at the initiation of sporulation. J. Bacteriol. *173*, 2625-2632.

Cordell, S.C., and Lowe, J. 2001. Crystal structure of the bacterial cell division regulator MinD. FEBS Lett. *492*, 160-165.

Cutting, S., Roels, S., and Losick, R. 1991. Sporulation operon SpoIVF and the characterization of mutations that uncouple mother cell from forespore gene expression in *Bacillus subtilis*. J. Mol. Biol. *221*, 1237-1256.

deBoer, P.A.J., Crossley, R.E., Hand, A.R., and Rothfield, L.I. 1991.The MinD protein is a membrane ATPase required for the correct placement of the *Escherichia coli* division site. EMBO J. *10*, 4371-4380.

deBoer, P.A.J., Crossley, R.E., and Rothfield, L.I. 1992. Roles of MinC and MinD in the site-specific septation block mediated by the MinCDE system. J. Bacteriol. *174*, 63-70.

Dubnau, D. 1999. DNA uptake in bacteria. Ann. Rev. Microbiol. *53,* 217-244.

Duncan, L., Alper, S., Arigoni, F., Losick, R., and Stragier, P. 1995. Activation of cell-specific transcription by a serine phosphatase at the site of asymmetric division. Science *270,* 641-644.

Duncan, L., Alper, S., and Losick, R. 1996. SpoIIAA governs the release of the cell-type specific transcription factor σ^F from its anti-sigma factor SpoIIAB. J. Mol. Biol. *260,* 147-164.

Duncan, L., and Losick, R. 1993. SpoIIAB is an anti-sigma factor that binds to and inhibits transcription by regulatory protein σ^F from *Bacillus subtilis.* Proc. Natl. Acad. Sci. USA *90,* 2325-2329.

Dworkin, J., and Losick, R. 2001. Differential gene expression governed by chromosomal spatial asymmetry. Cell *107,* 339-346.

Edwards, D.H., and Errington, J. 1997. The *Bacillus subtilis* DivIVA protein targets to the division septum and controls the site specificity of cell division. Mol. Microbiol. *24,* 905-915.

Eichenberger, P., Fawcett, P., and Losick, R. 2001. A three-protein inhibitor of polar septation during sporulation in *Bacillus subtilis.* Mol. Microbiol. *42,* 1147-1162.

Errington, J. 1993. *Bacillus-subtilis* sporulation-regulation of gene-expression and control of morphogenesis. Microbiol. Rev. *57,* 1-33.

Errington, J. 2001. Septation and chromosome segregation during sporulation in *Bacillus subtilis.* Curr. Opin. Microbiol. *4,* 660-666.

Errington, J. 2003. Regulation of endospore formation in *Bacillus subtilis.* Nature Rev. *1,* 117-126.

Errington, J., Daniel, A., and Scheffers, D.J. 2003. Cytokinesis in bacteria. Microbiol. Mol. Biol. Rev. 52-65.

Fawcett, P., Eichenberger, P., Losick, R., and Youngman, P. 2000. The transcriptional profile of early to middle sporulation in *Bacillus subtilis.* Proc. Natl. Acad. Sci. USA *97,* 8063-8068.

Feucht, A., Daniel, R.A., and Errington, J. 1999. Characterization of a morphological checkpoint coupling cell-specific transcription to septation in *Bacillus subtilis.* Mol. Microbiol. *33,* 1015-1026.

Feucht, A., Magnin, T., Yudkin, M.D., and Errington, J. 1996. Bifunctional protein required for asymmetric cell division and cell-specific transcription in *Bacillus subtilis.* Genes Dev.*10,* 794-803.

Frandsen, N., Barak, I., KarmazynCampelli, C., and Stragier, P. 1999. Transient gene asymmetry during sporulation and establishment of cell specificity in *Bacillus subtilis.* Genes Dev. *13,* 394-399.

Frandsen, N., and Stragier, P. 1995. Identification and characterization of the *Bacillus subtilis spoIIP* locus. J. Bacteriol. *177,* 716-722.

Fu, X.L., Shih, Y.L., Zhang, Y., and Rothfield, L.I. 2001. The MinE ring required for proper placement of the division site is a mobile structure that changes its cellular location during the *Escherichia coli* division cycle. Proc. Natl. Acad. Sci. USA *98,* 980-985.

Fujita, M., and Losick, R. 2003. The master regulator for entry into sporulation in *Bacillus subtilis* becomes a cell-specific transcription factor after asymmetric division. Genes Dev. *17,* 1166-1174.

Glaser, P., Sharpe, M.E., Raether, B., Perego, M., Ohlsen, K., and Errington, J. 1997. Dynamic, mitotic-like behavior of a bacterial protein required for accurate chromosome partitioning. Genes Dev. *11,* 1160-1168.

Gonzalez-Pastor, J.E., Hobbs, E.C., and Losick, R. 2003. Cannibalism by sporulating bacteria. Science *301,* 510-513.

Graumann, P.L., and Losick, R. 2001. Coupling of asymmetric division to polar placement of replication origin regions in *Bacillus subtilis.* J. Bacteriol. *183,* 4052-4060.

Harry, E.J. 2001. Bacterial cell division: regulating Z-ring formation. Mol. Microbiol. *40,* 795-803.

Harry, E.J., Rodwell, J., and Wake, R.G. 1999. Co-ordinating DNA replication with cell division in bacteria: a link between the early stages of a round of replication and mid-cell Z ring assembly. Mol. Microbiol. *33,* 33-40.

Hoa, N.T., Brannigan, J.A., and Cutting, S.M. 2002. The *Bacillus subtilis* signaling protein SpoIVB defines a new family of serine peptidases. J. Bacteriol. *184,* 191-199.

Hoch, J.A. 1993. The phosphorelay signal transduction pathway in the initiation of *Bacillus subtilis* sporulation. J. Cell. Biochem. *51,* 55-61.

Hoch, J.A. 1998. Initiation of bacterial development. Curr. Opin. Microbiol. *1,* 170-174.

Hofmeister, A.E.M., Londono-Vallejo, A., Harry, E., Stragier, P., and Losick, R. 1995. Extracellular signal protein triggering the proteolytic activation of a developmental transcription factor in *Bacillus subtilis.* Cell *83,* 219-226.

Hu, Z.L., and Lutkenhaus, J. 1999. Topological regulation of cell division in *Escherichia coli* involves rapid pole to pole oscillation of the division inhibitor MinC under the control of MinD and MinE. Mol. Microbiol. *34,* 82-90.

Hu, Z.L., and Lutkenhaus, J. 2001. Topological regulation of cell division in *Escherichia coli*: Spatiotemporal oscillation of MinD requires stimulation of its ATPase by MinE and phospholipid. Mol. Cell *7,* 1337-1343.

Illing, N., and Errington, J. 1991. Genetic regulation of morphogenesis in *Bacillus subtilis* – roles of σ^E and σ^F in prespore engulfment. J. Bacteriol. *173,* 3159-3169.

Ju, J.L., Luo, T.Q., and Haldenwang, W.G. 1 997. *Bacillus subtilis* pro-σ^E fusion protein localizes to the forespore septum and fails to be processed when synthesized in the forespore. J. Bacteriol. *179,* 4888-4893.

Ju, J.L., Luo, T.Q., and Haldenwang, W.G. 1998. Forespore expression and processing of the σ^E transcription factor in wild-type and mutant *Bacillus subtilis.* J. Bacteriol. *180,* 1673-1681.

Kellner, E.M., Decatur, A., and Moran, C.P. 1996. Two-stage regulation of an anti-sigma factor determines developmental fate during bacterial endospore formation. Mol. Microbiol. *21,* 913-924.

Kenney, T.J., Kirchman, P.A., and Moran, C.P. 1988. Gene encoding σ^E is transcribed from a σ^A – like promoter in *Bacillus subtilis.* J. Bacteriol. *170,* 3058-3064.

Khvorova, A., Chary, V.K., Hilbert, D.W., and Piggot, P.J. 2000. The chromosomal location of the *Bacillus subtilis* sporulation gene *spoIIR* is important for its function. J. Bacteriol. *182,* 4425-4429.

Khvorova, A., Zhang, L., Higgins, M.L., and Piggot, P.J. 1998. The *spoIIE* locus is involved in the Spo0A-dependent switch in the location of FtsZ rings in *Bacillus subtilis.* J. Bacteriol. *180,* 1256-1260.

King, G.F., Shih, Y.L., Maciejewski, M.W., Bains, N.P.S., Pan, B.L., Rowland, S.L., Mullen, G.P., and Rothfield, L.I. 2000. Structural basis topological specificity function of MinE. Nature Struc. Biol. *7,* 1013-1017.

King, N., Dreesen, O., Stragier, P., Pogliano, K., and Losick, R. 1999. Septation, dephosphorylation, and the activation of σ^F during sporulation in *Bacillus subtilis.* Genes Dev. *13,* 1156-1167.

Kunkel, B., Losick, R., and Stragier, P. 1990. The *Bacillus subtilis* gene for developmental transcription factor σ^K is generated by excision of a dispensable element containing a sporulation recombinase gene. Genes Dev. *4,* 525-535.

Ladds, J.C., Muchova, K., Blaskovic, D., Lewis, R.J., Brannigan, J.A., Wilkinson, A.J., and Barak, I. 2003. Response regulator Spo0A phosphorylated from *Bacillus subtilis* is efficiently in *Escherichia coli.* FEMS Microbiol. Lett. *223,* 153-157.

Lee, C.S., Lucet, I., and Yudkin, M.D. 2000. Fate of the SpoIIAB*-ADP liberated after SpoIIAB phosphorylates SpoIIAA of *Bacillus subtilis.* J. Bacteriol. *182,* 6250-6253.

Levin, P.A., and Losick, R. 1996. Transcription factor Spo0A switches the localization of the cell division protein FtsZ from a medial to a bipolar pattern in *Bacillus subtilis.* Genes Dev. *10,* 478-488.

Lewis, R.J., Brannigan, J.A., Muchova, K., Barak, I., and Wilkinson, A.J. 1999. Phosphorylated aspartate in the structure of a response regulator protein. J. Mol. Biol. *294,* 9-15.

Lewis, R.J., Krzywda, S., Brannigan, J.A., Turkenburg, J.P., Muchova, K., Dodson, E.J., Barak, I., and Wilkinson, A.J. 2000a. The *trans*-activation domain of the sporulation response regulator Spo0A revealed by X-ray crystallography. Mol. Microbiol. *38,* 198-212.

Lewis, R.J., Muchova, K., Brannigan, J.A., Barak, I., Leonard, G., and Wilkinson, A.J. 2000b. Domain swapping in the sporulation response regulator Spo0A. J. Mol. Biol. *297,* 757-770.

Lewis, R.J., Scott, D.J., Brannigan, J.A., Ladds, J.C., Cervin, M.A., Spiegelman, G.B., Hoggett, J.G., Barak, I., and Wilkinson, A.J. 2002. Dimer formation and transcription activation in the sporulation response regulator Spo0A. J. Mol. Biol. *316,* 235-245.

Lin, D.C.H., Levin, P.A., and Grossman, A.D. 1997. Bipolar localization of a chromosome partition protein in *Bacillus subtilis.* Proc. Natl. Acad. Sci. USA *94,* 4721-4726.

Losick, R., and Stragier, P. 1992. Crisscross regulation of cell-type specific gene expression during development in *Bacillus subtilis.* Nature *355,* 601-604.

Lucet, I., Feucht, A., Yudkin, M.D., and Errington, J. 2000. Direct interaction between the cell division protein FtsZ and the cell differentiation protein SpoIIE. EMBO J. *19,* 1467-1475.

Margolin, W. 2001. Bacterial cell division: A moving MinE sweeper boggles the MinD. Curr. Biol. *11,* R395-R398.

Marston, A.L., and Errington, J. 1999a. Dynamic movement of the ParA-like soj protein of B-subtilis and its dual role in nucleoid organization and developmental regulation. Molecular Cell 4, 673-682.

Marston, A.L., and Errington, J. 1999b. Selection of the midcell division site in Bacillus subtilis through MinD-dependent polar localization and activation of MinC. Mol. Microbiol. 33, 84-96.

Marston, A.L., Thomaides, H.B., Edwards, D.H., Sharpe, M.E., and Errington, J. 1998. Polar localization of the MinD protein of Bacillus subtilis and its role in selection of the mid-cell division site. Genes Dev. 12, 3419-3430.

McCaul, T.F. 1991. The developmental cycle of Coxiella burnetii. In: Q fever: The biology of Coxiella burneti. Williams, J.C., and Thompson, H.A., ed. CRS Press London, p. 223-258.

Min, K.T., Hilditch, C.M., Diederich, B., Errington, J., and Yudkin, M.D. 1993. σ^F, the first compartment-specific transcription factor of Bacillus subtilis, is regulated by an anti-sigma factor that is also a protein kinase. Cell 74, 735-742.

Molle, V., Fujita, M., Jensen, S.T., Eichenberger, P., Gonzalez-Pastor, J.E., Liu, J.S., and Losick, R. 2003. The Spo0A regulon of Bacillus subtilis. Mol. Microbiol. 50, 1683-1701.

Muchova, K., Kutejova, E., Scott, D.J., Brannigan, J.A., Lewis, R.J., Wilkinson, A.J., and Barak, I. 2002. Oligomerization of the Bacillus subtilis division protein DivIVA Microbiol.-SGM 148, 807-813.

Najafi, S.M.A., Harris, D.A., and Yudkin, M.D. 1997. Properties of the phosphorylation reaction catalyzed by SpoIIAB that help to regulate sporulation of Bacillus subtilis. J. Bacteriol. 179, 5628-5631.

Ohlsen, K.L., Grimsley, J.K., and Hoch, J.A. 1994. Deactivation of the sporulation transcription factor Spo0A by the Spo0E protein phosphatase. Proc. Natl. Acad. Sci. USA 91, 1756-1760.

Oswald, W., and Thiele, D. 1993. A sporulation gene in Coxiella burnetii. J. Vet. Med., Series B 40, 366-370.

Pan, Q., Garsin, D.A., and Losick, R. 2001. Self-reinforcing activation of a cell-specific transcription factor by proteolysis of an anti-sigma factor in Bacillus subtilis. Mol. Cell 8, 873-883.

Perego, M. 1998. Kinase-phosphatase competition regulates Bacillus subtilis development. Trends Microbiol. 6, 366-370.

Piggot, P., and Coote J.G. 1973. Genetic aspects of Bacillus subtilis: a minimal estimate of the number of sporulation operons. J. Bacteriol. 114, 908-962.

Prepiak, P., Chromikova, Z., and Barak, I. 2001. Use of yeast two-hybrid system for detection of Bacillus subtilis FtsZ protein partners. Folia Microbiologica 46, 292-296.

Quisel, J.D., Lin, D.C.H., and Grossman, A.D. 1999. Control of development by altered localization of a transcription factor in Bacillus subtilis. Mol. Cell 4, 665-672.

Raskin, D.M. and deBoer, P.A.J. 1997. The MinE ring: An FtsZ-independent cell structure required for selection of the correct division site in Escherichia coli. Cell 91, 685-694.

Raskin, D.M., and de Boer, P.A.J. 1999a. Rapid pole-to-pole oscillation of a protein required for directing division to the middle of Escherichia coli. Proc. Natl. Acad. Sci. USA 96, 4971-4976.

Raskin, D.M., and deBoer, P.A.J. 1999b. MinDE-dependent pole-to-pole oscillation of division inhibitor MinC in Escherichia coli. J. Bacteriol. 181, 6419-6424.

Ricca, E., Cutting, S., and Losick, R. 1992. Characterization of bofA, a gene involved in intercompartmental regulation of pro-σ^K processing during sporulation in Bacillus subtilis. J. Bacteriol. 174, 3177-3184.

Rong, S., Rosenkrantz, M.S., and Sonenshein, A.L. 1986. Transcriptional control of the Bacillus subtilis spoIID gene. J. Bacteriol. 165, 771-779.

Rowland, S.L., Fu, X., Sayed, M.A., Zhang, Y., Cook, W.R., and Rothfield, L.I. 2000. Membrane redistribution of the Escherichia coli MinD protein induced by MinE. J. Bacteriol. 182, 613-619.

Satola, S., Kirchman, P.A., and Moran, C.P. 1991. Spo0A binds to a promoter used by σ^A RNA-polymerase during sporulation in Bacillus subtilis. Proc. Natl. Acad. Sci. USA 88, 4533-4537.

Seavers, P.R., Lewis, R.J., Brannigan, J.A., Verschueren, K.H.G., Murshudov, G.N., and Wilkinson, A.J. 2001. Structure of the Bacillus cell fate determinant SpoIIAA in phosphorylated and unphosphorylated forms. Structure 9, 605-614.

Seshadri, R. Paulsen, I.T., Eisen, J.A., et al. 2003. Complete genome sequence of the Q-fever pathogen Coxiella burnetii. Proc. Natl. Acad. Sci. 100, 5455-5460.

Shapiro, L., and Losick, R. 1997. Protein localization and cell fate in bacteria. Science 276, 712-718.

Smith, K., and Youngman, P. 1992. Use of an integrational vector to investigate compartment specific expression of the Bacillus subtilis spoIIM gene. Biochimie 74, 705-711.

Sonenshein, A.L. Hoch, A.J., and Losick, R. 2002. Bacillus subtilis: From Cells to Genes and from Genes to Cells. ASM Press, Washington, D.C.

Stephenson, K. and Hoch, J.A. 2002. Evolution of signalling in the sporulation phosphorelay. Mol. Microbiol. 46, 297-304.

Stragier, P. 2002. A gene odyssey: Exploring the genomes of endospore-forming bacteria. In: Bacillus subtilis: From Cells to Genes and from Genes to Cells. Sonenshein, A.L. Hoch, A.J., and Losick, R., ed. ASM Press, Washington, D.C. p. 519-525.

Stragier, P., Bonamy, C., and Karmazyn-Campelli, C. 1988. Processing of a sporulation sigma factor in Bacillus subtilis – how morphological structure could control gene expression. Cell 52, 697-704.

Stragier, P., and Losick, R. 1996. Molecular genetics of sporulation in Bacillus subtilis. Ann. Rev. Gen. 30, 297-341.

Strauch, M., Webb, V., Spiegelman, G., and Hoch, J.A. 1990. The Spo0A protein of Bacillus subtilis is a repressor of the abrB gene. Proc. Natl. Acad. Sci. USA 87, 1801-1805.

Strauch, M.A., Trach, K.A., Day, J., and Hoch, J.A. 1992. Spo0A activates and represses its own synthesis by binding at its dual promoters. Biochimie 74, 619-626.

Strauch, M.A., Wu, J.J., Jonas, R.H., and Hoch, J.A. 1993. A positive feedback loop controls transcription of the Spo0F gene, a component of the sporulation phosphorelay in Bacillus subtilis. Mol. Microbiol. 7, 967-974.

Thomaides, H.B., Freeman, M., El Karoui, M., and Errington, J. 2001. Division site selection protein DivIVA of Bacillus subtilis has a second distinct function in chromosome segregation during sporulation. Genes Dev. 15, 1662-1673.

Trach, K., Burbulys, D., Strauch, M., Wu, J.J. and others 1991. Control of the initiation of sporulation in Bacillus subtilis by a phosphorelay. Res. Microbiol. 142, 815-823.

Wakeley, P.R., Dorazi, R., Hoa, N.T., Bowyer, J.R., and Cutting, S.M. 2000. Proteolysis of SpoIVB is a critical determinant in signalling of pro-σ^K processing in Bacillus subtilis. Mol. Microbiol. 36, 1336-1348.

Weir, J., Predich, M., Dubnau, E., Nair, G., and Smith, I. 1991. Regulation of spo0H, a gene coding for the Bacillus subtilis σ^H factor. J. Bacteriol. 173, 521-529.

Woldringh, C.L., Mulder, E., Huls, P.G., and Vischer, N. 1991. Toporegulation of bacterial division according to the nucleoid occlusion model. Res. Microbiol. 142, 309-320.

Wu, J.J., Piggot, P.J., Tatti, K.M., and Moran, C.P. 1991. Transcription of the Bacillus subtilis spoIIA locus. Gene 101, 113-116.

Wu, J.J., Schuch, R., and Piggot, P.J. 1992. Characterization of a Bacillus subtilis sporulation operon that includes genes for an RNA-polymerase sigma-factor and for a putative dd-carboxypeptidase. J. Bacteriol. 174, 4885-4892.

Wu, L.J., and Errington, J. 1994. Bacillus subtilis SpoIIIE protein required for DNA segregation during asymmetric cell division. Science 264, 572-575.

Wu, L.J., and Errington, J. 1997. Septal localization of the SpoIIIE chromosome partitioning protein in Bacillus subtilis. EMBO J. 16, 2161-2169.

Wu, L.J., and Errington, J. 1998. Use of asymmetric cell division and spoIIIE mutants to probe chromosome orientation and organization in Bacillus subtilis. Mol. Microbiol. 27, 777-786.

Wu, L.J., and Errington, J. 2003. RacA and the Soj-Spo0J system combine to effect polar chromosome segregation in sporulating Bacillus subtilis. Mol. Microbiol. 49, 1463-1475.

Wu, L.J., Feucht, A., and Errington, J. 1998. Prespore-specific gene expression in Bacillus subtilis is driven by sequestration of SpoIIE phosphatase to the prespore side of the asymmetric septum. Genes Dev. 12, 1371-1380.

York, K., Kenney, T.J., Satola, S., Moran, C.P., Poth, H., and Youngman, P. 1992. Spo0A controls the σ^A dependent activation of Bacillus subtilis sporulation specific transcription unit spoIIE. J. Bacteriol. 174: 2648-2658.

Zhang, Y., Rowland, S., King, G., Braswell, E., and Rothfield, L. 1998. The relationship between hetero-oligomer formation and function of the topological specificity domain of the Escherichia coli MinE protein. Mol. Microbiol. 30: 265-273.

Zhao, H., Msadek, T., Zapf, J., Madhusudan, Hoch, J.A., and Varughese, K.I. 2002. DNA complexed structure of the key transcription factor initiating development in sporulating bacteria. Structure 10: 1041-1050.

Zupancic, M.L., Tran, H., and Hofmeister, A.E.M. 2001. Chromosomal organization governs the timing of cell type specific gene expression required for spore formation in Bacillus subtilis. Mol. Microbiol. 39: 1471-1481.

Chapter 6

The Functional Architecture and Assembly of the Spore Coat

Adriano O. Henriques, Teresa V. Costa, Lígia O. Martins, and Rita Zilhão

SUMMARY

Bacterial endospores are shielded against noxious chemicals and peptidoglycan-breaking enzymes by a protein coat that is also required for efficient spore germination. In most *Bacillus* and *Clostridia* species, the coat obeys to a common architecture, with three main layers: an amorphous undercoat that contacts the underlying cortex peptidoglycan, a lamellar lightly-staining inner layer, and a striated electrodense outer layer. Most species to species structural variation relates to the presence or absence of exosporia and of various surface appendages. The coat is mainly composed of proteins whose assembly is controlled at several levels: *i*) by the time, level and site of expression of the various genes involved; *ii*) by the action of a group of morphogenetic proteins, that guide assembly of the structural components; *iii*) by various post-transcriptional and post-translational mechanisms. The result is a stable yet dynamic structure that senses and responds to nutritional cues. Transcriptional profiling studies, proteomics, cell biological techniques, and the structural characterization of selected components is illuminating the assembly process, and expanding the repertoire of tools to manipulate the composition, structure, and properties of the spore coat in view of specific applications in biotechnology or biomedicine.

INTRODUCTION

Most genetic and biochemical studies on the structure, assembly, and function of the spore coat have been confined to the model organism *B. subtilis*, in part because of the large array of sophisticated genetic tools available for its manipulation, and also because its genome was the first from a spore-forming organism to be sequenced and annotated (Kunst et al., 1997). However, other paradigms for studies of coat assembly are emerging, most notably *B. anthracis* (Liu et al., 2004; Lai et al., 2003; see below).

The genomes of six other endospore formers are now known, *B. anthracis* (Read et al., 2002; Read et al., 2003), *B. cereus* (Ivanova et al., 2003), *B. halodurans* (Takami et al., 2000), *Oceanobacillus iheyensis* (Takami et al., 2002), *Clostridium acetobutylicum* (Nolling et al., 2001), and *C. perfringens* (Shimizu et al., 2002). The discussion below is mostly dedicated to a review of the literature available for *B. subtilis* but references to other spore-formers are made, when appropriate. No attempts were made, however, to provide an exhaustive comparative genomics analysis. The reader is also referred to other recent reviews on coat assembly (e.g., Driks, 1999; Henriques and Moran, 2000; Driks, 2003). Spore germination is also important from the point of view both of its connection to the process of spore construction, and for certain applications of spores. Although references are made to several germination loci, this topic is also not covered exhaustively here (see Moir, 2003; Moir et al., 2002, for recent reviews).

COMPOSITION AND STRUCTURE

Overall Structure

Spores consist of a central compartment, the spore core, which contains a copy of the genome, and is delimeted by the prespore inner membrane (Piggot and Coote, 1976; Driks, 1999; Henriques and Moran, 2000) (Figure 1A). The inner prespore membrane is the site of assembly of a thin layer of peptidoglycan called the primordial germ cell-wall, which will originate the cell wall of the newly-formed vegetative cell upon spore germination. The germ cell is surrounded by a much thicker and chemically distinct form of peptidoglycan (PG), called the cortex, which is essential for the development and maintenance of heat resistance. The outermost spore structure is a multilayered protein shell, or coat, which protects the cortex from enzymatic attack, confers protection against organic solvents, and oxygen-reactive species, UV light, is essential for spore germination and probably has a number of other roles in the survival and ecological interactions of spores in natural ecosystems (Driks, 1999; Henriques and Moran, 2000; Nicholson et al., 2000; see also Chapter by W. Nicholson, this volume). Mutants that form heat-resistant but lysozyme-susceptible spores have specific lesions of the coat. Some coat mutants also show various degrees of deficiency in spore germination (reviewed by Driks, 1999; Henriques and Moran, 2000). When spores are viewed by thin-sectioning transmission electron microscopy, the coat evidences three layers: an amorphous, darlkly-staining undercoat in contact with the underlying cortex, an intermediate lamellar layer called the inner coat and, closely apposed to it, a thick, electrodense and striated outer coat (Figure 1A). *Bacillus* and *Clostridia* spores all show these main spore structures, but differ to varying extents in the thickness and appearance of the cortex and coat protective layers (Barbosa et al., this volume). They also differ in additional structures that decorate the spore surface, which may include an exosporium, and/or other appendages (see also Chapter by A. Driks).

The coat is mainly formed of protein, but also contains some polysaccharide (Pandey and Aronson, 1979). The collection of proteins that can be solubilized from purified spores varies greatly from species to species, from over 40 in the case of *B.*

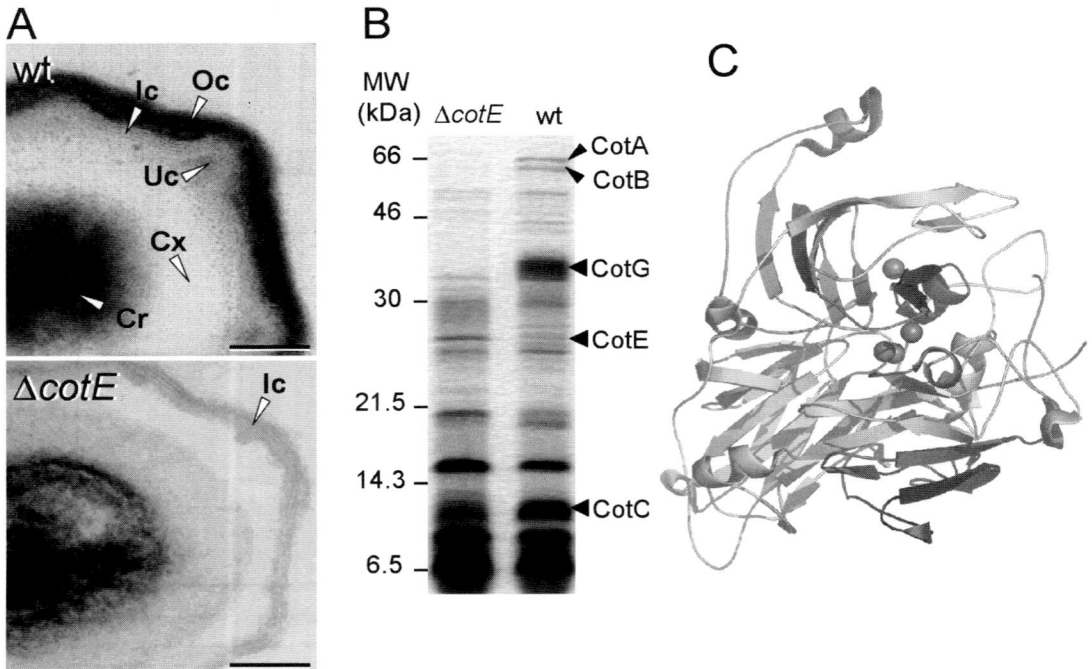

Figure 1. The figure illustrates the ultrastructure (Panel A) of wild type and *cotE* mutant spores. Note the absence of the outer coat structure in the mutant. Panel B shows the collection of proteins that can be extracted from wild type or *cotE* mutant spores following treatment with detergents and reducing agents, and resolved in one-dimensional SDS-PAGE gels. The position of several known outer coat components is indicated. Panel C shows the crystal structure of the outer coat protein CotA, a highly thermo stable laccase. Cr, spore core; Cx, spore cortex; Uc, under coat; Ic, inner coat; Oc, outer coat. Scale bar, 0.2 μm.

subtilis (see below), to only a few major proteins. All *Bacillus* spores are heat resistant, although to various extents (Barbosa et al., this volume), and resistant to lysozyme treatment, which raises intriguing questions regarding the need for the more complex coats. This complexity suggests that the coats have yet to be discovered functions. Possibly, the coat composition reflects in part the complex and for the most part unknown ecological interactions of spores in different environments (Nicholson et al., 2000; Nicholson, 2000; see also the Chaper by W. Nicholson in this volume). It is worthy of note that for a given species the spore coat composition appears to vary with the nutritional conditions during which sporulation occurred, or with other environmental factors such as temperature (Melly et al., 2002). This suggests that the spore surface can be adjusted in response to particular environments, a notion that is also supported by an emerging pattern of complex transcriptional regulation of the coat genes, as well as by the apparent extreme metabolic plasticity of the mother cell, which nurtures spore formation (see below).

In sharp contrast with the traditional view of spores as static structures, waiting for the restoration of growth-promoting conditions, spores have recently been shown to have a remarkable dynamic behaviour. Spores of *Bacillus thuringiensis* are capable or rapid contraction and expansion in response to changes in relative humidity without breaking dormancy, a behavior that may prepare the spore's reponse to germination signals (Westphal et al., 2002; Driks, 2002).

In *B. subtilis* and possibly other spore-forming organisms, the differentiation between inner coat and outer coat has a loose but operationally useful genetic basis. Spores of a *cotE* mutant fail to assemble the outer coat structure (Figure 1A) and are highly susceptible to lysozyme treatment, on which basis the outer coat has been viewed as the main barrier against peptidoglycan-breaking enzymes (Zheng et al., 1988). However, *cotE* spores are also deficient in germination, and have subtle lesions in the inner coat layers, presumably because CotE also recruits proteins to the inner coat. Nevertheless, proteins missing from *cotE* spores have been often assigned to the outer coat layers (Figure 1B). Conversely, cells mutant for *gerE*, which encodes a key transcriptional regulator of late coat gene expression, form lysosyme-sensitive spores that lack all the morphological features normally associated with the inner coat (Moir et al., 1981). On this basis, *gerE*-dependent assembly has been often interpreted as indicating a inner coat localization, provided that the corresponding locus is not dependent on *gerE* for expression. However, *gerE* mutants are deficient in the transcription of several genes encoding abundant outer coat (i.e., CotE-dependent) proteins, and hence *gerE* spores have greatly altered outer coats. A *gerE/cotE* double mutant lacks all discernible signs of inner and outer coat, but may still retain some of the undercoat material around the mostly exposed cortex (Driks, 1999).

THE COAT COMPONENTS

In *B. subtilis*, the coat is formed by the ordered assembly of over 40 polypeptides, ranging in size from about 6 to more that 70 kDa (Driks, 1999; Henriques and Moran, 2000) (Figure 1B, and Table 1). *B. anthracis* seems to produce a coat with a very

Table 1. *B. subtilis* coat proteins and orthologs in other spore-formers.

Protein[a]	Comment	Accession number	No. of residues	MW (kDa)	Paralogs[b]	Orthologs[b]								
						B. a.[c]	*B. c.*[d]	*B. h.*[e]	*C. p.*[f]	*C. a.*[g]	*C. t.*[h]	*C. th.*[i]	*O. i.*[j]	*T. t.*[k]
CotA	OC	BG10490	513	58.3		+	+	+	-	-	-	-	+	-
CotB	OC	BG10491	380	42.8		+	+	-	-	-	-	-	-	-
CotC	OC	BG10492	118	14.6	CotU	-	-	-	-	-	-	-	-	-
CotD	IC	BG10493	75	8.7		+	+	+	-	-	-	-	+	-
CotE	OC	BG10494	181	20.8		+	+	+	-	-	-	-	+	-
CotF	OC	BG10012	160	18.6	YraD; YraF	+	-	+	-	+	-	-	+	-
CotG	OC	BG11017	195	23.8		-	+	-	-	-	-	-	-	-
CotH	IC/OC	BG11791	362	42.6	YisJ	+	+	-	-	-	-	+	+	-
CotI (YtaA)	-	BG13821	357	41.1	CotS; YutH	+	+	+	+	+	+	+	+	-
CotJA	IC	BG11799	82	9.6		+	+	-	-	-	-	-	-	-
CotJB	-	BG11800	100	11.6		+	+	-	-	+	-	-	+	+
CotJC	IC	BG11801	189	21.6	YdbD; YjqC	+	+	+	-	+	+	+	+	+
CotM	-	BG11822	130	15.1		+	+	-	-	-	-	-	-	-
CotP	-	BG12167	143	16.9		-	-	-	-	-	-	-	-	-
CotQ (YvdP)	-	BG12424	447	49.9	YgaK	+	+	-	+	-	-	-	+	-
CotR (YvdO)	-	BG12423	320	35.2		+	-	-	-	-	-	-	+	-
CotS	IC	BG11380	351	40.9	CotI; YutH	+	+	+	+	+	+	+	-	-
CotSA	-	BG11381	337	42.8	YtcC; YveN; YqgM; YpjH	+	+	+	+	+	+	+	+	+
CotT	IC	BG10495	107	12.9		-	-	-	-	-	-	-	-	-
CotU (YnzH)	-	BG13471	86	11.4	CotC	+	-	-	-	-	-	-	-	-
CotV	OC (Ins)	BG10496	128	14.1		-	-	+	-	-	-	-	+	-
CotW	OC (Ins)	BG10497	105	12.2		-	-	-	-	-	-	-	+	-
CotX	OC (Ins)	BG10500	172	18.5		-	-	+	-	-	-	-	+	-
CotY	OC (Ins)	BG10498	162	17.7	CotZ	+	-	-	-	-	-	-	+	-
CotZ	OC (Ins)	BG10499	148	16.4	CotY	+	-	-	-	-	-	-	+	-
CwlJ	-	BG11172	142	16.2	SleB; YkvT	+	+	+	-	+[l]	+	+	+	+
GerQ (YwdL)	OC?	BG10608	181	20.1		+	+	+	-	-	-	-	+	-
OxdD	IC	BG13484	392	43.4	OxdC	+	+	-	-	-	-	+	-	-
SafA	Cortex/coat	BG13781	387	43.1	YhdD[l]; YojL[l]	+[l]	+	+	+[l]	+[l]	+[l]	+[l]	+	+[l]
SodA	-	BG11676	202	22.4	SodF	+	+	+	+	+	+	+	+	+
SpoIVA	OFM	BG10275	492	55		+	+	+	+	+	+	+	+	+
SpoVID	Cortex/coat	BG10346	575	64.8		+	+	+	-	-	-	-	+	-
TasA	-	BG11697	261	28.2		+	+	+	-	-	-	-	+	-
Tgl	-	BG10946	245	28.2		+	+	+	-	-	-	-	-	-
YaaH	-	BG10080	427	48.5	YdhD; YvbX; YkvQ; XlyA	+	+	+	-	+	+	+	+	+
YabG	-	BG10106	290	33.2		+	+	+	+	+	+	+	+	+
YdhD	-	BG12181	439	48.8	YaaH; YvbX; YkvQ; YhdD[l]; YojL[l] YvfQ; YtxM	+	+	+	-	+	+	+	+	+
YkuD	-	BG13288	164	17.6	YqjB; YhdD[l]	+	+	+	-	+	+	+	-	-
YkvP	-	BG13318	399	46.1	CgeB; YhdD[l]	+[l]	-	+	-	+	-	+	+[l]	+[l]

[a]) The list is not exhaustive.
[b]) NCBI BLASTP, E value ≤ 10^{-4}
[c]) *Bacillus anthracis* A2012
[d]) *Bacillcus cereus* ATCC14579
[e]) *Bacillus halodurans*
[f]) *Clostridium perfringens* str.13
[g]) *Clostridium acetobutylicum* ATCC13124
[h]) *Clostridium tetani* E88
[i]) *Clostridium thermocellum* ATCC27405 (unfinished genome)
[j]) *Oceanobacillus iheyensis* HTE831
[k]) *Thermoanaerobacter tengcongensis*
[l]) Similarity restricted to the CWB motif
OC, outer coat; IC, Inner coat; Cortex/coat, interface between cortex and coat; OFM, outer forespore membrane; Ins, coat insoluble fraction.

similar protein composition (Lai et al., 2003), while in other species the coat appears simpler. For example, a single main protein of about 20 kDa is extracted from the coats of *B. clausii* spores, and *B. megaterium* spores have a major protein of some 26 kDa (see Barbosa et al., this volume). Most spore coat proteins have been identified by reverse genetics, following extraction and electrophoretic resolution of coat proteins, either by N-terminal sequence analysis or by mass spectrometry (MS)-based approaches (Donovan et al., 1987; Zheng et al., 1988; Zhang et al., 1993; Cutting et al., 1991; Aronson et al., 1989; Abe et al., 1993; Henriques et al., 1997; Henriques et al., 1998; Kuwana et al., 2002; Driks et al., 2003; Liu et al., 2004; Costa et al., 2004; see the following section). Both proteins from wild type spores, or proteins that are misassembled and more extractable from spores with specific coat lesions have been identified in this manner (Bauer et al., 1999; Little and Driks, 2001; Serrano et al., 1999).

Genetic screens for sporulation mutants (Piggot and Coote, 1992; Roels et al., 1992; Stevens et al., 1991), or screens for mother cell or prespore-specific loci (Beall et al., 1993; Henriques et al., 1995; Bagyan et al., 1996; Feucht et al., 2003; Eichenberger et al., 2003), together with their functional analysis and subcellular localization studies have also lead to the identification of several genes coding for coat structural components, or genes involved in the control of the assembly process. Lastly, coat genes have also been found by biopanning random peptide libraries for tight binders to purified coat proteins (Ozin et al., 2000).

The role of most individual coat structural proteins in assembly and function of the spore coat is unclear, as null mutations often do not cause a measurable phenotypic effect at least under Laboratory conditions (Driks, 1999; Henriques and Moran, 2000; Nicholson et al., 2000; Nicholson, 2002; see also the Chapter by W. Nicholson, this volume). This suggests extensive redundancy or minor contributions of the various components to the structure and function of the coat layers as normally assessed. The relative harsh conditions needed to extract proteins from the spore coats (Driks, 1999; Henriques and Moran, 2000) have precluded detailed biochemical studies, but ultimately, a description of the assembly process that also accounts for the coat properties will require the detailed functional and structural characterization of selected components. For instance CotA (Donovan et al., 1987), a component of the spore outer coat layers (Zheng et al., 1988), is a highly thermostable laccase (Hullo et al., 1999; Martins et al., 2000; see also below). Remarkably the enzymatic properties of the recombinant enzyme expressed and purified from *E. coli* cells, or of the coat-associated enzyme are indistinguishable (Martins et al., 2000). The crystal structure of CotA was determined (Figure 1C), and will serve as a platform for the detailed analysis of its assembly (Enguita et al., 2003; see also below).

Structural Proteomics

Proteomics has been recently applied to the study of spore proteins in both *B. subtilis* and *B. anthracis* (Kuwana et al., 2002; Lai et al., 2003). In their study, Kuwana et al. (2002) have used a combination of one-dimensional SDS-PAGE resolution of proteins extracted from purified spores, and liquid chromatography coupled to tandem mass MS. They carried out the systematic identification of sequences present in gel slices ranging from 6 to 66 kDa, and found a total of 154 spore proteins, of which 69 were novel. Of the previously identified proteins, many are known to be associated with the spore core, spore membranes or cortex (Kuwana et al., 2002). Based on the reasonable assumption that proteins synthesized under the control of the prespore-specific regulators σ^F and σ^G should localize in the spore core, the prespore inner membrane or the cortex, whereas genes under the control of σ^E and σ^K should code for cortex, prespore outer membrane, or coat proteins (however, see below), these authors examined the expression of several genes, using *lacZ* fusions. They found 10 σ^E-dependent genes, 4 of which coded for proteins with N-terminal signal sequences, and 8 σ^K-controlled genes, and therefore a total of 14 putative new coat proteins. Among the genes in the σ^K-dependent group were *ytaA*, which was found to code for a coat protein in an independent study and re-named *cotI* (Lai et al., 2003), *ytxO*, which is the third gene in the *cotS* operon (Takamatsu et al., 1998; Takamatsu et al., 1999), and *ywrJ*, which is downstream of, and co-transcribed with the gene for a well characterized coat protein (Donovan et al., 1987; Eichenberger et al., 2003; Zilhão et al., 2004).

In another study, Lai et al (2003) have used a combination of one- and two-dimensional gel electrophoresis of spore extracts, followed by matrix-assisted laser desorption ionization-time of flight and/or dual MS to identify spore proteins in *B. subtilis* and *B. anthracis*. They identified 38 spore proteins in *Bacillus subtilis*, 12 of which were known components of the spore coat. They proposed that 3 of the new proteins CotI (YtaA), CotQ (YvdP), and CotU (YnzH) should be considered bona-fide coat components, and 14 other as potential coat proteins. With the exception of CotI (YtaA), none of these prospective coat proteins were found by Kuwana et al. (2002). Therefore, the two studies combined allowed the identification of no less than 29 potential new coat proteins. Lai et al. (2003) also identified 11 spore proteins in *B. anthracis*, 6 of which associated with the coat or exosporium. Their analysis indicates that both species have very similar coats, which share a core group of proteins, including the major coat morphogenetic proteins known for *B. subtilis* (see also below). Some of the putative structural components identified in *B. anthracis* were homologues of inner coat (CotJC), outer coat (CotB), or proteins of the insoluble, cross-linked coat fraction of *B. subtilis* (CotZ) (see below), suggesting that the composition of these main coat fractions may to some extent be conserved (Lai et al., 2003). Both the studies of Kuwana et al. (2002), and Lai et al. (2003) provide evidence for the involvement of processing and cross-linking in the posttranslational modification of coat proteins (see also below).

A recent study has related global changes in gene expression during growth and sporulation of *B. anthracis* to its spore proteome (Liu et al., 2004). Proteins were identified by multidimensional chromatography and tandem MS in several spore fractions, including a total spore fraction, a fraction enriched for the exosporium, and a nude spore fraction (i.e., a spore stripped of the exosporium), which was further divided into a soluble and an insoluble fraction (Liu et al., 2004). A total of 744 proteins were identified in the nude spore fraction, and 137 in the exosporium, including 21 that may be unique to this structure. Some of the most abundant proteins among these were two homologues of *B. subtilis* CotZ, a CotJC-like catalase, CotJA and CotJB, superoxide dismutase (SodA), alanine racemase, and hydrolases, including a homologue of the *B. subtilis* cortex-lytic enzyme CwlJ (below). In *B. subtilis* all these proteins are associated with the spore coat (see below). Other studies have also found evidence for the presence of other homologues of *B. subtilis* coat proteins in the exosporium of *B. cereus* and *B. anthracis*. For instance CotY, which is functionally related to CotZ, CotE, SodA, CotB (Steichen, et al., 2003; Baillie et al., 2004; Todd et al., 2003; see also Chapter by A. Driks). It is still early to fully comprehend and explore the data in the monumental study of Liu et al. (2004), but their work will evidently be used as an important guide not only for studies on the control of gene expression and spore morphogenesis in other spore-formers, but also in detection, prevention, and treatment of anthrax.

SYNTHESIS OF THE COAT PROTEINS

Transcriptional Control

The Main Transcriptional Regulators of Coat Gene Expression

Expression of genes involved in coat assembly spans a period of some 6-8 hours during the process of spore formation, starting with the asymmetric division of the cell at the onset of sporulation and culminating with the lysis of the mother cell and release of the mature spore (Driks, 1999; Henriques and Moran, 2000; Figure 2A). During this period, the temporal window and

level of expression of a large number of genes that contribute to the process must be carefully controlled. With so far only two possible exceptions (Serrano et al., 1999; Kakeshita et al., 2000; Kakeshita et al., 2001), expression of the genes involved in spore coat assembly is driven by a cascade of mother cell-specific regulators, in the order σ^E, SpoIIID, σ^K, and GerE (reviewed by Driks, 1999; Henriques and Moran, 2000; Errington, 2003, Chapter 5, this volume). σ^E becomes active in the mother cell soon after asymmetric division of the sporulation cell, whereas σ^k is activated only following the complete engulfment of the prespore by the mother cell (Cutting et al., 1990; reviewed by Errington, 2003) (Figure 2A). The cascade is hierarchical, in that each regulator drives expression of the next transcription factor (Cutting et al., 1989; Zheng and Losick, 1990). SpoIIID and GerE are small DNA-binding proteins that modify the specificity of σ^E- and σ^K-containing RNA polymerase, respectively, and both function to activate or repress transcription (Zheng et al., 1992; Ichikawa et al., 1999; Kroos et al., 1989; Halberg et al., 1994; Halberg et al., 1995). The effects of SpoIIID and GerE on σ^E- and σ^K-dependent gene expression divide the mother cell line of gene expression into four temporally distinct classes (Zheng and Losick, 1990) (Figure 2B). Feedback mechanisms help driving the proper temporal progression in the cascade. For example, GerE negatively regulates expression of the gene *sigK*, encoding σ^K, and of genes encoding structural components, e.g. *cotA*, which are expressed under σ^K control in the previous temporal class of gene expression (Zheng et al., 1992; Sandman et al., 1988). Also, activation of σ^K generates a signal that represses transcription of the gene for σ^E, and hence the production of SpoIIID (Zhang and Kroos, 1997). σ^E drives expression of several important morphogenetic regulators of coat assembly (see below), as well as the genes for some structural components of the coat, whereas σ^K drives expression of most coat structural genes (Figure 2B). GerE controls the production of the CotB, CotC and CotG proteins, which are

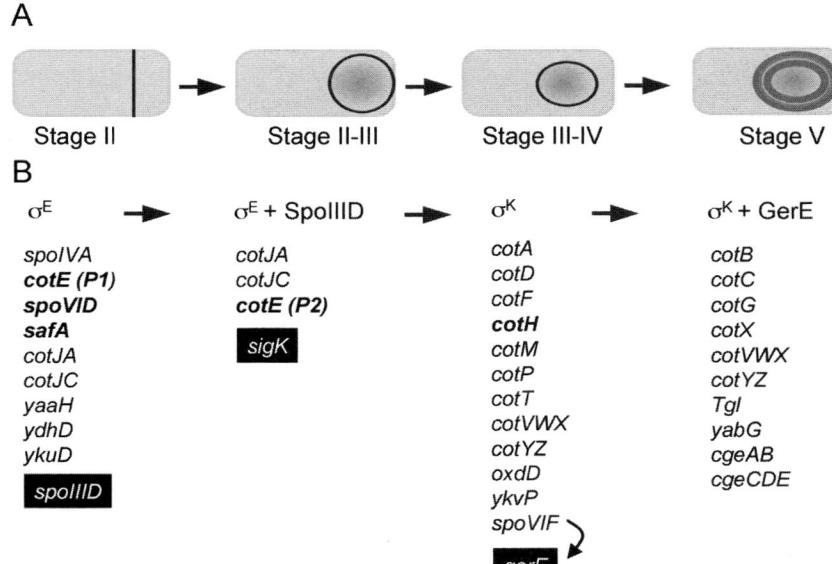

Figure 2. Synthesis of the coat proteins is initiated soon after the asymmetric division at the onset of sporulation that produces the prespore (PS) and the larger mother cell (MC). With the possible exception of TasA and FtsY (see text), synthesis of the coat proteins is governed by the early and late mother cell-specific RNA polymerase sigma factors σ^E and σ^K, respectively. Gene expression in the σ^E and σ^K regulons is further devided into temporal classes by the action of SpoIIID and GerE. Transcriptional regulators are indicated by the black boxes, and the genes for the morphogenetic proteins are in bold. SpoVIF negatively regulates the expression of *gerE*, but it is not known whether it acts at the transcriptional level. Representatives of the four temporal classes of coat gene expression are shown.

among the most abundant outer coat components (Naclerio et al., 1996; Zhang et al., 1990; Sacco et al., 1995). The crystal structure of GerE has been solved (Ducros et al., 2001), and the regions in σ^K and GerE important for promoter activation have been defined (Wade et al., 1999; Crater and Moran, 2001; Crater and Moran, 2002).

Additional Regulators

One study using microarrays has placed a putative transcription factor of the Myb family, encoded by *ylbO*, in the σ^E regulon (Eichenberger et al., 2003; see below). YlbO has a counterpart in the prespore, the RsfA protein, which was shown to modulate gene expression in this compartment (Wu and Errington, 2000). A different microarray analysis (Feucht et al., 2003) evidenced that the expression of a gene (*ytzE*) encoding a putative transcription factor of the DeoR family is σ^E-dependent. The analysis of inactivating mutations in the *ylbO* and *ytzE* genes has not yet been reported.

The recent analysis of the *spoVIF* (*yjcC*) gene, adds complexity to the story (Kuwana et al., 2003; Kuwana et al., 2004). *spoVIF* mutants form spores that lack both the inner and outer coat layers (Kuwana et al., 2003). Overall, the spores resembled those produced by a *gerE* mutant, and as *gerE* spores they were susceptible to heat and lysozyme (Kuwana et al., 2003). Like *gerE*, the *spoVIF* gene is under σ^K control (Kuwana et al., 2003). In *spoVIF* mutants, transcription of the *gerE*-dependent genes *cgeA*, *cotG*, and *cotS* was reduced, as was the accumulation of GerE (Kuwana et al., 2004). These observations have led to the suggestion that SpoVIF may directly control the transcription of *gerE* (Kuwana et al., 2004), adding a further level of control to the late stages of mother cell-specific gene expression (Figure 2B). Further work is needed to verify whether SpoVIF is a transcription factor, or whether it affects *gerE*-dependent gene expression by some other mechanism. The mother cell appears to have great metabolic plasticity (Eichenberber et al., 2003; Feucht et al., 2003; Mekjian et al., 1999). It could be that the existence of additional regulators helps coordinating the mother cell metabolism and morphogenesis to the available resources. One well-documented example of this regards the utilization of hexuronate in *B. subtilis*. Expression of the *exu* operon is under the control the main vegetative σ factor, σ^A and is repressed by galacturonate via the ExuR repressor (Mekjian et al., 1999). However, the *exu* operon is also transcribed form an internal σ^E-dependent promoter (Mekjian et al., 1999). The σ^E promoter could ensure that expression of ExuR, so that the operon would only be expressed in the presence of galacturonate (Mekjian et al., 1999). It is not known whether ExuR controls the expression of additional loci.

Proper coat formation additionally requires SpoVT, a prespore-specific regulator that modulates expression of σ^G-controlled genes (Bagyan et al., 1996). *spoVT* mutants have a thin poorly defined outer coat, which forms swirls of material that remains attached to the spore, but project into the mother cell cytoplasm (Bagyan et al., 1996). The spores also presented reduced heat and lysozyme resistance (Bagyan et al., 1996). Possibly, SpoVT somehow controls the expression of mother cell-specific genes required for coat formation. Alternatively,

the expression of certain genes involved in coat assembly takes place in the prespore, and is altered in the *spoVT* mutant. If so, these genes could influence the assembly process from the prespore, or their products could somehow be transported out of the prespore to the site of coat assembly.

The Importance of Temporal Control

The number of proteins that are assembled into the spore coat at different stages in the process is high, and growing everyday. Several reports have shown that the coat proteins are to some extent capable of self-assembly (e.g., Aronson and Fitz-james, 1971; Goldman and Tipper, 1978; Aronson et al., 1992). This raises the question of how the cells manage to prevent inappropriate and unproductive interactions among the various coat proteins during assembly. One member of the α-crystallin family of small heat shock chaperons, YocM is expressed as part of the σ^E regulon (Eichenberger et al., 2003), and two more, CotM and CotP, are under σ^K control (Henriques et al., 1997; Reischl et al., 2001). However, even though CotM has been shown to be spore-associated (Lai et al., 2003), and to participate in outer coat formation, neither single mutant nor the double mutant shows a phenotype that supports their role as chaperones (Henriques et al., 1997; Reisch et al., 2001).

In principle, inappropriate interactions could be prevented if only specific complexes can undergo assembly, or if the proteins only assume their final form as the result of posttranslational modifications occurring only at the spore surface. Indeed, certain proteins which are encoded in operons or clusters of functionally-related genes appear to be co-assembled as pre-formed complexes (Henriques et al., 1995; Seyler et al., 1997; Zhang et al., 1993; Takamatsu et al., 1998; Tkamatsu et al., 1999), and in several cases cross-linking or processing has been detected at the spore surface (discussed below). Yet another mechanism is to ensure that the critical level of a particular component is only attained following the assembly of another, and this could be achieved by carefully adjusting the time and level of expression of each of the intervening genes. The assembly process may take place mostly in a sequential manner, with some proteins being recruited only after the assembly of others (Jenkinson et al., 1981). The following examples serve to stress the importance of the transcriptional control for proper assembly and function of the spore coat. In cells bearing *bof* alleles, the activation of σ^K is uncoupled from the activity of σ^G, and occurs about 30 to 60 minutes earlier than normal (Cutting et al., 1990). As a consequence, expression of the σ^K regulon is premature, and the resulting spores show deficient germination (Cutting et al., 1990). Another example comes from studies on the assembly of CotB (see also below). Most of *cotB* transcription is dependent on σ^K and GerE, and hence *cotB* belongs to the last known temporal class of coat gene expression (Eichenberger et al., 2003; Zheng and Losick, 1990). A strain in which *cotB* is expressed from the *cotE* P1 promoter, and hence as part of the first wave of coat gene expression, forms spores which germinate slowly, even though spores of a *cotB* insertional mutant responded normally to the germinant (Donovan et al., 1987; Zilhão et al., 2004).

Microarray Analysis

The composition of the σ^E regulon of *B. subtilis* has been recently investigated using DNA microarrays analysis (Eichenberger et al., 2003; Feucht et al., 2003). Eichenberger et al. (2003) found 253 genes in the σ^E regulon, organized in 157 operons, 181 of which had not been previously assigned to σ^E control. Feucht et al. (2003) found 178 genes in the σ^E regulon, 130 of which in operons. In confirmation of earlier studies, expression of the genes for coat proteins CotE, CotJABC, CwlJ, SafA, YaaH, SpoVID, and YdhD, as well as that of the CotZ, previously shown to be expressed from a σ^K-dependent promoter (Zhang et al., 1994) was found to be under σ^E control (Figure 2B).

Eichenberger and co-workers (2003) found that fusions of PrkA, YhbB, YybI, YhaX, YuzC, and YjbX to GFP, localized as ring-like structures or caps around the developing spore, presumably in the inner coat, whereas YjbX-GFP could be in the outer coat. Their assignement was based on the dependencies of known morphogenetic reguultors of coat assembly (see below). The proteomic analysis of Kuwana et al. (2002) confirms that at least YjbX, YuzC and YybI are spore proteins. Presumably, more of the large collection of new σ^E-dependent genes will be found to encode additional coat proteins. With the exception of *Bacillus anthracis* (Liu et al., 2004; see above), genome-wide transcription profiling studies have not yet been extended to the late sporulation regulons.

CONTROL OF THE ASSEMBLY PROCESS BY MORPHOGENETIC PROTEINS

The earliest stages in coat assembly involve the action of a group of morphogenetic proteins that appear to act by preparing the surface of the developing spore for the synthesis and attachment of the coat. These proteins are thought to act by establishing a morphological imprint that guides the assembly of the coat components. Regardless of their association with the final structure, they guide the assembly of several structural components. The morphogenetic proteins synthesized early under the control of σ^E affect the assembly process more dramatically than those produced late, under the control of σ^K. The topological plan established by the σ^E-controlled proteins is essential for the correct targeting of the early- and late-produced components to the inner and outer coat layers (Figure 3).

SpoIVA and CotE

The morphogenetic protein SpoIVA is produced in the mother cell early in sporulation, under σ^E control, soon after the asymmetric division of the sporulating cell (Stevens et al., 1992; Roels et al., 1992). At this time it localizes at or very close to the mother cell surface of the division septum (Driks et al., 1994; Pogliano et al., 1995; Price and Losick, 1999). At later stages, following engulfment of the prespore by the mother cell, SpoIVA is detected at or close to the prespore outer

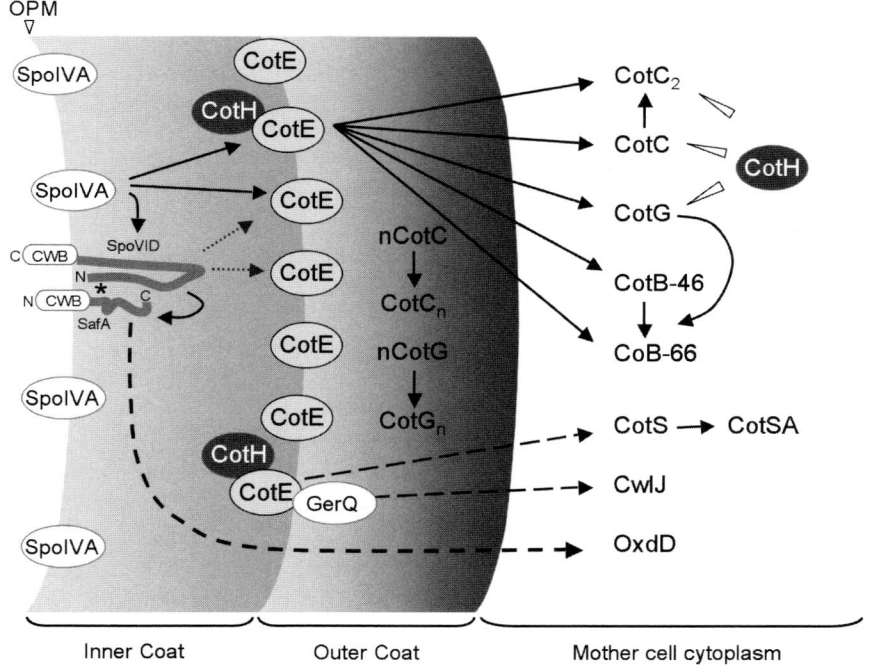

Figure 3. Structural organization of the coat layers and roles of the SpoIVA, CotE, SpoVID, and SafA morphogenetic proteins. Assembly of the coat requires the localization of the SpoIVA protein near or at the outer prespore membrane (OPM). Localization of SpoIVA is required for the assembly of CotE in a ring-like structure at a distance of about 75 nm from the site defined by the localization of SpoIVA. Maintenance of the CotE ring around the developing spore at a time (following activation of σ^K) when synthesis of most of the coat proteins is triggered requires SpoVID. The localization of SpoVID around the prespore is SpoIVA-dependent, but independent of CotE. SafA localizes around the prespore in a SpoIVA-, and SpoVID-dependent manner, and SafA and SpoVID interact (asterisk) via their N-terminal regions. The cell wall binding domains of SafA and SpoVID may promote localization of both proteins close to the cortex. CotE recruits CotH, CotG, and all forms of CotB and CotC to the outer coat, and also seems to recruit at least two inner coat proteins (CwlJ, and CotS). Assembly of CotSA requires CotS, and hence CotE. CotH appears to act in the mother cell cytoplasm to stabilize two forms (monomer and dimer) of CotC, and CotG (24 and 36 kDa forms), CotG in turn is required for the formation of CotB-66 from CotB-46 (note that it is not known whether CotB-66 is formed in the mother cell or at the spore surface). Following assembly, CotG and the forms of CotC undergo further multimerization.

membrane by immunogold labeling, and the use of a GFP fusion and deconvolution microscopy reveals that it forms a shell that completely surrounds the prespore (Driks et al., 1994; Price and Losick, 1999). Localization of SpoIVA requires another σ^E-controlled sporulation protein, SpoVM, as well as another unidentified factor (Price and Losick, 1999). The last 5 amino acid residues of the 492-residues SpoIVA protein, and at least the region centered around amino acid 393, are required for its targeting to the developing spore (Price and Losick, 1999; Catalano et al., 2001). Assembly also requires unspecified sequences elsewhere in the protein that promote correct direct or indirect interactions between SpoIVA monomers (Price and Losick, 1999). SpoIVA has a pivotal role in the assembly of both the spore cortex and coat. *spoIVA* mutants fail to form the spore cortex, and the coat accumulates as swirls of partially structured material dispersed in the mother cell cytoplasm (Piggot and Coote, 1976; Stevens et al., 1992; Roels et al., 1992). SpoIVA marks the prespore outer membrane for both its use in the synthesis of the cortex and as the site of coat attachment, but these functions may not be genetically separable (Catalano et al., 2001). The localization of SpoIVA close to the prespore outer membrane is a prerequisite for the localization of another morphogenetic protein, CotE (Zheng et al., 1988; Driks et al., 1994) (Figure 3). CotE is produced early, first from a promoter, P1, recognized by σ^E, and later from a promoter, P2, that additionally requires SpoIIID (Zheng et al., 1992) (Figure 2B). Thus, transcription of *cotE* is mostly confined to the period prior to engulfment completion and therefore, prior to the σ^K-driven production of most of the coat structural components. CotE is also targeted to the surface of the developing spore immediately after formation of the asymmetric septum, and later it fully encircles the prespore (Driks et al., 1994; Webb et al., 1995). However, CotE assembles as a ring-like structure about 75 nm from the position marked by SpoIVA (Driks et al., 1994). The space delimited by the positions of SpoIVA and CotE forms a compartment of unknown composition (but see below), in which presumably the inner coat will develop, whereas the CotE ring may serve as the site for nucleation of outer coat assembly (Driks et al., 1994) (Figure 3). Accordingly, a *cotE* mutant forms spores that still retain at least part of the inner coat, but from which the outer coat is absent (Zheng et al., 1988). These spores are susceptible to lysozyme treatment, suggesting that the outer coat is an important barrier against peptidoglycan-breaking enzymes, and are also impaired in germination (Zheng et al., 1988; see below). CotE appears to be highly modular, with regions required for its targeting to the developing coat, and for multimerization, as well as regions (towards its C-terminus) involved in the recruitment of specific structural components (Bauer et al., 1999; Little and Driks, 2001). Presumably, CotE sits at the inner coat/outer coat interface, from where it recruits several proteins via direct protein-protein interactions (Little and Driks, 2001). Among the proteins whose recruitment was shown to require specific regions of CotE are prominent outer coat components, such as CotA, but also inner coat proteins such as CotS (Little and Driks, 2001; see below). It is unclear if CotE directly recruits proteins to the inner coat layers (see also below), but the structural alterations in the coat of *cotE* spores are not

restricted to the outer coat; *cotE* mutant spores have a reduced undercoat, and a greatly expanded cortical region (Aronson et al., 1992; Serrano et al., 1999; Driks, 1999).

SpoVID and SafA

A third morphogenetic protein produced under σ^E control is SpoVID (Beall et al., 1993). SpoVID is also targeted to the surface of the developing spore soon after formation of the septum, and later encircles the prespore (Driks et al., 1994; Ozin et al., 2001). Localization of SpoVID is SpoIVA-dependent, but CotE-independent (Driks et al., 1994; Ozin et al., 2001) (Figure 3). SpoVID is not required for the formation of the CotE ring, but is needed for its maintenance around the forespore following engulfment completion, when synthesis of most of the coat structural components begins (Driks et al., 1994). In the absence of SpoVID the coat misassembles as swirls of material dispersed throughout the mother cell cytoplasm (Beall et al., 1993). This phenotype is reminiscent of the effect caused by the elimination of *spoIVA*, but *spoVID* mutants still form the cortex layer. However, because the cortex is exposed, the resulting spores are highly susceptible to lysozyme treatment (Beall et al., 1993). SpoVID is a long (575 residues) highly acidic protein that shares some similarity to neurofilament proteins. To investigate its function in the assembly process, a random peptide library was biopanned for SpoVID tight binders (Ozin et al., 2000). These studies have led to the identification of a peptide motif, PYYH, present in the C-terminal half of the 45 kDa SafA protein, which is also produced in the mother cell under σ^E control (Ozin et al., 2000; Takamatsu et al., 1999). A *safA* mutant forms spores with abnormal coat layers, which are lysozyme susceptible and impaired in germination (Ozin et al., 2000; Takamatsu et al., 1999). SpoVID and SafA form complexes during the early stages in coat assembly, and directly interact (Ozin et al., 2000; Ozin et al., 2001a). SafA was also found to interact with itself, and with a form of the protein that corresponds to its C-terminal half (SafA-C$_{30}$), which is mainly produced by internal translation (Ozin et al., 2001b; see below). Only the full-length form of SafA (SafA-FL) is required for lysozyme resistance (Ozin et al., 2001b).

SafA has a modular design. It has a cell wall-binding (CWB) motif at its N-terminus, whereas the rest of the protein is proline-rich and bears sequence similarity to inner coat proteins such as CotD, CotT, and CotJA. Interestingly, a CWB motif is also present in the C-terminus of SpoVID, and both SafA and SpoVID appear to localize to the cortex/coat interface (Driks et al., 1994; Ozin et al., 2000) (Figure 3). It seems plausible that SafA and SpoVID form complexes at the cortex/coat interface that interact with the cortex peptidoglycan via their CWB motifs, while the remaining regions of both proteins interact with other coat components (Ozin et al., 2001a). SpoVID does not interact with itself, at least as accessed using a yeast two-hybrid assay (Ozin et al., 2001a). Therefore, it is possible that several SpoVID molecules are brought together by SafA (Ozin et al., 2001a). The interactions between SpoVID and SafA may help forming the matrix that dictates assembly of the CotE ring at a distance from the prespore surface. Since the phenotype of a *safA* mutant is less drastic than that of a *spoVID* mutant, it follows that SpoVID does not act solely via SafA. At least three other proteins with CWB motifs that localize to the coats (YaaH, YdhD, and YkuD) are synthesized under σ^E control, and one (YkvP) under the control of σ^K (Kodama et al., 1999; Kodama et al., 2000). The analysis of the mutants indicates minor roles in coat assembly or spore germination (Kodama et al., 1999; Kodama et al., 2000). However, it is not known whether any of these proteins interacts with SafA or SpoVID, and the analysis of multiple CWB protein mutants has not yet been reported.

Assembly of SafA occurs in two stages. The initial targeting of SafA to the spore surface is independent of SpoIVA and SpoVID. Then, SafA encircles the prespore in a SpoIVA- and SpoVID-dependent, but CotE-independent manner (Ozin et al., 2001a). Targeting of SpoVID does not require SafA. Presumably, the localization of SafA around the prespore requires its interaction with SpoVID (Ozin et al., 2001a). Recent work has shown that the interaction of SafA with SpoVID involves the PYYH motif initially found in the screen for strong peptide binders (Ozin et al., 2000), but also a short region of 12 residues just downstream of the CWB motif in SafA. Deletion of this region results in the production of lysozyme-sensitive spores, suggesting that it constitutes the main site of interaction with SpoVID. Since SafA interacts with the first 200 amino acids of SpoVID (our unpublished results), SafA-SpoVID complex formation seems to occur mainly via the N-terminal region of both proteins, and does not seem to involve the CWB in either SafA or SpoVID. In agreement with these observations, targeting of SafA-C_{30} (which only includes the PYYP motif) to the spore surface requires full length SafA (Ozin et al., 2001b). Consistent with the notion that the interaction of SafA with SpoVID is required for its localization around the spore, both the determinants for the interaction and localization of SafA are within the N-terminal half of the protein.

Even though the interaction with SpoVID does not involve the CWB domain, the localization of SafA may be facilitated by this structural motif. However, it has not been shown directly whether the CWB domain of either SafA or SpoVID interacts with the cortex peptidoglycan. If so, the mechanism by which their CWB domains directly contact the cortex, which develops on the outer surface of the prespore outer membrane remains mysterious. Neither SafA nor SpoVID contain signal sequences for secretion.

CotH

Conclusion of the engulfment process triggers activation of σ^K, and the expression of most of the genes that code for coat structural components (Driks, 1999; Henriques and Moran, 2000). Transcription of the *cotH* gene is under the control of σ^K (Naclerio et al., 1996) (Figure 2B). The 42.8 kDa CotH protein is a minor coat component, but the coats of *cotH* spores lack several abundant coat proteins, including CotG, CotB, and CotC, and show abnormal morphology at the both the inner and outer coat levels (Naclerio et al., 1996; Zilhão et al., 1999). Moreover, a *cotH* mutation exacerbates the germination phenotype of a *cotE* mutant (Naclerio et al., 1996; Zilhão et al., 1999). Because the assembly of CotH depends on both *cotE* and *gerE*, and because of the functional linkage between the *cotE* and *cotH* loci, it has been proposed that CotH localizes at

the inner coat/outer coat interface, from where it could recruit to both coat layers (Naclerio et al., 1996; Zilhão et al., 1999). Recent results indicate that CotH may act in part in the mother cell cytoplasm, to stabilize otherwise unstable coat proteins (Isticato et al., 2004; Zilhão et al., 2004; see also the section on protein-protein interactions). CotH shows similarity to a structural component of the cellulosome (Zverlov et al., 2003), and has a paralog in *B. subtilis* (*yisJ*; Table 1) that like *cotH*, is expressed under σ^K control (our unpublished results), but whose function if any in coat assembly remains to be determined. CotH has a clear homologue in *B. anthracis* (Read et al., 2003).

POST-TRANSCRIPTIONAL AND POST-TRANSLATIONAL REGULATION OF COAT ASSEMBLY

Translational Control

Several *safA*-derived polypeptides are found in extracts of sporulating cells, including SafA-FL (45 kDa), SafA-C_{30} (30 kDa), and a form of 21 kDa (or SafA-N) derived from the N-terminal region of the full-length protein (Takamatsu et al., 1999; Ozin et al., 2000; Ozin et al., 2001b). However, only SafA-FL and or SafA-C_{30} are present or at least extractable from the spore coats. SafA-C_{30} starts with a methionine coded for by codon 164 of the *safA* gene (Takamatsu et al., 1999; Ozin et al., 2001b). This polypeptide is produced primarily by internal translation initiating at codon 164 of the *safA* transcript. First, expression of *safA* is driven from a single promoter. Second, introduction of a stop codon upstream of a predicted ribosome-binding site close to codon 164, or an alanine substitution at the first codon of *safA*, abolished formation of SafA-FL, but not of SafA-C_{30}. Lastly, accumulation of SafA-C_{30} was prevented by an alanine substitution at codon 164 (Ozin et al., 2001b). A form of SafA of about 30 kDa but distinct from SafA-C_{30} is also produced by proteolysis (Takamatsu et al., 2000; see below). Production of SafA-N presumably arises from ribosome dissociation prior to re-assembly at the internal ribosome binding-site (Ozin et al., 2001b).

The function of SafA-C_{30} is unclear, as a strain unable to produce SafA-C_{30} does not show any apparent phenotype. However, its production must be under precise regulation. A multicopy allele of *safA* bearing the alanine substitution at the first methionine codon, caused the overproduction of SafA-C_{30} in the absence of other forms of SafA. The resulting strain was severely impaired in spore formation, showed decreased σ^E-dependent gene transcription, and virtually undetected expression of the *sigK* gene (encoding σ^K), or of the σ^K-dependent *gerE* gene (Ozin et al., 2001b). In contrast, forespore-specific gene expression was not severely affected. This suggested that accumulation of SafA-C_{30} interfered with the synthesis or activity of σ^K. One possibility is that the accumulation of SafA-C_{30} in the absence of some of its interactors, such as full-length SafA, may titrate a factor that is required for the normal expression of σ^K activity. Since pro-σ^K is activated by a dedicated protease at the membrane level (see chapter by I. Barak), it is possible that abnormal assembly of SafA-C_{30} interferes with this reaction. In any event, these observations suggest a link between proper coat morphogenesis and the normal deployment of the mother cell-specific program of gene expression (Ozin et al., 2001b). The generation of multiple coat polypeptides by internal translation has so far been detected only in the case of *safA*.

Protein-Protein Interactions

The classical work of Jenkinson and co-authors (1981), produced the first dynamic view of the assembly of the coat proteins, in relation to their time of synthesis, and the development of spore resistance properties. Their work showed that synthesis of several coat proteins started at early time in sporulation, presumably following the asymmetric division of the sporulating cell, but that some of these appeared to be deposited onto the spore surface much later in the process. In addition, the assembly process appeared to be sequential, since proteins surface exposed at intermediate times during the process were successively overlaid at later times (Jenkinson et al., 1981). The basis for the assembly of a growing group of coat proteins is now beginning to be described in a molecular language. This includes their interactions and dependencies for assembly. However, in no case have the sequence or structural requirements for assembly been precisely defined. Assembly is often accompanied by posttranslational modifications such as proteolytical processing or cross-linking. In all cases documented, e.g., those of CotT, CotC, or CotG (see below), the modifications appear to occur at the surface of the developing spore.

Assembly of Proteins into the Inner Coat Layers

Direct protein-protein interactions between coat proteins have only been reported in a few cases, but so far it has not been shown directly that the interaction is required for assembly. The first example involved the product of the *cotJC* gene, the third and last cistron of an operon found in a screen for σ^E-controlled genes (Henriques et al., 1995). *cotJC* codes for a protein of 21.7 kDa with significant similarity to non-heme (type II) Mn^{2+}-dependent catalases (or pseudocatalases), found in many microorganisms in microaerophilic environments (Barynin et al., 2001 and references therein). The first (*cotJA*) and second (*cotJB*) cistrons of the *cotJ* operon encode small proteins of about 9.7 and 11.7 kDa, respectively (Henriques et al., 1995). Expression of *cotJA* is necessary for the assembly of CotJC, and in turn *cotJC* is required for the assembly of CotJA (Seyler et al., 1007). CotJA and CotJC are found in complexes at the time of coat assembly, and using a yeast two-hybrid assay, CotJA and CotJC interact with themselves and with each other (Seyler et al., 1997). Presumably, CotJA and CotJC form a complex in the mother cell cytoplasm, which is then recruited to the surface of the developing spore, while neither CotJA nor CotJC per se can undergo assembly. CotJB is also targeted to the developing spore (Kuwana et al., 2002), but is not required for the assembly of CotJA or CotJC (Seyler et al., 1997) (Figure 5). The crystal structure of the manganese catalase holoenzyme from *Lactobacillus plantarum* has recently been solved, revealing a hexamer stabilized by extensive intersubunit contacts (Barynin et al., 2001). The N- and C-terminal regions (residues 1-18, and 191-260), which flank a central domain carrying the active

site, are important in the packing of each monomer against the neighboring catalytic domains. The conserved catalytic core is conserved in CotJC, but the C-terminal domain involved in subunit interactions in the *Lactobacillus plantarum* enzyme is missing (not shown). In view of the interactions found between CotJA and CotJC, it is tempting to speculate that CotJC also forms a hexamer, and that CotJA even though unrelated at the sequence level, substitutes for the missing C-terminal domain. Possibly, the CotJA polypeptide has evolved as a module for both the multimerization of CotJC, and for the targeting of the multimeric enzyme to the developing coat. Since its assembly is detected in cells doubly mutant for *cotE* and *gerE*, the CotJA-CotJC complex is likely to be targeted to the inner coat layers, close to the outer prespore membrane (Seyler et al., 1997). At least one other type II catalase is produced during sporulation (our unpublished results), but this enzyme does not seem to need the prior interaction with a smaller polypeptide for assembly. It is not known whether this second catalase is associated with the coat, or whether the double mutant is impaired in spore resistance.

Often, interactions have been inferred from genetic data, but not verified by complimentary approaches. For example, the *cotS* operon encodes three coat components, CotSA (43 kDa), CotS (41 kDa), and Ytx0 (17 kDa), which are produced under the dual control of σ^K and GerE (Abe et al., 1993; Abe et al., 1995; Takamatsu et al., 1999; Kuwana et al., 2002). Assembly of CotSA requires expression of the downstream gene *cotS*, but not of the third cistron of the operon, *ytx0* (Takamatsu et al., 1999). Very little CotSA accumulates in cells of a *cotS* mutant, suggesting instability of the protein in the absence of assembly. In contrast to the reciprocal relationship found between CotJA and CotJC (see above), assembly of CotS does not requires CotSA (Takamatsu et al., 1999). Conceivably, the interaction of CotS with CotSA is responsible for the assembly of CotSA into the spore coat (Takamatsu et al., 1999). Both CotSA and CotS are missing from the coats of a *cotE* mutant (Takamatsu et al., 1998; Takamatsu et al., 1999). However, immunogold labeling studies have shown that CotS localizes to the inner coat (Takamatsu et al., 1998). Assembly of CwlJ is also CotE-dependent but the protein is thought to localize in the inner coat (Bagyan and Setlow, 2002; see also below). It is possible that CotE is also involved in the recruitment of proteins to this coat layer. Alternatively, some inner coat proteins are not retained in spores of a *cotE* mutant, or are degraded (Bagyan and Setlow, 2003). Studies on the assembly of the inner coat protein OxdD show that its assembly around the prespore in sporulating cells is *cotE*-independent, but that the protein is not retained in the released spores (Costa et al., 2004) (Figure 4). Presumably the absence of CotS or CwlJ from *cotE* spores is also due to their inability to remain associated with the coat of the mutant spores.

The OxdD protein, which has oxalate decarboxylase activity, is produced from a promoter recognized by σ^K and negatively regulated by GerE (Costa et al., 2004). The detailed analysis of the assembly of an enzymatically active OxdD-GFP fusion protein evidenced two phases in its assembly. First, the fusion protein was targeted to both poles of the spore in a *safA*-

dependent manner (Figure 4) and its later migration around the prespore required at least one *gerE*-dependent factor (Costa et al., 2004) (Figure 4B). This two-step pathway may be followed by other inner coat proteins. Assembly of SafA also occurs in two stages (Ozin et al., 2001a; see above). It may be that the inner coat proteins recognize specific receptors or morphological features at the spore poles and then spread around the prespore as the inner coat develops. Note that the morphological features associated with the inner coat layers are absent in *gerE* mutant spores (Moir, 1981; Driks, 1999). Spores of the *oxdD* mutant did not show any perceptible phenotype. A second oxalate decarboxylase, OxdC, is found in spores (Kuwana et al., 2002). It is not yet known whether OxdC also localizes to the coat, and whether OxdD and OxdC have redundant functions.

SafA may have a more general role in the recruitment of proteins to the inner coat layers. It is unknown whether CotT and CotD, which share similarity with the C-terminal half of SafA, CotJA-CotJC or CotS are recruited to the inner coat layers in a *safA*-dependent manner.

Assembly of Proteins into the Outer Coat Layers

Some of the most abundant coat proteins, including CotB, CotG, and CotC, appear to be components of the outer coat, on the basis of their absence from the coats of *cotE* spores (Donovan et al., 1987; Sacco et al., 1995; Zheng et al., 1988), and at least CotB is exposed at the spore surface (Duc et al., 2003; Isticato et al., 2001). Synthesis of CotB, CotG, and CotC, requires σ^K and GerE (Naclerio et al., 1996; Sacco et al., 1995; Zheng and Losick, 1990; Figure 2B). Expression of the σ^K-directed *cotH* gene is a pre-requisite for the assembly of CotB, CotG, and CotC (Sacco et al., 1995; Naclerio et al., 1996). Based on the analysis of Coomassie-stained gels of coat protein extracts, *cotH*, *cotG*,

cotB, and *cotC* were found to define a cascade of functional or direct interactions: in addition to CotH, *cotH* mutants lack, CotG, CotB, and CotC, whereas *cotG* mutants lack CotG and CotB, and have reduced amounts of CotC (Sacco et al., 1995; Naclerio et al., 1996). Lastly, only CotB appears to be missing from *cotB* spores (Donovan et al., 1987). The basis for these dependencies is now known in some detail. CotB is synthesized as a species of 46 kDa (CotB-46) that is rapidly converted into a form of 66 kDa (CotB-66), which accumulates in the spore coat (Donovan et al., 1987; Zilhão et al., 2004) (Figure 3). Formation of CotB-66 requires expression of *cotG*, which is under σ^K and GerE control (Sacco et al., 1995). In a *cotG* mutant, CotG-46 accumulates in sporulating cells and is the only form of CotB found in the spore coats. CotG is found in complexes with CotB-46 at the time when formation of CotB-66 is detected (Zilhão et al., 2004). Presumably, CotG interacts with CotB-46 to promote formation of CotB-66, a conclusion supported by the finding that CotB interacts with itself and with CotG in a yeast two-hybrid assay (Zilhão et al., 2004). CotB-66 is presumably a cross-linked, stable homodimer. The nature of the cross-linking is presently unknown, but we note that cysteine residues are absent from the CotB primary structure. The C-terminal half of the 389 amino acids-long CotB (residues 252 through 330) contains four repeats of a serine-rich sequence (Kunst et al., 1997). Unpublished work from our laboratory suggests that the interaction of CotB with CotG may involve the serine repeats in CotB. Interestingly, the spore coat of *B. anthracis* contains a homologue of CotB, but this protein is shorter (149 residues; 19 kDa), and does not contains the serine-rich repeats present in the C-terminal half of *B. subtilis* CotB (Lai et al., 2003). A similarly short version of CotB is encoded by the *B. cereus* genome (Ivanova et al., 2003). This suggests that the details of

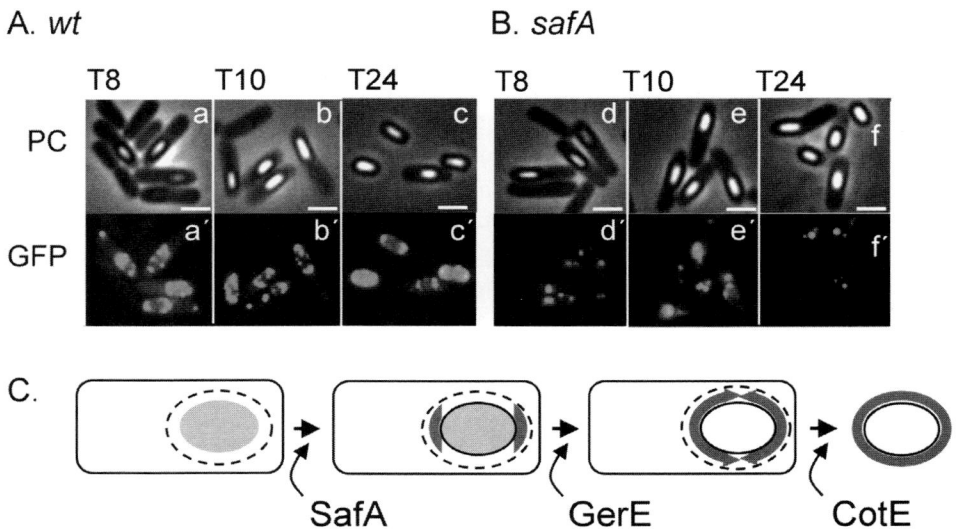

Figure 4. Assembly as monitored by using functional fusions of coat proteins to GFP. The upper part of the figure depicts the stages in assembly of a fusion of OxdD-GFP in a wild type strain (Panel A), and in a *safA* mutant (Panel B). Note that in the mutant, OxdD-GFP forms aggregates in the mother cell cytoplasm, occasionally close to the surface of the spore, and does not encircles the prespore. Other mutants are not shown for simplicity (Costa et al., 2004). Panel C depicts a model for the assembly of OxdD in the inner coat layers. The initial targeting of OxdD-GFP to the spore surface is *safA*-dependent, but then the product of at least one *gerE*-dependent gene is required for OxdD-GFP to migrate around the prespore. Maintenance of the fusion protein around the prespore following its release is *cotE*-dependent. Scale bar, 2 µm.

the assembly and contribution of CotB for coat structure and function varies among species. The GerE-dependent promoter of *cotB* also drives expression of a downstream gene, *ywrJ* (Eichenberger, et al., 2003), which codes for a spore-associated protein (Kuwana et al., 2002). The region delimited by residues 83 to 222 of YwrJ is highly similar to the first 247 residues of CotB, but disruption of *ywrJ* does not prevent formation of CotB-66 (see above) or otherwise interferes with assembly of the spore coat (Zilhão et al., 2004).

The requirement for *cotH* for the assembly of CotB-66 is indirect. Expression of *cotH* is required for the accumulation of CotG, suggesting that CotH or a protein under the control of CotH is required for the stabilization of the otherwise unstable CotG (Zilhão et al., 2004). CotH also stabilizes CotC (Figure 3). When made in *Escherichia coli* cells, or in the mother cell of sporulating cells of *B. subtilis*, CotC (12 kDa) forms a homodimer (21 kDa) that resists detergent treatment and reducing conditions (Isticato et al., 2004). The capacity of CotC to dimerize was also detected using a Gal4-based yeast two-hybrid system (Isticato et al., 2004). The 12 kDa (monomer) and the 21 kDa (dimer) forms undergo assembly and are further modified at the spore surface (Isticato et al., 2004; see below). Expression of *cotH* is also required for the stabilization of CotC monomers and homodimers in the absence of their assembly onto the developing coat (as in a *cotE* mutant) (Isticato et al., 2004) (Figure 3). The mechanism by which CotH achieves the stabilization of CotC and CotG is unknown.

CotG is detected in Coomassie-stained coat protein gels as an abundant, diffuse polypeptide of about 36 kDa (Sacco et al., 1995). However, immunoblot analysis reveals that CotG is first synthesized as a species of about 30 kDa, closer to the size predicted from the gene's sequence (24 kDa), and rapidly converted to the major form of CotG, of 36 kDa, found in the coat; further multimerization of CotG then occurs at the spore surface (Zilhão et al., 2004; our unpublished results; Figure 3). CotG is a key factor determining the striated pattern of the outer coat (Henriques et al., 1998). CotG has a central region (residues 35 to 155) formed by 9 tandem repeats of a lysine, serine, and arginine-rich motif, the SKR region (Sacco et al., 1995). We suspect that CotG forms long filaments, which may serve as a platform for the assembly and structural organization of other outer coat proteins, including CotB (Sacco et al., 1995). Possibly, the SKR repeats play a central role in the establishing this pattern. A caveat with this interpretation is that *B. anthracis*, which forms a coat structurally very similar to that of *B. subtilis* (apart from the presence of the exosporium), does not seem to encode a clear homologue of CotG but this could be due to the presence of long regions of low complexity in CotG (Read et al., 2003) (Table 1). However, the genome of the related species *B. cereus* appears to encode CotG homologues (Ivanova et al., 2003) (Table 1).

Outer coat proteins CotA, CotB, and CotE make an important contribution to the surface topology of spores (Chada et al., 2003; Driks, this volume).

Protein Cross-Linking

About 30% of the total coat protein forms an insoluble fraction, resistant to extraction by a variety of treatments, which has a high cysteine content, and appears to contain other modified residues, suggesting extensive cross-linking (Pandey and Aronson, 1979; Zhang et al., 1993). Coat protein cross-linking has been detected immunologically in several cases, e.g., for CotB, CotC, CotE, CotG, and CotT (Zilhão et al., 1994; Isticato et al., 2004; Zhang et al., 1991; Aronson et al., 1992). Multimeric, highly insoluble forms of CotE and CotT have been found, and since multimerization of CotE was somewhat dependent on *cotT*, the two proteins may be cross-linked together (Aronson et al., 1992; see above).

Recent proteomics studies have provided further evidence for the cross-linking of at least CotB, CotC, CotD, CotE, CotS, and CotY (Kuwana et al., 2002; Lai et al., 2003). Since both CotT and CotD appear to be in the inner coat (Bourne et al., 1991; Zheng et al., 1988), cross-linking doe not seems to be restricted to the outer layers of the coat. At least three types of cross-have been found, or proposed to contribute to proper coat assembly: disulfide bonds, *o,o*-dityrosine bonds, and ε-(γ-glutamyl)lysil cross-links (Pandey and Aronson, 1979; Zhang et al., 1993; Kobayashi et al., 1991; Henriques et al., 1997). Formation of cross-links between the γ-amino group of lysine residues and L-aspartate has also been reported in *B. spahericus* (Tipper and Gauthier, 1972). Unlike sulfhydryl oxidation or sulfhydryl exchange, which can be enzymatically facilitated but can occur spontaneously, formation of *o,o*-dityrosine bonds, and ε-(γ-glutamyl)lysil bonds within or between polypeptide chains involves enzyme catalysis.

Cross-linking is thought to be a key factor in controlling the dynamic behavior of spores (Westphal et al., 2002; Driks, 2002), and is likely to be judiciously controlled.

Disulfide Bonds

The cysteine-rich coat insoluble fraction is essential for the functional integrity of the coat, as upon treatment with reducing agents spores of various species become susceptible to lysozyme and H_2O_2 (Gould et al., 1970). The *cotVWXYZ* cluster was identified by reverse genetics, following the characterization of a peptide solubilized by formic acid hydrolysis of the isolated insoluble material (Zhang et al., 1993). The CotY protein (16.5 kDa) contains 15 cysteine residues, and CotZ (16.5 kDa) contains 10 cysteines (8.6 and 6.2% of the total mass, respectively). CotY and CotZ are minor components of the coat soluble fraction, of 26 and 18 kD, respectively, but CotY also exists as 52 (dimers) and 76 kDa forms (trimers), possibly containing CotZ. Since the multimeric forms of CotY, can be solubilized by high concentrations of reducing agents (Zhang et al., 1993), they appear to contain disulfide bonds. That elimination of *cotXYZ* provokes a reduction of the outer coat, altered surface properties, and increased accessibility to germinants, highlights the importance of disulfide-bond formation for correct assembly and function of the spore coat (Zhang et al., 1993).

Transcription of the *cotVWXYZ* cluster is achieved by means of three promoters, designated P_{VWX}, P_X, and P_{YZ}, in accordance with the genes they control, all of which are recognized by the σ^K form of RNA polymerase (Zhang et al., 1994). P_X has a strict requirement for GerE, whereas GerE enhances transcription from both P_{VWX} and P_{YZ} (Zhang et al., 1994) (Figure 2B). Thus, proteins of the insoluble fraction are made at the end of the assembly process. The disintegration of the mother cell may provide the oxidative conditions required for their efficient cross-linking. Note however that in two recent microarray analysis suggest that some transcription of the *cotYZ* occurs as part of the σ^E regulon (Eichenberger et al., 2003; Feucht et al., 2003). It is unknown whether disulfide-bond formation in the coat is enzymatically facilitated, and if so, in what cellular compartment do those reactions take place. The CcdA protein of *B. subtilis*, resembles the central part of the DsbD protein of *Escherichia coli*, which seems to transfer reducing equivalents to protein disulfide isomerases in the periplasm (Schiöt and Hederstedt, 2000). CcdA is required for a late stage in sporulation, probably the synthesis of the cortex and/or spore coat layers (Schiöt and Hederstedt, 2000). It is tempting to speculate that some proteins may cycle between the mother cell and the compartment defined by the forespore inner and outer membranes, for the introduction of disulfide bonds, and that this pathway could be facilitated by components of the secretion machinery.

Surprisingly, recent results suggest that transglutaminases have protein disulfide isomerase activity (see the following section).

A Coat Transglutaminase

Formation of the covalent ϵ-(γ-glutamyl)lysil cross-link is catalyzed by transglutaminases (TGases) (Griffin, et al., 2002; Lorand and Graham, 2003). Transglutaminases (protein-glutamine γ-glutamyl transferases) catalyze an acyl transfer reaction in which the γ-carboxyamide groups of peptide-bound glutamines act as acyl donors, the most common acceptors being the ϵ-amino groups of lysine residues in proteins or the primary amino groups in some polyamines (Griffin, et al., 2002; Lorand and Graham, 2003). When the acceptors are lysines in polypeptide chains, intra or intermolecular cross-links are formed, and the result is protein polymerization. Transglutaminases play important roles in a variety of processes such as cytoskeletal scaffolding, the control of protein-protein interactions, blood clotting, wound healing, apoptosis, development and morphogenesis, and the keratinization of skin (Lorand and Graham, 2003; Griffin et al., 2002). Since the ϵ-(γ-glutamyl)lysil bond is stable and resistant to proteolysis, it makes an important contribution to the stabilization of tissues or cellular structures, increasing their rigidity and resistance to chemical, enzymatic, physical and mechanical dilapidation (Griffin, et al., 2002; Lorand and Graham, 2003). One of the best-studied enzymes is human coagulation factor XIII, which catalyzes the formation of cross-links between fibrin molecules (Griffin, et al., 2002; Lorand and Graham, 2003; Yee et al., 1994). The crystal structures of factor XIII (75 kDa) and of another factor XIII-like enzyme, a tissue TGase of fish origin, are known. The structure of factor XIII reveals a four-domain organization with a thiol protease-type (papain-lyke) active site involving a cysteine, a histidine and an aspartate in the catalytic domain (Yee et al., 1994).

Recently, the structure of a microbial TGase from *Streptoverticillium mobaraense* was solved at 2.4 Å resolution (Kashiwagi et al., 2002). The enzyme is much smaller (37.9 kDa), shows litlle sequence similarity to the factor XIII-type enzymes and has a different overall fold (Kashiwagi et al., 2002). However, the relative positions of the residues forming the catalytic triad can be superimposed on the structure of the factor XIII-type enzymes, and some elements of secondary structure around the active site are conserved (Kashiwagi et al., 2002). It has been suggested that the two types of enzymes have undergone convergent evolution (Kashiwagi et al., 2002). TGase activity is detected in sporulating cells of *B. subtilis*, and ϵ-(γ-glutamyl)lysil cross-links have been detected in both intact spores and purified coat material (Ramanujam and Hageman, 1990; Kobayashi et al., 1996). The gene (*tgl*) coding for the sporulation-specific TGase was cloned and expressed in an active form in *E. coli* (Kobayashi et al., 1998). The σ^K- and GerE-controlled *tgl* gene encodes a protein of only 28 kDa, which shows no significant similarity to factor XIII or to other transglutaminases of microbial origin (Yee et al., 1994; Makarova et al., 1999). However, a cysteine at position 116 of the Tgl primary structure is essential for enzyme activity (Kobayashi et al., 1998), and the protein was found in spores (Kuwana et al., 2002).

The physiological substrates for Tgl are not known, but several candidates exist. One, the CotM protein, has similarity to the vertebrate eye lens α-crystallins, which can be subjected to transglutaminase-mediated cross-linking (Henriques et al., 1997, and refs. therein). Other Tgl substrates could be encoded by genes in the *cotVXWYZ* cluster. CotW (12.3 kDa) has no cysteines and CotV (14.4 kDa) and CotX (18.6 kDa) bear a single cysteine residue, but in contrast are rich in glutamine and lysine residues. Low levels of 24 kDa and 48 kDa forms of CotX, as well as higher order multimers that fail to enter a resolving SDS-PAGE gel can be solubilized from the coats, and immunologically detected (Zhang et al., 1993). The CotX multimers are not solubilized by excess reducing agents, and the multimerization of CotY and CotZ is partly dependent on CotX (Zhang et al., 1993).

Recently, distant homologues of factor XIII-type TGases were found in several archaeal, bacterial, and eukaryotic genomes, including that of *B. subtilis* (Makarova et al., 1999). Only one member of this group has been functionally characterized, a protein from an archaeal phage, which was shown to have protease activity (Makarova et al., 1999). This suggest that the animal TGases evolved from ancestral proteases, and also that the other members of this new superfamily may also have protease activity. The *B. subtilis* gene has not yet been characterized in detail. TGase-type proteins may also bear other activities. For example, a protein disulfide isomerase (PDI) from *Caenorhabditis elegans* was shown to have transglutaminase activity and to be involved in the control of body shape (Eschenlauer and Page, 2003). Conversely, a tissue-type TGase was shown to have PDI activity (Hasegawa et al., 2003). It is not known whether the smaller microbial-type enzymes are also able to carry PDI reactions.

Evidence for Peroxidase-mediated Cross-Linking

The H_2O_2-dependent formation of *o,o*-dityrosine bonds by peroxidases has been implicated during hardening of the nematode cuticule, the insect egg chorion, the sea urchin fertilization membrane, or during elicitor- and wound-induced oxidative cross-linking of plant cell wall proteins (Henriques et al., 1998 and refs. therein). In a classical study, Pandey and Aronson (1979) found a high content of tyrosine as well as dityrosine cross-links in the coat fraction, but this later observation was not confirmed by the subsequent work of Goldman and Tipper (1981). However, more recently, the application of HPLC techniques has confirmed the presence of dityrosine cross-links in spores, suggesting that tyrosine-rich coat proteins such as CotC, CotU, and CotG, may be subjected to peroxidase-mediated cross-linking (Driks, 1999; Henriques and Moran, 2000). CotC (66 residues) and CotU (86 resideues) are unusual proteins, with only three amino acids, aspartate, lysine, and tyrosine accounting for 75% of the total mass of CotU, and 60% of that of CotC. The 195-amino acids long CotG protein is also tyrosine-rich (15% of the total mass), even though the most represented amino acid is lysine (39% of the total mass). As stated above, multiple forms of CotC, CotU, and CotG are present in the spore coat (Isticato et al., 2004; Zilhão et al., 2004).

Following synthesis, both the CotC monomer and dimeric forms are rapidly assembled onto the developing coat. Some two hours later, two additional forms of 12.5 and 30 kDa are detected, probably resulting from posttranslational modification of the two early forms that occur at the surface of the spore during its maturation (Isticato et al., 2004) (Figure 3). While the 12.5 kDa form may result from proteolysis, the 30 kDa species is likely to result from cross-linking. CotC is highly similar to another coat protein, CotU, particularly in their N-terminal regions, which differ in only one out of 24 amino acid residues (Donovan et al., 1987; Kunst et al., 1997; Lai et al., 2003). Two forms of CotU, which has a predicted size of 18 kDa, are detected in the coat: a form of 17 kDa, presumably the monomer, and a form of 23 kDa (Isticato et al., 2004). The later requires expression of both *cotC* and *cotU*, suggesting that it may represent a cross-linked CotC-CotU heterodimer (Isticato et al., 2004). The nature of the cross-links in CotC or CotU has not been elucidated.

Several multimeric forms of CotG, ranging from 36 kDa to over 100 kDa, are immunologically detected in purified coat material, again suggesting extensive cross-linking of CotG (Zilhão et al., 2004). Multimerization of CotG is also supported by the observation that CotG interacts with itself in a yeast two-hybrid system (Zilhão et al., 2004) but again, the nature of the cross-links holding the CotG multimers is not known. However, the form of CotG of 36 kDa is more extractable in spores produced by a *sodA* mutant, which fail to produce a Mn^{2+}-dependent superoxide dismutase (Henriques et al., 1998). This led to the hypothesis that the Mn^{2+}-dependent SodA could be involved in the conversion of superoxide radicals to H_2O_2, which in turn could be used by a peroxidase for the oxidative cross-linking of CotG (Henriques et al., 1998). SodA appears to associate with the *B. subtilis* spore (Henriques et al., 1998; Kuwana et al., 2002), and the formation of CotG multimers occurs at the spore surface (our unpublished results). Interestingly, a second, Cu^{2+}/Zn^{2+}-dependent superoxide dismutase, SodF, is expressed in the σ^E regulon (Feucht et al., 2003).

In *B. cereus*, a peroxide activity that increased just prior to the accumulation of dipicolinic acid by sporulating cells (presumably at the onset of cortex and coat deposition), was localized to the prespore membranes (Ishida et al., 1987). In *B. subtilis*, a coat-associated chloride peroxidase, the product of the *yisY* gene, was identified by MS techniques (Lai et al., 2003), and the microarray analysis of Eichenberger et al. (2003), assigns this gene to the σ^E regulon. The functional analysis of this gene has not been reported, but it could be involved in detoxification, rather than cross-linking. Other enzymes may be involved in formation of *o,o*-dityrosine bonds. For example, laccases have been implicated in the cross-linking of tyrosine-containing proteins, and the architecture of the CotA active site appears designed to accomodate high molecular weight apolar compounds, such as the tyrosine-rich CotC and CotU proteins (Enguita et al., 2003; see also below). At present, there is no evidence for the involvement of CotA in coat protein cross-linking.

Protein secretion

The mechanisms involved in sorting polypeptides to cellular compartments lacking protein synthesis are to a considerable extent conserved across kingdoms (Tjalsma et al., 2000, and refs. therein). In particular, several similarities exist between the export of proteins across the bacterial plasma membrane and the eukaryotic endoplasmic reticular (ER) membrane, including the signal peptide, the signal recognition particle, signal peptidases (SPases), and several components of the protein translocation channel (Tjalsma et al., 2000). SPases are of two main classes. The P-class enzymes are exclusive to eubacteria, mitochondria, and chloroplasts, whereas ER-type SPases are confined to archae, eukaryotes, and sporulating Gram-positive bacteria, including *B. subtilis*, *B. amyloliquefacies*, *B. anthracis*, *B. cereus*, and *Clostridium perfrigens* (Tjalsma et al., 2000). The first eubacterial ER-type SPase was identified in *B. subtilis* and found to be involved in the targeting of a protein with antibacterial activity, TasA, to the spore coat (Serrano et al., 1999; Stöver et al., 1999a). TasA was found as the predominant polypeptide extracted from the coats of *gerE* spores by alkali treatment (Serrano et al., 1999). The polypeptide found in the coats has 30 kDa and lacks a 27 amino acid signal peptide (Stöver and Driks, 1999a; Serrano et al., 1999). TasA is encoded by the last of a three-gene operon transcribed in the pre-divisional sporangium from a σ^H-type promoter that also requires Spo0A, a response regulator essential for entry into sporulation, and is negatively regulated by the transition stage regulator AbrB (Serrano et al., 1999; Stöver and Driks, 1999b). The second gene in the operon codes for an ER-type SPase, SipW, which is required for processing of TasA (Serrano et al., 1999; Stöver et al., 1999a). TasA is secreted into the culture medium at the onset of sporulation, and thus the presence of TasA in the coat could be due to its association with the spore following its release from the mother cell. However, TasA is also detected intracellularly throughout sporulation, and spores of a *tasA* mutant show a greatly expanded undercoat, in which unstructured material tends to accumulate close to one of the poles of the elipsoidal spore, as well as abnormal inner and outer coat layers (Serrano et al., 1999). Moreover, spores of a *gerE/tasA* double mutant show a synergistic decrease in L-alanine-induced germination, suggesting that TasA somehow contributes to the normal spore physiology (Serrano et al., 1999). In light of these observations it has been suggested that TasA is also secreted to the septum following asymmetric division at the onset of sporulation, and acts from the septal compartment to influence assembly of the coat (Serrano et al., 1999). The initial targeting of TasA to the septum, from which one of the spore poles derives, could explain the asymmetric accumulation of material in the undercoat region of the mutant. In contrast to the P-type enzymes, which use a Ser-Lys catalytic dyad, the ER-type SPases appear to employ a Ser-His-Asp catalytic triad or a Ser-His diad (Tjaslma et al., 2000). Importantly, alanine substitutions in the catalytic residues impaired processing and incorporation of mature TasA into spores (Tjalsma et al., 2000). Possibly, sporulating bacteria employ ER-type SPases to target proteins to the specialized compartment delimited by the prespore inner and outer membranes, that results from the asymmetric division of the sporulating cell and the subsequent engulfment of the prespore by the mother cell.

Another component of the *B. subtilis* secretion machinery is the homologue of the α subunit of the mammalian signal-recognition particle (SRP) encoded by *ftsY* gene (Kakeshita et al., 2000). *ftsY* is the third and last cistron of an operon transcribed in growing cells from a σ^A-type promoter (Kakeshita et al., 2000). Studies with a conditional mutant have shown that *ftsY* is essential for protein secretion and growth (Kakeshita et al., 2000). During sporulation, expression of *ftsY* occurs monocistronically under the control of σ^K and GerE (Kakeshita et al., 2000). Depletion studies using an *ftsY*-inducible allele have shown that the spores formed in the absence of inducer have reduced amounts of CotA and CotE, and a shorter form of CotB-66 (Kakeshita et al., 2001). In agreement with these findings, the outer coat was reduced and disorganized, and the spores susceptible to lysozyme (Kakeshita et al., 2001). FtsY was found in both the cytoplasm and the membrane of growing cells but interestingly, in spores it appeared to associate predominantly with the cortex/inner coat and inner/outer coat interfaces (Kakeshita et al., 2001). The localization of FtsY suggests both a role in secretion of proteins into or out of the cortical region, as well as a structural role. The requirements for the localization of FtsY to the spore are not known, and the exact role of FtsY in spore morphogenesis is unclear.

Proteolysis

Several protease activities have been detected in sporulating cells of *B. subtilis* and *B. cereus*, and were implicated in spore coat morphogenesis (Sastry et al., 1983; Jenkinson et al., 1981; Goldman and Tipper, 1978; Jenkinson and Lord, 1983; James and Mandelstam, 1985). For example, a 15 kDa activity appears just prior to spore release, and addition of the serine protease

inhibitor PMSF to sporulating cultures at the time of coat deposition prevents assembly of an abundant 12 kDa component, as well as spore release (Jenkinson et al., 1980; Jenkinson et al., 1981; James and Mandelstam, 1985). Another activity was associated with a polypeptide of about 30 kDa produced in a *gerE*-dependent manner and found in mature spores (Jenkinson and Lord, 1983; James and Mandelstam, 1985). Proteases associated with the spore coat may be involved in processing of coat protein precursors, and in the optimization of the coat structure so that it interacts adequately with germinants (see below). Since the addition of serine protease inhibitors to mature spores blocks germination in both *B. cereus* T and *B. subtilis* 168, it seems that coat-associated proteases may also be involved in breaking of the coat during germination germination (Boschwitz et al., 1991).

Several spore coat components are subjected to proteolytical processing. The *cotT* gene was identified by reverse genetics, as encoding an abundant polypeptide of 8 kDa, which corresponds to the last two thirds of the gene (Aronson et al., 1989). A precursor of 12 kDa accumulates in coat extracts of both a *cotT* multicopy mutant, implying that processing occurrs at the spore surface, and of a *gerE* mutant, suggesting that processing depends on the integrity of the coat. That processing is not a pre-requisite for assembly is also supported by the observation that the precursor molecule accumulates in the coats of a mutant bearing a single amino acid substitution of argine 19 (the presumed processing site, deduced from the amino acid sequence of the mature form) to isoleucine (Bourne et al., 1991). Consistent with the deduced site of cleavage at arginine 19, the 12 kDa precursor could be converted into the 8 kDa form by incubation of coat extracts with trypsin (Aronson et al., 1989). CotT is most likely an inner coat protein. First, the CotT precursor was present although not fully processed in a *cotE* mutant (Bourne et al., 1991). This also suggests that CotE or assembly of the outer coat is required for efficient processing. Second, accumulation of the 12 kDa precursor in spores of the multicopy mutant led to a thickening of the inner coat, and caused slow germination. Since spores of a *cotT* insertional mutant show deficient germination (see below), it appears that both synthesis and proper processing of the CotT precursor are required for germination (Bourne et al., 1991). The 19 kDa product of the *cotF* gene is also processed to polypeptides of about 5 and 8 kDa which accumulate in the coats (Cutting et al., 1991). The 5 kDa polypeptide, which derives from the 5' end of *cotF*, undergoes a second cleavage event that removes the first five amino acid residues from its N-terminus. *cotF* gene was also found by reverse genetics, and shown to be under the control of σ^K and GerE (Cutting et al., 1991). Deletion of the gene causes no discernible phenotype. There is also evidence that the N-terminal end of SpsC (first 5 amino acid residues) is cleaved by a trypsin-like protease, prior to binding of the protein to the spore (Knurr et al., 2003; see below).

Recently, several proteases have been shown to associate with spores of *B. subtilis*. One is the 30 kDa product of the *yabG* gene, which is transcribed under the control of σ^K (Takamatsu et al., 1999). *yabG* spores are unaffected in their resistance properties, but have altered coats from which several proteins were more extractable, including YeeK, YxeE, and SpoIVA; the later is not normally found in extracts of sporulating cells at late times in sporulation, or in spore coat extracts (Takamatsu et al., 1999). Moreover, precursor forms of SafA, CotT and CotF accumulated in spore coat extracts, suggesting that YabG is directly or indirectly involved in processing of these proteins (see also above). SafA polypeptides of about 42 and 30 kDa, accumulated in a *yabG*-dependent manner, and appeared to derive from SafA-FL (45 kDa), or the SafA-C$_{30}$ form produced by internal translation, respectively (Takamatsu et al., 2000; Ozin et al., 2001; see above). In agreement with this notion, purified YabG could cleave SafA in vitro, and a cocktail of protease inhibitors prevented proteolysis of SafA in vivo (Takamatsu et al., 2000). In contrast to the insertional mutation, a multicopy allele of *yabG* resulted in the formation of heat-sensitive spores (Takamtsu *et al.*, 2000). The recent analysis of Kuwana et al. (2002) confirmed the association of YabG with spores. Presumably, YabG is the 30 kDa *gerE*-dependent, spore-associated protease reported by Jenkinson and Lord (1983) and James and Mandelstam (1985). Three other proteases, the membrane-bound metalloprotease FtsH, the alkaline serine protease AprX, and the extracellular metalloprotease Mpr were also found in spores of *Bacillus subtilis* (Kuwana et al., 2002; Henriques and Moran, unpublished results). Mpr is synthesized as a secretory pre-pro protein precursor in pre-divisional cells at the onset of sporulation, but expression of *mpr* was also found to be under σ^E control in the microarray analysis of Feucht et al. (2003). It is not known how it reaches the coat compartment (see also section on secretion). The roles of these proteases in coat assembly, if any remain to be determined. The application of proteomics methodology to the analysis of the coat in both *B. subtilis* and *B. anthracis* confirms that many coat proteins undergo proteolysis (Kuwana et al., 2002; Lai et al., 2003).

Glycosylation

At least three operons in *B. subtilis* encode enzymes that appear to be involved in the glycosylation of spore coat proteins. Two form a cluster of divergently oriented operons, *cgeAB* and *cgeCDE*, both of which under the control of σ^K and GerE (Roels and Losick, 1995). Spores of a *cgeAB* deletion strain, or those from a strain bearing a deletion of *cgeD* to *cgeB*, have altered surface properties, tended to clump and adhere to glass and plastic surfaces. The first 258 amino acid residues of CgeD are highly similar to the SpsA protein, which is encoded by the first cistron of the *spsA-K* operon (Kunst et al., 1997), and both proteins share sequence similarity with enzymes involved in the biosynthesis of extracellular polysaccharides (Roels and Losick, 1995). The *spsA-K* operon is under the control of σ^K (our unpublished results), but microarray analysis indicates it may also be expressed albeit weakly under σ^E-control (Eichenberger et al., 2003; Feucht et al., 2003). Deletion of the *spsA-K* operon also alters the spore surface properties (Henriques and Moran, 2000). SpsA is a member of glycosyltransferase family 2, that use nucleotide-diphospho-α-D-sugars to generate β-linked products (Charnock and Davies, 1999). Members of glycosyltransferase family 2 are involved in the synthsesis of some of the most abundant polymers found in nature, cellulose and chitin, and

also include several lipopolysaccharide and bacterial O-antigen synthases (Charnock and Davies, 1999; Tarbouriech et al., 2001). The structure of SpsA has been determined both in the native form and in complex with sugar nucleotides (Charnock and Davies, 1999). Mass spectroscopic techniques have shown that at least one protein encoded by the *spsA-K* operon, SpsI, is spore-associated (Kuwana et al., 2002). Also, the sequence of a peptide found to bind tightly to spores of *B. subtilis* (Knurr et al., 2003), matches exactly residues 6-10 of SpsC, encoded by the third cistron of the *spsA-K* operon (Kunst et al., 1997). Since the peptide only binds to spores when at the N-terminus of a fusion protein, it may be that the first 5 amino acids of SpsC, are cleaved-off at a lysine or argine residue (residues 4 and 5), by a trypsin-like protease for the protein to bind to spores (Knurr et al., 2003). Processing may be a timing event, to ensure the correct binding and contribution of SpsC for spore maturation, presumably via glycosylation (Knurr et al., 2003). The expression of the *cge* and *sps* genes, at the time of coat maturation is consistent with the biochemical analyis of the coat fraction by Pandey and Aronson (1979). These authors found that about 6% of the dry weight of the material solubilized from spores, was polysaccharide, and two abundant spore coat components of about 8 and 9 kDa, appear to be glycoproteins (Pandey and Aronson, 1979; Jenkinson et al., 1981). The identity of these proteins is still unknown.

STRUCTURE AND FUNCTION

The main functions of the spore coat at least when accessed under laboratory conditions relate to the protection of the spore against noxious chemicals, peptidoglycan-breaking enzymes, and physical stresses such as UV light, and in spore germination (Driks, 1999; Henriques and Moran, 2000; Nicholson et al., 2000). However, some *Clostridium* species are part of the gut microflora, and at least some *Bacillus* species cycle between the soil and the gastrointestinal tract of various animals (Barbosa et al., this volume). In both cases the robustness of the spore makes a decisive contribution to survival in both environments, but spores interact in more complex ways with the environment (Nicholson et al., 2000; Nicholson, 2002; see als chapter by W. Nicholson, this volume). Also, the spore is the infectious particle of *B. anthracis*, and is designed to survive the encouter with and phagocytosis by alveolar macrophages. Central to this process, is the robustness of the spore, which cannot be destroyed by the macrophage, as well as its capacity to germinate efficiently in the phagosomal compartment of bronchoalveolar macrophages (Guidi-Rontani, et al., 1999; Guidi-Rontani et al., 2001; Mock and Fouet, 2001). Newly-formed vegetative cells can reach the macrophage cytoplasm and propagate (Guidi-Rontani, et al., 1999; Guidi-Rontani et al., 2001; Mock and Fouet, 2001). In contrast, germinated spores or vegetative cells of *B. subtilis* are rapidly destroyed in the macrophage (Duc et al., 2003a).

The Coats and Germination

Resistance and germination are in part dictated by the ability of the coat to act as either a physical barrier, or an interaction surface for compounds in the medium (including germinants). For example, mutations in the hexacistrocic *gerP* operon in both *B. subtilis* (*yisH* to *yisC*) and *B. cereus* cause a block in the early stages of germination, before the later activation of cortex lytic enzymes, cortex hydrolysis, and spore rehydration (Behravan et al., 2000; see Moir et al., 2003, and Moir et al., 2002 for recent reviews on germination). Permeabilization of the spore by extraction of the coat or by the introduction of a *cotE* mutation (in *B. subtilis*), bypasses the early germination block (Behravan et al., 2000). Presumably, the *gerP*-encoded proteins have a role in producing a normal, permeable coat structure, which is important for small hydrophilic germinants such as L-alanine or inosine to interact with the spore, and to gain access to their receptors (Behravan et al., 2000). The *gerP* operon is located close to the *yisJ* gene, encoding a paralog of CotH, and is expressed from a σ^K-dependent promoter negatively regulated by GerE. Although *gerP* mutants do not show an evident coat defect, it is highly likely that they make subtle contributions to the assembly of coat proteins that allow the passage of germinants (Behravan et al., 2000).

Germination also results from the assembly and activity of enzymes at the spore surface. For example, CwlJ is a cortex lytic enzyme activated during spore germination by the release of dipicolinic acid (DPA) from the spore core (Ragkousi et al., 2003). Activation of CwlJ can also be achieved by the addition of Ca^{2+}-DPA (Ragkousi et al., 2003). CwlJ is recruited by CotE and GerQ to the spore coat, in part explaining why *cotE* spores germinate poorly in response to exogenous Ca^{2+}-DPA (Bagyan and Setlow, 2002; Ragkousi et al., 2003). Also, an alanine racemase is part of the spore coat and appears to control the extent of spore germination by modulating the levels of the germinant L-ala in contact with the spores (Kanda-Nambu et al., 2000).

Coat Enzymes

Some of the coat proteins have enzymatic counterparts that were extensively studied in plants or fungi. This group of proteins includes a laccase (CotA), an oxalate decarboxylase (OxdD), a reticuline-oxidase (YvdP or CotQ), and a patatin-like phospholipase (YvdO or CotR) (Bauer et al., 1999; Little and Driks, 2001; Kuwana et al., 2002; Lai et al., 2003; Costa et al., 2004; Hullo et al., 1999; Martins et al., 2000; our unpublished results). For some, enzyme activity within the coat structure has been demonstrated. For instance, oxalate decarboxylase activity was detected in spores of a mutant bearing a multicopy allele of *oxdD* but not in wild type spores (Costa et al., 2004). In another example, CotA is a highly thermostable laccase whose assembly into the coat is required for spore resistance against hydrogen peroxide, for the synthesis of a melanin-like pigment and for resistance to UV light (Donovan et al., 1987; Hullo et al., 2001; Martins et al., 2002). The crystal structure of CotA was determined at 1.7 Å resolution (Enguita et al., 2003) (Figure 1C). This was the first structure of an endospore coat component, and also the first structure of a bacterial laccase. Similarly to other multicopper oxidases, the overall fold of CotA reveals three cupredoxin-like domains, and includes one mononuclear and one trinuclear copper center (Enguita et al., 2003). The three cupredoxin domains in CotA are linked by external inter-domain loops, which increase the packing level of the structure,

and may contribute to the remarkable thermostability of CotA. Other distinctive features of CotA include the largest substrate-accepting cavity of any known laccase, and a flexible lid-like region close to the substrate-binding site that may mediate substrate accessibility or be involved in its assembly within the coat (Figure 1C). We suspect that the thermostability of CotA is an indirect consequence of evolutionary constraints imposed by its incorporation into the highly ordered coat structure, because other heat stable spore-associated activities (catalases) have been reported (Lawrence and Halvorson, 1954; Norris and Baillie, 1964). In any event, the intrinsic properties of CotA, together with the observation that spores can be engineered to display increased levels of the enzyme with no consequences for their structural or resistance properties (Martins et al., 2002), opens the way for its use in biotechnological applications.

Assembly of CotA may involve a large number of contacts dispersed along its surface, a specific region of the protein, or a combination of the two mechanisms. Residues 155 to 158 of CotE are part of a highly acidic region (151-DWEEDDEEDWEDELDEE-166) that is important for the assembly of CotA (Little and Driks, 2001), but a different region of CotA may be involved in its assembly, as evidenced by studies in which peptide ligands able to bind tightly to *B. subtilis* spores were isolated (Knurr et al., 2003). Screening of the *B. subtilis* genome sequence (Kunst et al., 1997), revealed a close match between a sequence in CotA (Thr-His-Phe-Leu-Pro, residues 79 to 83) and the sequence of a peptide known to bind tightly to spores (Knurr et al., 2003). It remains to be determined whether this or other sequences in CotA are involved in its targeting to the spore coat.

The presence of a reticuline oxidase-like protein in the spore coat is intriguing. Reticuline oxidase (also known as the berberine bridge enzyme) catalyzes the formation of (*S*)-scoulerine in the pathway leading to the biosynthesis of benzophenanthridine alkaloids (Dittrich and Kutchan, 1991). These compounds have cytotoxic effects, and accumulate in certain plants in response to pathogenic attack. Synthesis of the benzophenanthridine core is initiated by two molecules of L-tyrosine, and requires 12 enzymatic steps. Reticuline oxidase functions in the branching point of the pathway, to form (*S*)-scoulerine from (*S*)-reticuline, which can then be used for the synthesis of various benzophenanthridine alkaloids (Dittrich and Kutchan, 1991). Thus, it appears that reticuline oxidase cannot function independently of other enzymes in the pathway (Dittrich and Kutchan, 1991). It is possible that the coat-associated protein has a different activity, or that it serves only a structural role within the coat. Alternatively, YvdP and other coat-associated enzymes confer upon the spore the capacity to interact with other organisms in natural environments (Nicholson et al., and Barbosa et al., this volume).

CONCLUDING REMARKS

The extreme endurance of the spore, in part conferred by its coat layers is central to many of the emerging applications of spores in biotechnology and biomedicine. The utilization of spores as probiotics appears to rely on the survival of spores to extreme acidity and to bile salts (Duc et al., 2003a and b), conditions found at different levels of the gastrointestinal tract, and this capacity can be attributed at least in part to the robustness of the spore coat. Also, the germination of spores in the gut may contribute to their probiotic effect, and again germination is influenced by the status of the coat layers. In this respect, it is worth a note that spores of *B. subtilis* strains isolated from poultry feaces do not germinate in the Laboratory in response to L-alanine (Barbosa et al., this volume). It is possible that in the gastrointestinal tract, where spores are known to germinate (Duc et al., 2003a), or in other ecosystems the spore germinates in response to different stimuli.

Spore coat proteins are also useful markers for the rapid identification of spores by using peptide ligands (Knurr et al., 2003) or MS-based protein profiling (Dickinson et al., 2004). Another line of applications involves the use of coat proteins to display toxin subunits or other proteins at the spore surface, with potential uses as vaccines (Isticato et al., 2001; Duc et al., 2003b). Here again, a detailed knowledge of the structure and assembly of the native proteins used as fusion partners, may help designing more efficient strategies for the display of heterologous polypeptides at the spore surface (Isticato et al., 2004; Zilhão et al., 2004).

There is a vast array of enzymes associated with the spore coat, and possibly several other waiting to be discovered in *B. subtilis* and in the coats of other sporeformers (Read et al., 2002; Read et al., 2003; Ivanova et al., 2003; Takami et al., 2000; Takami et al., 2002; Nolling et al., 2002; Shimizu et al., 2002; Kunst et al., 1997). The panoply of native or imported enzymes, in combination with the tools for the genetic manipulation of *B. subtilis*, can be explored in several ways. For example, spores of *B. subtilis* and *B. cereus* are already being used as biosensors (Rotman, 2001). The biochemical and structural characterization of CotA (Hullo et al., 1999; Martins et al., 2000; Enguita et al., 2003) already suggests that some may have unique properties. The level of native enzymes normally present in the outer coat can be increased without altering its structure and properties (Martins et al., 2002).

A more complete description of the enzymes and other proteins associated with the coat will also surface from detailed investigations of the ecological roles of spores of *B. subtilis* and other species, and of their interactions with other organisms in different ecosystems. These studies will be facilitated by the knowledge of the genome sequences of various spore-forming organisms. They will increase our ability to make use of the repertoire of coat structural proteins and enzymes, to manipulate the spore surface in view of specific applications.

REFERENCES

Abe, A., Koide, H., Kohno, T., and Watabe, K. 1995. A *Bacillus subtilis* spore coat polypeptide gene, *cotS*. Microbiol. *141*, 1433-1442.

Abe, A., Ogawa, S., Kohno, T., and Watabe, K. 1993. Purification of *Bacillus subtilis* spore coat protein by electrophoretic elution procedure and determination of NH2-terminal amino acid sequences. Microbiol. Immunol. *37*, 809-812.

Aronson, A.I., and Fitz-James, P.C. 1971. Reconstitution of bacterial spore coat layers *in vitro*. J. Bacteriol. *108*, 571-578.

Aronson, A.I., Ekanayake, L., and Fitz-James, P.C. 1992. Protein filaments may initiate the assembly of the *Bacillus subtilis* spore coat. Biochimie *74*, 661-667.

Aronson, A.I., Song, H.-Y., and Bourne, N. 1989. Gene structure and precursor processing of a novel *Bacillus subtilis* spore coat protein. Mol. Microbiol. *3*, 437-444.

Bagyan, I., and Setlow, P. 2002. Localization of the cortex lytic enzyme CwlJ in spores of *Bacillus subtilis*. J. Bacteriol. *184*, 1219-1224.

Bagyan, I., Hobot, J., and Cutting, S. 1996. A compartmentalized regulator of developmental expression in *Bacillus subtilis*. J. Bacteriol. *178*, 4500-4507.

Barynin, V.V., Whittaker, M.M., Antonyuk, S.V., Lamzin, V.S., Harrison, P.M., Artymiuk, P.J., and Whittaker, J.W. 2001. Crystal structure of manganese catalase from *Lactobacillus plantarum*. Structure *9*, 725-738.

Bauer, T., S. Little, A. G. Stover, and A. Driks. 1999. Functional regions of the *Bacillus subtilis* spore coat morphogenetic protein CotE. J Bacteriol *181*, 7043-51.

Beall, B., A. Driks, R. Losick, and C. P. Moran, Jr. 1993. Cloning and characterization of a gene required for assembly of the *Bacillus subtilis* spore coat. J Bacteriol *175*, 1705-16.

Beall, B., Driks, A., Losick, R., and Moran, C.P. Jr. 1993. Cloning and characterization of a gene required for assembly of the *Bacillus subtilis* spore coat. J. Bacteriol. *175*, 1705-1716.

Behravan, J., Chirakkal, H., Masson, A., and Moir. A. 2000. Mutations in the *gerP* locus of *Bacillus subtilis* and *Bacillus anthracis* affect access of germinants to their targets in spores. J. Bacteriol. *182*, 1987-1994.

Boschwitz, H., Gofshtein-Gandman, L., Halvorson, H.O., Keynan, A., and Milner, Y. 1991. The possible involvement of trypsin-like enzymes in germination of spores of *Bacillus cereus* T, and *Bacillus subtilis* 168. J. Gen. Microbiol. *137*, 1145-1153.

Bourne, N., Fitz-James, P.C., and Aronson, A.I. 1991. Structural and germination defects of *Bacillus subtilis* spores with altered contents of a spore coat protein. J. Bacteriol. *173*, 6618-6625.

Catalano, F.A., Meador-Parton, J., Popham, D.L., and Driks, A. 2001. Amino acids in the Bacillus subtilis morphogenetic protein SpoIVA with roles in spore coat and cortex formation. J. Bacteriol. *183*, 1645-1654.

Chada, V.G.R., Sanstad, E.A., Wang, R., and Driks, A. 2003. Morphogenesis of *Bacillus* spore surfaces. J. Bacteriol. *185*, 6255-6261.

Charnock, S.J., and Davies, G.J. 1999. Structure of the nucleotide-diphospho-sugar transferase, SpsA from *Bacillus subtilis*, in native and nucleotide-complexed forms. Biochem. *38*, 6380-6385.

Costa, T.V., Martins, L.O., Voelker, U., and Henriques, A.O. 2004. Assembly of an oxalate decarboxylase produced under σ^K control into the *Bacillus subtilis* spore coat. J. Bacteriol. *186*, 1462-1474.

Crater, D.L., and Moran, C.P.Jr. 2001. Identification of a DNA binding region in GerE from *Bacillus subtilis*. J. Bacteriol. *183*, 4183-4189.

Crater, D.L., and Moran, C.P.Jr. 2002. Two regions of GerE required for promoter activation in *Bacillus subtilis*. J. Bacteriol. *184*, 241-249.

Cutting, S., Oke, V., Driks, A., Losick, R., Lu, S., and Kroos, L. 1990. A forespore checkpoint for mother cell gene expression during development in *Bacillus subtilis*. Cell *62*, 239-250.

Cutting, S., Panzer, S., and Losick, R. 1989. Regulatory studies on the promoter for a gene governing synthesis and assembly of the spore coat in *Bacillus subtilis*. J. Mol. Biol. *207*, 393-404.

Cutting, S.M., Zheng, L., and Losick, R. 1991. Gene encoding two alkali-soluble components of the spore coat from *Bacillus subtilis*. J. Bacteriol. *173*, 2915-2919.

Dickinson, D.N., La Duc, M.T., Haskins, W.E., Gornushkin, I., Winefordner, J.D., Powell, D.H., and Venkateswaran. 2004. Species differentiation of a diverse suite of *Bacillus* spores by mass spectrometry-based protein profiling. Appl. Environm. Microbiol. *70*, 475-482.

Dittrich, H., and Kutchan, T.M. 1991. Molecular cloning, expression, and induction of berberine bridge enzyme, an enzyme essential to the formation benzophenanthridine alkaloids in the response of plants to pathogenic attack. Proc. Natl. Acad. Sci. USA *88*, 9969-9973.

Donovan, W., Zheng, L.B., Sandman, K., and Losick, R. 1987. Genes encoding spore coat polypeptides from *Bacillus subtilis*. J. Mol. Biol. *196*, 1-10.

Driks, A. 1999. *Bacillus subtilis* spore coat. Microbiol. Mol. Biol. Rev. *63*, 1-20.

Driks, A. 2003. Maximum shields: the armor plating of the bacterial spore. Trends Microbiol *10*, 251-154.

Driks, A. 2003. The dynamic spore. Proc. Natl. Acad. Sci. USA *100*, 3007-3009.

Driks, A., Roels, S., Beall, B., Moran, C.P.Jr., and Losick, R. 1994. Subcellular localization of proteins involved in the assembly of the spore coat of *Bacillus subtilis*. Genes Dev. *8*, 234-244.

Duc, L.H., Hong, H.A., and Cutting, S.M. 2003a. Germination of the spore in the gastrointestinal tract provides a novel route for heterologous antigen delivery. Vaccine *21*, 4215-4224.

Duc, L.H., Hong, H.A., Fairweather, N., Ricca, E., and Cutting, S.M. 2003b. Bacterial spores as vaccine vehicles. Infect. Immun. *71*, 2810-2818.

Ducros, V.M., Lewis, R.J., Verma, C.S., Dodson, E.J., Leonard, G., Turkenburg, J.P., Murshudov, G.N., Wilkinson, A.J., and Brannigan. J.A. 2001. Crystal structure of GerE, the ultimate transcriptional regulator of spore formation in *Bacillus subtilsi*. J. Mol. Biol. *306*, 759-771.

Eichenberger, P., S. T. Jensen, E. M. Conlon, C. van Ooij, J. Silvaggi, J. E. Gonzalez-Pastor, M. Fujita, S. Ben-Yehuda, P. Stragier, J. S. Liu, and R. Losick. 2003. The sigma E regulon and the identification of additional sporulation genes in *Bacillus subtilis*. J Mol Biol *327*, 945-72.

Enguita, F.J., Martins, L.O., Henriques, A.O., and Carrondo, M.A. 2003. Crystal structure of a bacterial endospore coat component: a laccase with enhanced thermostability properties. J. Biol. Chem. *278*, 19416-19425.

Errington, J. 2003. Regulation of endospore formation in *Bacillus subtilis*. *Nature Rev. Microbiol 1*, 117-126.

Eschenlauer, S.C., and Page, A.P. 2003. The *Caenohabditis elegans* Erp60 homolog protein sisulfide isomerase-3 has disulfide isomerase and transglutaminase-like cross-linking activity and is involved in the maintenance of body morphology. J. Biol. Chem. *278*, 4227-4237.

Goldman, R.C., and Tipper, D.J. 1978. *Bacillus subtilis* spore coats: complexity and purification of a unique polypeptide component. J. Bacteriol. *135*, 1091-1106.

Goldman, R.C., and Tipper, D.J. 1981. Coat protein synthesis during sporulation of *Bacillus subtilis*: immunological detection of soluble precursors to the 12,200-Dalton spore coat protein. J. Bacteriol. *147*, 1040-1048.

Gould, G.W., Stubbs, J.M., and King, W.L. 1970. Structure and composition of resistant layers in bacterial spore coats. J. Gen. Microbiol. *60*, 347-355.

Griffin, M., Casadio, R., and Bergamini, C.M. 2002. Transglutaminases: nature's biological glues. Biochem. J. *368*, 377-396.

Guidi-Rontani, C., Levy, M., Ohayon, H., and Mock, M. 2001. Fate of germinated *Bacillus anthracis* spores in primary murine macrophages. Mol. Microbiol *42*, 931-938.

Guidi-Rontani, C., Weber-Levy, M., Labruyère, E., and Mock, M. 1999. Germination of *Bacillus anthracis* spores within alveolar macrophages. Mol. Microbiol. *31*, 9-17.

Halberg, R., and Kroos, L. 1994. Sporulation regulatory protein SpoIIID from *Bacillus subtilis* activates and represses transcription by both mother-cell-specific forms of RNA polymerase. J. Mol. Biol. *243*, 425-436.

Halberg, R., Oke, V., and Kroos, L. 1995. Effects of *Bacillus subtilis* sporulation regulatory protein SpoIIID on transcription by σ^K RNA polymerase in vivo and in vitro. J. Bacteriol. *177*, 1888-1891.

Hasegawa, G., Suwa, M., Ichikawa, Y., Ohtsuka, T., Kumagi, S., Kikuchi, T., Sato, Y., and Saito, Y. 2003. A novel function of tissue-type transglutaminase: protein disulfide isomerase. Biochem. J. *373*, 793-803.

Henriques A.O., and Moran, C.P. Jr. 2000. Structure and assembly of the bacterial endospore coat. Methods *20*, 95-110.

Henriques, A. O., Beall, B. W., and Moran, C. P. Jr. 1997. CotM of *Bacillus subtilis*, a member of the α-crystallin family of stress proteins, is induced during development and participates in spore outer coat formation. J. Bcateriol. *179*, 1887-1897.

Henriques, A. O., Beall, B.W., Roland, K., and Moran, C.P. Jr. 1995. Characterization of *cotJ*, a σ^E-controlled operon affecting the polypeptide composition of the coat of *Bacillus subtilis* spores. *J. Bacteriol.*, *177*, 3394-3406.

Henriques, A.O., Melsen, L.R., and Moran, C.P. Jr. 1998. Involvement of superoxide dismutase in spore coat assembly in *Bacillus subtilis*. J. Bacteriol. *180*, 2285-2291.

Hullo, M. F., I. Moszer, A. Danchin, and I. Martin-Verstraete. 2001. CotA of *Bacillus subtilis* is a copper-dependent laccase. J Bacteriol *183*, 5426-30.

Ichikawa, H., and L. Kroos. 2000. Combined action of two transcription factors regulates genes encoding spore coat proteins of *Bacillus subtilis*. J Biol Chem *275*, 13849-55.

Ichikawa, H., R. Halberg, and L. Kroos. 1999. Negative regulation by the *Bacillus subtilis* GerE protein. J. Biol. Chem. *274*, 8322-7.

Ishida, A., Futamura, N., and Matsusaka, T. 1987. Detection of peroxidase activity and its localization in the forespore envelopes of *Bacillus cereus*. J. Gen. Appl. Microbiol. *33*, 27-32.

Isticato, R., Cangiano, G., Tran, T.-H., Ciabattini, A., Medaglini, D., Oggioni, M.R., De Felice, M., Pozzi, G., and Ricca, E. 2001. Surface display of recombinant proteins on *Bacillus subtilis* spores. J. Bacteriol. *183*, 6294-6301.

Isticato, R., Esposito, G., Zilhão, R., Nolasco, S., Cangiano, G., De Felice, M., Henriques, A.O., and Ricca, E. 2004. Assembly of multiple CotC forms into the *Bacillus subtilis* spore coat. J. Bacteriol. *186*, 1129-1135.

Ivanova, N., Sorokin, A., Anderson, I., Galleron, N., Candelon, B., Kapatral, V., Bhattacharyya, A., Reznik, G., Mikhailova, N., Lapidus, A., Chu, L., Mazur, M., Goltsman, E., Larsen, N., D'Souza, M., Walunas, T., Grechkin, Y., Push, G., Haselkorn, R., Fonstein, M., Dusko-Ehrlich, S., Overbeek, R., and Kyrpides, N. 2003. Genome sequence of *Bacillus cereus* and comparative analysis with *Bacillus anthracis*. Nature *423*, 87-91.

James, W., and Mandelstam, J. 1985. Protease production during sporulation of germination mutants of *Bacillus subtilis* and the cloning of a functional *gerE* gene. J. Gen. Microbiol. *131*, 2421-2430.

Jenkinson, H.F., and Lord, H. 1983. Protease deficiency and its association with defects in spore coat structure, germination and resistance properties in a mutant of *Bacillus subtilis*. J. Gen Microbiol. *129*, 2727-2737.

Jenkinson, H.F., Sawyer, W.D., and Mandelstam, J. 1981. Synthesis and order of assembly of spore coat proteins in *Bacillus subtilis*. J. Gen. Microbiol. *123*, 1-16.

Kakeshita, H., Oguro, A., Amikura, R., Nakamura, K., and Yamane, K. 2000. Expression of the *ftsY* gene, encoding a homologue of the α subunit of mammalian signal recognition particle receptor, is controlled by different promoters in vegetative and sporulating cells of *Bacillus subtilis*. Microbiol. *146*, 2595-2603.

Kakeshita, H., Takamatsu, H., Amikura, R., Nakamura, R., Watabe, K., and Yamane, K. 2001. Effects of depletion of FtsY on spore morphology and the protein composition of the spore coat layer in *Bacillus subtilis*. FEMS Microbiol Lett. *195*, 41-46.

Kanda-Nambu, K., Yasuda, Y., and Tochikubo, L. 2000. Isozymic nature of spore coat-associated alanine racemase of *Bacillus subtilis*. Amino Acids *18*, 375-387.

Kashiwagi, T., Yokoyama, K., Ishikawa, K., Ono, K., Ejima, D., Matsui, H., and Suzuki, E. 2002. Crystal structure of microbial transglutaminase from *Streptoverticillium mobaraense*. J. Biol. Chem. *277*, 44252-44260.

Knurr, J., Benedek, O., Heslop, J., Vinson, R.B., Boydston, J.A., McAndrew, J., Kearney, J.F., and Turnbough, C.L.Jr. 2003. Peptide ligands that bind selectively to spores of *Bacillus subtilis* and closely related species. J. Bacteriol. *69*, 6841-6847.

Kobayashi, K., Hashiguchi, K., Yokozeki, K., and Yamanaka, S. 1998. Molecular cloning of the transglutaminase gene from *Bacillus subtilis* and its expression in *Escherichia coli*. Biosci. Biotechnol. Biochem. *62*, 1109-1114.

Kobayashi, K., Kumazawa, Y., Miwa, K., and Yamanaka, S. 1996. ε-(γ-glutamyl)lysine cross-links of spore coat proteins and transglutaminase activity in *Bacillus subtilis*. FEMS Microbiol. Lett. *144*, 157-160.

Kodama, T., Takamatsu, H., Asai, K., Kobayashi, K., Ogasawara, N., and Watabe, K. 1999. The *Bacillus subtilis yaaH* gene is transcribed by SigE RNA polymerase during sporulation, and its product is involved in germination of spores. J. Bacteriol. *181*, 4584-4591.

Kodama, T., Takamatsu, H., Asai, K., Ogasawara, N., Sadaie, Y., and Watabe, K. 2000. Synthesis and characterization of the spore proteins of *Bacillus subtilis* YdhD, YkuD, and YkvP, which carry a motif conserved among cell wall binding proteins. J. Biochem. *128*, 655-663.

Kroos, L., Kunkel, B., and Losick, R. 1989. Switch protein alters specificity of RNA polymerase containing a compartment-specific sigma factor. Science *243*, 526-529.

Kunst, F., Ogasawara, N., Moszer, I., Albertini, A.M., Alloni, G., Azevedo, V., Bertero, M.G., Bessieres, P., Bolotin, A., Borchert, S., Borriss, R., Boursier, L., Brans, A., Braun, M., Brignell, S.C., Bron, S., Brouillet, S., Bruschi, C.V., Caldwell, B., Capuano, V., Carter, N.M., Choi, S.K., Codani, J.J., Connerton, I.F., Danchin, A., and et al. 1997. The complete genome sequence of the gram-positive bacterium *Bacillus subtilis*. Nature *390*, 249-56.

Kuwana, R., Ikejiri, H., Yamamura, S., Takamatsu, H., and Watabe, K. 2004. Functional relationship between SpoVIF and GerE in gene regulation during sporulation of *Bacillus subtilis*. Microbiol. *150*, 163-170.

Kuwana, R., Kasahara, Y., Fujibayashi, M., Takamtsu, H., Ogasawara, N., and Watabe, K. 2002. Proteomics characterization of novel spore proteins of *Bacillus subtilis*. Microbiol. *148*, 3971-3982.

Kuwana, R., Yamamura, S., Ikejiri, H., Kobayashi, K., Ogasawara, N., Asai, K., Sadaie, Y., Takamatsu, H., and Watabe. K. 2003. *Bacillus subtilis spoVIF* (*yjcC*) gene, involved in coat assembly and spore resistance. Microbiol. *149*, 3011-3021.

Lai, E.-M., Phadke, N.D., Kachman, M.T., Giorno, R., Vasquez, S., Vazquez, J.A., Maddock, J.R., and Driks, A. 2003. Proteomic analysis of the spore coats of *Bacillus subtilis* and *Bacillus anthracis*. J. Bacteriol. *185*, 1443-1454.

Little, S., and A. Driks. 2001. Functional analysis of the *Bacillus subtilis* morphogenetic spore coat protein CotE. Mol. Microbiol. *42*, 1107-1120.

Liu, H., Bergman, N.H., Thomason, B., Shallom, S., Hazen, A., Crossno, J., Rasko, D.A., Ravel, J., Read, T.D., Peterson, S.N., Yates III, J., and Hanna, P.C. 2003. Formation and composition of the *Bacillus anthracis* endospore. J. Bacteriol. *186*, 164-178.

Lorand, L., and Graham, R.M. 2003. Transglutaminases: cross-linking enzymes with pleiotropic functions. Nat. Rev. Mol. Cell Biol. *4*, 140-156.

Makarova, K.S., Aravind, L., and Koonin, E.V. 1999. A superfamily of archaeal, bacterial, and eukaryotic proteins homologous to animal transglutaminases. Prot. Sci. *8*, 1714-1719.

Martins, L.O., Soares, C.M., Pereira, M.M., Teixeira, M., Costa, T., Jones, G.H. and Henriques, A.O. 2002. Molecular and biochemical characterization of a highly stable bacterial laccase that occurs as a structural component of the *Bacillus subtilis* endospore coat. J. Biol. Chem. *277*, 18849-18859.

Melly E., Genest, P.C., Gilmore, M.E., Little S., Popham, D.L., Driks, A., and Setlow, P. 2002. Analysis of the properties of spores of *Bacillus subtilis* prepared at different temperatures. J. Appl. Microbiol. *92*, 1105-15.

Mock, M., and Fouet, A. 2001. Anthrax. Annu. Rev. Microbiol. *55*, 647-671.

Moir, A. 1981. Germination properties of a spore coat-defective mutant of *Bacillus subtilis*. J Bacteriol *146*, 1106-16.

Moir, A. 2003. Bacterial spore germination and protein mobility. Trends Microbio. *11*, 452-454.

Moir, A., Corfe, B.M., and Behravan, J. 2002. Spore germination. Cell Mol. Life Sci. *59*, 403-409.

Naclerio, G., Baccigalupi, L., Zilhão, R., de Felice, M., and Ricca, E. 1996. *Bacillus subtilis* spore coat assembly requires *cotH* gene expression. J. Bacteriol. *178*, 4375-4380.

Nicholson, W. L. 2002. Roles of *Bacillus* endospores in the environment. Cell Mol Life Sci *59*, 410-6.

Nicholson, W. L., N. Munakata, G. Horneck, H. J. Melosh, and P. Setlow. 2000. Resistance of *Bacillus* endospores to extreme terrestrial and extraterrestrial environments. Microbiol Mol Biol Rev *64*, 548-72.

Nolling, J., Breton, G., Omelchenko, M.V., Makarova, K.S., Zeng, Q., Gibson, R. et al. 2001. Genome sequence and comparative analysis of the solvent-producing bacterium *Clostridium acetobutylicum*. J. Bacteriol. *183*, 4823-4838.

Ozin, A. J., Costa, T., Henriques, A. O., and Moran, C. P. Jr. 2001a. Alternative translation initiation produces a short form of a spore coat protein in *Bacillus subtilis*. J. Bacteriol. *183*, 2032-2040.

Ozin, A.J., Henriques, A.O., Yi, H., and Moran, C.P. Jr. 2000. Morphogenetic proteins SpoVID and SafA form a complex during assembly of the *Bacillus subtilis* spore coat. J. Bacteriol. *182*, 1828-1833.

Ozin, A.J., Samford, C.S., Henriques, A.O., and Moran, C.P. Jr. 2001b. SpoVID guides SafA to the spore coat in *Bacillus subtilis*. J. Bacteriol. *183*, 3041-3049.

Piggot, P.J., and Coote, J.G. 1976. Genetic aspects of bacterial endospore formation. Bacteriol. Rev. *40*, 908-962.

Pogliano, K., Harry, E., and Losick, R. 1995. Visualization of the subcellular location of sporulation proteins in *Bacillus subtilis* using immunofluorescence microscopy. Mol. Microbiol. *18*, 459-470.

Price, K.D., and Losick, R. 1999. A four-dimensional view of the assembly of a morphogenetic proteins during sporulation in *Bacillus subtilis*. J. Bacteriol. *181*, 781-790.

Ragkousi, K., Eichenberger, P., van Ooij, C., and Setlow, P. 2003. Identification of a new gene essential for spore germination of *Bacillus subtilis* spores with Ca^{2+}-dipicolinate. J. Bacteriol. *185*, 2315-2319.

Ramanujam, M.V., and Hageman, J.H. 1990. Intracellular transglutaminase (EC 2.3.2.13) in a prokaryote: evidence from vegetative and sporulating cells of *Bacillus subtilis*. FASEB J. *4*, A2321.

Read, T.D., Peterson, S.N., Tourasse, N., Baillie, L.W., Paulsen, I.T., Nelson, K.E., Tettelin, H., Fouts, D.E., Eisen, J.A., Gill, S.R., Holtzapple, E.K., Oa, O.K., Helgason, E., Rilstone, J., Wu, M., Kolonay, J.F., Beanan, M.J., Dodson, R.J., Brinkac, L.M., Gwinn, M., DeBoy, R.T., Madpu, R., Daugherty, S.C., Durkin, A.S., Haft, D.H., Nelson, W.C., et al. 2003. The genome sequence of *Bacillus anthracis* Ames and comparison to related bacteria. Nature *423*, 81-86.

Read, T.D., Salzberg, S.L., Pop, M., Shumway, M., Umayam, L., Jiang, L., et al. 2002. Comparative genome sequencing for discovery of novel polymorphisms in *Bacillus anthracis*. Science *296*, 2028-2033.

Redmond, C., Baillie, L.W., Hibbs, S., Moir, A.J., and Moir, A. 2004. Identification of proteins in the exosporium of *Bacillus anthracis*. Microbiol. *150*, 355-363.

Reischl, S., Thake, S., Homuth, G., and Shumann, W. 2001. Transcriptional analysis of three *Bacillus subtilis* genes coding for proteins with the α-crystallin domain characteristic of small heat shock proteins. FEMS Microbiol Lett. *194*, 99-103.

Riesenman, P.J., and Nicholson, W.L. 2000. Role of the spore coat layers in *Bacillus subtilis* resistance to hydrogen peroxide, artificial UV-C, UV-B, and solar UV radiation. Appl. Environm. Microbiol. *66*, 620-626.

Roels, S., and Losick, R. 1995. Adjacent and divergently oriented operons under the control of the sporulation regulatory protein GerE in *Bacillus subtilis*. J. Bacteriol. *177*, 6263-6275.

Roels, S., Driks, A., and Losick, R. 1992. Characterization of *spoIVA*, a sporulation gene involved in coat morphogenesis in *Bacillus subtilis*. J. Bacteriol. *174*, 575-585.

Rotman, B. Analytical system based on spore germination. U.S. Patent No. 6,228,574. May 8, 2001.

Sacco, M., Ricca, E., Losick, R., and Cutting, S. 1995. An additional GerE-controlled gene encoding an abundant spore coat protein from *Bacillus subtilis*. J. Bacteriol. *177*, 372-377.

Sandman, K., Kroos, L., Cutting, S., Youngman, P. and Losick, R. 1988. Identification of the promoter for a spore coat protein gene in *Bacillus subtilis* and studies on the regulation of its induction at a late stage of sporulation. J. Mol. Biol. *200*, 461-473.

Sastry, K.J., Srivastava, O.P., Millet, J., Fitz-James, P.C., and Aronson, A.I. 1983. Characterization of *Bacillus subtilis* mutants with a temperature-sensitive intracellular protease. J. Bacteriol. *153*, 511-519.

Serrano, M., Zilhão, R., Ricca, E., Ozin, A.J., Moran, C.P. Jr., and Henriques, A.O. 1999. A *Bacillus subtilis* secreted protein with a role in endospore coat assembly and function. J. Bacteriol. *181*, 3632-3643.

Seyler, R., Henriques, A.O., Ozin, A., and Moran, C.P. Jr. 1997. Interactions and assembly of *cotJ*-encoded products, constituents of the inner layers of the *Bacillus subtilis* spore coat. Mol. Microbiol., *25*, 955-966.

Shimizu, T., Ohtani, K., Hirakawa, H., Ohshima, K., Yamashita, A., Shiba, T. et al. 2002. Complete genome sequence of *Clostridium perfringens*, an anaerobic flesh-eater. Proc. Natl. Acad. Sci. USA *99*, 996-1001.

Steichen, C., Chen, P., Kearney, J.F., and Turnbough, C.L.Jr. 2003. Identification of the immunodominant protein and other proteins of the *Bacillus anthracis* exosporium. J. Bacteriol. *185*, 1903-1910.

Stevens, C.M., Daniel, R., Illing, N., and Errington, J. 1992. Characterization of a sporulation gene, *spoIVA*, involved in spore coat morphogenesis in *Bacillus subtilis*. J. Bacteriol. *174*, 586-594.

Stöver, A., and Driks, A. 1999a. Secretion, localization, and antibacterial activity of TasA, a *Bacillus subtilis* spore-associated protein. J. Bacteriol. *181*, 1664-1672.

Stöver, A., and Driks, A. 1999b. Regulation of synthesis of the *Bacillus subtilis* transition-phase spore-associated antibacterial protein TasA. J. Bacteriol. *181*, 5476-5481.

Takamatsu, H., Chikahiro, Y., Kodama, T., Koide, H., Kozuka, S., Tochikubo, K., and Watabe, K. 1998. A spore coat protein, CotS, of *Bacillus subtilis* is synthesized under the regulation of σ^K and GerE during development and is located in the inner coat layers of spores. J. Bacteriol. *180*, 2968-2974.

Takamatsu, H., Imamura, A., Kodama, T., Asai, K., Ogasawara, N., and Watabe, K. 2000. The *yabG* gene of *Bacillus subtilis* encodes a sporulation specific protease which is involved in the processing of several coat proteins. FEMS Microbiol. Lett. *192*, 33-38.

Takamatsu, H., Kodama, T, Nakayama, T., and Watabe, T. 1999. Characterization of the *yrbA* gene of *Bacillus subtilis* involved in the resistance and germination of spores. J. Bacteriol. *181*, 4986-4994.

Takamatsu, H., Kodama, T., and Watabe, K. 1999. Assembly of the CotSA coat protein into spores requires CotS in *Bacillus subtilis*. FEMS Microbiol. Lett. *174*, 201-206.

Takamatsu, H., Kodama, T., Imamura, A., Asai, K., Kobayashi, K., Nakayama, T., Ogasawara, N., and Watabe, K. 2000. The *Bacillus subtilis yabG* gene is transcribed by SigK RNA polymerase, and *yabG* mutant spores have altered coat protein composition. J. Bacteriol. *182*, 1883-1888.

Takami, H., Nakasone, K., Takaki, Y., Maeno, G., Sasaki, R., Masui, N., et al. 2000. Complete genome sequence of the alkaliphilic bacterium *Bacillus halodurans* and genome sequence comparison with *Bacillus subtilis*. Nucl. Acids Res. *28*, 4317-4331.

Takami, H., Takaki, Y., and Uchiyama, I. 2002. Genome sequence of *Oceanobacillus iheyensis* isolated from the Iheya ridge and its unexpected adaptive capabilities to extreme environments. Nucl. Acids Res. *30*, 3927-3935.

Tarbouriech, N., Dharnock, S.J., and Davies, G.J. 2001. Three-dimensional structures of the Mn and Mg dTDP complexes of the family GT-2 glycosyltransferase SpsA: a comparison with related NDP-sugar glycosyltransferases. J. Mol. Biol. *34*, 655-661.

Tipper, D.J., and Gauthier, J.J. 1972. Structure of the bacterial endospore. p. 3-12. *In* Halvorson, H.O., Hanson, R., and Campbell, L.L. (ed.). Spores-V. Amreican Society for Microbiology, Wahihngton, D.C.

Tjalsma, H., Stöver, A.G., Driks, A., Venema, G., Bron, S., and van Dijl, J.M. 2000. Conserved serine and histidine residues are critical for activity of the ER-type signal peptidase SipW of *Bacillus subtilis*. J. Biol. Chem. *275*, 25102-25108.

Todd, S.J., Moir, A.J.G., Hohnson, M.J., and Moir, A. 2003. Genes of *Bacillus cereus* and *Bacillus anthracis* encoding proteins of the exosporium. J. Bacteriol. *185*, 3373-3378.

Wade, K.H., Schyns, G., Opdyke, J.A., and Moran, C.P.Jr. 1999. A region of σ^K involved in promoter activation by GerE in *Bacillus subtilis*. J. Bacteriol. *181*, 4365-4373.

Webb, C. D., A. Decatur, A. Teleman, and R. Losick. 1995. Use of green fluorescent protein for visualization of cell-specific gene expression and subcellular protein localization during sporulation in *Bacillus subtilis*. J. Bacteriol. *177*, 5906-11.

Yee, V.C., Pedersen, L.C., Trong, I.L., Bishop, P.D., Stenkamp, R.E., and Teller, D.C. 1994. Three-dimensional structure of a transglutaminase: human blood coagulation factor XIII. Proc. Natl. Acad. Sci. USA *91*, 7296-7300.

Zhang, B., and Kroos, L. 1997. A feedback loop regulates the switch from one sigma factor to the next in the cascade controlling *Bacillus subtilis* mother cell gene expression. J. Bacteriol. *179*, 6138-6144.

Zhang, J., Fitz-James, P.C., and Aronson, A.I. 1993. Cloning and characterization of a cluster of genes encoding polypeptides present in the insoluble fraction of the spore coat of *Bacillus subtilis*. J. Bacteriol. *175*, 3757-3766.

Zhang, J., Ichikawa, H., Halberg, R., Kroos, L., and Aronson, A.I. 1994. Regulation of the transcription of a cluster of *Bacillus subtilis* spore coat genes. J. Mol. Biol. *240*, 405-415.

Zheng, L., and Losick, R. 1990. Cascade regulation of spore coat gene expression in*Bacillus subtilis*. J. Mol. Biol. *212:* 645-660.

Zheng, L., Donovan, W.P., Fitz-James, P.C., and Losick, R. 1988. Gene encoding a morphogenetic protein required in the assembly of the outer coat of the *Bacillus subtilis* endospore. Genes Dev. *2:* 1047-1054.

Zheng, L., R. Halberg, S. Roels, H. Ichikawa, L. Kroos, and R. Losick. 1992. Sporulation regulatory protein GerE from *Bacillus subtilis* binds to and can activate or repress transcription from promoters for mother-cell-specific genes. J. Mol. Biol. *226:* 1037-50.

Zilhão, R., Serrano, M., Isticato, R., Ricca, E., Moran, C.P. Jr., and Henriques, A.O. 2004. Interactions among CotB, CotG, and CotH during assembly of the *Bacillus subtilis* spore coat. J. Bacteriol. *186:* 1110-1119.

Zverlov, V.V., Velikodvorskaya, G.A., and Schwartz, W.H. 2003. Two new cellulosome components downstream of celI in the genome of *Clostridium thermocellum*: the non-processive endoglucanase CelN and the possibly structural protein CseP. Microbiol. *149:* 515-524.

Chapter 7

The Spore Surface

Adam Driks

SUMMARY

Bacterial spores are encased in a protein coat and, for some species, an additional structure called the exosporium. The best-studied features of the spore, namely its resistance properties and the ability to revive after long periods of dormancy depend, to a large degree, on structures inside the spore. Less well-studied though is the spore surface. Among other functions, the spore surface is home to important enzymatic activities and may play essential roles in attachment and signaling to other organisms. Recent results stemming from basic studies of the coat reveal that the spore surface can be genetically manipulated, leading to important biotechnological applications. Finally, studies of the spore surface are providing clues to the dynamic nature of the spore and, as a result, shedding light on how the structural flexibility of the coat may be fundamental to proper spore function.

INTRODUCTION

The spore surface is the site of contact with and attachment to environmental substrates, the entry point for germinants on their way to receptors located deep within the spore and the frontline defense against environmental assault. Given its obvious importance to spore function it may come as a surprise that the spore surface is so poorly understood. Currently, as the relevance of spores to biological weapons and biotechnology occupies an increased share of the scientific community's attention, the spore surface is beginning to receive much more serious study. In this chapter I will review what is known and attempt to outline promising areas of future investigation.

The extraordinary resistance properties of bacterial spores are well documented (Nicholson et al., 2000 and Chapter 1). Understanding how these remarkable features arise has been a central motivation for research into spore formation and function. However, a variety of observations suggest that spores are not just inert capsules that passively protect DNA between periods of nutrient availability. Rather, spores are likely to be relatively active participants in the environments in which they persist. Certainly, they are, for the most part, biochemically dormant and highly resilient; but this does not necessarily prevent them from interacting with other organisms or from performing chemical reactions. Most likely, the ability of a dormant spore to influence its environment derives, to a large degree, from molecules on, or very close to, the surface. Likely functions for these proteins include adhesion, enzymatic processing of minerals and toxic compounds and, perhaps, signaling to other organisms. This chapter will discuss some of these potential natural functions of the surface as well as the utility of spore surfaces in biotechnological applications.

COMPOSITION OF SPORE SURFACES

All bacterial spores possess a coat (see Chapter 6) (Figure 1). In some species, this is the outermost structure and, therefore, the spore surface is the outer surface of the coat. In other species, the spore is encased in an additional structure called the exosporium. As the coat and exosporium are biochemically and structurally distinct, I will consider their surfaces separately.

The coat

In a number of strains, including at least certain isolates of *B. megaterium*, *B. licheniformis* and *B. subtilis* (DesRosier and Lara, 1981) (but see Holt and Leadbetter, 1969), the spore surface is that of the coat. In many if not most species, the coat is biochemically complex. The coats of *B. subtilis*, *B. cereus* *B. anthracis* and *B. licheniformis* may possess as many as 60 protein species (Eichenberger et al., 2003; Giorno and Driks, unpublished results; Kuwana et al., 2002; Lai, 2003). Even in *B. subtilis*, many of these remain poorly characterized (Driks, 2002b; Henriques and Moran, 2000; Takamatsu and Watabe, 2002) (Chapter 6). Transmission electron microscopy (TEM) reveals that coats possess morphologically distinct layers, whose number and thickness vary between species (Aronson and Fitz-James, 1976; Driks, 1999a; Holt and Leadbetter, 1969; Warth et al., 1963). These layers are assembled under the control of a subset of coat proteins with specific roles in

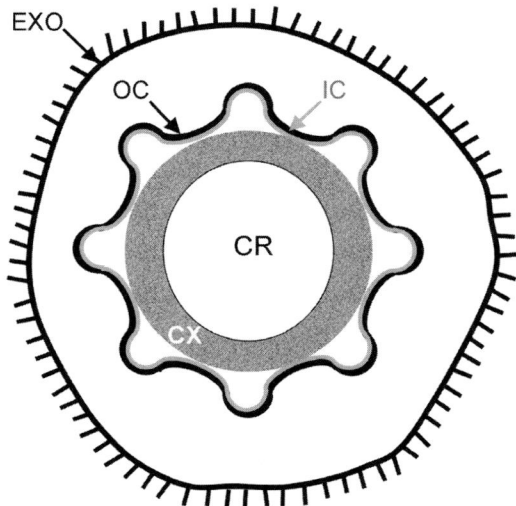

Figure 1. Cartoon of a spore, based on *B. anthracis*, showing the major structural features in cross section. EXO, exosporium; OC, outer coat (black line); IC, inner coat (gray line); CX, cortex; CR, core.

coat morphogenesis (Driks, 1999b; Driks, 2002b; Henriques and Moran, 2000; Takamatsu and Watabe, 2002). Presumably, assembly of the coat surface is also under the control of this morphogenetic program.

The single known coat surface protein is a previously identified *B. subtilis* protein called CotB (Donovan et al., 1987; Isticato et al., 2001; Zheng and Losick, 1990). Importantly, it has been shown that CotB can display heterologous protein species when these are fused to it (see Chapter 17). As a result, genetic manipulation of CotB permits addition of novel proteins to the spore surface. These heterologous species are in an active form in the sense that when recombinant spores are used as immunogen, antisera reacting against the fusion partner were readily obtained. This has significant implications for biotechnology, as will be discussed below and elsewhere in this volume (see, especially, Chapters 17 and 18). The ability to derivatize CotB has been used to characterize the immunological response of mice after exposure to spores by the oral and intranasal routes (Duc et al., 2003). These results clearly demonstrate the utility of spores as vehicles for antigen delivery, and underscore how the identification of even a single spore surface protein opens up important areas of investigation. Although only one coat surface protein has been identified so far, morphological studies indicate that several proteins impact on surface structure (Chada et al., 2003) (see below). One interpretation of this finding is that multiple protein species are present on the surface. The identification of several *B. subtilis* coat proteins with known or possible oxidase activity in *B. subtilis* (including CotA (Hullo et al., 2001; Martins et al., 2002), CotQ, and YisY (Lai, 2003)) raises the question of whether any of these proteins are also surface located. If they are, they could act on substrates that are larger than the small molecules, such as amino acids and sugars, which are presumed to diffuse through the porous coat to reach the germinant receptors located inside the spore (Moir et al., 2002).

It is likely that at least some coat surface proteins are glycosylated, based on biochemical evidence, as well as the presence of candidate carbohydrate biosynthetic genes in *B. subtilis*, which has no exosporium (Driks, 1999a; Fox et al., 2003; Wunschel et al., 1994). One of these genes, *cgeD*, is a member of a gene cluster whose expression is under the control of the late-acting mother-cell transcription factor GerE (Roels and Losick, 1995), which also directs expression of a large set of coat protein genes (Driks, 2002b; Kroos and Yu, 2000). Deletion of genes in the cluster alters spore surface properties, such as the ability to pellet after centrifugation or to clump, consistent with a possible role in modification of the coat surface (Roels and Losick, 1995). Glycosylation could also impact on the ability of germinants to penetrate the coat, as well as on attachment to environmental surfaces. However, no other consequence of these mutations was discovered, so the actual role, if any, of these genes in spore function remains unknown.

The exosporium

The exosporium, which encases the entire spore in those species that possess it, has a unique appearance in TEM. It consists of a membrane-like basal layer and, often, fine hair-like projections that extend from this layer (as in *B. cereus* and *B. anthracis*) (Figure 1). Strikingly, a significant gap separates the coat and the exosporium, at least in *B. cereus* and *B. anthracis*. Although the exosporium encases the spore, the contours of the exosporium do not necessarily follow those of the coat. Instead, the shape of the exosporium varies considerably from spore to spore within each field of view. This has led it to sometimes be described, perhaps misleadingly, as loosely connected to the spore. TEM does not reveal any electron-dense material connecting the coat and exosporium. Nonetheless, it seems intuitive that some connecting material should exist. Probably, an as yet uncharacterized substance, capable of significant changes in shape which fills the gap and connects the coat and exosporium.

With renewed interest in *B. anthracis*, biochemical characterization of the exosporium surface in this organism as well as in *B. cereus* is proceeding rapidly. Initial experiments in *B. cereus* identified a variety of spore-associated proteins, including several candidate exosporium proteins (Charlton et al., 1999). However, whether any of these proteins are surface-located remained unclear. In contrast, the first exosporium species to be identified and characterized in *B. anthracis*, the glycoprotein BclA (Steichen et al., 2003; Sylvestre et al., 2002), is a very strong candidate for an exosporium-surface protein. A *bclA* deletion mutant version of the Sterne strain of *B. anthracis* lacks the hair-like projections (Sylvestre et al., 2002). Additionally, in strains bearing partially truncated alleles of *bclA*, these hair-like structures are reduced in length, in proportion to the size of the deletion (Sylvestre et al., 2003). Further evidence for the surface location of BclA comes from analysis of rabbit anti-sera generated using intact *B. anthracis* spores. In these experiments, BclA is the immunodominant species. These studies raise the possibility that BclA is the most abundant and, perhaps, only protein on the *B. anthracis* exosporium surface. A homologue of *bclA* is present in the *B. cereus* genome and may play a role similar to that in *B. anthracis*.

The best studied example of an exosporium-surface protein is provided by the marine bacillus species SG-1, (Bacillus SG-1) whose spores harbor an oxidase called MnxG that catalyzes the oxidation of manganese(II) and, as a result, directs formation of a manganese precipitate that encases spores of this organism (van Waasbergen et al., 1996; van Waasbergen et al., 1993). This and other biochemical evidence argues that MnxG is an exosporium-surface-located oxidase (Francis et al., 2002). Most likely, MnxG activity allows SG-1 spores to participate in geochemical events in marine environments. More recently, it has become clear that this type of activity is present in several marine bacilli other than SG-1, as well as spore formers from non-marine environments (Francis and Tebo, 2002). These results argue that, in general, spore surface activities likely have a wider and more complex impact on the environment than has been appreciated so far. In this regard, it is informative to note that spores are ubiquitous, being present in widely ranging niches (Chapter 1 and Nicholson, 2002). In some cases spores, or at least sporulating bacteria, appear to engage in symbiotic relationships (Dunn and Handelsman, 2002; Rhee et al., 2004). The mounting evidence of spore-host interactions and the presence of spores in so many environments suggest the possibility that species-specific differences in surface proteins account for some important niche-specific adaptations (Driks, 1999a).

Comparison of the genomes of sequenced bacilli indicates that *B. subtilis*, *B. cereus* and *B. anthracis* share at least certain candidate carbohydrate biosynthetic genes, as already alluded to, and raises the possibility that the surfaces of spores that lack exosporia (i.e., those of *B. subtilis*) and those that possess them (such as *B. anthracis* and *B. cereus*) are modified in similar ways (Fox et al., 2003; Steichen et al., 2003) Although too little is known about the surfaces of spores from these species to reliably interpret this observation, one could speculate that, at least in part, a common mechanism is used to modify the coat and the exosporium. It should be noted that, in spite of the possibility that some aspects of coat and exosporium glycosylation are shared, their surfaces are clearly significantly different, given that BclA has no homologue in *B. subtilis* and the structures of the surfaces of spores of species with and without exosporia are morphologically quite distinct (Chada et al., 2003; Driks, 2002a) (see below). In addition, the chemical properties of the surfaces of *B. cereus* and *B. subtilis* spore are also very different (Faille et al., 2002) (see below).

The finding that only a subset of species possess exosporia begs the question of what differs between the life styles of bacilli with and without these structures. However, too little is known about the natural ecology of spores to allow more than speculation at this stage. Probably, exosporia enhance the functionality and survival of the spore in multiple ways, and are not simply an adaptation to any single feature of the environment.

SPORE SURFACE MORPHOLOGY

The surfaces of spore coats of a variety of species have been extensively characterized in early studies by a variety of microscopic methods, especially TEM (Aronson and Fitz-James, 1976; Holt et al., 1975; Holt and Leadbetter, 1969; Piggot and Coote, 1976; Santo and Doi, 1974; Warth et al., 1963), which suggested that the spore surface possesses ridges or folds. However, this is not true of every spore in a field and, furthermore, the number and curvature of the folds varies significantly between and within spores. In contrast to TEM analyses, scanning electron microscopy (SEM) suggested that ridges are a more consistent feature, but one that still varies to a significant degree (see, for example, (Comas-Riu and Vives-Rego, 2002; Franklin and Bradley, 1957; Roth and Williams, 1963)). SEM shows that most ridges extend along the long axis of the spore and are approximately equally spaced. There are typically about three ridges visible on a single spore face. In contrast, freeze-etch microscopic images do not show deep ridges (Aronson and Fitz-James, 1976; Holt and Leadbetter, 1969). Rather, the spore surface is relatively flat, although undulations in the spore surface are present that are probably flattened ridges. Furthermore, otherwise undetected fine details are apparent in these images. The coats of *B. subtilis*, *B. cereus* and other species are covered by fine striations of about 5 nm in

diameter, which are not oriented in any particular direction. The differences between TEM and SEM images, on the one hand, and freeze etch microscopy data, on the other hand, could potentially be explained, at least in part, by the significant chemical fixation and other preparative procedures that accompany the former techniques. Freeze-etch methods use cryo-fixation in an aqueous environment. From this perspective, the freeze-etch microscopic data could be considered more reliable.

Recent analysis of the spore surface by atomic force microscopy (AFM) suggests that important features seen by TEM and SEM, in particular the ridges, are real, in spite of the significant possibility of fixation-related artifacts in electron microscopy. AFM uses an extremely fine probe to scan surfaces and record topographical features (Morris et al., 1999). Typically, no specimen preparation at all is required. AFM images of air-dried spores clearly show ridges, similar to those seen in previous SEM studies (Chada et al., 2003). Furthermore, AFM reveals a series of small roughly circular bumps that populate both the ridges and the regions between them. The interfaces between these bumps may be sites of openings in the coat surface that could permit entry of small germinant molecules, such as sugars and amino acids, which must penetrate the coat. If the ridges that span the coat are real, why are they not seen by freeze-etch methods? One possibility is that the freeze-etch fracture plane removes a structure needed for appearance of the ridges. This seems unlikely, given that TEM indicates that the ridges are not solely a surface feature but rather are present throughout the thickness of the coat. A second possibility is that the lack of ridges and the fine surface striations are a result of the hydrated conditions of freeze-etch microscopy. If so, then the spore surface may be capable of significant macromolecular rearrangement.

The possibility of rearrangements of coat-surface proteins under hydrated conditions is particularly important when one considers that the coat appears to be a flexible shell. Coat flexibility can be inferred from two lines of study. First, it is well established that as the spore swells during germination, the coat alters morphologically to accommodate the increase in core volume (Santo and Doi, 1974). Apparently, the coat permits this increase in volume by unfolding the ridges. As would be expected, AFM analysis also reveals that the surfaces of germinated spores are smooth (Chada et al., 2003). Second, spore dimensions alter in response to changes in relative humidity (Driks, 2003; Westphal et al., 2003). The notion that an accordion pleat-like architecture underlies the ability to accommodate volume changes is appealing because of its apparent simplicity. However, when considered in the context of the complex ultrastructure of the coat, the molecular details of this type of global flexibility are difficult to envision. A logical first step in addressing this would be careful characterization of the coat surface in the folded and unfolded states. In this regard, the surface details of dormant spores in a wet environment, seen so far only in freeze-etch studies, may be especially useful, if these features are indeed the result of hydration (and, therefore, unfolding of the ridges), as seems likely.

AFM analysis reveals several additional interesting features of the coat surface (Chada et al., 2003). First, comparative

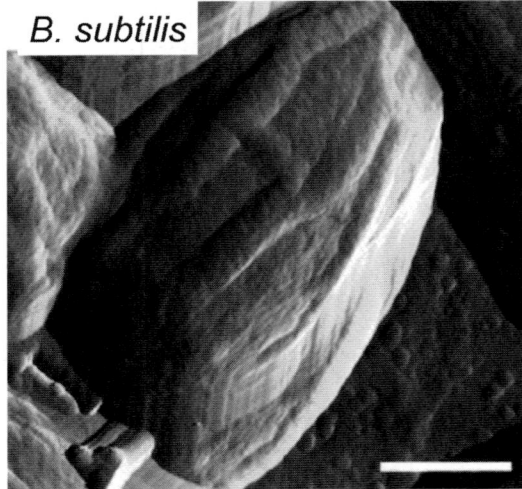

Figure 2. AFM analysis of *B. subtilis* spores. Spores were imaged in tapping mode and amplitude information collected. Scale bar = 375 nm. Reproduced, with permission from from Chada et al., 2003.

analysis of the surfaces of individual strains of *B. anthracis*, *B. cereus* and *B. subtilis* spores shows that each of these organisms has solved the problem of building ridges somewhat differently (see Figure 2 and Figure 3). In *B. subtilis*, the ridges are relatively smooth (Figure 2) hemi-cylinders that run from pole to pole along the long axis of the spore. The same is true for *B. anthracis*, except that in many cases each ridge appears to be composed of two hemi-cylinders with consistent diameter, running side by side (Figure 3). The ridges in *B. cereus* resemble those of *B. subtilis* in that they do not appear to be double structures (Figure 3). However, they are distinct from ridges in both *B. subtilis* and *B anthracis* in that each ridge takes a zig-zag course as it transits from pole to pole. The angular bends in the *B. cereus* ridges may alter the dynamic flexibility of the coat in this species. What is particularly striking is that

B. anthracis and *B. cereus*, which are especially closely related (Helgason et al., 2000), have such different ridge morphologies. Although these observations are extremely suggestive, they will be more robust when a larger number of isolates from each species has been analyzed. As the data from *B. cereus* and *B. anthracis* indicate, even genetically similar strains can have very different surface morphologies, indicating that deeper analysis of otherwise taxonomically indistinguishable strains can still yield surprising results.

A second finding from AFM is the impact of two previously identified coat proteins, CotA and CotB, on coat surface morphology in *B. subtilis* (Chada et al., 2003). Spores bearing a null allele of *cotA* still possess ridges, but they are significantly disorganized. They appear branched and of varying diameter. *cotB* null mutant spores also possess ridges, but they tend to be in closely apposed pairs that are connected by a series of very fine ridges, running at right angles to the major ridges. Deletion of both *cotA* and *cotB* in a single strain eliminates the ridges, leaving the spore surface relatively smooth, and divided into several roughly square patches. CotA is the multi-copper oxidase and CotB is the coat surface protein, already discussed above. The fact that absence of either CotA or CotB alters but does not eliminate the ridges indicates that ridges are probably not composed of a single protein species. One possibility is that the coat is approximately homogenous (i.e. the major coat proteins are equally distributed within each coat layer and not restricted to the regions of the ridges) and that ridge formation is a global property of the coat. Given that the ridges are not, however, entirely uniform, local deviations from homogeneity are very likely. These deviations could be the result of variations in coat composition that arise during assembly or in the process of core contraction.

Structural studies of the exosporium are less advanced than those of the coat. AFM analysis indicates that the exosporium is relatively smooth, being comprised of a series of very small bumps (Chada et al., 2003). These bumps may correspond to the hair-like projections seen by TEM. However, this initial

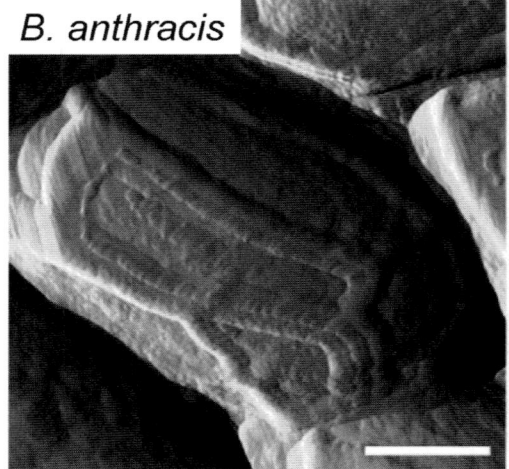

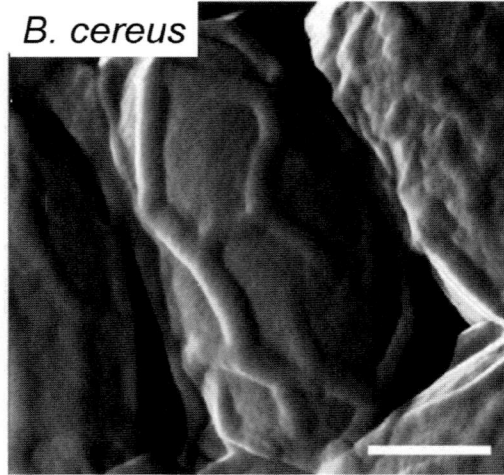

Figure 3. AFM analysis of *B. anthracis and B. cereus* spores. Spores were imaged in tapping mode and amplitude information collected. Scale bar = 375 nm for *B. anthracis* and 382nm for *B. cereus*. Reproduced, with permission from from Chada et al., 2003.

AFM study needs to be extended with analysis of strains bearing exosporium protein mutations, particularly alleles of *bclA*, and with analysis of additional species.

TECHNOLOGICAL USE OF SPORE SURFACES

A major challenge for nanotechnology is identification of suitable platforms for presentation of technologically useful biomolecules. Appropriate platforms will often need to be highly stable, readily manufactured, easily derivatized and possess significant useable surface area. Clearly, spores readily meet these criteria . They possess an additional feature as well; they are essentially self-replicating. The recent demonstration of the ease of presenting immunologically visible fusion partners to CotB on the spore surface clearly shows that the spore can be readily reengineered as a useful nanomachine (Duc et al., 2003; Isticato et al., 2001). Several applications of surface-modified spores are obvious: delivery of drugs, vaccine antigens and other immunomodulating activities (perhaps leading to improved probiotics). Other applications might include trapping enzymatic activities for improved bioreactors, use as bioactive filters, and as environmental monitors. Potentially the spore surface might also be used for display a series of enzymes involved in a reaction cascade thus minimizing diffusion distances between substrate and enzyme ensuring enhanced enzyme catalysis. This short, and very incomplete, list is meant to underscore the enormous range of technological applications still to be investigated. As our understanding of the spore surface increases, as it surely will in the next few years, new technological applications for spores will appear.

FUTURE DIRECTIONS

Clearly, reductionist dissection of the spore surface will continue, as coat and exosporium proteins continue to be identified, and their functions are clarified. However, because important spore functions appear to be emergent properties (Thomas, 1993) (Chada et al., 2003), these studies are necessarily limited and should be complemented by holistic approaches that characterize global spore surface properties. As should be clear from the discussion above, AFM has only just begun to be used to analyze spore surfaces and much remains to be done. In particular, using AFM to quantify biophysical features of the spore surface, such as hardness and electrostatic charge, will be important. Methodologies from more physical sciences may also be very useful in holistic analyses. For example, the field of colloid chemistry has an array of sensitive assays to measure surface characteristics of small particles, including electrostatic charge, which would be particularly relevant to spores (Morrison and Ross, 2002).

Another group of approaches come from efforts to combat bacterial biofouling of industrial equipment, work that has directed significant attention to the mechanisms of adhesion of cells to artificial surfaces. A by-product of this effort has been investigation of some of the physical characteristics of spore surfaces (Faille et al., 2002). A particularly interesting result from this study is that, based on solvent affinity, the *B. cereus* spore surface appears to be hydrophobic, whereas the *B. subtilis* surface is hydrophilic. These differing surface characteristics could have very significant consequences for interactions between spores and their environments. The possibility that the hydrophobic nature of the *B. cereus* spore surface is due to its exosporium could be a clue to the *in vivo* role of this structure. However, these results stand in apparent contrast to an earlier study, using electrophoretic mobility, indicating that *B. cereus* and *B. subtilis* spore surfaces are hydrophilic (Douglas, 1957). Resolving this issue will require analysis of otherwise well studied strains, using multiple methods to characterize the chemical nature of the spore surface. An informative example of such an analysis was carried out on SG-1 spores that, as already discussed, possess an exosporium (He and Tebo, 1998). This study showed that SG-1 spores are hydrophilic. If *B. cereus* spores are indeed hydrophobic, then the surfaces of exosporia may differ significantly between species.

Other physical and chemical analyses will also be important to bring to bear on the study of the spore surface. The analysis of the SG-1 spore surface, cited above, includes measurement of the spore surface area. Strikingly, the results suggest that when wet, the spore expands and becomes porous, whereas in the dry state pores are not evident (He and Tebo, 1998). It is not clear whether these pores are present in the exosporium, the coat, or both. Nonetheless, this approach may have significant potential as a tool for characterization of ultrastructural dynamics during dormancy under varying environmental conditions (such as the changes in hydration already described: Driks, 2003; Westphal et al., 2003) and during germination.

Ultimately, understanding the role of the spore surface will require correlating its global characteristics (such as charge, permeability or hardness) with the ability to survive in the environment. These experiments will be difficult, but they hold the promise of helping us to understand how spore coat and exosporium molecules, whose roles are often so cryptic when analyzed individually, function as a unit to generate structures with such varied and extraordinary properties.

ACKNOWLEDGEMENTS

Thanks to David Garcia for helpful conversations. Work in the author's laboratory is supported by grants GM 53989 and AI53365 from the NIH. We thank the American Society for Microbiology (ASM) for permission to reproduce Figures 2 and 3.

REFERENCES

Aronson, A. I., and Fitz-James, P. (1976). Structure and morphogenesis of the bacterial spore coat. Bacteriol Rev *40*, 360-402.

Chada, V. G., Sanstad, E. A., Wang, R., and Driks, A. (2003). Morphogenesis of *bacillus* spore surfaces. J Bacteriol *185*, 6255-6261.

Charlton, S., Moir, A. J., Baillie, L., and Moir, A. (1999). Characterization of the exosporium of *Bacillus cereus*. J Appl Microbiol *87*, 241-245.

Comas-Riu, J., and Vives-Rego, J. (2002). Cytometric monitoring of growth, sporogenesis and spore cell sorting in *Paenibacillus polymyxa* (formerly *Bacillus polymyxa*). J Appl Microbiol *92*, 475-481.

DesRosier, J. P., and Lara, J. C. (1981). Isolation and properties of pili from spores of *Bacillus cereus*. J Bacteriol *145*, 613-619.

Donovan, W., Zheng, L., Sandman, K., and Losick, R. (1987). Genes encoding spore coat polypeptides from *Bacillus subtilis*. J Mol Biol *196*, 1-10.

Douglas, H. W. (1957). Electrophoretic studies on spores and vegetative cells of certain strains of *B. megaterium*, *B. subtilis* and *B. cereus*. J Appl Bacteriol *20*, 390-403.

Driks, A. (1999a). The *Bacillus subtilis* spore coat. Microbiol Mol Biol Rev *63*, 1-20.

Driks, A. (1999b). Spatial and temporal control of gene expression in prokaryotes. In Development: genetics, epigenetics and environmental regulation, V. E. A. Russo, D. J. Cove, L. G. Edgar, R. Jaenisch, and F. Salamini, eds. (Berlin, Springer).

Driks, A. (2002a). Maximum shields: the armor plating of the bacterial spore. Trends Microbiol *10*, 251-254.

Driks, A. (2002b). Proteins of the spore core and coat. In *Bacillus subtilis* and its closest relatives, A. L. Sonenshein, J. A. Hoch, and R. Losick, eds. (Washington, American Society for Microbiology), pp. 527-536.

Driks, A. (2003). The dynamic spore. Proc Natl Acad Sci U S A *100*, 3007-3009.

Duc, L.H., H., Hong, H. A., Fairweather, N., Ricca, E., and Cutting, S. M. (2003). Bacterial spores as vaccine vehicles. Infect Immun *71*, 2810-2818.

Dunn, A. K., and Handelsman, J. (2002). Toward an understanding of microbial communities through analysis of communication networks. Antonie Van Leeuwenhoek *81*, 565-574.

Eichenberger, P., Jensen, S. T., Conlon, E. M., van Ooij, C., Silvaggi, J., Gonzalez-Pastor, J. E., Fujita, M., Ben-Yehuda, S., Stragier, P., Liu, J. S., and Losick, R. (2003). The sigmaE regulon and the identification of additional sporulation genes in *Bacillus subtilis*. J Mol Biol *327*, 945-972.

Faille, C., Jullien, C., Fontaine, F., Bellon-Fontaine, M. N., Slomianny, C., and Benezech, T. (2002). Adhesion of *Bacillus* spores and *Escherichia coli* cells to inert surfaces: role of surface hydrophobicity. Can J Microbiol *48*, 728-738.

Fox, A., Stewart, G. C., Wallera, L. N., Fox, K. F., Harley, W. M., and Price, R. L. (2003). Carbohydrates and glycoproteins of *Bacillus anthracis* and related bacilli: targets for biodetection. J Microbiol Meth *54*, 143– 152.

Francis, C. A., Casciotti, K. L., and Tebo, B. M. (2002). Localization of Mn(II)-oxidizing activity and the putative multicopper oxidase, MnxG, to the exosporium of the marine *Bacillus* sp. strain SG-1. Arch Microbiol *178*, 450-456.

Francis, C. A., and Tebo, B. M. (2002). Enzymatic manganese(II) oxidation by metabolically dormant spores of diverse *Bacillus* species. Appl Environ Microbiol *68*, 874-880.

Franklin, J. G., and Bradley, D. E. (1957). A further study of the spores of species of the genus *bacillus* in the electron microscope using carbon replicas, and some preliminary observation on *Clostridium welchii*. J Appl Bacteriol *30*, 467-472.

He, L. M., and Tebo, B. M. (1998). Surface charge protperties of and Cu(II) absorbtion by spore of the marine *Bacillus* sp. strain SG-1. Appl Environ Microbiol *64*, 1123-1129.

Helgason, E., Okstad, O. A., Caugant, D. A., Johansen, H. A., Fouet, A., Mock, M., Hegna, I., and Kolsto (2000). *Bacillus anthracis, Bacillus cereus, and Bacillus thuringiensis*-one species on the basis of genetic evidence. Appl Environ Microbiol *66*, 2627-2630.

Henriques, A. O., and Moran, C. P., Jr. (2000). Structure and assembly of the bacterial endospore coat. Methods *20*, 95-110.

Holt, S. C., Gauther, J. J., and Tipper, D. J. (1975). Ultrastructural studies of sporulation in *Bacillus sphaericus*. J Bacteriol *122*, 1322-1338.

Holt, S. C., and Leadbetter, E. R. (1969). Comparative ultrastructure of selected aerobic spore-forming bacteria: a freeze-etching study. Bacteriol Rev *33*, 346-378.

Hullo, M. F., Moszer, I., Danchin, A., and Martin-Verstraete, I. (2001). CotA of *Bacillus subtilis* is a copper-dependent laccase. J Bacteriol *183*, 5426-5430.

Isticato, R., Cangiano, G., Tran, H. T., Ciabattini, A., Medaglini, D., Oggioni, M. R., De Felice, M., Pozzi, G., and Ricca, E. (2001). Surface Display of Recombinant Proteins on *Bacillus subtilis* Spores. J Bacteriol *183*, 6294-6301.

Kroos, L., and Yu, Y. N. (2000). Regulation of sigma factor activity during *Bacillus subtilis* development. Curr Opin Microbiol *3*, 553-560.

Kuwana, R., Kasahara, Y., Fujibayashi, M., Takamatsu, H., Ogasawara, N., and Watabe, K. (2002). Proteomics characterization of novel spore proteins of *Bacillus subtilis*. Microbiology *148*, 3971-3982.

Lai, E.-M., Phadke, N. D., Kachman, M. T., Giorno, R., Vazquez. S., Vazquez, J. A., Maddock, J. R. and Driks, A. (2003). Proteomic Analysis of the Spore Coats of *Bacillus subtilis* and *Bacillus anthracis*. J Bacteriol *185*, 1443-1454.

Martins, L. O., Soares, C. M., Pereira, M. M., Teixeira, M., Costa, T., Jones, G. H., and Henriques, A. O. (2002). Molecular and biochemical characterization of a highly stable bacterial laccase that occurs as a structural component of the *Bacillus subtilis* endospore coat. J Biol Chem *277*, 18849-18859.

Moir, A., Corfe, B. M., and Behravan, J. (2002). Spore germination. Cell Mol Life Sci *59*, 403-409.

Morris, V. J., Gunning, A. P., and Kirby, A. R. (1999). Atomic Force Microscopy for Biologists, Imperial College Press).

Morrison, I. D., and Ross, S. (2002). Colloidal Dispersions : Suspensions, Emulsions, and Foams, Wiley-Interscience).

Nicholson, W. L. (2002). Roles of *Bacillus* endospores in the environment. Cell Mol Life Sci *59*, 410-416.

Nicholson, W. L., Munakata, N., Horneck, G., Melosh, H. J., and Setlow, P. (2000). Resistance of *Bacillus* endospores to extreme terrestrial and extraterrestrial environments. Microbiol Mol Biol Rev *64*, 548-572.

Piggot, P. J., and Coote, J. G. (1976). Genetic aspects of bacterial endospore formation. Bacteriol Rev *40*, 908-962.

Rhee, K.-J., Sethupathi, P., Driks, A., Lanning, D.K., and Knight, K.L. (2004). Role of commensal bacteria in development of gut-associated lymphoid tissues and preimmune antibody repertoire. J. Immunol. *172*, 1118-1124.

Roels, S., and Losick, R. (1995). Adjacent and divergently oriented operons under the control of the sporulation regulatory protein GerE in *Bacillus subtilis*. J Bacteriol *177*, 6263-6275.

Roth, I. L., and Williams, R. P. (1963). Comparison of the fine structure of virulent and avirulent spores of *Bacillus anthracis*. Texas Rept Biol Med *21*, 394-399.

Santo, L. Y., and Doi, R. H. (1974). Ultrastructural analysis during germination and outgrowth of *Bacillus subtilis* spores. J Bacteriol *120*, 475-481.

Steichen, C., Chen, P., Kearney, J. F., and Turnbough, C. L., Jr. (2003). Identification of the immunodominant protein and other proteins of the *Bacillus anthracis* exosporium. J Bacteriol *185*, 1903-1910.

Sylvestre, P., Couture-Tosi, E., and Mock, M. (2002). A collagen-like surface glycoprotein is a structural component of the *Bacillus anthracis* exosporium. Mol Microbiol *45*, 169-178.

Sylvestre, P., Couture-Tosi, E., and Mock, M. (2003). Polymorphism in the collagen-like region of the *Bacillus anthracis* BclA protein leads to variation in exosporium filament length. J Bacteriol *185*, 1555-1563.

Takamatsu, H., and Watabe, K. (2002). Assembly and genetics of spore protective structures. Cell Mol Life Sci *59*, 434-444.

Thomas, J. H. (1993). Thinking about genetic redundancy. Trends Genet *9*, 395-399.

van Waasbergen, L. G., Hildebrand, M., and Tebo, B. M. (1996). Identification and characterization of a gene cluster involved in manganese oxidation by spores of the marine *Bacillus* sp. strain SG-1. J Bacteriol *178*, 3517-3530.

van Waasbergen, L. G., Hoch, J. A., and Tebo, B. M. (1993). Genetic analysis of the marine manganese-oxidizing *Bacillus* sp. strain SG-1: protoplast transformation, Tn917 mutagenesis, and identification of chromosomal loci involved in manganese oxidation. J Bacteriol *175*, 7594-7603.

Warth, A. D., Ohye, D. F., and Murrell, W. G. (1963). The composition and structure of bacterial spores. J Cell Biol *16*, 579-592.

Westphal, A. J., Price, P. B., Leighton, T. J., and Wheeler, K. E. (2003). Kinetics of size changes of individual *Bacillus thuringiensis* spores in response to changes in relative humidity. Proc Natl Acad Sci USA *100*, 3461-3466.

Wunschel, D., Fox, K. F., Black, G. E., and Fox, A. (1994). Discrimination among the *Bacillus cereus* group, in comparison to *B. subtilis*,by structural carbohydrate profiles and ribosomal RNAspacer region PCR. Syst Appl Microbiol *17*, 625– 635.

Zheng, L., and Losick, R. (1990). Cascade regulation of spore coat gene expression in *Bacillus subtilis*. J Mol Biol *212*, 645-660.

Chapter 8

Safety of Aerobic Endospore-Forming Bacteria

Niall A. Logan

SUMMARY

The safety of aerobic endospore-forming bacteria to humans and other animals is a matter of increasing concern, and this occurs at a time of rapid changes in the classification and nomenclature of these organisms. Given the increasing interest in *Bacillus* species as probiotic supplements for both human and animal use, it seems pertinent to consider what is presently known about the safety issues that arise from their use. The pathogenicities of *Bacillus anthracis* and *Bacillus cereus* are well established, but less is generally known about the risks associated with other aerobic endospore-forming species and their products. This review provides a comprehensive survey of current knowledge about the involvement of species of *Bacillus* and related genera in human opportunistic infections, and the association of *Bacillus* enzymes with occupational asthma. Also covered are the issues of revised species nomenclature and product mislabelling, both of which are areas of potential misunderstanding or even abuse by commercial manufacturers of probiotic supplements.

INTRODUCTION

The production of resistant endospores in the presence of oxygen remains the practically-useful, defining feature for *Bacillus* and the new genera derived from it (*Bacillus sensu lato*). However, this definition has been undermined by the discoveries of *Bacillus infernus* and *B. arseniciselenatis,* which are strictly anaerobic (Boone et al., 1995; Switzer Blum et al., 1998), and spores have not been observed in *B. infernus* and several other species. Nonetheless, those species of *Bacillus sensu lato* likely to be used as probiotics are rod-shaped, endospore-forming organisms which may be aerobic or facultatively anaerobic. They are usually Gram-positive (in young cultures), but sometimes Gram-variable or frankly Gram-negative. They are mostly catalase-positive, and may be motile by means of peritrichous flagella. Most species are mesophilic, and although the group contains some thermophiles and psychrophiles, they need not detain us in the present context.

As will become apparent later in this chapter, nomenclature is of critical importance to any discussion of the use of aerobic endosporeformers as probiotics. *Bacillus* has been subdivided on the basis of 16S rRNA sequencing studies, and ten new genera have been proposed so far: *Alicyclobacillus* (Wisotzkey et al., 1992), which contains eight species of thermoacidophiles; *Paenibacillus* (Ash et al., 1993), containing 44 species and including organisms formerly called *B. polymyxa, B. macerans, B. alvei,* and the honey bee pathogens *B. larvae* and *B. pulvifaciens* (now both subspecies of *P. larvae*); *Brevibacillus*

(Shida et al., 1996), containing 11 species and including organisms formerly called *B. brevis* and *B. laterosporus; Aneurinibacillus* (Shida et al., 1996), with *A. aneurinilyticus* and two other species; *Virgibacillus* (Heyndrickx et al., 1998) with *V. pantothenticus* and four other species; *Gracilibacillus* and *Salibacillus* (Wainø et al, 1999), which each contain two species of halophiles; *Geobacillus* (Nazina et al., 2001) with ten species of thermophiles including *Bacillus stearothermophilus*; *Ureibacillus* (Fortina et al., 2001) with two round-spored, thermophilic species, and single-membered *Marinibacillus* (Yoon et al., 2001b). *Salibacillus* was subsequently merged with *Virgibacillus* (Heyrman et al., 2003). *Sporosarcina* contains the motile, sporeforming coccus *S. ureae,* which is closely related to *B. sphaericus,* and seven rod-shaped, round-spored species, including *Bacillus pasteurii*, that have now been placed in this genus (Yoon et al., 2001a). Several other genera of aerobic endospore formers, not derived from *Bacillus* (but mostly with generic names containing the suffix 'bacillus' to indicate that they are sporeformers), have also been proposed; most of them based upon one or a very few isolates. They are: *Ammoniphilus, Amphibacillus, Filobacillus, Halobacillus, Jeotgalibacillus , Lentibacillus, Oceanobacillus, Sulfobacillus* and *Thermobacillus.*

Bacillus still contains most of the familiar species such as: *B. subtilis* (the type species) and its relatives *B. licheniformis* and *B. pumilus*; members of the *B. cereus* group, which include *B. anthracis* and *B. thuringiensis*, all of which are really pathovars of a single species (Turnbull et al. 2002); and *B. megaterium* and *B. sphaericus*. It continues to be a large genus, with over 80 species, because proposals for new *Bacillus* species have kept pace with transfers of species to other genera.

Regrettably, this rapid taxonomic progress has not been accompanied by the discovery of salient characters that might allow a distinctive definition for each of the new genera, and so permit the ready recognition of their members. Members of these genera may show wide ranges of sporangial morphologies and other phenotypic profiles, and many recently-described species are based upon very few strains, frequently only single isolates. Furthermore, new species and subspecies often represent genomic groups disclosed by DNA-DNA pairing experiments, so that routine phenotypic characters for distinguishing many of them may be few in number and of unproven value.

The identification of species of *Bacillus sensu lato* remains a challenge for many routine diagnostic laboratories (Logan, 2002). The chief difficulties are the ever-expanding numbers of species (which often represent the splitting of an established species following molecular analysis), the need for special test kits and media, the incompleteness of commercially-available and published databases, and considerable between-strain variation in many familiar species such as *B. subtilis*. A further complication is that reference strains held in culture collections around the world are not always authentic strains of the species named on the label (Logan, 2002), a problem also encountered with the labelling of some probiotic preparations.

HABITATS AND IMPORTANCE OF THE AEROBIC ENDOSPOREFORMERS

Most aerobic endosporeformers are saprophytes widely distributed in the natural environment, but some species are opportunistic or obligate pathogens of animals, including humans, other mammals, and insects; however, the majority of known species of *Bacillus* and related genera are heterotrophic organisms that have been isolated on complex organic media. The main habitats are soils of all kinds, ranging from acid to alkaline, hot to cold, and fertile to desert, and the water columns and bottom deposits of fresh and marine waters. Their spores readily survive distribution in soils, dusts and aerosols from these natural environments to a wide variety of other habitats, and Nicholson (2002) has considered the roles of *Bacillus* spores in the natural environment. This aspect is discussed further in Chapter 1. In studies of indoor and outdoor air and dusts, *Bacillus* species commonly dominate the cultivable flora or form a large part of it; the commonly isolated species, such as *B. subtilis* and *B. cereus* are very widely distributed worldwide, and *B. thuringiensis* has been isolated from all continents, including Antarctica (Forsyth and Logan, 2000). Some species appear to be ubiquitous contaminants of man, other animals, their foodstuffs, water and environments: natural, domestic, industrial and hospital. The resistance of the spores to heat, radiation, disinfectants and desiccation results in their being troublesome contaminants in the operating room, on surgical dressings, and in pharmaceutical products. Spores are typically more resistant to heat than vegetative cells by a factor of 10^5 or more, while resistance to UV and ionizing radiation may be 100-fold or more and, if protected from radiation, spores may show remarkable longevities. These organisms are also important as deleterious agents and contaminants in the food industry; dried foods such as spices, milk powders and farinaceous products are often quite heavily contaminated with spores.

Aerobic endosporeformers have been and continue to be important in many fields of basic research, and long-term research into the sporulation process in *B. subtilis* has led to its being probably the best understood developmental system. The genomes of four *Bacillus* species, *B. subtilis*, *B. halodurans*, *B. anthracis* and *B. cereus*, have been sequenced (Kunst et al., 1997; Takami et al., 2000; Read et al., 2003; Ivanova et al., 2003). The restriction endonucleases and DNA polymerases of several *Bacillus* species are of considerable importance as research tools for better understanding of disease and improved diagnosis. *Bacillus* species lend themselves well to host-vector systems for production of bio-engineered therapeutic products.

Bacillus species are of considerable commercial importance in the production of industrial enzymes; the detergent industry is the largest consumer and it accounts for one third of the global

market in these products. The proteases and amylases it uses are all of *Bacillus* origin, and allergies to such widely-used proteins can occur. *Bacillus* amylases, proteases and glucanases are also widely used by the food and potable alcohol industries (Outtrup and Jørgensen, 2002). *Bacillus thuringiensis* is preeminent as a biopesticide, and even though the global market for *Bacillus* insecticides is quite small, representing less than 2% of the global insecticide market (Bishop, 2002), the intentional distribution of this close relative of the established pathogen, *B. cereus*, has potential implications for health. The genes for the δ-endotoxins of *B. thuringiensis* have been in the forefront of the development of transgenic crop plants (Van Rie, 2002), which would avoid the need to apply *B. thuringiensis* itself to crops.

Nitrogen-fixing *Bacillus* and *Paenibacillus* growing in the rhizosphere may help to promote plant growth; and aerobic endosporeformers may promote plant growth in other ways (Chanway, 2002), including the production of phytohormones, increasing nutrient availability, interactions with symbiotic bacteria and fungi, enhancement of root nodulation, and biological control of plant pathogens by various mechanisms including the production of antibiotics.

Most species of aerobic endospore-forming bacteria appear to have little or no pathogenic potential and are rarely associated with disease, but *B. anthracis* is, to all intents and purposes, an obligate pathogen of animals and humans. Some authorities consider that microcycling of this organism in soil is very rare, and that the highly conserved nature of the species is consistent with it having infrequent multiplication phases in animals, but others believe that multiplication may occur in certain soils (Cherkasskiy, 1999 and see Chapter 3). Its close relative, *B. cereus*, is now well established as an opportunistic pathogen, and other aerobic endosporeformers, particularly *B. licheniformis*, can also be opportunistic pathogens from time to time, or agents of food-borne illness; the use of these or related organisms in therapeutic preparations must therefore be very carefully considered. Six organisms are important as insect pathogens: *B. thuringiensis* (another close relative of *B. anthracis*), *B. sphaericus*, the two *P. larvae* subspecies, *P. lentimorbus* and *P. popilliae*. L-form *Bacillus* cells have been reported from humans, other animals and plants (see *B. licheniformis* and *B. subtilis* below).

Aside from the pathogenicity of some species, several aerobic endosporeformers are of clinical or health importance in very positive ways: in the production of antibiotics (bacitracin from *B. licheniformis* or *B. subtilis*, gramicidin from *Brevibacillus brevis*, and polymyxin from *Paenibacillus polymyxa*) or vitamins (e.g. vitamins B12 and B2 from *B. megaterium*, biotin and riboflavin from *B. subtilis*); as the bases of antibiotic assays (*B. cereus*, *B. circulans*, *B. megaterium*, *B. pumilus*, *B. subtilis* and *G. stearothermophilus*); in the validation of disinfectants (*B. cereus*), or the monitoring of fumigation (*B. subtilis*), heat sterilization (*G. stearothermophilus*) or radiation (*B. pumilus*) processes; and in various clinical tests (such as a uric acid assay using *B. fastidiosus*, a *Chlamydia* detection assay employing a variant of subtilisin from *B. subtilis*, or a blood screening test for phenylketonuria utilizing *B. subtilis*). Spores of *B. subtilis* have been evaluated as vehicles for vaccines (Duc et al., 2003; and Chapter 17). Finally, of course, several species of *Bacillus sensu lato* are the active ingredients of probiotics for animals and humans, and the safety implications of such applications are an important theme of this chapter.

PATHOGENICITY OF AEROBIC ENDOSPOREFORMERS

Bacillus anthracis

Anthrax is a disease of great antiquity, and convincing descriptions of it are found in ancient literature such as Virgil's *Georgics* (36-29 BC). It is primarily a disease of herbivores, and before a veterinary vaccine became available in the late 1930s, it was one of the foremost causes worldwide of mortality in domesticated herbivores. Veterinary and human vaccines, improvements in factory hygiene, sterilization of imported animal products, and man-made alternatives to animal hides or hair have resulted in a marked decline in the incidence of the disease. However, the disease continues to be endemic in several countries of Africa, Asia, and central and southern Europe, particularly those that lack an efficient vaccination policy, and non-endemic regions must be constantly on the alert for the arrival of *B. anthracis* in imported animal products.

In a typical natural infectious cycle a grazing animal ingests the spores which may gain access to the lymphatics, and so the spleen, though gut abrasions. The organism multiplies and produces toxin in the spleen for a few days, and then the animal collapses with a sudden and fatal septicaemia, and haemorrhagic exudates escape from the mouth, nose and anus and so contaminate the soil, where the vegetative cells sporulate. Not all the animals develop clinical disease, and mild and symptomless infections may occur (Turnbull et al., 1992). The spores may remain viable in soil for many years and so the persistence of the organism does not depend on animal reservoirs; *B. anthracis* is therefore exceedingly difficult to eradicate from an endemic area. Direct animal-to-animal transmission within a species is very rare; however, but carnivorous scavengers can show transient carriage of the organism or occasional overt disease (Lindeque and Turnbull, 1994).

Human anthrax

In public consciousness, *B. anthracis* is associated more with warfare and terrorism than with a disease of herbivores, and it is feared accordingly. It has long been considered as a potential weapon, and its first use is believed to have been against livestock during World War I (Christopher et al., 1997); it has been included in various development and offensive programmes in several countries since, and it has also been used in terrorist attacks (Lane and Fauci, 2001; Barnaby, 2002; Takahashi et al., 2004.). Apart from such artificial attacks, humans almost invariably contract anthrax directly or indirectly from animals, and direct human-to-human transmission is exceptionally rare. Naturally-acquired human anthrax may result from close contact with infected animals or their carcasses after death from the disease, or be acquired by those employed in processing wool ('wool sorter's disease'),

hair, hides, bones, or other animal products. The vast majority of cases are cutaneous infections, but inhalational, meningitic, and gastrointestinal anthrax are occasionally reported (for more information, and details of pathogenesis, see Logan and Turnbull, 2003). Anthrax weapons are normally intended to cause the inhalational disease but are likely to cause cutaneous cases as well; the former type of infection also used to occur, rarely, in the wool industry from the inhalation of spore-laden dust. There were a few laboratory-acquired infections in the past (Collins, 1988); 66 anthrax deaths occurred in April 1979 in the city of Sverdlovsk (now Yekaterinburg) in the Urals following the accidental release of spores from a military production facility (Meselson et al., 1994).

Anthrax is not highly contagious, and circumstantial evidence shows that humans are moderately resistant to anthrax as compared with obligate herbivores; infectious doses in the human inhalational and intestinal forms are generally very high (LD$_{50}$ 2 500 to 55 000 spores). Cutaneous anthrax is readily treated and is only life-threatening in exceptional cases. The infectious doses in the human inhalational and intestinal forms (also treatable if recognized early) are generally very high, and, generally speaking, precautions need to be sensible, not extreme. However, non-clinical materials associated with an attempt at deliberate release of *B. anthracis* spores may be very hazardous and no attempt to process them should be made without the appropriate instructions from the correct authorities. The laboratory examination of clinical specimens from suspected or potential victims of such an attack will be as described for cases of naturally-acquired anthrax. When collecting environmental specimens related to suspected natural anthrax, disposable gloves, disposable apron or overalls, and boots which can be disinfected after use should be worn. For dusty samples that might contain many spores, the use of head-gear and dust masks should be considered.

For details of pathogenesis, symptoms, diagnosis and treatment see Logan and Turnbull (2003). For further information on safety aspects see Turnbull et al. (1998), and visit www.who.int, www.bt.cdc.gov and www.hopkins-biodefense.org.

Aerobic endosporeformers as opportunistic pathogens

With the obvious exception of *B. anthracis*, the majority of aerobic endosporeformers are common environmental contaminants, and isolation from a single clinical specimen is generally not enough for incriminating one of these organisms as the aetiological agent. When an aerobic sporeformer is isolated in a clinical laboratory it is necessary to ask: was it isolated in pure culture or at least apparently dominating the flora? was it isolated in large numbers? was it isolated more than once? However, moderate or heavy growth of aerobic endosporeformers from wounds is usually significant, and repeated isolation in large numbers in pure culture, particularly from blood cultures, is unlikely to be mere contamination. *B. cereus* infections of the eye are always emergencies (see below).

Low-level contamination of foodstuffs by aerobic endosporeformers is common, as is asymptomatic, transient, faecal carriage by human and animal populations. Therefore, in foodborne illness investigations, qualitative isolation tests are insufficient. The ideal criteria for establishing that an aerobic endospore former is the aetiological agent are the isolation of significant numbers ($>10^5$ cfu/g) of the organism from the incriminated food together with recovery of the same strain in significant numbers from the patient's faeces or vomitus; also, detection of emetic toxin and/or enterotoxin in the case of suspected *B. cereus* food poisoning. Clinical specimens for isolation of species other than *B. anthracis* can be handled without special precautions, and organisms will normally survive transport in freshly collected specimens or in a standard transport medium. All these species of aerobic endospore-forming bacteria that may be isolated from clinical specimens can be handled safely on the open bench. All the clinically-significant isolates reported to date are of species that grow, and often sporulate, on routine laboratory media at 37 °C. It seems unlikely that many clinically important, but more fastidious, strains are being missed for the want of special media or growth conditions.

Not only is *B. cereus* a close relative of *B. anthracis*, it is also next in importance as a pathogen of humans and other animals, and its ubiquity ensures that cases occur not infrequently. Reports of infections with non-*B. cereus* group species are comparatively rare, but very diverse (Logan, 1988; Berkeley and Logan, 1997), and there have been several hospital pseudoepidemics associated with contaminated blood culture systems.

Bacillus cereus and food-borne illness

B. cereus is the aetiological agent of two distinct food poisoning syndromes (Kramer and Gilbert, 1992):

(i) a diarrhoeal-type, characterized by abdominal pain with diarrhoea 8 to 16 h after ingestion of the contaminated food, and associated with a diversity of foods from meats and vegetable dishes to pastas, desserts, cakes, sauces, and milk, and

(ii) an emetic-type that is characterized by nausea and vomiting 1 to 5 h after eating the offending food, predominantly oriental rice dishes, although occasionally other foods such as pasteurized cream, milk pudding, pastas, and reconstituted infant-feed formulas have been implicated.

Recovery is usually complete in 24 hours, although there have been two reports of death associated with diarrhoeal illness (Logan, 1988). An emetic outbreak followed the mere handling of contaminated rice in a children's craft activity (Briley et al., 2001), and fulminant liver failure associated with the emetic toxin has been reported (Mahler et al., 1997). Both syndromes arise as a direct result of the fact that *B. cereus* spores can survive normal cooking procedures. Under improper storage conditions after cooking, the spores germinate and the vegetative cells multiply. In diarrhoeal illness, the toxin(s) responsible are produced by organisms in the small intestine (infective doses 10^4-10^9 cells per gram of food), while the emetic toxin is preformed and ingested in food (about 10^5-10^8 cells per gram

in order to produce sufficient toxin). Variations in the infective dose of cases of diarrhoeal illness reflect the proportion of ingested cells that are sporulated, and so can survive the acid barrier of the stomach. The capacity of the strain concerned to produce toxin(s) will, of course, influence the infective or intoxicating dose in both types of illness. It is probable that cases with diarrhoeal and emetic symptoms occurring together are caused by organisms producing both diarrhoeal and emetic toxins.

B. cereus is known to produce six toxins - five enterotoxins, and the emetic toxin:

(i) Haemolysin BL (Hbl), a 3-component proteinaceous toxin which also has dermonecrotic and vascular permeability activities, and causes fluid accumulation in ligated rabbit ileal loops; Hbl is produced by about 60% of strains tested (Granum, 2002), and it has been suggested that it is the primary virulence factor in *B. cereus* diarrhoea, but the mechanism of its enterotoxic activity is unclear (Granum, 2002);

(ii) Non-haemolytic enterotoxin (Nhe), which is another 3-component proteinaceous toxin, produced by most strains tested (Granum, 2002), and whose components show some similarities to Hbl;

(iii) and (iv) Enterotoxin T (BceT) and Enterotoxin FM (EntFM), which are single-component proteinaceous toxins whose roles and characteristics are not known;

(v) Enterotoxin K (EntK), that is similar to the β-toxin of *Clostridium perfringens* and was associated with a French outbreak of necrotic enteritis in which three people died (Lund et al., 2000). The genetics of toxin production are summarized by Granum (2002).

(vi) The emetic toxin, cereulide, is a dodecadepsipeptide comprising a ring of four amino- and/or oxy-acids: [D-*O*-Leu-D-Ala-L-*O*-Val-L-Val] thrice repeated; chemically speaking, it is closely related to the potassium ionophore valinomycin (Agata et al., 1994). It is resistant to heat, pH and proteolysis, but it is not antigenic (Kramer and Gilbert, 1989).

Cereulide is probably an enzymatically synthesized peptide rather than a direct genetic product; it is produced in larger amounts at lower incubation temperatures, and its production does not appear to be connected with sporulation (Finlay et al., 2000), and it is not produced in anaerobic conditions (Finlay et al, 2002). Its mechanism of action is unknown, but it has been shown to stimulate the vagus afferent through binding to the 5-HT$_3$ receptor (Agata et al., 1995). The earliest detection system for emetic toxin involved monkey-feeding tests (Logan et al., 1979), but a semi-automated metabolic staining assay has now been developed (Finlay et al., 1999), and an assay based upon loss of motility of boar spermatozoa has been described (Andersson et al., 1999). Two commercial kits are available for the detection in foods and faeces of the enterotoxin complex responsible for the diarrhoeal type of *B. cereus* food-poisoning: the Oxoid BCET-RPLA (Oxoid Ltd, Basingstoke, UK, Product Code TD950) and the TECRA VIA

(TECRA Diagnostics, Roseville, NSW, Australia, Product Code BDEVIA48). However, these kits detect different antigens, and there is some controversy about their reliabilities. Other assays, based on tissue culture, have also been developed (Fletcher and Logan, 1999).

Other illnesses owing to *B. cereus*

Bacillus cereus is one of the most virulent and destructive of ocular pathogens. The most serious of these conditions is panophthalmitis, a rapidly developing infection which may follow penetrating trauma of the eye (commonly by a metal fragment in an environment such as a farm or a garage), intraocular surgery, or haematogenous dissemination of the organism from another site (typically in intravenous drug abusers). Either way, the condition usually evolves so rapidly that irreversible damage occurs before effective treatment can be started; vision is therefore lost, and loss of the eye is normal (Davey and Tauber, 1987; Das et al., 2001). *B. cereus* keratitis associated with contact lens wear has also been reported (Pinna et al., 2001).

Other *B. cereus* infections occur mainly, though not exclusively, in persons predisposed by neoplastic disease, immunosuppression, alcoholism and other drug abuse, or some other underlying condition, and fatalities occasionally result. Reported conditions include bacteraemia, septicaemia, fulminant sepsis with haemolysis, meningitis, brain haemorrhage, ventricular shunt infections, endocarditis, pneumonia, empyema, pleurisy, lung abscess, brain abscess, osteomyelitis, salpingitis, urinary tract infection, and primary cutaneous infections. Wound infections, mostly in otherwise healthy persons, have been reported following surgery, road traffic and other accidents, scalds, burns, plaster fixation, drug injection (including a case associated with contaminated heroin; Dancer et al., 2002), and close-range gunshot and nail bomb injuries; some became necrotic and gangrenous. A fatal inflammation was caused by a blank firearm injury; blank cartridge propellants are commonly contaminated with the organism (Rothschild and Leisenfeld, 1996). Neonates also appear to be particularly susceptible to *B. cereus*, especially with umbilical stump infections, and respiratory tract infections associated with contaminated ventilation systems have occurred also (Van der Zwet et al., 2000).

Bacillus cereus also causes infections in domestic animals. It is a well-recognized agent of mastitis and abortion in cattle, and can cause these conditions in other livestock (Blowey and Edmondson, 1995).

Bacillus thuringiensis, including its application as a pesticide

This species is closely related to *B. cereus*, and is distinguished principally by the ability of its members to produce crystalline, proteinaceous, parasporal bodies within the sporangia; the insecticidal activities of many of these δ-endotoxins have made the organism (often referred to as Bt) one of the most widely produced and studied bacteria in biotechnology. The applications and development of Bt pesticides have been reviewed by Bishop (2002), while Van Rie (2002) has reviewed the development of transgenic crop plants. Strains of *B. thuringiensis* may also produce the *B. cereus* diarrhoeal toxin, and *B. thuringiensis* has indeed been implicated in cases of gastroenteritis (Damgaard et al., 1997). There have also been reports of wound, burn, and ocular infections with *B. thuringiensis* (Damgaard et al., 1997), Cases of illness caused by *B. thuringiensis* may have been diagnosed as caused by *B. cereus*, as the former may not produce its characteristic insecticidal toxin crystals when incubated at 37°C, owing to the loss of the plasmids carrying the toxin genes. The safety of using *B. thuringiensis* as a biopesticide on crop plants has been reviewed by Bishop (2002); Bishop et al. (1999) found that the main pesticide strains that they assayed produced low titres of enterotoxin, and there is as yet no evidence of infections associated with the use of this organism as an insecticide.

A further general point to be made with respect to the safety of Bt transgenic crop plants is that they allow huge reductions in the use of chemical insecticides, and this in itself is beneficial to human health, especially in those countries where crop spraying is usually done by hand (Van Rie, 2002; Thomson, 2003), and is also beneficial, of course, to harmless invertebrates. Although concerns have been raised about the potential danger of such transgenic plants to non-target species, including beneficial insects, all studies reported to date have found such toxicities to be undetectable or insignificant (Van Rie, 2002; Thomson, 2003).

Bacillus licheniformis

This species has been reported from ventriculitis following the removal of a meningioma, cerebral abscess after penetrating orbital injury, septicaemia following arteriography, bacteraemia associated with indwelling central venous catheters (Blue et al., 1995), bacteraemia during pregnancy with eclampsia and acute fibrinolysis, peritonitis in a CAPD patient and in a patient with volvulus and small-bowel perforation, ophthalmitis, and corneal ulcer after trauma. There have also been reports of L-form organisms, phenotypically similar to *B. licheniformis,* occurring in the blood of normal and arthritic persons in association with erythrocytes (Bisset and Bartlett, 1978), in other body fluids such as synovial fluids of arthritic patients (Pease, 1969), in association with neoplasms (Livingston and Alexander-Jackson, 1970), and in chickens and turkeys with infectious synovitis. These organisms may revert to small, acid-fast diphtheroids, and on prolonged culture or subculture in the presence of agents known to stimulate reversion of L-forms, some increase in size and become Gram-positive endospore-forming rods. The fully reverted isolates of Bisset and Bartlett (1978) were phenotypically similar to *B. licheniformis* in other respects, and they named them "*B. licheniformis* var. *endoparasiticus*". Although some authors have claimed a relationship between these organisms and diseases with postulated immunological elements, and higher isolations from the synovial fluids and membranes of arthritic patients have been reported, Bartlett and Bisset (1981) were unable to confirm the latter association.

B. licheniformis can cause food-borne diarrhoeal illness, and has been associated with an infant fatality (Mikkola et al., 2000). A toxin possibly associated with *B. licheniformis* food poisoning has been identified (Mikkola et al., 2000) but, in general, toxins or virulence factors widely accepted as responsible for symptoms periodically associated with *Bacillus* species other than those in the *B. cereus* group have not been identified. *B. licheniformis* is frequently associated with bovine abortion and has been reproduced by experimental infection of cows, which demonstrated the tropism of the organism for the bovine placenta (Agerholm et al., 1997); this species has also been associated with abortion in water buffalo (Galiero and De Carlo, 1998), and is occasionally associated with bovine mastitis (Blowey and Edmondson, 1995). Many of these types of *B. licheniformis* and *B. cereus* infections are associated with wet and dirty conditions during winter housing, particularly when the animals lie in spilled silage (Blowey and Edmondson, 1995); in one outbreak, a water tank contaminated with *B. licheniformis* was implicated (Parvanta, 2000).

Bacillus subtilis

The name *B. subtilis* was often used to mean any aerobic, endospore-forming organism, but since 1970 there have been reports of infection in which identification of this species appears to have been made accurately. Cases associated with neoplastic disease include: fatal pneumonia and bacteraemia, a septicaemia, and an infection of a necrotic axillary tumour in breast cancer patients. Breast prosthesis and ventriculo-atrial shunt infections, endocarditis in a drug abuser, meningitis following a head injury, cholangitis associated with kidney and liver disease, and isolations from surgical wound-drainage sites have also been reported. This species has also been associated with cases of bovine mastitis and of ovine abortion (Logan, 1988; Berkeley and Logan, 1997). A probiotic preparation labelled as containing strains of this species led to a fatal septicaemia in an immunocompromised patient (Oggioni et al., 1998); subsequently, the organisms concerned were identified as *B. clausii* (Spinosa et al., 2000). These authors reported another *B. clausii* infection, cholangitis in polycystic kidney disease in a 15-year-old French boy who had undergone renal transplant. The original authors (Wallet et al., 1996) had identified the organism as *B. subtilis*; their patient had not been taking a probiotic preparation and the source of the infecting *B. clausii* was unclear (Spinosa et al., 2000).

B. *subtilis* has been implicated in food-borne illness: vomiting has been the commonest symptom, but with accompanying diarrhoea frequently reported, the onset periods have been short (ranging from 10 min to 14 h; median 2.5 h), the bacterial loads of the organism were high (10^5 to 10^9 cfu/g), and the implicated foods were often prepared dishes in which meat or fish were served with cereal-based components such as bread, pastry, rice or stuffing (Kramer and Gilbert, 1989). No toxic agents were identified in any of these cases, but although the high bacterial load might be a factor in such rapid-onset illnesses, other members of the species and related organisms are regularly swallowed in large numbers in probiotic preparations without symptoms of food poisoning being reported, indeed, probiotics comprising *B. subtilis* spores are intended to treat antibiotic therapy-associated diarrhoea (Mazza, 1994), and large doses administered to mice in vaccine vehicle investigations elicited no adverse reactions (Duc et al., 2003).

Symbiotic associations between L-form bacteria and plants have been observed (Paton and Innes, 1991), and a stable L-form of *B. subtilis* in Chinese cabbage seedlings has been shown to inhibit the germination of *Botrytis cinerea* conidia (Walker et al., 2002).

Other species of aerobic endosporeformers

Organisms identified as *B. circulans* have been isolated from cases of meningitis, a cerebrospinal fluid shunt infection, endocarditis, endophthalmitis (Tandon et al., 2001), a wound infection in a cancer patient, and a bite wound (but the strains involved may not have been authentic *B. circulans*, as this very heterogeneous species continues to undergo radical taxonomic revision, with some strains being assigned to species within *Paenibacillus*). *B. coagulans* has been isolated from corneal infection, bacteraemia and bovine abortion. *B. pumilus* has been found in cases of pustule and rectal fistula infection, and in association with bovine mastitis. *B. sphaericus* has been implicated in a fatal lung pseudotumour, and meningitis (Logan, 1988). As noted above, a probiotic strain labelled as *B. subtilis* and implicated in a fatal septicaemia in an immunocompromised patient (Oggioni et al., 1998); was later identified as *B. clausii* (Spinosa et al., 2000).

Bacillus brevis has been isolated from corneal infection and has been implicated in several incidents of food poisoning; since these reports, the species was split (see Introduction, above) and transferred to the new genus *Brevibacillus*. Strains of the new species, *Brevibacillus agri*, have been isolated in association with an outbreak of waterborne illness in Sweden (Logan et al., 2002), and other *Brevibacillus* species have been found in human blood and bronchio-alveolar lavage specimens. *Br. laterosporus* has been reported in association with a severe case of endophthalmitis (Logan, 1988).

Paenibacillus alvei has been isolated from cases of meningitis, a prosthetic hip infection in a patient with sickle cell anaemia, a wound infection, and, in association with *Clostridium perfringens,* a case of gas gangrene. *P. macerans* has been isolated from a wound infection following removal of a malignant melanoma, and from bovine abortion, and *P. polymyxa* from ovine abortion (Logan, 1988).

ALLERGIES TO AEROBIC ENDOSPOREFORMERS OR THEIR PRODUCTS

The detergent and food industries use large amounts of enzymes from *Bacillus* species, and although these enzymes do not normally appear to pose much of a danger to consumers nowadays , they can and do cause allergic reactions in workers at the production plants where they may be handled in large quantities.

The earliest 'biological' washing powders were marketed in the 1960s, and contained powdered proteases from *B. subtilis*. Nowadays the preferred protease is produced by the alkaliphilic species *B. clausii*, but proteases from another alkaliphile, *B. halodurans*, and from *B. licheniformis* are also used. The addition of other kinds of enzymes began in the 1990s: amylases from the alkalitolerant *B. halmapalus* and from *B. licheniformis*, and cellulases and lipases derived mainly from fungi (Outtrup and Jørgensen, 2002).

The earliest powdered enzyme preparations gave rise to numerous epidemics of occupational asthma (i.e. a Type I hypersensitivity reaction, or immediate hypersensitivity) among exposed workers, and to occasional IgE-associated allergies and skin irritation in some consumers (Pepys et al., 1969, 1973; Belin et al., 1970; Franz et al., 1971). Encapsulating the enzymes and the introduction of tighter engineering, better ventilation, additional protective clothing for the workers, and, later, product formulation controls led to a dramatic reduction of allergy problems in the industry. The industry's own association drew up progressively tighter recommendations on the manufacturing process and medical surveillance, and a seven-year study of over 1500 workers in a large washing-powder factory confirmed the effectiveness of those recommendations (Juniper et al., 1977). There have, however, been further outbreaks of occupational asthma associated with detergent enzymes, even in factories that use exclusively encapsulated enzymes; it was shown in one study that the capsules had become fissured, so allowing contact of workers with the allergen (Perdu et al., 1992). Cullinan et al. (2000) reported that sensitization to amylase was the most common type in their factory, and considered that their findings, along with the animal data of Sarlo et al. (1997b), imply that bacillary amylase is more allergenic than proteases, either innately or through potentiation.

It is clear that engineering and product formulation controls and enzyme encapsulation may not be sufficient to prevent all further outbreaks of occupational asthma in detergent factories, and that there is a need for the establishment of exposure guidelines for workers. Some way of evaluating the allergenicities of new enzymes as they are introduced to the detergent industry is therefore required, and two such approaches are the guinea pig intratracheal test (Sarlo et al., 1997a) and the mouse intranasal test (Robinson et al., 1998).

Following a survey of 2500 patients it was concluded that the addition of enzymes to laundry products is safe for consumer use, as even highly atopic users of such detergents did not become sensitized during the two-year period that these washing powders had been on the market in the UK (Pepys et al., 1973). A follow-up study of 88 individuals over a further three years showed that even though enzyme-specific radioallergosorbent test IgE levels might be raised in exposed consumers, there was no evidence that modern, non-dusty enzyme-containing detergents would lead to clinical sensitization (Pepys et al., 1985). However, Tripathi and Grammer (2001) reported a case of extrinsic allergic alveolitis (i.e. a Type III hypersensitivity reaction or immune complex disease) following exposure to proteolytic enzymes in a liquid cleaner. A safety assessment study on a prototype enzyme-containing personal cleansing bar showed that an IgE specific for the protease enzyme concerned was detectable by skin-prick test in a small number of the test subjects, and it was concluded that this kind of application represented an inappropriate use of enzymes in a consumer product (Kelling et al., 1998).

Microbial enzymes used in food manufacture are normally sold as 'enzyme preparations', because they contain substances other than the desired enzyme, including microbial metabolites, stabilizers and preservatives. A wide variety of enzymes used in food processing are derived from aerobic endospore formers: carbohydrases such as α-amylases, β-glucanases, maltogenic amylase, pullulanases and xylanases, and glutaminase and proteases from *B. amyloliquefaciens, B. circulans, B. licheniformis, B. naganoensis, B. subtilis* and *G. stearothermophilus,* genetically manipulated in some cases with related species as donors.

As with detergent enzymes, it is important to evaluate the safety of these food enzyme preparations, rather than just the enzymes themselves (Pariza and Johnson 2001). Although some *Bacillus* species may be Generally Regarded as Safe (GRAS) (Saunders et al., 2003) for the purposes of enzyme production, the GRAS petitions by the enzyme producers have tended to concentrate on the potential pathogenicities and toxicities of the organisms concerned, such as *B. subtilis*, or recombinant strains of *B. subtilis* producing *B. megaterium, B. stearothermophilus* or maltogenic amylases, and on the potential toxicities of the enzymes (Andersen et al., 1987; De Boer and Diderichsen, 1991). These enzymes have been deemed safe for use in food manufacture, and no published reports of their causing occupational asthma or other allergic reactions in the food industry have been traced; however, there have been several reports of sensitization to fungal amylases used in baking and other industries, where dry enzyme preparations can give rise to dusts or become components of dusts (Losada et al., 1992; Houba et al., 1996; Smith et al., 1997).

SAFETY OF AEROBIC ENDOSPOREFORMERS USED AS PROBIOTICS

The efficacies and modes of action of aerobic endosporeformers used in products for animals and humans are covered elsewhere in this volume, and the present review will focus only on the safety implications of probiotic usage of these organisms; it will not address the more ethical issue of whether or not the probiotic is of any health-enhancing effect in the application for which it is intended.

As with any safety assessment, there are several facets to consider:

(i) Does the probiotic product actually contain the organism(s) specified on the label, and if not, what does it contain?

(ii) Is any organism in the product a member or close relative of a species that has been isolated in association with infection?

(iii) Does any organism in the product carry toxins, or is it a member or close relative of a group of organisms known to produce or have the potential to produce such virulence factors? Has it been screened for toxicity?

(iv) Does any organism in the product show any adherence to or invasiveness for tissues of the animal species for which it is prescribed?

(v) Does any organism in the product harbour, or have the potential to harbour, transferable antibiotic resistance? Has it been screened for plasmids antibiotic-resistance plasmids?

(vi) Has any organism in the product been genetically modified, and if it has, has a specific environmental risk assessment

been carried out and reached a satisfactory conclusion?
(vii)Have contraindications to use been considered, and are these clearly listed on the product label?

Nomenclature

It must first be appreciated that the accuracy of identification and naming of the organism(s) used in a probiotic product is of fundamental importance, and that the difficulties of identifying *Bacillus sensu lato* species at a time of rapid taxonomic and nomenclatural change are no excuse for shoddiness. If the manufacturer does not know the true identity of the organism(s) he is marketing, how can quality control be achieved and how can claims of efficacy and modes of action be properly investigated and accurately reported?

According to product labelling, species used in probiotic preparations for administration to humans and other animals include *Bacillus cereus, B. clausii, B. coagulans, B. licheniformis, B. pumilis* and *B. subtilis*, and *Brevibacillus laterosporus* and *Paenibacillus polymyxa*. In some cases, many organisms are listed; one product claims to contain some 42 "probiotic complex ingredients" which include the last four species listed in the preceding sentence (Sanders et al., 2003). However, several studies have revealed that the species names given on the product label do not always match the contents. Green et al. (1999) examined two commercial products that claimed to contain *B. subtilis*, but found that one contained an organism more closely related to *B. pumilus* and that the other contained *B. clausii*; the latter identification was confirmed by Spinosa et al. (2000) and Senesi et al. (2001) and the product label has been amended. Hoa et al. (2000) examined five products, and found that in only one of them could the identity of the organism listed (*B. cereus*) be confirmed - of three products listing *B. subtilis*, one contained *B. cereus,* one contained *B. clausii*, and one was unidentifiable; the fifth product did contain *B. subtilis* but the organism was listed as "*Lactobacillus sporogenes*". This brings us to another problem of labelling, that of the use of invalid or obsolete nomenclature. At a time of unprecedented powers of communication and the rapid updating of freely-available nomenclatural and other taxonomic databases, there is no excuse for this. As pointed out by Sanders et al. (2003), the name "*Lactobacillus sporogenes*", proposed by Horowitz-Wlassowa and Nowotelnow (1932), has never been valid and is in fact a synonym of *Bacillus coagulans*, yet commercial preparations may still bear the invalid name. The implication that the products contain a *Lactobacillus* species, some of which are well known as being important for their protective roles in the gut and vagina, as opposed to *Bacillus coagulans* which is better known as a soil organism and food-spoilage agent, is frankly misleading. Another invalid name is "*B. toyoi*", thought to be a synonym of *B. cereus* (Sanders et al., 2003), and the enzyme and other industries regularly use invalid names such as "*B. polyfermenticus*" (Lee et al., 2001), "*B. acidopullulyticus*" and "*B. deramificans*" (Outtrup and Jørgensen, 2002). Bacterial nomenclature is carefully regulated for a good reason, and without close adherence to it we cannot make sound progress in evaluating the safety or efficacy of probiotic products as we are unable to communicate effectively.

Pathogenicity and toxicity

It will be clear from the review of infections given earlier in this chapter, that strains representing all of the aerobic endosporeforming species used in probiotics have been isolated at some time or another in association with human and/or animal infection or food-borne illness. This must be put in perspective. Most such cases occurred in persons debilitated by neoplastic disease, immunosuppression, alcoholism and other drug abuse, or some other underlying condition, and reports of infections with *B. clausii, B. coagulans, B. pumilus, Brevibacillus laterosporus* and *Paenibacillus polymyxa* are exceedingly rare. Infections with *B. subtilis* are also rare, but with some of the earlier reports we may entertain some doubts as to the accuracies of the identifications. These occasional cases do not seem to be sufficient grounds upon which to recommend the withdrawal of products containing these species; after all, some of these products have been widely used over many years, and the great between-strain variation of some species of *Bacillus sensu lato* has already been stressed. We encounter these soil organisms regularly in our everyday environments, and we inadvertently consume them in our food; if they had appreciable pathogenic potentials we would expect to see many more cases reported. The key concern with probiotic preparations is that the size of the inoculum is huge by comparison with casual, accidental consumption of, or wound contamination with, a few spores, and so we must be satisfied that all seven of the requirements set out earlier in this section are met. Without thorough, independent and scientifically controlled study of the actual organism to be marketed, we cannot be entirely confident of the safety of any of these products. When the product comprises mixtures of many different organisms, it seems unlikely that all have been tested for absence of the undesirable properties outlined above.

Attention to contraindications is just as important as it is with any other therapeutic product. A probiotic preparation labelled as containing *B. subtilis* led to a fatal septicaemia with the same organism (actually a *B. clausii*; Spinosa et al., 2000) in an immunocompromised patient (Oggioni et al., 1998). Both of the reports of *B. clausii* infection reviewed by Spinosa et al. (2000) occurred in immunocompromised patients: the French cholangitis patient had undergone a kidney transplant, while the Italian septicaemic patient was undergoing chemotherapy; there was no history, however, of probiotic treatment in the former case. On the other hand, Vacca et al. (1983) found that cell-mediated immunity parameters improved in a number of multiple myeloma patients treated with *B. subtilis* spores, and that several of the patients enjoyed a decrease in recurrent respiratory tract infections; the probiotic preparation used in that study was the same that Spinosa et al. (2000) subsequently found to contain *B. clausii*.

Cases of *B. cereus* infection are not unusual, of course, and an appreciable number have occurred in patients who were apparently in good health beforehand, and *B. cereus* food poisoning is commonplace, as is well known. Toxin production is so widespread among strains of *B. cereus* and its relatives (see, for example, Fletcher and Logan, 1999; Rivera et al., 2000), that this gives cause for concern at the use of strains of these species in probiotic preparations, even though they may be regarded as non-toxic strains. Indeed, the European Commission's Scientific Committee on Animal Nutrition (SCAN, 1999a, 2001a) considered that two animal feed additives containing *B. cereus* posed a risk to human and animal health, and recommended that: "for all future applications involving the addition of living organisms to animal feeds, the use of strains from the *B. cereus* taxonomic group be strongly discouraged" (SCAN, 2000a, 2003a), although a "*B. cereus* var. *toyoi*" strain could not be shown to produce *B. cereus* toxins (SCAN, 2001b). The same committee considered that an animal feed additive containing *B. licheniformis* and *B. subtilis* was safe for use and that no deleterious effects on customers were to be expected (SCAN, 2000b), while another product containing *B. licheniformis* was unsafe on account of its antibiotic resistance (SCAN, 2002). An animal feed additive product containing *B. clausii* was also considered to pose a risk to human or animal health, but this was not on account of strains of this species having been found in association with human infections (Spinosa et al., 2000), but because of concerns about the potential for dissemination of the drug resistance genes that the organisms carried (SCAN 1999b).

For specific applications, such as enzyme production, some species of aerobic endosporeformer are considered to be GRAS by the Federal Drug Administration (FDA) in the United States of America, but the FDA has not as yet granted GRAS status to any aerobic endosporeformer in a probiotic application (Sanders et al., 2003).

Antibiotic resistance and genetic modification

Some concerns about the use of aerobic endosporeformers in probiotic preparations for humans and animals arises not so much from their potential to cause opportunistic infections as from their potential to disseminate any drug resistance genes that they might carry. Strains of *B. clausii* that are used in an animal feed additive were derived from a pharmaceutical product registered for human use. The strains concerned were originally selected for their resistance to a range of clinically useful antibiotics, as the product was intended for use as an adjunct to antibiotic therapy in humans. Animals are believed to be a reservoir of drug resistance transferable to humans, so contributing to the development of resistance to clinically important antibiotics. SCAN was particularly concerned at the high levels of resistance to erythromycin and lincosamides in the product and so considered it to be unsafe for use as an animal feed additive (SCAN, 1999b, and see SCAN 2003b). Another product, containing *B. licheniformis* and intended for feeding to pigs, was considered to be unsafe because of the risk of disseminating a gene conferring resistance to erythromycin; although it was thought probable that the magnitude of this unquantifiable risk was low, a precautionary approach was adopted (SCAN, 2002).

Despite the well-established importance of *B. cereus* as an opportunistic pathogen, there have been rather few studies of its antibiotic sensitivity, and most information has to be gleaned from the reports of individual cases or outbreaks. *B. cereus* and *B. thuringiensis* produce a broad spectrum β-lactamase and are thus resistant to penicillin, ampicillin, and cephalosporins;

they are also resistant to trimethoprim. A plasmid carrying resistance to tetracycline in *B. cereus* has been transferred to a strain of *B. subtilis* and stably maintained (Bernhard et al., 1978). Although strains are almost always susceptible to clindamycin, erythromycin, chloramphenicol, vancomycin, and the aminoglycosides and are usually sensitive to tetracycline and sulphonamides, there have been several reports of treatment failures with some of these drugs (see Logan and Turnbull, 2003). Oral ciprofloxacin has been used successfully in the treatment of *B. cereus* wound infections, and clindamycin with gentamicin, given early, appears to be the best treatment for ophthalmic infections caused by *B. cereus*.

Information is sparse on treatment of infections with species outside the *B. cereus* group. Gentamicin was effective in treating a case of *B. licheniformis* ophthalmitis and cephalosporin was effective against *B. licheniformis* bacteraemia/septicaemia. Resistance to macrolides appears to occur naturally in *B. licheniformis*. *B. subtilis* endocarditis in a drug abuser was successfully treated with cephalosporin, and gentamicin was successful against a *B. subtilis* septicaemia. Penicillin, or its derivatives, or cephalosporins probably form the best first choices for treatment of infections attributed to other *Bacillus* species. Isolates of "*B. polymyxa*" and *B. circulans* were more likely to be resistant to the penicillins and cephalosporins than strains of the other species – it is possible that some or all of the strains identified as *B. circulans* might now be accommodated in *Paenibacillus*, along with "*B. polymyxa*". An infection of a human bite wound with an organism identified as *B. circulans* did not respond to treatment with amoxicillin and flucloxacillin, but was resolved with clindamycin (Goudswaard et al., 1995). The above-mentioned recurrent septicaemia with *B. subtilis* in an immunocompromised patient yielded two isolates, both of which could be recovered from the probiotic preparation that the patient had been taking. These strains were later identified as *B. clausii* (i.e., not *B. subtilis*); one isolate was resistant to penicillin, erythromycin, rifampin and novobiocin, while the other was sensitive to rifampin and novobiocin but resistant to chloramphenicol (Oggioni et al., 1998). As indicated above, the product concerned was intended to be taken as an adjunct to antibiotic therapy, and so the strains used were chosen for their drug resistances; however, in a case such as the one reported, some of the very properties for which the probiotic strains were chosen will limit the therapeutic choices for the clinician encountering an opportunistic infection arising from use of the product.

A strain of *B. circulans* showing vancomycin resistance has been isolated from an Italian clinical specimen (Ligozzi et al., 1998). Vancomycin resistance was reported for a strain of *B.* (now *Paenibacillus*) *popilliae* in 1965, and isolates of this species dating back to 1945 have been shown to carry a *vanA*- and *vanB*-like gene, that is to say a gene resembling those responsible for high-level vancomycin resistance in enterococci. Vancomycin resistant enterococci (VRE) were first reported in 1986, and so it has been suggested that the resistance genes in *B. popilliae* and VRE may share a common ancestor, or even that the gene in *B. popilliae* itself may have been the precursor of those in VRE; *B. popilliae* has been used for over 50 years as a biopesticide, and no other potential source of *vanA* and *vanB* has been identified (Rippere et al. 1998). Of two South African vancomycin-resistant clinical isolates, one was identified as *P. thiaminolyticus* and the other was unidentified but considered to be related to *B. lentus* (Forsyth and Logan, unpublished); the latter was isolated from a case of neonatal sepsis, and has been shown to have inducible resistance to vancomycin and teicoplanin; this is in contrast to the *B. circulans* and *P. thiaminolyticus* isolates mentioned above, in which expression of resistance was found to be constitutive (A. von Gottberg and W. van Nierop, personal communication).

There is very little information publicly available on the use of genetically modified strains of endosporeformers in probiotic applications. Products for human and animal use, and containing recombinant *Bacillus* strains, are manufactured in the Ukraine, but there is little reporting of any independent study on their efficacy and safety.

CONCLUSION

Over the decades since aerobic endospore formers and their products were introduced to biotechnological and therapeutic applications, including detergent enzymes, pest control and probiotics, and as their uses in these areas have widened, our understandings of allied matters such as allergy, drug resistance and opportunistic infection have increased. We also have a larger population of vulnerable individuals who are rendered especially susceptible to injury from some such agents, by virtue of suppressed immunity consequent upon chronic illness or its treatment.

Over the same period, our attitudes towards occupational and consumer health and safety have substantially changed. No longer can we use organisms such as "*B. subtilis* var. *globigii*" with impunity as experimental tracers in public environments, as the risk of harming someone is now appreciated (Barnaby, 2002). When *Bacillus* enzymes were introduced to washing powders in the 1960s, the problems of occupational asthma and consumer allergy that emerged were dealt with reactively, even though they might have been anticipated. Nowadays such a set of circumstances would be much less likely to occur because, more than ever before, emphasis is placed upon the protection of workers and consumers from the side effects of the manufacture and use of the majority of products – witness, as outlined above, the detailed considerations given by the European Commission to the safety of various animal feed additives. Such proactive approaches to occupational and consumer health and safety are, of course, to be welcomed, not least because they help biotechnological developments to proceed in an informed manner, rather than in a climate of public suspicion.

REFERENCES

Agata, N. Ohta, M., Mori, M. and Isobe M. 1995. A novel dodecadepsipeptide, cereulide, is an emetic toxin of *Bacillus cereus*. Int. J. Syst. Evol. Microbiol. *129*, 17-20.

Agata, N., Mori, M., Ohta, M., Suwan, S., Ohtani, I. and Isobe, M. 1994. A novel dodecadepsipeptide, cereulide, isolated from *Bacillus cereus* causes vacuole formation in Hep-2 cells. Int. J. Syst. Evol. Microbiol. *121*, 31-34.

Agerholm, J.S., N.E. Jensen, V. Dantzer, H.E. Jensen and F.M. Aarestrup. 1997. Experimental infection of pregnant cows with *Bacillus licheniformis* bacteria. Vet. Path. *36*, 191-201.

Andersen, J.R., B.K. Diderichsen, R.K. Hjortkjaer, A.S. de Boer, J. Bootman, H. West and R. Ashby. 1987. Determining the safety of maltogenic amylase produced by rDNA technology. J. Food Prot. *50*, 521-526.

Andersson, M.A., R. Mikkola, J. Helin, M.C. Andersson and M.S. Salkinoja-Salonen. 1999. A novel sensitive bioassay for detection of *Bacillus cereus* emetic toxin and related depsipeptide ionophores. Appl. Environ. Microbiol. *64*, 1338-1343.

Ash, C., F. G. Priest, and M. D. Collins. 1993. Molecular identification of rRNA group 3 bacilli (Ash, Farrow, Wallbanks and Collins) using a PCR probe test. Ant. van Leeuwenhoek *64L*, 253-260.

Barnaby, W. 2002. The plague makers: the secret world of biological warfare. Vision, London.

Bartlett, R. and K.A. Bisset. 1981. Isolation of *Bacillus licheniformis* var. *endoparasiticus* from the blood of rheumatoid arthritis patients and normal subjects. J. Med. Microbiol. *14*, 97-105.

Belin, L, E. Falsen, J. Hoborn and J. Andre. 1970. Enzyme sensitisation in consumers of enzyme containing washing powders. Lancet ii, 1153.

Berkeley, R.C.W. and N.A. Logan. 1997. *Bacillus, Alicyclobacillus* and *Paenibacillus. In* Emmerson, Hawkey, and Gillespie (Editors), Principles and Practice of Clinical Bacteriology. John Wiley, Chichester, pp. 185-204.

Bernhard, K., H. Schrempf and W. Goebel. 1978. Bacteriocin and antibiotic resistance plasmids in *Bacillus cereus* and *Bacillus subtilis*. J. Bact. 133, 897-903.

Bishop, A.H. 2002. *Bacillus thuringiensis* insecticides. *In* Berkeley, Heyndrickx, Logan and De Vos (Editors), Applications and Systematics of *Bacillus* and Relatives. Blackwell Science, Oxford, pp. 160-175.

Bishop, A.H., J. Johnson and M. Perani. 1999. The safety of *Bacillus thuringiensis* to mammals investigated by oral and subcutaneous dosage. Wld J. Biotech. 15, 375-380.

Bisset, K.A. and R. Bartlett. 1978. The isolation and characters of L-forms and reversions of *Bacillus licheniformis* var. *endoparasiticus* (Benedek) associated with the erythrocytes of clinically normal persons. J. Med. Microbiol. *11*, 335-349.

Blowey, R., and P. Edmondson. 1995. Mastitis Control in Dairy Herds. An Illustrated and Practical Guide. Farming Press Books, Ipswich.

Blue, S.R., V.R. Singh and M.A. Saubolle. 1995. *Bacillus licheniformis* bacteremia: five cases associated with indwelling central venous catheters. Clin. Inf. Dis. 20, 620-633.

Boone, D. R., Liu, Y., Zhao, Z.-J., Balkwill, D. L., Drake, G. R., Stevens, T. O. and H. C. Aldrich. 1995. *Bacillus infernus* sp. nov., an Fe(III)- and Mn(IV)-reducing anaerobe from the deep terrestrial subsurface. Int. J. Syst. Bacteriol. 45, 441-448.

Briley, R.T., Teel, J.H. and J.P. Fowler. 2001. Nontypical *Bacillus cereus* outbreak in a child care center. J. Environ. Hlth 63, 9-11.

Chanway, C.P. 2002. Plant growth promotion by *Bacillus* and relatives. *In* Berkeley, Heyndrickx, Logan and De Vos (Editors), Applications and Systematics of *Bacillus* and Relatives, Blackwell Science, Oxford, 219-235.

Cherkasskiy, B.L. 1999. A national register of historic and contemporary anthrax foci. J. Appl. Microbiol. *87*,192-195.

Christopher, G.W., T.J. Cieslak, J.A. Pavlin and E.M. Eitzen. 1997. Biological warfare; a historical perspective. J. Amer. Med. Assoc.. 278, 412-417.

Collins, C.H. 1988. *Laboratory Acquired Infections.* p. 16. 2nd ed. Butterworths, London, United Kingdom.

Damgaard, P.H., P.E. Granum, J. Bresciani, M.V. Torregrossa, J. Eilenberg, and L. Valentino. 1997. Characterization of *Bacillus thuringiensis* isolated from infections in burn wounds. FEMS Immunol. Med. Microbiol. *18*, 47-53.

Dancer, S.J., D. McNair, P. Finn and A.-B. Kolstø. 2002. *Bacillus cereus* cellulitis from contaminated heroin. J. Med. Microbiol. *51*, 278-281.

Das, T., K. Choudhury, S. Sharma, S. Jalali and R. Nuthethi. 2001. Clinical profile and outcome in *Bacillus* endophthalmitis. Ophthalmology *108*, 1819-1825.

Davey, R.T. Jr., and W.B. Tauber. 1987. Posttraumatic endophthalmitis: the emerging role of *Bacillus cereus* infection. Rev. Infect. Dis. 9, 110-123.

Franz, T., K.D. McMurrain, S. Brooks and I.L. Bernstein. 1971. Clinical, immunologic, and physiologic observations in factory workers exposed to *B. subtilis* enzyme dust. J. Allergy. *47*, 170-180.

Duc, L. H., Hong, H. A., Fairweather, N., Ricca, E. and Cutting, S. M. 2003. Bacterial spores as vaccine vehicles. Inf. Immun. *71*, 2810-2818.

Finlay, W. J. J., Logan, N. A. and A. D. Sutherland. 1999. Semiautomated metabolic staining assay for *Bacillus cereus* emetic toxin. Appl. Environ. Microbiol. *65*,1811-1812.

Finlay, W.J.J., N.A. Logan and A.D. Sutherland. 2000. *Bacillus cereus* produces most emetic toxin at lower temperatures. Lett. Appl. Microbiol. *31*, 385-389.

Finlay, W.J.J., N.A. Logan and A.D. Sutherland. 2002. *Bacillus cereus* emetic toxin production in relation to dissolved oxygen tension and sporulation. Food Microbiol. *19*, 423-430.

Fletcher, P. and N.A. Logan. 1999. Improved cytotoxicity assay for *Bacillus cereus* diarrhoeal enterotoxin. Lett. Appl. Microbiol. *28*, 394-400.

Forsyth, G. and N.A. Logan. 2000. Isolation of *Bacillus thuringiensis* from northern Victoria Land, Antarctica. Lett. Appl. Microbiol. *30*, 263-266.

Fortina, M. G., Pukall, R., Schumann, P., Mora, D., Parini., C., Manachini, P. L. and E. Stackebrandt. 2001. *Ureibacillus* gen. nov., a new genus to accommodate *Bacillus thermosphaericus* (Andersson et al. 1995), emendation of *Ureibacillus thermosphaericus* and description of *Ureibacillus terrenus* sp. nov. Int. J. Syst. Evol. Microbiol. *51*, 447-455.

Galiero, G. and E. De Carlo. 1998. Abortion in water buffalo (*Bubalis bubalis*) associated with *Bacillus licheniformis*. Vet. Record *143*, 640.

Goudswaard, W.B., M.H. Dammer and C. Hol. 1995. *Bacillus circulans* infection of a proximal interphalangeal joint after a clenched-fist injury caused by human teeth. Eur. J. Clin. Microbiol. Inf. Dis. *14*, 1015-1016.

Granum, P.E. 2002. *Bacillus cereus* and food poisoning. *In* Berkeley, Heyndrickx, Logan and De Vos (Editors), Applications and Systematics of *Bacillus* and Relatives, Blackwell Science, Oxford, pp. 37-46.

Green, D.H., P.R. Wakeley, A. Page, A. Barnes, L. Baccigalupi, E. Ricca and S.M. Cutting. 1999. Characterization of two *Bacillus* probiotics. Appl. Environ. Microbiol. *65*, 4288-4291.

Heyndrickx, M., L. Lebbe, M. Vancanneyt, K. Kersters, P. De Vos, G. Forsyth, and N. A. Logan. 1998. *Virgibacillus*: a new genus to accommodate *Bacillus pantothenticus* (Proom and Knight 1950). Emended description of *Virgibacillus pantothenticus*. Int. J. Syst. Bacteriol. *48*, 99-106.

Heyrman, J., N. A. Logan, H.-J. Busse, A. Balcaen, L. Lebbe, M. Rodriguez-Diaz, J. Swings and P. De Vos. 2003. *Virgibacillus carmonensis* sp. nov., *Virgibacillus necropolis* sp. nov. and *Virgibacillus picturae* sp. nov., three new species isolated from deteriorated mural paintings, transfer of the species of the genus *Salibacillus* to *Virgibacillus*, as *Virgibacillus marismortui* comb. nov. and *Virgibacillus salexigens* comb. nov., and emended description of the genus *Virgibacillus*. . Int. J. Syst. Evol. Microbiol. *53*, 501-511.

Hoa, N.T., L. Baccigalupi, A. Huxham, A. Smertenko, P.H. Van , S. Ammendola, E. Ricca, and S.M. Cutting. 2000. Characterization of *Bacillus* species used for oral bacteriotherapy and bacterioprophylaxis of gastrointestinal disorders. Appl. Environ. Microbiol. *66*, 5241-5247.

Horowitz-Wlassowa, L.M. and N.W. Nowotelnow. 1932. Über eine sporogene Milchsäurebakterienart, *Lactobacillus sporogenes* n. sp. Zentralbl. Bakteriol. Parasitenkd. Infektionskr. Hyg. Abt. II *87*, 331.

Houba. R., D.J. Heederik, G. Doekes and P.E. van Run. 1996. Exposure-sensitization relationship for alpha-amylase allergens in the baking industry. Am. J. Respir. Crit. Care Med. *154*, 130-136.

Ivanova, N., A. Sorokin, I. Anderson, N. Galleron, B. Candelon, V. Kapatral, A. Bhattacharyya, G. Reznik, N. Mikhilova, A. Lapidus, L. Chu, M. Mazur, E. Goltsman, N. Larsen, M. D'Souza, T. Walunas, Y. Grechkin, G. Pusch, R. Haselkorn, M. Fonstein, S.D. Ehrlich, R. Overbeek and N. Kyrpides. 2003. Genome sequence of *Bacillus cereus* and comparative analysis with *Bacillus anthracis*. Nature *423*, 87-91.

Juniper, C.P., M.J. How, B.F.J. Goodwin and A.K. Kinshott. 1977. *Bacillus subtilis* enzymes: a 7-year clinical, epidemiological and immunological study of an industrial antigen. J. Soc. Occup. Med. *27*, 3-12.

Kelling, C.K., R.G. Bartolo, K.D. Ertel, L.A. Smith, D.D. Watson and K. Sarlo. 1998. Safety assessment of enzyme-containing personal cleansing products: exposure characterization and development of IgE antibody to enzymes after a 6-month use test. J. Allergy Clin. Immunol. *101*, 179-187.

Kramer, J.M. and R.J. Gilbert. 1989. *Bacillus cereus* and other *Bacillus* species. *In* M.P. Doyle (Editor), Foodborne Bacterial Pathogens. Marcel Dekker, New York and Basel, pp. 21-70.

Kunst, F., N. Ogasawara, I. Moszer, A.M. Albertini, G. Alloni, V. Azevedo, M.G. Bertero, P. Bessières, A. Bolotin, S. Borchert, R. Borriss, L. Boursier, A. Brans, M. Braun, S.C. Brignell, S. Bron, S. Brouillet, C.V. Bruschi, B. Caldwell, V. Capuano, N.M. Carter, S.-K. Choi, J.-J. Codani, I.F. Connerton, N.J. Cummings, R.A. Daniel, F. Denizot, K.M. Devine, A. Düsterhöft, S.D. Ehrlich, R.T. Emmerson, K.D. Entian, J. Errington, C. Fabret, E. Ferrari, D. Foulger, C. Fritz, M. Fujita, Y. Fujita, S. Fuma, A. Galizzi, N. Galleron, S.-Y. Ghim, P. Glaser, A. Goffeau, E.J. Golightly, G. Grandi, G. Guiseppi, B.J. Guy, K. Haga, J. Haiech, C.R. Harwood, A. Hénaut, H. Hilbert, S. Holsappel, S.

Hosono, M.-F. Hullo, M. Itaya, L. Jones, B. Joris, D. Karamata, Y. Kasahara, M. Klaer-Blanchard, C. Klein, Y. Kobayashi, P. Koetter, G. Koningstein, S. Krogh, M. Kumano, K. Kurita, A. Lapidus, L. Lardinois, J. Lauber, V. Lazarevic, S.-M. Lee, A. Levine, H. Liu, S. Masuda, C. Mauël, C. Médigue, N. Medina, R.P. Mellado, M. Mizuno, D. Moestl, S. Nakai, M. Noback, D. Noone, M. O'Reilly, K. Ogawa, A. Ogiwara, B. Oudega, S.-H. Park, V. Parro, T.M. Pohl, D. Portetelle, S. Porwollik, A.M. Prescott, E. Presecan, P. Pujic, B. Purnelle, G. Rapoport, M. Rey, S. Reynolds, M. Rieger, C. Rivolta, E. Rocha, B. Roche, M. Rose, Y. Sadaie, T. Sato, E. Scanlan, S. Schleich, R. Schroeter, F. Scoffone, J. Sekiguchi, A. Sekowska, S.J. Seror, P. Serror, B.-S. Shin, B. Soldo, A. Sorokin, E. Tacconi, T. Takagi, H. Takahashi, K. Takemaru, M. Tacheuchi, A. Tamakoshi, R. Tanaka, P. Terpstra, A. Tognoni, V. Tosato, S. Uchiyama, M. Vandenbol, F. Vannier, A. Vassarotti, A. Viari, R. Wambutt, E. Wedler, H. Wedler, T. Weitzenegger, P. Winters, A. Wipat, H. Yamamoto, K. Yamane, K. Yasumoto, K. Yata, K. Yoshida, H.-F. Yoshikawa, E. Zumstein, H. Yoshikawa and A. Danchin. 1997. The complete genome sequence of the Gram-positive bacterium *Bacillus subtilis*. Nature *390*, 249-256.

Lane, H.C. and A.S. Fauci. 2001. Bioterrorism on the home front: A new challenge for American medicine J. Amer. Med. Assoc.. *286*, 2595.

Lee, K.H., K.D. Jun, W.S. Kim and H.D. Paik. 2001. Partial characterization of polyfermenticin SCD, a newly identified bacteriocin of *Bacillus polyfermenticus*. Lett. Appl. Microbiol. *32*, 146-151.

Ligozzi, M., G.L. Cascio and R. Fontana. 1998. *vanA* gene cluster in a vancomycin-resistant clinical isolate of *Bacillus circulans*. Antimicrob. Agents Chemother. *42*, 2055-2059.

Lindeque, P.M. and P.C.B. Turnbull. 1994. Ecology and epidemiology of anthrax in the Etosha National Park, Namibia. Onderspoort J. Vet. Res. *61*, 71-83.

Livingston, V.W. and E. Alexander-Jackson. 1970. A specific organism cultivated from malignancy: bacteriology and proposed classification. Ann. N. Y. Acad. Sci. *174*, 636.

Logan, N.A. 1988. *Bacillus* species of medical and veterinary importance. J. Med. Microbiol. *25*, 157-165.

Logan, N.A. 2002. Modern methods for identification. *In* Berkeley, Heyndrickx, Logan and De Vos (Editors), Applications and Systematics of *Bacillus* and Relatives, Blackwell Science, Oxford, pp. 123-140.

Logan, N.A. and P.C.B. Turnbull. 2003. *Bacillus* and other aerobic endospore-forming bacteria. *In* Manual of Clinical Microbiology, 8th Edn, Vol. 1, Murray, Baron, Jorgensen, Pfaller, Yolken (Editors). ASM, Washington, 445-460.

Logan, N.A., B.J. Capel, J. Melling and R.C.W. Berkeley. 1979. Distinction between emetic and other strains of *Bacillus cereus* using the API system and numerical methods. Int. J. Syst. Evol. Microbiol. *5*, 373-375.

Logan, N.A., G. Forsyth, L. Lebbe, J. Goris, M. Heyndrickx, A. Balcaen, A. Verhelst, E. Falsen, Å. Ljungh, H.B. Hansson and P. De Vos. 2002. Polyphasic identification of *Bacillus* and *Brevibacillus* strains from clinical, dairy, and industrial specimens and proposal of *Brevibacillus invocatus*, sp. nov. Int. J. Syst. Evol. Microbiol. *52*, 953-966.

Losada, E., M. Hinojosa, S. Quirce, M. Sanchez-Cano and I. Moneo. 1992. Occupational asthma caused by α-amylase inhalation: clinical and immunologic findings and bronchial response patterns. J. Allergy Clin. Immunol. *89*, 118-125.

Lund, T., De Buyser, M.L. and Granum, P.E. 2000. A new enterotoxin from *Bacillus cereus* that can cause necrotic enteritis. Mol. Microbiol. *38*, 254-261.

Mahler, H., A. Pasi, J.M. Kramer, P. Schulte, A.C. Scoging, W. Bär, and S. Krähenbühl. 1997. Fulminant liver failure in association with the emetic toxin of *Baillus cereus*. New Eng. J. Med. *336*, 1142-1148.

Mazza, P. (1994). The use of *Bacillus subtilis* as an antidiarrhoeal microorganism. Boll. Chim. Farmaceutico. *133*, 3-18.

Meselson, M., J. Guillemin, M. Hugh-Jones, A. Langmuir, I. Popova, A. Shelokov, and O. Yampolskaya. 1994. The Sverdlovsk anthrax outbreak of 1979. Science *266*, 1202-1208.

Mikkola, R., M. Kolari, M.A. Andersson, J. Helin,. and M.S. Salkinoja-Salonen. 2000. Toxic lactonic lipopeptide from food poisoning isolates of *Bacillus licheniformis*. Eur. J. Biochem. *267*, 4068-4074.

Nazina, T. N., Tourova, T. P., Poltaraus, A. B., Novikova, E. V., Grigoryan, A. A., Ivanova, A. E., Lysenko, A. M., Petrunyaka, V. V., Osipov, G. A., Belyaev, S. S. and M. V. Ivanov. 2001. Taxonomic study of aerobic thermophilic bacilli: descriptions of *Geobacillus subterraneus* gen nov, sp. nov. and *Geobacillus uzenensis* sp. nov. from petroleum reservoirs and transfer of *Bacillus stearothermophilus, Bacillus thermocatenulatus, Bacillus thermoleovorans Bacillus kaustophilus, Bacillus thermoglucosidasius, Bacillus thermodenitrificans* to *Geobacillus* as *Geobacillus stearothermophilus,*

Geobacillus thermocatenulatus, Geobacillus thermoleovorans Geobacillus kaustophilus, Geobacillus thermoglucosidasius, Geobacillus thermodenitrificans. Int. J. Syst. Evol. Microbiol. *51*, 433-446.

Nicholson, W.L. 2002. Roles of *Bacillus* endospores in the environment. *Cell. Mol. Life Sci.* 59, 410-416.

Oggioni, M. Pozzi, G., Valensis, P.E., Galieni, P. and C. Bigazzi. 1998. Recurrent septicemia in an immunocompromised patient due to probiotic strains of *Bacillus subtilis*. J. Clin. Microbiol. *36*, 325-326.

Outtrup, H. and S. T. Jørgensen. 2002. The importance of *Bacillus* species in the production of industrial enzymes. *In* R.C.W. Berkeley, M. Heyndrickx, N.A. Logan, and P. de Vos (ed) *Applications and Systematics of Bacillus and Relatives*. Blackwell Science, Oxford, United Kingdom, 206-218.

Parvanta, M.F. 2000. Abortion in a dairy herd associated with *Bacillus licheniformis*. Tierarztliche Umschau 55, 126.

Paton, A.M. and C.M.J. Innes. 1991. Methods for the establishment of intracellular associations of L-forms with higher plants. J. Appl. Bacteriol. *71*, 59-64.

Pease, P.E. 1969. Bacterial L-forms in the blood and joint fluids of arthritic subjects. Ann. Rheum. Dis. *28*, 270.

Pepys, J., J.L. Longbottom, F.E. Hargreave and J. Faux. 1969. Allergic reactions of the lungs to enzymes of *Bacillus subtilis*. Lancet i, 1181-1184.

Pepys, J., I.D. Wells, M.F. D'Souza and M. Greenberg. 1973. Clinical and immunological responses to enzymes of *Bacillus subtilis* in factory workers and consumers. Clin. Allergy *3*, 143-160.

Pepys, J. J. Mitchell, R. Hawkins and J.L. Malo. 1985. A longitudinal study of possible allergy to enzyme detergents. Clin. Allergy *15*, 101-115.

Perdu, D., F. Lavaud, C. Cossart, S. Legrele, P. Passemard, G. Deltour and J.M. Dubois de Montreynaud. 1992. Enzymes des lessives: le risque de sensibilisation professionelle a-t-il disparu? Rev. Mal. Resp. *9*, 443-448.

Pinna A, L.A. Sechi, S. Zanetti, D. Esai, G. Delogu, P. Cappuccinelli and F. Carta. 2001. *Bacillus cereus* keratitis associated with contact lens wear. Ophthalmology, *108*, 1830-1834.

Read, T.D., S.N. Peterson, N. Tourasse, L.W. Baillie, I.T. Paulsen, K.E. Nelson, H. Tetellin, D.E. Fouts, J.A. Eisen, S.R. Gill, E.K. Holtzapple, O.A. Økstad, E. Helgason, J. Rilstone, M. Wu, J.F. Kolonay, M.J. Beanan, R.J. Dodson, L.M. Brinkac, M. Gwinn, R.T. DeBoy, R. Madpu, S.C Daugherty, A.S. Durkin, D.H. Haft, W.C. Nelson, J.D. Peterson, M. Pop, H.M. Khouri, D. Radune, J.L. Benton, Y. Mahamoud, L. Jiang, I.R. Hance, J.F. Weidman, K.J. Berry, R.D. Plaut, A.M. Wolf, K.L. Watkins, W.C. Nierman, A. Hazen, R. Cline, C. Redmond, J.E. Thwaite, O. White, S.L. Salzberg, B. Thomason, A.M. Friedlander, T.M. Koehler, P.C. Hanna, A.-B. Kolstø and C. Fraser. 2003. The genome sequence of *Bacillus anthracis* Ames and comparison to closely related bacteria. Nature *423*, 81-86.

Rippere, K., R. Patel, J.R. Uhl, K.E. Piper, J.M. Steckelberg, B.C. Kline, F.R. Cockerill III and A.A. Yousten. 1998. DNA sequence resembling *vanA* and *vanB* in the vancomycin-resistant biopesticide *Bacillus popilliae*. J. Infect. Dis. *178*, 584-588.

Rivera, A.M.G., P.E. Granum and F.G. Priest. 2000 Common occurrence of enterotoxin genes and enterotoxicity in *Bacillus thuringiensis*. Int. J. Syst. Evol. Microbiol. *190*, 151-155.

Robinson, M.K., P.A. Horn, T.T. Kawabata, L.S. Babcock, E.R. Fletcher and K. Sarlo. 1998. Use of the mouse intranasal test (MINT) to determine the allergenic potency of detergent enzymes: comparison to the guinea pig intratracheal (GPIT) test. Toxicol. Sci. *43*, 39-46.

Rothschild, M.A. and O. Leisenfeld. 1996. Is the exploding powder from blank cartridges sterile? Forensic. Sci. Int. *83*, 1-13.

Sanders, M.E., L. Morelli and T.A. Tompkins. 2003. Sporeformers as human probiotics: *Bacillus, Sporolactobacillus* and *Brevibacillus*. Comp. Rev. Food Sci. Food Safety 2, 101-110.

Sarlo, K., E.R. Fletcher, W.G. Gaines and H.L. Ritz. 1997a. respiratory allergenicity of detergent enzymes in the guinea pig intratracheal test: association with sensitization of occupationally exposed individuals. Fund. Appl. Toxicol. *39*, 44-52.

Sarlo, K., H.L. Ritz, R. Fletcher, K.R. Schrotel, and E.D. Clark. 1997b. Proteolytic detergent enzymes enhance the allergic antibody responses of guinea pigs to nonproteolytic detergent enzymes in a mixture: implications for occupational exposure. J. Allergy Clin. Immunol. *100*, 480-487.

[SCAN] Scientific Committee on Animal Nutrition. 1999a. Assessment by the Scientific Committee on Animal Nutrition (SCAN) of a micro-organisms product: Esporafeed Plus®. European Commission, Health and Consumer Protection Directorate-General. [SCAN] Scientific Committee on Animal Nutrition. http://europa.eu.int/comm/food/fs/sc/scan/out39_en.pdf

[SCAN] Scientific Committee on Animal Nutrition. 1999b. Assessment by the Scientific Committee on Animal Nutrition (SCAN) of a micro-organisms product: Neoferm BS-10®. European Commission, Health and Consumer Protection Directorate-General. [SCAN] Scientific Committee on Animal Nutrition. http://europa.eu.int/comm/food/fs/sc/scan/out28_en.pdf

[SCAN] Scientific Committee on Animal Nutrition. 2000a. Report of the Scientific Committee on Animal Nutrition, Opinion of the Scientific Committee on Animal Nutrition on the safety of use of *Bacillus* species in animal nutrition. European Commission, Health and Consumer Protection Directorate-General. [SCAN] Scientific Committee on Animal Nutrition. http://europa.eu.int/comm/food/fs/sc/scan/out41_en.pdf

[SCAN] Scientific Committee on Animal Nutrition. 2000b. Report of the Scientific Committee on Animal Nutrition on product Bioplus 2B® for use as a feed additive. European Commission, Health and Consumer Protection Directorate-General. [SCAN] Scientific Committee on Animal Nutrition. http://europa.eu.int/comm/food/fs/sc/scan/out49_en.pdf

[SCAN] Scientific Committee on Animal Nutrition. 2001a. Report of the Scientific Committee on Animal Nutrition of the safety of product Paciflor® for use as a feed additive. European Commission, Health and Consumer Protection Directorate-General. [SCAN] Scientific Committee on Animal Nutrition. http://europa.eu.int/comm/food/fs/sc/scan/out62_en.pdf

[SCAN] Scientific Committee on Animal Nutrition. 2001b. Report of the Scientific Committee on Animal Nutrition on product Toyocerin® for use as a feed additive. European Commission, Health and Consumer Protection Directorate-General. [SCAN] Scientific Committee on Animal Nutrition. http://europa.eu.int/comm/food/fs/sc/scan/out72_en.pdf

[SCAN] Scientific Committee on Animal Nutrition. 2002. Opinion of the Scientific Committee on Animal Nutrition on the use of *Bacillus licheniformis* NCTC 13123 in feedingstuffs for pigs (Product AlCare™). European Commission, Health and Consumer Protection Directorate-General. [SCAN] Scientific Committee on Animal Nutrition. http://europa.eu.int/comm/food/fs/sc/scan/out79_en.pdf

[SCAN] Scientific Committee on Animal Nutrition. 2003a. Opinion on the use of certain micro-organisms as additives in feedingstuffs. European Commission, Health and Consumer Protection Directorate-General. [SCAN] Scientific Committee on Animal Nutrition. http://europa.eu.int/comm/food/fs/sc/scan/out93_en.pdf

[SCAN] Scientific Committee on Animal Nutrition. 2003b. Opinion of the Scientific Committee on Animal Nutrition, on the criteria for assessing the safety of micro-organisms resistant to antibiotics of human clinical and veterinary importance. European Commission, Health and Consumer Protection Directorate-General. [SCAN] Scientific Committee on Animal Nutrition. http://europa.eu.int/comm/food/fs/sc/scan/out108_en.pdf

Senesi, S., F. Celandroni, A. Tavanti and E. Ghelardi. 2001. Molecular characterization and identification of *Bacillus clausii* strains marketed for use in oral bacteriotherapy. Appl. Environ. Microbiol. *67*, 834-839.

Shida, O., H. Takagi,. K. Kadowaki, and K. Komagata. 1996. Proposal for two new genera, *Brevibacillus* gen. nov. and *Aneurinibacillus* gen. nov. Int. J. Syst. Bacteriol. *46*, 939-946.

Smith, T.A., K.P. Lumley and E.H. Hui 1997. Allergy to flour and fungal amylase in bakery workers. Occup. Med. (Lond.) *47*, 21-24.

Spinosa, M.R., F. Wallet, R.J. Courcol and M.R. Oggioni. 2000. The trouble in tracing opportunistic pathogens: cholangitis due to *Bacillus* in a French hospital caused by a strain related to an Italian product? Microb. Ecol. Health. Dis. *12*, 99-101.

Switzer Blum, J., A. Burns Bindi, J Buzzelli, J. F. Stolz and R. S. Oremland. 1998. *Bacillus arsenicoselenatis*, sp. nov., and *Bacillus selenitireducens*, sp. nov.: two haloalkaliphiles from Mono Lake, California that respire oxyanions of selenium and arsenic. Arch. Microbiol. *171*, 19-30.

Takahashi, H., P. Keim, A.F. Kaufmann, C. Keys, K.L. Smith, K. Taniguchi, S. Inouye and T. Kurata. 2004. *Bacillus anthracis* incident, Kameido, Tokyo, 1993. Emerg. Infect. Dis. *10*, 117-120.

Takami, H., K. Nakasone, Y. Takaki, G. Maeno, R. Sasaki, N. Masui, F. Fuji, C. Hirama, Y. Nakamura, N. Ogasawara, S. Kuhara and K. Horikoshi. 2000. Complete genome sequence of the alkaliphilic bacterium *Bacillus halodurans* and genomic sequence comparison with *Bacillus subtilis*. Nuc. Acids Res. *28*, 4317-4331.

Tandon, A., M.L. Tay-Kearney, C. Metcalf, and I. McAllister. 2001. *Bacillus circulans* endophthalmitis. Clin. Exp. Ophthalmol. *29*, 92-93.

Thomson, J. 2003. Genetically modified crops for improving agricultural practice and their effects on human health. Trends Food Sci. Technol. *14*, 210-228.

Tripathi, A. and L.C. Grammer. 2001. Extrinsic allergic alveolitis from a proteolytic enzyme. Ann. Allergy Asthma Immunol. *86*, 425-427.

Turnbull, P.C.B., M. Doganay, P.M. Lindeque, B. Aygen, and J. McLaughlin. 1992. Serology and anthrax in humans, livestock and Etosha National Park wildlife. Epidemiol. Infect. *108*, 299-313.

Turnbull, P.C.B., P.J.Jackson, K.K. Hill, A-B.Kolstø, P. Keim, and D.J. Beecher. 2002. Longstanding taxonomic enigmas with the 'Bacillus cereus group' are on the verge of being resolved by far-reaching molecular developments. Forecasts on the possible outcome by an *ad hoc* team. *In* R.C.W. Berkeley, M. Heyndrickx, N.A. Logan, and P. de Vos (ed) *Applications and Systematics of Bacillus and Relatives*. Blackwell Science, Oxford, United Kingdom, 23-36.

Turnbull, P.C.B., R. Böhm, O. Cosivi, M. Doganay, M.E. Hugh-Jones, D.D. Joshi, M.K. Lalitha, and V. de Vos. 1998. Guidelines for the surveillance and control of anthrax in humans and animals. World Health Organization, Geneva, WHO/EMC/ZDI/98.6.

Vacca, A., G. Pantaleo, M. Ronco and F. Dammacco. 1983. Chemoimmunotherapy for multiple myeloma using an intermiteent combination drug schedule (Mephalan + Prednisone) and alternating courses of *Bacillus subtilis* spores. Chemioterapia II, 300-306.

Van der Zwet, W.C., G.A. Parlevliet, P.H. Savelkoul, J. Stoof, A.M. Kaiser, A.M. Van Furth and C.M. Vandenbroucke-Grauls. 2000. Outbreak *of Bacillus cereus* infections in a neonatal intensive care unit traced to balloons used in manual ventilation. J. Clin. Microbiol. *38*, 4131-4136.

Van Rie, J. 2002. Bt crops – a novel insect control tool. *In* Berkeley, Heyndrickx, Logan and De Vos (Editors), Applications and Systematics of *Bacillus* and Relatives, Blackwell Science, Oxford, pp. 177-189.

Wainø, M., Tindall, B. J., Schumann, P. and K. Ingvorsen. 1999. *Gracilibacillus* gen. nov., with description of *Gracilibacillus halotolerans* gen. nov., sp. nov.: transfer of *Bacillus dipsosauri* to *Gracilibacillus dipsosauri* comb. nov., and *Bacillus salexigens* to the genus *Salibacillus* gen. nov., as *Salibacillus salexigens* comb. nov. Int. J. Syst. Bact. *49*, 821-831.

Walker, R., C.M.J. Ferguson, N.A. Booth and E.J. Allan. 2002. The symbiosis of *Bacillus subtilis* L-forms with Chinese cabbage seedlings inhibits conidial germination of *Botrytis cinerea*. Lett. Appl. Microbiol. *34*, 42-45.

Wallet, F., V. Crunelle, A. Roussel-Delvallez, A. Furchard, P. Saunier and R.J. Courcol. 1996. *Bacillus subtilis* as a cause of cholangitis in polycystic kidney and liver disease. Am. J. Gastroenterol. *91*, 1477-1478.

Wisotzkey, J. D., P. Jurtshuk, Jr., G. E. Fox, G. Deinhard, and K. Poralla. 1992. Comparative sequence analyses on the 16S rRNA (rDNA) of *Bacillus acidocaldarius, Bacillus acidoterrestris,* and *Bacillus cycloheptanicus* and proposal for creation of a new genus, *Alicyclobacillus* gen. nov. Int. J. Syst. Bacteriol. *42*, 263-269.

Yoon, J.-H., Lee, K.-C., Weiss, N., Kho, Y. H., Kang, K. H. and Y.-H. Park. 2001a. *Sporosarcina aquimarina* sp. nov., a bacterium isolated from seawater in Korea, and transfer of *Bacillus globisporus* (Larkin and Stokes 1967) , *Bacillus psychrophilus* (Nakamura 1984), and *Bacillus pasteurii* Chester 1898) to the genus *Sporosarcina* as *Sporosarcina globispora* comb. nov., *Sporosarcina psychrophila* comb. nov. and *Sporosarcina pasteurii* comb. nov., and emended description of the genus *Sporosarcina*. Int. J. Syst. Evol. Microbiol. *51*, 1079-1086.

Yoon, J.-H., N. Weiss, K.-C. Lee, I.-S. Kho, K.H. Kang and Y.-H. Park. 2001b. *Jeotgalibacillus alimentarius* gen. nov., sp. nov., a novel bacterium isolated from jeotgal with L-lysine in the cell wall, and reclassification of *Bacillus marinus* Rüger 1983 as *Marinibacillus marinus* gen. nov., comb. nov. Int. J. Syst. Evol. Microbiol. *51*, 2087-2093.

Chapter 9

The Fate of Ingested Spores

Huynh A. Hong and Le H. Duc

SUMMARY

Bacillus spores when ingested are not treated simply as a food. Mounting studies, reported here, show that spores have a complex interaction with the GALT. The most important finding is that a proportion of ingested spores can germinate in the intestine and undergo limited rounds of growth before sporulating again. Spores can disseminate to the GALT and are likely to even persist briefly within phagocytic cells resident in the Peyers' Patches. This persistence may elicit cellular responses against spores as well as enhancing systemic humoral and local responses. Immune responses against spores when given at high dose is of course important when considering their use as probiotics. We propose in this chapter the novel idea that *Bacillus* spores, although not long term residents of the gut, may actually use the animal gut (including humans) as part of their natural life cycle.

INTRODUCTION

Bacillus spores are being used as probiotics for humans, for animals and for aquaculture. The rationale behind their use as probiotics is based upon a combination of circumstantial as well as factual evidence suggesting that ingested spores enhance the gut environment making it beneficial for commensal bacteria and prohibitive for the growth of harmful and possibly pathogenic bacteria (Mazza, 1994; Sanders et al., 2003). In some cases, scientific evidence supports some of the claimed probiotic effects, for example, *B. subtilis* spores have been shown to suppress both *E. coli* 078:K80 infection in poultry (La Ragione et al., 2001) and infection of the Black Tiger Shrimp (*Penaeus monodon*) with *Vibrio harveyi* (Rengpipat et al., 1998; Vaseeharan and Ramasamy, 2003). However, few if any proper human trials have been performed showing clear cut benefits (Sanders et al., 2003). Meanwhile, the lucrative health food market enables products to be marketed based on unsubstantiated, if not, dubious claims.

Most probiotics being marketed today are commensal bacteria normally found in the gastrointestinal tract (GIT), e.g., *Lactobacillus* species and *Bifidobacteria*. These gut bacteria exist only in the vegetative state and can not form a dormant life form. The commercial products carrying these bacteria are usually packaged in a lyophilised state and must have a short shelf-life. By contrast, *Bacillus* spores are used in the dessicated state and should, in theory, last indefinitely at ambient temperature. This may offer some advantages in storage and distribution of a probiotic product. What is important though, is the fact that the spore is being used and not the vegetative cell. Since established probiotics based on commensal bacteria are being used how

then can a spore from a soil-organism, also produce a probiotic effect? If the probiotic effect is due solely to the spore then it would imply that spores and vegetative commensal bacteria can each produce probiotic effects. A logical extension, if this is the case, is whether *any* microparticle or biological agent could exert a probiotic effect when ingested. This point, if true, is important and would devalue further research on specific 'probiotic' organisms. Alternatively, the spore could germinate within the GIT and exert its beneficial effect in the vegetative form. In this scenario the spore would enable safe transit across the stomach barrier. There are two arguments against this model. First and foremost, most *Bacillus* species are considered resident soil-microorganisms so how could they survive in the vegetative state within the GIT? Second, the GIT is thought to be anoxic and most *Bacillus* species are considered aerobic.

We will show in this chapter evidence that suggests bacterial spores are able to germinate within the GIT, they can undergo limited rounds of growth and replication and may have a much more intimate association with the GALT than was originally assumed. Finally, we make the case that *Bacillus* spore formers may actually use the GIT of their host as part of their natural life cycle and should not be considered solely as soil-microorganisms.

DIRECT COUNTING OF INGESTED SPORES IN FAECES: EVIDENCE FOR SPORE GERMINATION

A straightforward approach to examining the fate of ingested spores is to dose an animal with a fixed number of spores and then examine the transit of spores through the GIT by counting spores in faeces. Since spores are heat resistant, faeces can be heated to remove the substantial numbers of resident commensal bacteria. In one such study by Hoa and coworkers (Hoa et al., 2001) groups of mice were given a single dose of *B. subtilis* spores (ranging between 10^8-10^{11} depending on the experiment). In all cases spores were detectable in the faeces between 3-6 hours after dosing. By 18-24h the total number of spores recovered in the faeces was equivalent to the initial dose. However, continued sampling showed spores in the faeces for up to 7 days after dosing. Moreover, the total cumulative counts were greater than the original dose by up to a factor of seven. The only way to explain these results was to assume that spores must have germinated and undergone limited rounds of growth and replication and then re-sporulated in the GIT before being excreted. There are a number of important issues regarding these results that are worthwhile elaborating on. First, this experiment was performed using both inbred and outbred mice and the cumulative counts did not always exceed that of the original inoculum. This would suggest that the ability of spores to germinate and replicate may be dependant upon a number of physiological factors, for example, the diet and physiological condition of the mouse. Second, experimental error might be the first criticism of these experiments yet consideration of all possible errors would always reduce the faecal counts. For example, failure to homogenise the faeces before plating would lower the counts not increase them, similarly, dosing mice is at best approximate and it is difficult to ensure a mouse receives the entire inoculum without regurgitation even when using a gavage. Another question regards the ability of *Bacillus* to survive in the anoxic conditions of the GIT. However, the GIT is unlikely to be completely anoxic and *B. subtilis* has recently been shown to grow anaerobically if the appropriate nutrients are provided (Nakano and Zuber, 1998). Is it unrealistic for spores to germinate and grow within the GIT? Spores should be sufficiently robust to survive the stomach barrier and below we show evidence for this. Entry into the small intestine would provide the spores with an abundance of nutrients which could trigger germination of the spore. In the small intestine, particularly in the jejunum and ileum the lumen should be enriched with bile salts and this might be toxic preventing or retarding cell growth. This issue is discussed in detail below but with a large inoculum of spores sufficient numbers of germinated spores could be expected to survive as they transit the small intestine. Indeed, the data of Hoa et al. (Hoa et al., 2001) show that spores can transit the GIT in as little as 3h so the inhibitory effects of bile salts would be quickly diluted as the cells pass through the GIT. Germinated cells might be expected to undergo only a few rounds of replication after which they would re-sporulate presumably as a response to the harsh conditions present in the large intestine. For now the issue of re-sporulation is conjecture and has not been proven, but is the only rational explanation for these experiments. Interestingly, some work in other laboratories does support these findings including germination of *B. cereus var toyoi* in poultry (Jadamus et al., 2002; Jadamus et al., 2001), germination of *B. subtilis var natto* in pigs (cited in (Mazza, 1994) but unretrievable) and germination in ligated ileum loops from rabbits (Hisanga, 1980). Finally, Chapter 11 provides evidence of spore germination in the GIT of humans.

Interestingly, a recent study examining the fate of ingested spores has 'concluded' that spores do not germinate in the GIT since analysis of faecal counts showed a steady decline in CFU (Spinosa et al., 2000). This study is important because it was cited in a recent review as evidence that spores can not germinate (Sanders et al., 2003). However, a careful examination of the experimental approach shows that the conclusions are flawed and not supported by the experimental evidence. In this study mice were given a fixed dose of spores and were sacrificed at specific time points after dosing and individual faecal pellets removed and analysed for CFU. Thus, the *total* number of CFU in the faeces was not determined so when plotted over time the counts only show a steady decline. Moreover, different mice were sampled for analysis of faeces, not the same mice. To perform this experiment properly, groups of mice must be dosed and then total faeces collected between time points

and examined. Only, using this regime, can the true CFU be determined.

PROOF OF GERMINATION

Conclusive proof of spore germination has come recently using a molecular method (Casula and Cutting, 2002). First, a gene encoding a chimera was constructed by fusing the 5'-region of the *ftsH* gene to the *lacZ* gene of *E. coli*. *ftsH* is a vegetatively expressed gene (Deuerling et al., 1995; Lysenko et al., 1997) and therefore *ftsH-lacZ* mRNA would not be produced during differentiation. Spores carrying the *ftsH-lacZ* chimera were used to orally dose mice and at appropriate time points sections (duodenum, jejunum and ileum) were excised and total mRNA extracted. Using specific primers RT-PCR (reverse transcriptase PCR) was used to detect for the presence of *ftsH-lacZ* in the dissected sections. *ftsH-lacZ* was found to be detectable in the jejunum and ileum and using a second chimeric gene, *rrnO-lacZ*, in the ileum too. These experiments prove that spores must have germinated and the level of spore germination was estimated as below 1% of the original inoculum ($6 \times 10^8 \times$ CFU).

SURVIVAL OF SPORES IN THE GIT

Bacillus spores are considered extremely robust life forms, able to withstand extremes of temperature, desiccation and exposure to solvents and noxious chemicals. The most formidable barrier in the GIT would be the stomach that has a pH of between 2-3 and contains enzymes such as trypsin. A recent study has shown the effect of the GIT on spores and vegetative cells using both *in vivo* and *in vitro* methods (Duc et al., 2003a). Dosing of mice with 2.1×10^8 spores of *B. subtilis* has shown that *in vivo* most spores survive transit across the stomach and can be recovered from the small intestine (Duc et al., 2003a). In contrast, when mice were orally dosed with vegetative cells (2.4×10^{10}) of the same strain of *B. subtilis* only a tiny fraction of the original inoculum could be recovered alive in the small intestine (Duc et al., 2003a). For example, after 6h only 0.00016% of the original dose was present in the small intestine and at other times even lower numbers of viable cells could be recovered. Similarly, when vegetative cells were used to dose mice only 0.00025% of the original inoculum was recovered in the faeces. These results show that vegetative cells had little chance of surviving the stomach barrier. To account for those that do survive one must assume that aggregation of bacteria either with themselves or with food particles within the stomach must enable a low level of survival in transit.

Using simulated conditions representing the GIT environment the resistance properties of spores and of vegetative cells has been accurately defined (Duc et al., 2003a). In simulated gastric fluid *B. subtilis* spores are unaffected by gastric fluid. In contrast, vegetative cells were almost completely destroyed in gastric fluid within 1h of incubation with the viable population being reduced to 0.001%. Interestingly, in these studies a small proportion of cells appeared able to survive the simulated gastric conditions and perhaps this represents a subpopulation of cells with an intrinsic resistance or is due to cell aggregation. Either way, this low level of resistance supports the *in vivo* experiments described above where a fraction of cells can survive the stomach barrier. In the study of Duc et al. (Duc et al., 2003a) other enteric bacteria, notably, *E. coli* and *Citrobacter rodentium*, were tested in simulated gastric conditions and they also were killed under similar conditions with a small population surviving. In conditions simulating the small intestine (0.2% bile salts) vegetative *B. subtilis* was also destroyed with a reduction in viability to 0.0002%. In contrast, *E. coli* and *C. rodentium* were either unaffected or could actually grow. Spores on the other hand were unaffected with no loss in viability, although, interestingly, bile salts were shown to partially inhibit spore germination.

Taken together the *in vivo* and *in vitro* experiments of Duc et al. (2003a) reveal the likely fate of *B. subtilis* when ingested. Spores will survive essentially intact in the stomach. Entry into the duodenum and small intestine will provide a rich supply of nutrients and a rise in pH and it is here that the spore could germinate. The presence of bile salts would serve to inhibit germination but this effect is not absolute and would only be transient. As spores or germinated spores are moved through the small intestine the effect of bile salts would be diminished and presumably germinated cells could grow and replicate. Of course, this would apply only to a subpopulation of spores. Many, if not the vast majority, would simply pass through the GIT, unaffected, and be excreted in the faeces. Many spores that germinate would be killed by the effect of bile salts but a proportion (perhaps less than 1% (in accordance with the results of Casula and Cutting, 2002) would germinate and a sub-population of these be able to replicate.

One interesting discovery made in this laboratory (Duc et al., 2004a) is that spores of some *Bacillus* species (or strains) are inactivated and destroyed in simulated gastric fluid. Specifically, two commercial, probiotic, *B. cereus* strains were found to be extremely sensitive to gastric fluid with viability reduced to 0.02% or the inoculum within 1h of incubation. These spores were completely heat and solvent resistant and we have speculated that the acidity of gastric fluid actually induces or 'activates' germination, Acid induced activation (as opposed to heat-induced activation) of spore germination is not a new phenomenon and is documented for *B. cereus* strains (Faille et al., 2002; Keynan and Evenchik, 1969). Thus, spores can be killed in the stomach and for some species they must therefore be unable to survive the GIT even in the spore form.

DISSEMINATION IN THE GIT

Animal studies described above have shown mounting evidence that *Bacillus* spores (at least *B. subtilis*) can germinate to some degree in the GIT. Germination would presumably occur in the lumen of the GIT but it is also possible that spores could disseminate across the mucosal epithelium of the small intestine and then germinate. Normally, invasion into the gut associated lymphoid tissue (GALT) occurs by specialised epithelial cells known as M cells that transport bacteria across the epithelim into the Peyers' Patches (PP) a region rich in B cells and phagocytes (primarily dendritic cells). Further dissemination of bacteria can occur from the PP to the efferent lymph nodes including the mesenteric lymph nodes (MLN), submandibular glands (SMG) and cervical lymph nodes (CLN). Normally, dissemination studies are restricted to the study of pathogens

such as *Salmonella* species where dissemination plays an important part of the ensuing disease. Three studies have been made examining the dissemination of spores dosed orally in mice. In the most thorough study, inbred (Balb/c) mice were given 5 consecutive daily doses of 10^9 spores of *B. subtilis* and dissemination examined in 4 mice sacrificed on days 1, 2, 3, 5, 7 and 9 (Duc et al., 2003b). This study showed that significant numbers of viable bacteria were recovered from the PP and MLN as well as counts in the peritoneal macrophages. These counts represented spores but also vegetative cells implying that spores may have germinated within the dissected tissues. No counts were found in the spleen, liver of kidneys showing that although dissemination occurs it is not highly invasive. In another smaller study (Hoa et al., 2001) groups of four mice were sacrificed 12, 84 and 180 hours following dosing with 1.8 X 10^8 *B. subtilis* spores. In this study dissemination to the MLN, spleen and liver was found (the PP were not examined). Finally, in the study of Spinosa et al. (2000) bacteria (spores and vegetative cells) were recovered from the MLN and spleen of one mouse dosed orally with *B. clausii*.

In conclusion these studies suggest that a small proportion of spores can disseminate to the PP and lymphoid tissue. To do so they must presumably be taken up and translocated by M cells. The size of spores (~1.2 μm) is within the limitations of particle uptake by M cells (<5 μm). In the PP spores could interact with phagocytic cells as well as B cells. In the case of the latter this interaction would serve to prime humoral responses (ie, anti-spore Ig responses). Similarly, phagocytosis of spores could enhance both humoral responses as well as prime a cellular response. In the studies reported above the presence of vegetative cells has been found in the GALT. To achieve this, either spores had germinated in the GIT and then been taken up by M cells or, alternatively, the spore had entered the PP and then subsequently germinated. As discussed in more detail below *in vitro* studies have shown that spores of *B. subtilis* can efficiently germinate in macrophages so this event is likely to occur within the GALT. Importantly, dissemination by spores in the host is not extensive and spores appear to have a transient residence within the GALT unlike pathogenic bacteria that persist. These studies are important since probiotics consist of significant doses of spores that are often taken daily and the regular ingestion of spores could mimic the dissemination that has been shown from these experiments.

INTERACTION WITH THE GALT

Dissemination studies described above show that a small proportion of spores orally administered to the GIT can germinate, either in the lumen of the small intestine, following transit into the GALT or a combination of both. If spores or indeed vegetative cells enter the GALT then it is probable that antibody responses would be raised against the spore or germinated spore while persistence in the GALT might also lead to cell mediated immune responses.

Antibody responses

Several studies have examined the nature of antibody responses against spores. In the study of Duc et al. (Duc et al., 2003b) inbred C57/Black mice were dosed with 1.67 X 10^{10} spores

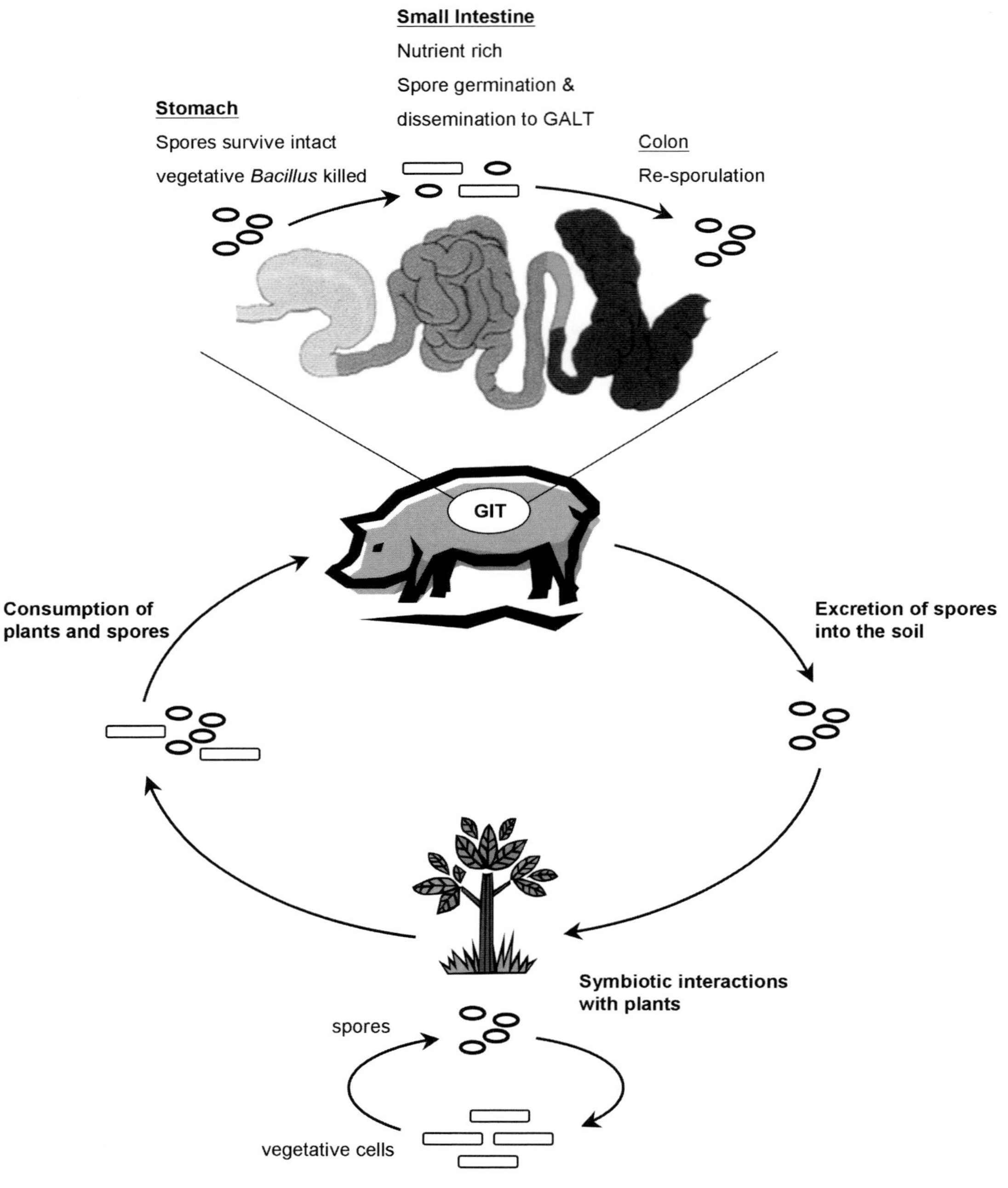

Small Intestine

Nutrient rich

Spore germination &

dissemination to GALT

Stomach

Spores survive intact

vegetative *Bacillus* killed

Colon

Re-sporulation

GIT

Consumption of
plants and spores

Excretion of spores
into the soil

Symbiotic interactions
with plants

spores

vegetative cells

Figure 1. The Natural Life Cycle of Spore Formers. The figure shows a hypothetical model for how *Bacillus* species might exploit the gut of a host animal. Most *Bacillus* species are considered soil organisms because they are found in the soil either as spores or in the vegetative form. Normally, in the latter they are found associated with plants probably in a symbiotic state (see Chapter 3). Consumption of plant matter by animals could introduce spores or vegetative cells into the GIT of the host animal. Vegetative cells would probably be killed in the stomach but spores are robust enough to survive. Entry into the small intestine would permit germination of a subpopulation of spores where they are able to grow and replicate. As the bacteria are passed through the GIT they would enter the harsher conditions present in the colon including overwhelming numbers of commensal bacteria (e.g., *Lactobacilli* and *Bifidobacteria*) that would out-compete the *Bacilli*. Sporulation would provide a suitable means of escape allowing spores to be excreted in the faeces and return to the soil. With some *Bacillus* species (e.g., *B. cereus* and *B. thuringiensis*) they can colonise the gut environment and are pathogenic representing a further evolutionary refinement.

(days 0, 2, 4, 18, 20, 22, 34, 35 and 36) of a strain isogenic to the laboratory strain PY79 (related to 168; (Youngman et al., 1984)). Significant IgG responses in the serum against spore coat proteins were determined by ELISA. Also, in the same study, secretory IgA (sIgA) responses against spore coat proteins were detected in the faeces. These studies show that spores dosed orally can elicit both systemic as well as local (mucosal) responses. In another study inbred Balb/c mice were dosed with 1.5×10^{10} spores of strain PY79 using an essentially identical dosing regime of 3 separate doses spaced by 20 days (Duc et al., 2003b). Systemic IgG responses against spores were high showing seroconversion as well as faecal sIgA. In this work, the IgG subclasses IgG1, IgG2a and IgG2b were examined and an early increase of spore-specific IgG2a was observed with a delayed increase in IgG2b and IgG1. This predominance of the IgG2a subclass over IgG1 during the early stages of immunisation suggests involvement of cellular responses. Compelling evidence shows that a predominance of the IgG2a subclass is indicative of a type 1 (Th1) T-cell response leading to CTL recruitment as well as enhancing IgG synthesis (Balloul et al., 1987; Isaka et al., 2001; Roberts et al., 1998; Robinson et al., 1997; VanCott et al., 1996). The increase in IgG2b during the later stages of immunisation indicates a type 2 (Th2) T-cell response and would account for the sIgA/IgG1 response.

Cellular responses

Other studies have provided support for cellular responses including *in vivo* analysis of cytokine mRNA. These studies showed an early induction of a major effector of cellular immunity, IFNγ, in the MLNs, submandibular glands (SMG) and liver following oral dosing of inbred mice with PY79 *B. subtilis* spores (Duc et al., 2003b). These early responses suggest an innate immune response with secretion of IFN-γ by peripheral blood mononuclear cells (PMBC). It is too early to predict which cell type is responsible for IFN-γ synthesis but CD4+ natural killer (NK) T cells have been shown to produce early IFN-γ induction following infection with *Mycobacterium bovis BCG* (Emoto et al., 1999) and *Listeria monocytogenes* (Andersson et al., 1998). Both NK cells and peritoneal macropohages have been shown to produce IFN-γ in experimentally induced bacterial peritonitis in mice (Sekl et al., 1998). Interestingly, these *in vivo* studies also showed early induction of TNF-α which is a pro-inflammatory cytokine whose production by macrophages has been linked with chronic infections (Dornand et al., 2002; Melby et al., 1994). Similarly, *in vitro* infection of a murine macrophage cell line RAW264.7 with *B. subtilis* spores of a strain isogenic to PY79 revealed early induction of the pro-inflammatory cytokine IL-6 as well as IL-1α and TNF-α. Together, these results suggest involvment of cellular responses when spores come into the contact with the GALT. Cytokines are probably produced early both to enhance the antibody responses to spores but also to enhance the destruction of spores. In simple terms, the production of the pro-inflammatory cytokines suggests that the host is mounting a significant response to clear the invading spores and although this may reflect the high dose of spores used in immunisation experiments it is worth noting that probiotics are used at similar doses. Finally, these studies have shown that spores are not regarded by the host as a food or as a transient passenger in the gut but instead as foreign and ingestion of high doses appears to switch on the full armoury of the host's immune responses.

Germination in macrophages

All the evidence discussed so far suggests that spores enter the GALT and elicit both antibody and cellular responses. Recent *in vitro* studies have revealed the fate of spores in the GALT using an *in vitro* macrophage cell line (Duc et al., 2004b). *B. subtilis* spores carrying a hybrid gene, *rrnO-lacZ*, that is expressed only in vegetative cells and expresses β-galactosidase, were incubated with the macrophages. Using a polyclonal antiserum specific against the spore coat the fate of spores was followed. Approximately 5h following uptake most of the spores were no longer detectable within the macrophage. However, using an anti- β-galactosidase polyclonal serum β-galactosidase was clearly expressed within the macrophage. This could only arise if the spore had germinated within the macrophage. No evidence for growth and replication of the germinated spore was found and the germinated cell did not grow but was retained in the phagolysosome. However, synthesis of β-galactosidase proves that for at least a short period of time the *B. subtilis* cell is able to undergo gene expression and protein synthesis. This is remarkably similar to the fate of *B. anthracis* spores when phagocytosed (Guidi-Rontani et al., 1999). Here, phagocytosis of the *B. anthracis* spore is a prelude to growth and replication and ultimate destruction of the phagocyte and the *B. anthracis* cells produces a capsule to protect it from the toxic conditions found within the macrophage. *B. subtilis* is of course non-pathogenic but clearly the spore itself can persist within the macrophage and, as with intracellular pathogens, this may induce cellular responses. For now, we can only speculate since the precise nature or importance of the cellular response still needs to be established yet we can state that the spore has a far more intimate interaction with the GALT than would initially have been conceived.

ARE SPORE FORMERS NORMAL RESIDENTS OF THE GIT?

Spores, or rather spore forming species are generally regarded as soil organisms. This implies that they only live in the soil. We believe that while the spore is undoubtedly present in large numbers in the soil its presence there may form just part of an intricate life cycle using the gut of host animals to complete the cycle. Chapter 1 outlines the diversity of spore formers and there natural habitats but as mentioned in this chapter there are mounting reports showing that spores and spore forming species are present in the gut of many animals. In this laboratory, we have found numerous different *Bacillus* spore forming species in the gut of humans and shrimps (unpublished). Similarly, *B. licheniformis. B. cereus, B. subtilis, B. sphaericus, B. circulans, B. megaterium, B. alvei*, and *B. pumilus* have all been found in the gut of insects (Gilliam, 1985; Gilliam et al., 1984; Gilliam et al., 1990a; Gilliam et al., 1990b). We believe, and propose here, that substantial numbers of spores accumulate in the soil by excretion from the gut of animals. Over time, the number of spores would accumulate to a level where they are abundant and of course the robustness of the spore would provide an excellent mechanism for long term survival in adverse conditions. However, where associated or close to the roots of plants spores could germinate and perhaps develop a symbiotic relationship explaining their plant growth-enhancing properties (see Chapter 3, 14 and Appendix III). Being in close association with plants would also enable spores to be ingested by animals which would enable the spore to re-enter the gut. Again, the spore would provide an excellent mechanism to survive the stomach barrier and being able to germinate in the intestine and grow before forming spores again would be an effective way to adapt to a changing environment. Studies reported in this chapter show that the spore does germinate in the gut of an animal and supports our hypothesis which is shown in Figure 1. Thus, although the spore is found in the soil it should not be considered an organism that lives only in the soil. Rather, its life cycle exploits both the soil and the animal gut.

REFERENCES

Andersson, A., Dai, W. J., Di Santo, J. P., and Brombacher, F. (1998). Early IFN-γ production and innate immunity during *Listeria monocytogenes* infection in the absence of NK cells. J. Immunol. *161*, 5600-5606.

Balloul, J.-M., Grzych, J.-M., Pierce, R. J., and Capron, A. (1987). A purified 28,000 dalton protein from *Schistosoma mansoni* adult worms protects rats and mice against experimental schistosomiasis. J. Immunol. *138*, 3448-3453.

Casula, G., and Cutting, S. M. (2002). *Bacillus* probiotics: spore germination in the gastrointestinal tract. App. Environ. Microbiol. *68*, 2344-2352.

Deuerling, E., Paeslack, B., and Schumann, W. (1995). The *ftsH* gene of *Bacillus subtilis* is transiently induced after osmotic and temperature upshift. J. Bacteriol. *177*, 4105-4112.

Dornand, J., Gross, A., Lafont, V., Liautard, J., Oliaro, J., and Liautard, J.-P. (2002). The innate immune response against *Brucella* in humans. Vet. Microbiol. *90*, 383-394.

Duc, L.H., Hong, H.A., and Cutting, S.M. (2003a). Germination of the spore in the gastrointestinal tract provides a novel route for heterologous antigen presentation. Vaccine 21: 4215-4224..

Duc, L.H., Hong, H.A., Fairweather, N., Ricca, E., and Cutting, S.M. (2003b). Bacterial spores as vaccine vehicles. Infect. Immun. *71*, 2810-2818.

Duc, L.H., Hong, H.A., Barbosa, T.M., Henriques, A.O., and Cutting, S.M. (2004a). Characterization of *Bacillus* probiotics available for human use. Appl. Environ. Microbiol. *70*, 2161-2171.

Duc, L.H., Hong, H.A., Uyen, N.Q., and Cutting, S.M. (2004b). Immunogenicity and intracellular fate of *B. subtilis* spores. Vaccine *22*, 1873-1885.

Emoto, M., Emoto, Y., Buchwalow, I. B., and Kaufmann, S. H. (1999). Induction of IFN-γ-producing CD4+ natural killer T cells by *Mycobacterium bovis* bacillus Calmette Guerin. Eur. J. Immunol. *29*, 650-659.

Faille, C., Membre, J., Kubaczka, M., and Gavini, F. (2002). Altered ability of *Bacillus cereus* spores to grow under unfavorable conditions (presence of nisin, low temperature, acidic pH, presence of NaCl) following heat treatment during sporulation. J. Food Prot. *65*, 1930-1936.

Gilliam, M. (1985). Microbes from apiarian sources: *Bacillus* spp. in frass of the greater wax moth. J. Invert. Pathol. 45, 218-224.

Gilliam, M., Buchmann, S. L., and Lorenz, B. J. (1984). Microbial flora of the larval provisions of the solitary bees, *Centris pallida* and *Anthophora* sp. Apidologie *15*, 1-10.

Gilliam, M., Buchmann, S. L., Lorenz, B. J., and Schmalzel, R. J. (1990a). Bacteria belonging to the genus *Bacillus* associated with three species of solitary bees. Apidologie *21*, 99-106.

Gilliam, M., Roubik, D. W., and Lorenz, B. J. (1990b). Microorganisms associated with pollen, honey, and brood provisions in the nest of a stingless bee, *Melipona fasciata*. Apidologie *21*, 89-98.

Guidi-Rontani, C., Weber-Levy, M., Labruyere, E., and Mock, M. (1999). Germination of *Bacillus anthracis* spores within alveolar macrophages. Mol. Microbiol. *31*, 9-17.

Hisanga, S. (1980). Studies on the germination of genus *Bacillus* spores in rabbit and canine intestines. J. Nagoya City Med. Assoc. *30*, 456-469.

Hoa, T. T., Duc, L. H., Isticato, R., Baccigalupi, L., Ricca, E., Van, P. H., and Cutting, S. M. (2001). Fate and dissemination of *Bacillus subtilis* spores in a murine model. Appl. Environ. Microbiol. *67*, 3819-3823.

Isaka, M., Yasuda, Y., Mizokami, M., Kozuka, S., Taniguchi, T., Matano, K., Maeyama, J., Mizuno, K., Morokuma, K., Ohkuma, K., et al. (2001). Mucosal immunization against hepatitis B virus by intranasal co-administration of recombinant hepatitis B surface antigen and recombinant cholera toxin B subunit as an adjuvant. Vaccine *19*, 1460-1466.

Jadamus, A., Vahjen, W., Schafer, K., and Simon, O. (2002). Influence of the probiotic strain *Bacillus cereus* var. *toyoi* on the development of enterobacterial growth and on selected parameters of bacterial metabolism in digesta samples of piglets. J. Anim. Physiol. Anim. Nutr. (Berl) *86*, 42-54.

Jadamus, A., Vahjen, W., and Simon, O. (2001). Growth behaviour of a spore forming probiotic strain in the gastrointestinal tract of broiler chicken and piglets. Arch. Tierernahr. *54*, 1-17.

Keynan, A., and Evenchik, Z. (1969). Activation. In The bacterial spore, G.W. Gould, and A. Hurst, eds. (London, Academic Press), pp. 359-396.

La Ragione, R. M., Casula, G., Cutting, S. M., and Woodward, M. J. (2001). *Bacillus subtilis* spores competitively exclude *Escherichia coli* O78:K80 in poultry. Vet. Microbiol. *79*, 133-142.

Lysenko, E., Ogura, T., and Cutting, S. M. (1997). Characterization of the *ftsH* gene of *Bacillus subtilis*. Microbiology *143* (Pt 3), 971-978.

Mazza, P. (1994). The use of *Bacillus subtilis* as an antidiarrhoeal microorganism. Boll. Chim. Farm. *133*, 3-18.

Melby, P., Andrade-Narvaez, F. J., Darnell, B. J., Valencia-Pacheco, G., Tryon, V. V., and Palomo-Cetina, A. (1994). Increased expression of proinflammatory cytokines in chronic lesions of human cutaneous leishmaniasis. Infect. Immun. *62*, 837-842.

Nakano, M. M., and Zuber, P. (1998). Anaerobic growth of a "strict aerobe" (*Bacillus subtilis*). Ann. Rev. Microbiol. *52*, 165-190.

Rengpipat, S., Phianphak, W., Piyatiratitivorakul, S., and Menasveta, P. (1998). Effects of a probiotic bacterium on black tiger shrimp *Penaeus monodon* survival and growth. Aquaculture *167*, 301-313.

Roberts, M., Li, J., Bacon, A., and Chatfield, S. (1998). Oral vaccination against tetanus: comparison of the immunogenicities of *Salmonella* strains expressing fragment C from the *nirA* and *htrA* promoters. Infect. Immun. *66*, 3080-3087.

Robinson, K., Chamberlain, L. M., Schofield, K. M., Wells, J. M., and Le Page, R. W. F. (1997). Oral vaccination of mice against tetanus with recombinant *Lactococcus lactis*. Nat. Biotechnol. *15*, 653-657.

Sanders, M. E., Morelli, L., and Tompkins, T. A. (2003). Sporeformers as human probiotics: *Bacillus*, *Sporolactobacillus*, and *Brevibacillus*. Comprehensive Rev. Food Sci. Food Safety *2*, 101-110.

Sekl, S., Osada, S.-I., Ono, S., Aosasa, S., Habu, Y., Nishikage, T., Mochizuki, H., and Hiraide, H. (1998). Role of liver NK cells and peritoneal macrophages in gamma interferon and interleukin-10 production in experimental bacterial peritonitis in mice. Infect. Immun. *66*, 5286-5294.

Spinosa, M. R., Braccini, T., Ricca, E., De Felice, M., Morelli, L., Pozzi, G., and Oggioni, M. R. (2000). On the fate of ingested *Bacillus* spores. Res. Microbiol. *151*, 361-368.

VanCott, J. L., Staats, H. F., Pascual, D. W., Roberts, M., Chatfield, S. N., Yamamoto, A., Coste, M., Carter, P. B., Kiyono, H., and McGhee, J. R. (1996). Regulation of mucosal and systemic antibody responses by T helper cell subsets, macrophages, and derived cytokines following or al immunisation with live recombinant *Salmonella*. J. Immunol. *156*, 1504-1514.

Vaseeharan, B., and Ramasamy, P. (2003). Control of pathogenic *Vibrio* spp. by *Bacillus subtilis* BT23, a possible probiotic treatment for black tiger shrimp *Penaeus monodon*. Lett. Appl. Microbiol. *36*, 83-87.

Youngman, P., Perkins, J., and Losick, R. (1984). Construction of a cloning site near one end of Tn*917* into which foreign DNA may be inserted without affecting transposition in *Bacillus subtilis* or expression of the transposon-borne *erm* gene. Plasmid *12*, 1-9.

Chapter 10

Mechanisms of Gene Transfer and the Spread of Antibiotic Resistance in Spore-Forming Organisms in the GI Tract

Peter Mullany, Teresa Barbosa, Karen Scott, and Adam P. Roberts

SUMMARY

In this chapter we will consider the mechanisms used for gene transfer in the spore forming bacteria that normally inhabit and have access to the gastrointestinal tract (GIT), these being principally of the genera *Bacillus* and *Clostridia*. The most abundant group of anaerobes inhabiting the GIT are low % DNA G+C Gram-positive bacteria closely related to Clostridial species and which fall in the Clostridial Cluster XIVa subphylum. Very little is known about the ability of these bacteria to sporulate, although some isolates have recently been found to contain homologues of sporulation genes (P. Louis, per. comun.). Since these bacteria may yet be shown able to form spores, gene transfer between them will also be considered. We will consider both the genetic elements responsible for mediating gene transfer and the environmental influences on gene transfer.

Major bacterial population densities in the GIT:

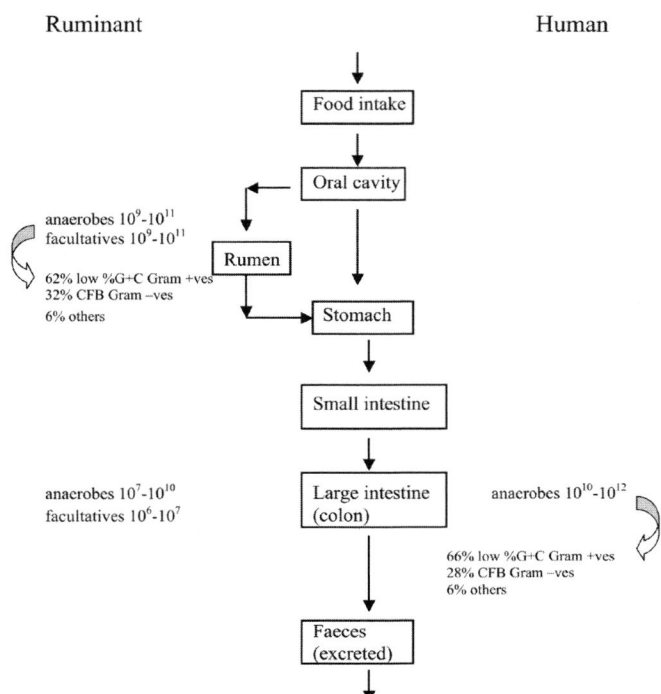

Figure 1. Schematic comparing the gastrointestinal tracts found in ruminants and humans, or other hindgut fermentors. The locations containing the greatest microbial biomass are indicated, and the bacterial population breakdown is shown.

BACTERIAL GENE TRANSFER IN THE GI TRACT

In terms of microbial population densities, mammals have one of two types of GIT (Figure 1). In ruminants the major site of microbial biomass is found in the rumen, a large organ preceding the acidic stomach that contains bacteria (average $10^9 – 10^{11}$ anaerobic cells per ml), fungi and protozoa. The large intestine, after the stomach, also has a large bacterial population ($10^7 – 10^{10}$ anaerobic cells per ml) but does not contain fungi or protozoa. In hindgut fermentors (including humans and pigs) the main microbial population occurs in the large intestine (or colon), and in healthy individuals consists only of bacteria ($10^{10} – 10^{12}$ anaerobic cells per ml). These different organisations mean that in the ruminant any incoming bacteria or food-derived DNA intermixes directly with a huge microbial population without first passing through the acidic barrier of the stomach, and thus the potential for direct horizontal gene transfer from incoming bacteria is greater in ruminants than in hindgut fermentors. However any bacteria passing through the GIT in the form of spores would probably survive passage through the acidic stomach. Recently 16S rRNA sequencing has facilitated bacterial profiling of GIT communities, and this has shown that the general population profiles of the human colon and rumen are similar (Suau et al., 1999; Tajima et al. 1999; Hold et al. 2002; Figure 1).

In general the variety and density of the resident GIT microbial populations, often existing as mixed biofilms on the gut epithelial surface or on the surface of food particles, make the GIT an ideal location for gene transfer events. Since gut environments are often exposed to low levels of antibiotics, shown to stimulate gene transfer in certain cases (Stevens et al., 1990; Showsh and Andrews, 1992), the likelihood of gene transfer occurring is increased even further. The contribution of sub-therapeutic levels of antibiotic to gene transfer between bacteria associated with farm animals is particularly relevant, where antibiotics, such as tetracycline, have been in long-term use as growth promoters. The low concentrations of the antibiotic involved are often supplied in drinking water, also accessible to birds and small rodents that inhabit farmyards.

The most widely described mechanism for horizontal gene transfer between bacteria, in any ecosystem, is conjugation. Most of the conjugation experiments that have been described involve *in vitro* filter-mating experiments between defined donor and recipient bacteria, and there is comparatively little evidence on *in vivo* gene transfer. Circumstantial evidence also exists that gene transfer occurs between bacteria; specifically the presence of identical genes on unrelated, phylogenetically or geographically, bacterial species. Recently, knowledge of

A. Conjugative transposons

Tn*5397* (*C. difficile*; tetracycline resistance)

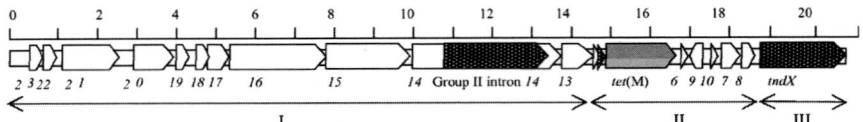

Tn*916* (*E. faecalis, C. difficile* plus others; tetracycline resistance)

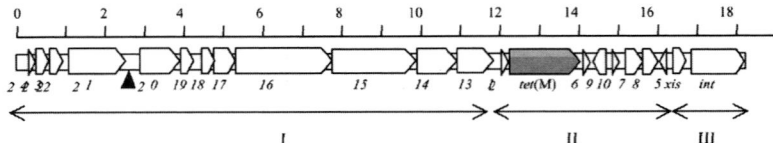

B. Mobilisable transposons

Tn*5398* (*C. difficile*; erythromycin resistance)

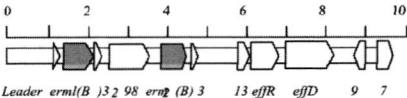

Tn*4451* (*C. perfringens*; chloramphenicol resistance), Tn*4453a* & b (*C. difficile*)

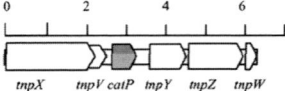

C. Other, non-mobile elements

CW459*tet*(M) (*C. perfringens*; tetracycline resistance)

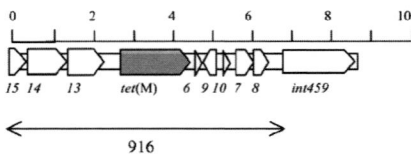

Figure 2. Structure of chromosomally located elements from the clostridia. For all elements the designated names are followed by the host the element was first discovered in and the resistance conferred by the element. The top line represents the scale in kilobases for all elements. The open reading frames (*orf*) for each element are indicated by the arrows pointing in the probable direction of transcription. The designated name of each *orf* is indicated below the arrow. Antibiotic resistance genes are shaded grey

A. The conjugative transposons Tn*5397* and Tn*916* are shown. The filled arrows in Tn*5397* represent *orfs* that are unique to that conjugative transposon. The filled triangle underneath Tn*916* marks the position of the *oriT*, based on nucleotide sequence comparison, this is also likely the case for Tn*5397*. The double headed arrows underneath both of these element represent the distinct modules, I; transfer (conjugation) module. II; resistance and regulatory module. III; excision and insertion (transposition) module.

B. Shows the structure of the mobilisable transposons Tn*5398* and Tn*4451/4453*. Interestingly Tn*5398* contains no genes thought to be involved in transposition or conjugation. The gene *tnpX* present on Tn*4451* and Tn*4453* are responsible for the excision and insertion of these elements.

C. The CW459*tet*(M) element has not been shown to be mobile to date. It is however highly related to Tn*916* (shown by the extent of the double headed arrow labeled 916), it does however appear to have acquired a completely different integrase gene when compared to Tn*916*.

complete bacterial genome sequences has also been used to prove that lateral gene transfer occurs between unrelated bacterial species. Although such gene transfer events are difficult to prove, they can be inferred by detailed sequence analysis, codon usage patterns and % DNA G+C content (see later section "Indirect evidence for widespread bacterial gene transfer").

GENE TRANSFER BY CONJUGATION

Conjugation is defined as the direct transfer of DNA from the cytoplasm of a donor to the cytoplasm of a recipient. The process requires intimate cell to cell contact and the structures involved in facilitating this process are almost always encoded by specialised genetic elements i.e. conjugative plasmids and conjugative transposons. These conjugal elements have also been shown to be capable of mobilising non-conjugative but mobilizable plasmids and transposons (these elements do not themselves contain the genetic information required to facilitate conjugation but can use the structures encoded by the self-conjugative element). Finally other non-conjugative mobile elements (transposons, integrative plasmids, mobile introns and IS sequences) can integrate into conjugative and mobiliseable elements and be transferred in *cis* (reviewed in Mullany 2000). In this section we will review what is known about the biology of the genetic elements that underpin conjugation in spore forming organisms and we will also consider the evidence for conjugative gene transfer in the GI tract.

CONJUGATION IN THE CLOSTRIDIA

The clostridia are a diverse group of spore forming anaerobic organisms. Members of this genus produce some of the most potent toxins known, e.g. *C. botulinium, C. tetani* and *C. difficile* (Stackebrandt 1997). They also contain species that are used in industry e.g. *C. acetobutylicum* (Nolling et al. 2001). The clostridia contain a rich array of mobile genetic elements that mediate gene transfer. Representatives of each of these will be discussed in this section.

CONJUGATIVE TRANSPOSONS

There has been some controversy in actually defining what a conjugative transposon is, therefore for the purposes of this article we will define them in the broadest possible terms. They are discrete DNA elements, usually integrated into the host genome, that are capable of mediating their own transfer from a donor to a recipient by conjugation. Some conjugative transposons have an exceptionally broad host range and the

donor and recipient do not necessarily have to be of the same genus of even the same Gram staining group (for a review of these elements see Mullany 2002 and accompanying articles).

TN*916*

Tn*916* was the first conjugative transposon to be discovered, and was identified on the chromosome of the *Enterococcus faecalis* strain DS16 (Franke and Clewell 1981). It is the most extensively investigated of the conjugative transposons, and elements that are indistinguishable to this have been found in, and have been transferred into, various species of clostridia (Roberts et al., 2001; Wang et al., 2000; Lin and Johnson, 1995; Volk 1988; Woolley et al., 1989).

PROPERTIES OF TN*916*

Tn*916* has subsequently been completely sequenced and extensively characterised (for recent reviews see Mullany et al 2002 and accompanying reviews). It is 18 kb in length (Figure 2) and encodes resistance to tetracycline. It is thought to consist of three distinct modules i.e. conjugation, recombination and regulation/resistance (see Figure 2). The initial step in conjugative transposition is excision from the chromosome followed by circularisation of the element. The circular molecule is then nicked at an origin of transfer, *oriT*, and a single strand is transferred to the recipient organism. In both donor and recipient the single strand is used as a template for second strand synthesis. The double stranded molecule then inserts into the recipient chromosome and can reinsert into the donor chromosome.

In the case of Tn*916* the products of the genes *int* and *xis* (Figure 2) mediate excision, circularisation and integration. Int

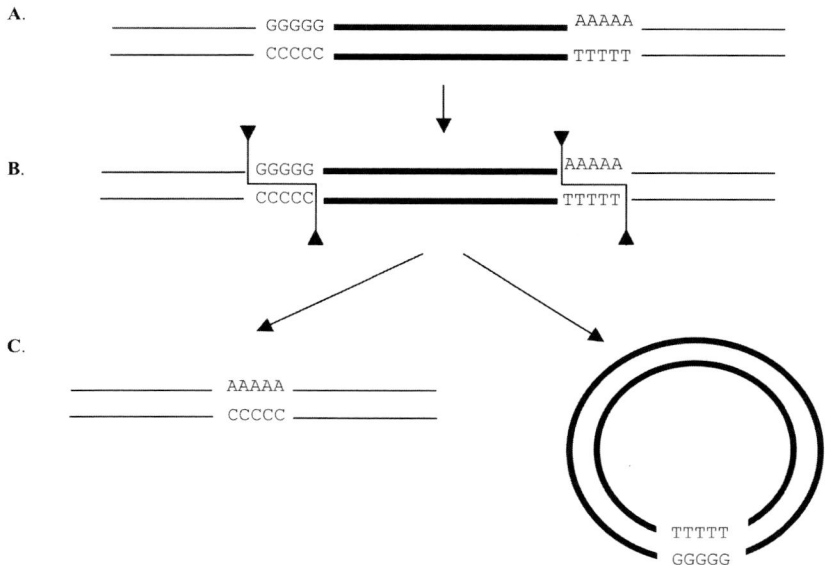

Figure 3. Excision and insertion of Tn*916*. A. Tn*916* (thick lines) is present in the chromosome (thin lines). The base pairing at each end of the element is shown by the G+C and the A+T groups.
B. Staggered endonucleolytic cuts are made at each end of the element (line with double inverted arrow heads) forming overhanging non-complementary regions at the end of Tn*916* and at the target site within the chromosome.
C. Strand exchange is completed resulting in a non-replicative circular form of Tn*916* containing a heteroduplex at the joint site (T+G groups) and a heteroduplex at the re-ligated target site (A+C groups). The insertion process is essentially the reverse of this mechanism.

is a member of the tyrosine recombinase family of site-specific recombinases and is required for both excision and integration. Xis is a small basic protein which greatly stimulates excision and in some hosts is absolutely required for this. At high concentrations Xis also inhibits integration. Prior to excision Int generates 5'-protruding, staggered, endonucleolytic cuts at each end of the element with one strand cut 5 bases from the end of the transposon and the other immediately adjacent to the other end. This generates single-strands that are ligated together; as Tn916 does not usually duplicate its target site on integration these sites form a heteroduplex at the junction of the ends of the transposon in the circular form (Figure 3). Insertion is essentially the reverse of excision (this is reviewed in Mullany et al., 2002).

Genetic analysis of Tn916 has indicated that the genes involved in conjugative transfer of the element were likely to be orfs 24 to 15 (Figure 2) and an oriT site between orf20 and orf21 (Flannagan et al., 1994; Senghas et al., 1988). Orf21 has a non-specific DNA nicking activity, this has led to speculation that this protein may be involved in generating a single strand nick at the oriT site, prior to transfer of the element (Jaworski and Clewell, 1995). The exact roles of the other orfs in this region in conjugation have not been determined although Orf15 and Orf16 have homologues in both pAD1 and pAM373 (conjugative plasmids from E. faecalis) and Orf18 has similarity with anti- restriction proteins, which could contribute to the extreme promiscuity of Tn916. Also in this regard the element has very few restriction sites this is also consistent with an element of very broad host range.

There is a regulatory region that encompasses orf7, orf8, orf9 and orf12. It has been postulated that the role of orf12 is to sense the presence of tetracycline via a transcriptional attenuation mechanism (Su et al., 1992). Northern blot analysis, promoter fusions, and primer extension analysis has allowed a model for regulation of Tn916 to be proposed which leads to increased transcription of the transfer genes when tetracycline is present (Celli and Trieu-Cuot 1998).

USE OF TN916 AS A TOOL IN THE CLOSTRIDIA

Due to the fact that some of the clostridia (e.g. C. difficile) are not able to be manipulated by conventional molecular methods, the use of naturally occurring conjugative elements as molecular tools is the only way to investigate the genetics of these organisms. To this end Tn916 has been used as an insertional mutagen in Clostridium perfringens (Briolat and Reysset 2002; Awad and Rood 1997). The element can either be introduced into this strain via conjugation from a suitable host or via electroporation of Tn916 contained on a suicide plasmid such as pAM120 (Gawron-Burke and Clewell 1984). Tn916 is not the ideal insertional mutagen as it has a tendency to preferentially insert into A+T rich regions, which are usually intergenic in low GC clostridia. Insertion at multiple sites in the same strain is also very common; this phenomenon complicates genetic analysis of mutants. Additionally there has been one example of Tn916 undergoing inversion within a particular target (O'Keeffe et al., 1999).

Despite these potential problems Tn916 has been successfully used to investigate genes involved in the stress response of C. perfringens and to investigate virulence properties of the organism. The element has also been used to generate insertion mutants in several other species of clostridia, including investigation of numerous fermentation pathways (Mattsson and Rogers 1994; Babb et al., 1993). The conjugative transposon Tn1545 (highly related to Tn916 with additional resistance genes to erythromycin and kanamycin) has also been used to generate mutants and as an aid to genetic mapping in different clostridial species (Jennert et al., 2000).

Tn916 has also been used for cloning foreign DNA in C. difficile (Mullany et al., 1994). This made use of the observation that there is a region within Tn916 (the tet(M) gene) into which DNA can be inserted without affecting the conjugation or transposition properties of the element. This technique has been used to introduce antisense constructs into C. difficile to investigate the role of adhesion in virulence (Roberts et al., 2003). It has also been used to investigate sporulation in the organism (Haraldsen and Sonenshein 2003).

TN5397

Tn5397 is a conjugative transposon that was originally isolated from C. difficile that also encodes resistance to tetracycline (Mullany et al., 1990). In fact it is the only conjugative transposon native to the clostridia that has been investigated in detail. Tn5397 is closely related to Tn916 but there are two major differences between the elements (Figure 2). Tn5397 contains a group II intron inserted into the homologue of orf14 (Figure 2). The intron is efficiently spliced in both C. difficile and B. subtilis (Roberts et al., 2001a). The second major difference is that Tn5397 integration and excision is mediated by the serine site-specific recombinase TndX instead of Int and Xis found in Tn916 (Wang et al., 2000a; Wang and Mullany 2000).

Tn5397 was originally found in C. difficile 630, and was transferable between C. difficile strains and to and from B. subtilis (Mullany et al., 1990). In C. difficile the element has a highly preferred insertion site but in B. subtilis it can insert into multiple sites.

The proposed model for TndX mediated insertion and excision of Tn5397, is described in detail elsewhere (Wang et al., 2000a; Wang and Mullany 2000). Briefly, in order to initiate excision of the element, TndX is thought to mediate 2 bp staggered cuts at the 3' ends of directly repeated GA dinucleotides that are present at each end of the transposon. Strand exchange then occurs which mediates the excision of the transposon as a circular form. Transfer of the element is thought to be similar to that of Tn916 as the two elements are highly homologous in the regions involved in conjugative transfer. Insertion of the element is thought to be the reverse of excision.

Tn5397 is the only conjugative transposon proven to be dependent on a serine recombinase for mobility, however DNA sequencing projects have revealed other genetic elements that could be conjugative transposons containing serine recombinases. Further work is required to understand the significance of this observation.

INTERACTION BETWEEN TN*5397* AND TN*916*

An interesting interaction between Tn*916* and Tn*5397* has been observed in *C. difficile* (Wang et al., 2000a). If a derivative of Tn*916*, Tn*916*ΔE (in which the tetracycline resistance gene has been replaced with an erythromycin resistance gene) is introduced into a strain containing Tn*5397* the majority of the transconjugants have been found to lose Tn*5397*. When Tn*5397* is introduced into a *C. difficile* strain containing Tn*916*ΔE the majority of the transconjugants were found to lose Tn*916*ΔE. In some transconjugants both elements are present together and in these cases they are both stably inherited. The molecular mechanism for this interaction is not completely clear, however, it presumably involves some *trans* acting factors that induce the resident element to excise. This phenomenon could be useful for curing strains of recombinant transposons and may provide insights into the possible interactions that occur between the different types of element.

CW459*TET*(M)

The CW459tet(M) element (found in *C. perfringens*) also has a modular structure, and although there is no evidence that this element is conjugative, it has regions that are very closely related to Tn*916* and Tn*5397* (Figure 2). However CW459*tet*(M) has a completely different recombinase to either of these transposons, the putative Int459 protein is a member of the integrase family of site-specific recombinases but is not closely related to Int from Tn*916*. Based on these studies it is concluded that the clostridial elements have a modular genetic organisation and were derived independently from distinct mobile genetic elements (see below).

CLOSTRIDIAL MOBILISEABLE TRANSPOSONS

Conjugative elements have all the required functions to initiate and complete conjugation. Mobilisable elements contain an *oriT* and a *mob* gene but do not contain all the other accessory factors required for conjugation, they rely on the conjugation functions of other genetic elements.

THE TN*4451*/TN*4453* FAMILY OF MOBILISEABLE TRANSPOSONS

The most extensively studied of the clostridial mobilisable transposons are Tn*4451* from *C. perfringens* and Tn*4453* from *C. difficile*. The elements are very closely related indicating recent gene transfer between the two clostridial species (Adams et al., 2002). The genetic organisation of Tn*4451* is shown in Figure 2B. The complete sequence of the 6338 bp element has been obtained. Like the conjugative transposon Tn*5397*, Tn*4451* is flanked by GA dinucleotides. Furthermore, the gene *tnpX*, which is responsible for insertion and excision of Tn*4451* encodes a protein that is a serine recombinase related to TndX.

TnpX mediates excision of the elements to form a circular molecule which results in the formation of a strong promoter upstream of *tnpX*. As the circular form cannot replicate, the formation of this strong promoter is thought to ensure that enough TnpX is produced to allow the transposon to reinsert before it is lost. When the element is inserted the promoter is disrupted thus allowing the transposon to remain stably integrated in the genome. TnpX is the only protein required for insertion and excision of Tn*4451*.

It has been shown that the *tnpZ* gene of Tn*4451* encodes a mobilisation protein that acts on a site termed RS$_A$ that is essentially an *oriT* site. TnpZ is required for nicking at this site prior to transfer of the element. The non-replicating circular form of Tn*4451* can be transferred from *E. coli* containing RP4 (a conjugative plasmid) to *E. coli* and *C. perfringens* strains where it integrates into the genome. The transposon is dependent on the RP4 functions for transfer. Therefore the Tn*4451*/*4453* family are proven mobiliseable transposons.

TN*5398*

In *C. difficile* 630 erythromycin resistance has been shown to be transferable at very low frequencies both between *C. difficile* strains and to *B. subtilis* strains, in the absence of detectable plasmid DNA (Adams et al., 2002). Further analysis of the mechanism of resistance showed that erythromycin resistance was encoded by an *ermB* gene (Mullany et al., 1995). This behaviour was consistent with resistance being contained within a conjugative transposon and the element was designated Tn*5398*.

However DNA sequencing of the element revealed that it did not have the structure of a classic conjugative transposon (see Figure 2). It was approximately 9.6 kb in length and contains two directly repeated *erm*(B) genes. The element contains *orf13*, *effR*, *effD* and *orf9* (Figure 2). The *effD* and *effR* have the potential to encode an efflux protein and its associated regulator respectively. The other named *orf*s are homologous to the similarly named *orf*s from Tn*916*. It is unlikely that any of these potential *orf*s are involved in mobility of this element as in Tn*916* the homologous *orf*s are involved in regulation (Adams et al., 2002). *Orf298* has the potential to produce a protein that has weak homology with the ParA and Soj families that are involved in plasmid and chromosome partitioning, however there is no obvious role for mobility in these proteins. The role of the other *orf*s in this region is unknown.

The ends of the element are highly AT rich and are very closely related to the putative target sequence of Tn*5398* in *C. difficile* CD37. Tn*5398* also contains at least two *oriT* sequences (Adams et al 2002). Therefore the element has the potential to be mobilised by another genetic element. The question as to how the element excises and integrates is not easily answered as there are no genes encoding recombinases within the element or any obvious binding sites for such a protein. However the homology with the target sequence in *C. difficile* CD37 indicates that homologous recombination could be involved. It was also suggested that a co-resident recombinase could be involved in excision/insertion although no evidence has been presented for this. Clearly more experimentation is required to understand the mobility of Tn*5398*.

PCW3 PLASMIDS

The plasmid pCW3 was found in *C. perfringens* where it was responsible for mediating the spread of tetracycline resistance (via the *tet*(P) gene). This plasmid also sometimes contains the transposon Tn*4451* (Abraham et al., 1985). This plasmid has a worldwide distribution in *C. perfringens* and is responsible for most of the conjugative transfer of tetracycline resistance in this organism. Interestingly the conjugative plasmid containing the *C. perfringens* enterotoxin is conjugative and is homologous to most of pCW3 the only regions that are not are those involved in plasmid replication and the fragment containing the *tet*(P) gene (Brynestad et al., 2001).

GENETIC ELEMENTS FROM CLOSTRIDIA-LIKE ORGANISMS

Mobile elements have been found in organisms that are phylogenetically close to the clostridia but have not yet been proved to sporulate.

The conjugative transposon Tn*B1230* was isolated from the rumen anaerobe *Butyrivibrio fibrisolvens* (Scott et al., 1997). This element is approximately 45 kb and mediates tetracycline resistance via the *tet*(W) gene. The element is related to the enterococcal conjugative transposon Tn*1549* in the region of the two elements thought to be involved in conjugation, but there are no other obvious similarities between the two elements (Melville et al., submitted). A feature of Tn*B1230* that is not often observed in conjugative transposons is that it can transfer at high frequencies between *B. fibrisolvens* strains, up to 10^{-2} transconjugants per donor (Scott et al., 1997). PFGE analysis of transconjugants indicated that the transposon had a preferential insertion site in the recipient genome (Barbosa et al., 1999).

A further new mobile genetic element identified in a human colonic anaerobe closely related to *Clostridium* sp. is Tn*K10* (Melville et al., 2001). This transposon, which has not yet been extensively characterised, also contains a novel tetracycline resistance gene, *tet*(32). Tn*K10* was transferable to two unrelated Clostridium Cluster XIVa bacteria, falling in different clusters in the phylogenetic tree: the human anaerobe, *Roseburia inulovorans* and the rumen anaerobe *B. fibrisolvens*.

Table 1. Examples of mobile elements found in *Bacillus* species.

Strain	Mobile element	Characteristics/location	Reference
Plasmids			
B. anthracis	pXO1	Toxin, mobilisable	Okinaka et al., 1999
B. anthracis	pXO2	Capsule, mobilisable	Pannucci et al., 2002
B. cereus	pBC16	Tc^R, mobilisable	Bernhard et al., 1978
B. thuringiensis	pXO12	ICP, conjugative	Battisti et al., 1985
B. thuringiensis	pXO16	aggregation, conjugative	Reddy et al., 1987
B. thuringiensis	pAW63	cryptic, conjugative	Wilcks et al., 1998
B. thuringiensis	pGI2	cryptic, mobilisable	Mahillon and Seurinck, 1988
B. thuringiensis	pTX14-1	cryptic, mobilisable	Andrup et al., 2003
B. subtilis (*natto*)	pLS20	cryptic, conjugative	Koehler and Thorne, 1987
B. coagulans	pL$_4$	bacteriocin, putative mobilisable	Hyronimus et al., 1998
Thermophilic *Bacillus*	pTB19	Tc^R, Km^R, Bleo^R, mobilisable	Imanaka et al., 1981
Conjugative transposons			
B. cereus group spp.	Tn*916*-like	Tc^R, *tet*(M)	Agerso et al., 2002
Transposons			
B. thuringiensis	Tn*4430*	cryptic, chromosomal and plasmid	Lereclus et al., 1984
B. thuringiensis	Tn*5401*	cryptic, chromosomal	Baum, 1994
B. cereus	Tn*5084*	Hg^R, chromosomal and plasmid	Bogdanova et al., 2001
B. megaterium	Tn*MERI*1	Hg^R, chromosomal	Huang et al., 1999
IS			
B. thuringiensis	IS*231*	IS*4* family	Mahillon et al., 1987
B. thuringiensis	IS*232*	IS*21*family	Menou et al., 1990
B. thuringiensis	IS*240*	IS*6* family	Delecluse et al., 1989
B. thuringiensis YBT-226	IS*Bt2*	IS*3* family	Mahillon et al., 1994
B. stearothermophilus IS*5376* CU21	IS*21*family	Xu et al., 1993	
B. stearothermophilus ISBs*1* CU21	IS*982*	family	Mahillon and Chandler, 1998
B. subtilis (*natto*)	IS*4Bsu*1	IS*4* family	Nagai et al., 2000
Introns			
B. anthracis	group I intron	chromosomal, *recA* gene	Ko et al., 2002
B. halodurans	group II intron	chromosomal	Takami et al., 2001
B. megaterium	group II intron	chromosomal, Tn*MERI*1	Huang et al., 1999

Again the conjugative transposon seemed to have a preferential insertion site in the recipient genome (Melville et al. 2001), although secondary insertions also occurred. These transfer experiments illustrate the potential for gene flow between bacteria inhabiting different environments.

GENE TRANSFER IN THE *BACILLUS*

The genus *Bacillus* comprises a diverse group of aerobic spore forming Gram positive organisms. Their presence in soil, water, dust and foodstuffs results in a considerable daily intake by animals and humans, reflected in their isolation from faecal material (Ghosh, 1978; Turnbull and Kramer, 1985; Barbosa and Henriques, unpublished). Most members of this genus are not harmful, however, there are some exceptions and these include *B. anthracis*, the etiological agent of anthrax, *B. cereus*, associated with food poisoning and opportunistic infections and *B. thuringiensis*, an insect pathogen (see Chapter 8).

The ubiquitous nature of these organisms suggests that there is considerable scope for them acting as both recipients and donors in gene transfer events in diverse ecological niches. This is particularly important given the fact that *Bacillus* spores are used as human probiotics, animal feed additives and as plant biopesticides. In the rest of this section we will discuss the genetic elements involved in gene transfer in the *Bacillus*.

BACILLUS MOBILE ELEMENTS

Naturally occurring plasmids are frequent among *Bacillus* isolates, with many strains harbouring multiple elements, most of which have not been characterised and remain cryptic. Only a few have been associated with a particular cell function, such as antibiotic resistance or synthesis of toxins and antibiotic compounds (Lovett et al., 1976; Bernhard et al., 1978; Mahler and Halvorson, 1980; Imanaka et al., 1981; Stahl, 1991). Nevertheless, many of these plasmids encode for conjugative or mobilisable properties (Table 1). Besides plasmids, other integrated mobile genetic entities, such as transposons and insertion sequences (IS elements) have also been identified in *Bacillus* species and will also be discussed here (Table 1).

PBC16

The tetracycline resistance plasmid pBC16 originally identified in *B. cereus* is one of the best characterised plasmids from *Bacillus* (Bernhard et al., 1978). Its 4630 bp nucleotide sequence has been completely determined and it was found to be almost identical to the kanamycin resistance plasmid pUB110 from *S. aureus* (Chopra et al., 1973, Gryczan et al., 1978). pBC16 is a rolling circle, non-conjugative but mobiliseable plasmid which has been mobilised by many *Bacillus* conjugative plasmids, such as pLS20, pXO11 and pXO12 (Battisti et al., 1985; Koehler and Thorne, 1987). The *B. subtilis* (natto) 55 kb conjugative plasmid pLS20 promoted transfer of pBC16 into a range of different *Bacillus* species including *B. subtilis* (natto), *B. megaterium*, *B. anthracis*, *B. cereus*, *B. thuringiensis*, *B. licheniformis*, *B. subtilis* and *B. pumilus*. ORF-b and a region located 5´ to this ORF, the RS$_A$ site (thought to be equivalent to the *mob* gene and *oriT*, respectively), of pBC16 were shown to be essential and sufficient for mobilisation of pBC16 by pLS20 in *B. subtilis*

(Koehler and Thorne, 1987; Selinger et al., 1990). pBC16 has been frequently used as a vector for investigating the genetics of *Bacillus* spp.

PLASMIDS OF THE *B. CEREUS* GROUP

B. cereus, *B. anthracis*, *B. thuringiensis* and *B. mycoides* all belong to the *B. cereus* group *sensu lato*, which is characterised by a close genetic and phenotypic relatedness of its species. Many authors contend that these should be considered as subspecies of a single generic *B. cereus* species (Ash et al., 1991; Leonard et al., 1997). Virulence factors are encoded on plasmids from this group including the *B. thuringiensis* insecticidal crystal proteins (ICPs) and the *B. anthracis* virulence determinants (Schnepf et al., 1998; Okinaka et al., 1999; Pannucci et al., 2002).

Isolates of different *B. thuringiensis* subspecies may carry multiple plasmids, with some containing as many as 12. These plasmids range from small mobilisable plasmids such as pGI1, pGI2, pTX14-1, pTX14-2, pTX14-3, to mega, self-transmissible plasmids of more than 200 kb, for example pXO11, pXO13, pXO14, pXO15, pXO16 and pAW63. Generally the information on the biology of these elements is limited (Mahillon and Seurinck, 1988; Wilcks et al., 1998; Andrup et al., 2003). Importantly, more than one conjugative plasmid can inhabit the same cell, even though in some cases an entry exclusion mechanism appears to exist (Battisti et al., 1985; Wilcks et al., 1998). Many of these conjugative plasmids are able to promote intra and interspecies transfer of mobilisable plasmids such as pBC16 and the *B. anthracis* virulence plasmids pXO1 and pXO2.

The *B. thuringiensis* ICPs are frequently encoded on large self-transmissible plasmids, such as pHT73 and pXO12 (Green et al., 1989; Wilcks et al., 1998). Importantly, pXO12 has been shown to be transferable to *B. cereus* and *B. anthracis* conferring the ability to produce toxin on the transconjugants (Green et al., 1989). *B. anthracis* transconjugants with plasmid pXO12 became efficient donors of resident plasmids pXO1, pXO2 and pBC16 (Green et al., 1989). pXO12-mediated transfer of pXO1 and pXO2 between *B. anthracis* isolates occurred by the formation of a co-integrate between the plasmids by the transposon Tn*4430* (Green et al., 1989). Usually the plasmids resolve into two plasmids in the transconjugants, but, less frequently, a large cointegrate plasmid was detected in the transconjugants. This plasmid was still efficient in mobilizing pBC16. The establishment of physical contacts between the plasmids was not necessary for the transfer of pBC16 which can be transferred at high frequencies presumably by factors acting *in trans*.

Many of the *B. thuringiensis* small plasmids, such as pGI2, pGI1, pTX14/1, are members of different families of rolling circle replicating plasmids presenting a modular structure, wherein the mobilisation modules comprise the *mob* gene and the *oriT* site (Andrup et al., 2003). Although the Mob proteins of these plasmids share conserved motifs they can belong to different Mob protein clusters and sequence analysis suggests the existence of subdomains among these proteins (Andrup et al., 2003). The Mob proteins are responsible for creating the specific single-stranded nick in the recognition sequence or

Table 2. Examples of intra and interspecies transfer of heterologous CTn in matings involving *Bacillus* isolates.

Donor	Recipient	Reference
Tn916 (Tc^R)		
E. coli	B. subtilis	Bertram et al., 1991
B. subtilis	E. faecalis	Storrs et al., 1991
B. subtilis	C. difficile	Mullany et al., 1991
B. subtilis	B. thuringiensis	Showsh and Andrews, 1992
E. faecalis	B. thuringiensis	Naglich and Andrews, 1988a
B. subtilis	Strepto. pyogenes	Norgren and Scott, 1991
B. subtilis	Thermus aquaticus	Sen and Oriel, 1990
Thermus aquaticus	B. subtilis	Sen and Oriel, 1990
B. sphaericus	B. thuringiensis	Showsh and Andrews, 1992
B. thuringiensis	B. subtilis	Showsh and Andrews, 1992
B. subtilis	Lactococcus lactis	Bringel et al., 1991
E. faecalis	B. licheniformis	Herzog-Velikonja et al., 1994
B. licheniformis	B. subtilis	Herzog-Velikonja et al., 1994
E. faecalis	B. pumilus	Hendrick et al., 1991
B. pumilus	B. pumilus	Hendrick et al., 1991
B. thuringiensis	E. faecalis	Naglich and Andrews, 1988
E. faecalis	B. anthracis	Ivins et al., 1988
B. anthracis	B. subtilis	Ivins et al., 1988
Tn925 (Tc^R)		
E. faecalis	B. subtilis	Christie et al., 1987
B. subtilis	E. faecalis	Christie et al., 1987
B. subtilis	B. thuringiensis	Showsh and Andrews, 1996
B. subtilis	B. firmus	Guffanti et al., 1991
B. firmus	B. subtilis	Guffanti et al., 1991
Tn919 (TcR)		
B. subtilis	Lactococcus lactis	Bringel et al., 1991
Tn1545 (Tc^R Em^R Km^R)		
B. subtilis	Clostridium acetobutylicum	Woolley et al., 1989
Tn5397 (Tc^R)		
C. difficile	B. subtilis	Mullany et al., 1990
B. subtilis	C. difficile	Mullany et al., 1990
B. subtilis	Strepto. acidominimus	Roberts et al., 1999
Tn5398 (MLS^R)		
C. difficile	B. subtilis	Mullany et al., 1995
B. subtilis	C. difficile	Mullany et al., 1995

promotes self-transfer and transfer of other mobilisable plasmids, and even plasmids that are not normally mobilisable, in liquid media. (Andrup et al., 1993; Jensen et al., 1995; Andrup et al., 1996). The aggregation phenotype is characterized by the formation of large macroscopic co-aggregates when exponential phase cultures of the donor (Agg+) and recipient (Agg-) strains are mixed in broth. After transfer of the aggregation phenotype to the recipient strains, which happens quickly at very high frequencies (~100%), the aggregate disperses. Recipients of *B. thuringiensis* and *B. cereus* that became Agg+ were able to act as efficient donors in subsequent matings demonstrating that the pXO16-encoded aggregation and conjugation properties can function in heterologous hosts (Jensen et al., 1996). Curiously, pXO16 was also able to transfer derivatives of pBC16 that lacked most of the required mobilisation sequences (Andrup et al., 1996). The mechanism by which this takes place has not been fully clarified.

Transfer of some of the conjugative plasmids described above has also been documented in insect larvae and other natural environments indicating that conjugation plays an important role in generating diversity in this *Bacillus* species (Jarret and Stephenson 1990; Thomas et al., 2000; Vilas-Boas et al., 2000; Thomas et al., 2001).

CONJUGATIVE TRANSPOSONS IN *BACILLUS*

There has been only one example of a conjugative transposon occurring naturally in a member of this genus. This was found in the chromosome of strains of the *B. cereus* group isolated from farm soil and manure (Agerso et al., 2002). The element appeared to be related to Tn916 although it was not characterised in detail. The tetracycline resistance gene, *tet*(M), associated with this putative conjugative transposon was transferable by conjugation to other *B. subtilis*, *S. aureus* and enterococci isolates. The lack of conjugative transposons in natural isolates of *Bacillus* is surprising, since conjugative transposons of the Tn916 family have been transferred into many different *Bacillus* species, which in turn became efficient donors, indicating that there are no particular biological barriers to the transfer of these elements (Table 2). However this may just represent a lack of a detailed investigation of these mainly non-pathogenic organisms.

Many insights into the molecular mechanisms of the conjugal transfer of conjugative transposons, such as the excision, transfer, transposition and integration processes have been gained in the *Bacillus* genetic background (Scott et al., 1988; Norgren and Scott, 1991; Storrs et al., 1991; Torres et al., 1991; Showsh and Andrews, 1992; Scott et al., 1994; Showsh and Andrews, 1996; Marra and Scott, 1999; Marra et al., 1999; Showsh and Andrews, 1999). In *Bacillus* these elements were able to mediate, not only their own transfer, but also transfer of non-conjugative plasmid DNA (Naglich and Andrews, 1988; Guffanti et al., 1991; Showsh and Andrews, 1996; Showsh and Andrews, 1999). In this way the use of, for example Tn1545 and Tn916, as tools to undertake mutagenesis and/or mutational cloning in a variety of *Bacillus* species, including *B. anthracis* has been explored (Ivins et al., 1988; Naglich and Andrews, 1988, 1988a; Natarajan and Oriel, 1991 ; Koehler, 2000).

oriT, where the transfer process is initiated. The *oriT* site in these plasmids is close to the lagging strand origin (*sso*) and is AT rich in comparison to the flanking regions as observed for the paradigm pMV158 (Andrup et al., 2003). Other plasmids like the large pAW63 have theta replicating origins (Wilcks et al., 1999).

The *B. thuringiensis* subsp. *israelensis* plasmid pXO16 (200 kb) encodes an aggregation phenotype that efficiently

A.

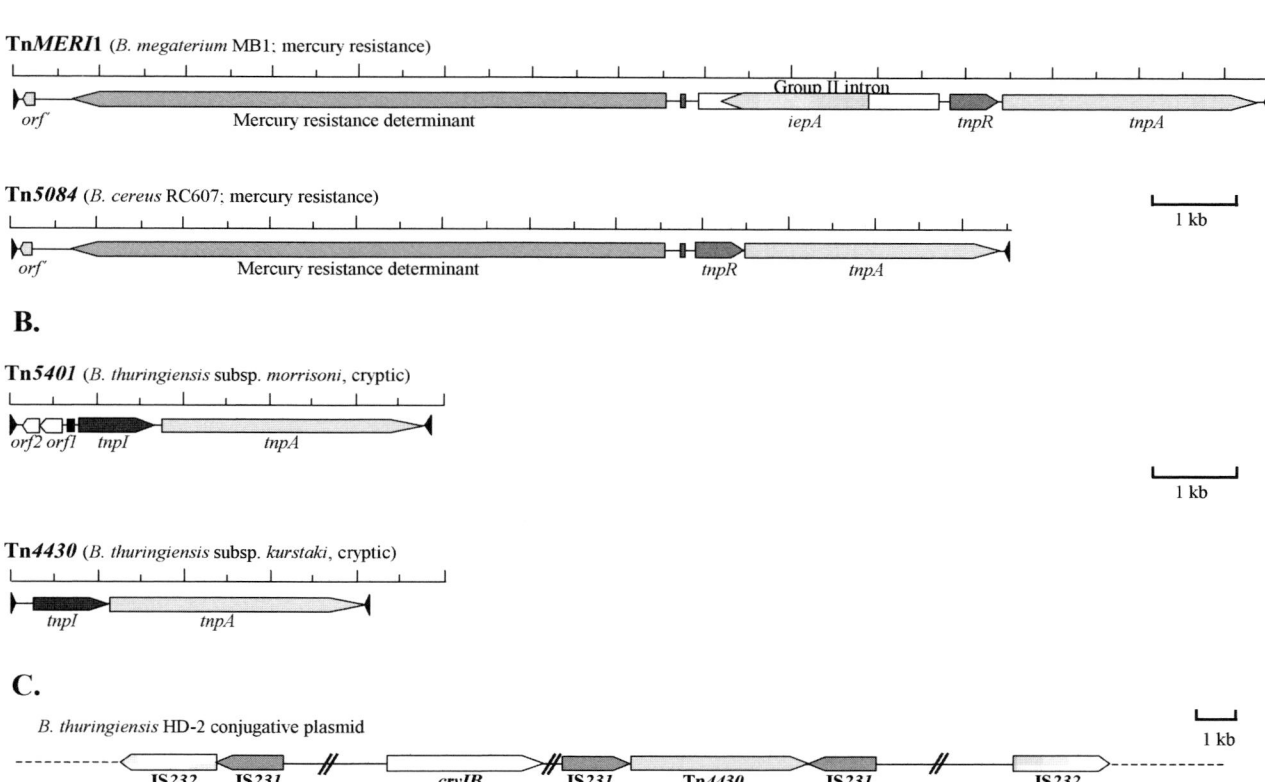

Tn*MERI*1 (*B. megaterium* MB1; mercury resistance)

orf´   Mercury resistance determinant   Group II intron   *iepA*   *tnpR*   *tnpA*

Tn*5084* (*B. cereus* RC607; mercury resistance)

1 kb

orf´   Mercury resistance determinant   *tnpR*   *tnpA*

B.

Tn*5401* (*B. thuringiensis* subsp. *morrisoni*, cryptic)

orf2 orf1   *tnpI*   *tnpA*

1 kb

Tn*4430* (*B. thuringiensis* subsp. *kurstaki*, cryptic)

tnpI   *tnpA*

C.

B. thuringiensis HD-2 conjugative plasmid

1 kb

IS*232*   IS*231*   *cryIB*   IS*231*   Tn*4430*   IS*231*   IS*232*

Figure 4. Schematic representation of the structure of different mobile elements from *Bacillus*.
Primary hosts are indicated. Arrowheads indicate the direction of transcription of the different open reading frames, operons or transposable elements. The scale is indicated.
A. Mercury resistance Class II transposons. Gray and hatched arrows represent *orf´s* related to transposition (*tnpA*, transposase and *tnpR*, resolvase) and mercury resistance, respectively. The Tn*MERI*1 group II intron is shown by a white box while the dotted arrow indicates the gene contained in the intron, *iepA*. *Orf´* indicates the 3´ end of a putative transposase gene. Black triangles and dark gray rectangles denote terminal inverted repeats (IRs) and putative *res*-regions.
B. Cryptic Class II transposons. Gray arrows represent *orf´s* related to transposition (*tnpA*, transposase and *tnpI*, recombinase). The black rectangle in Tn*5401* represents repeat elements with the consensus ATGTCCRCTAAY, which are recognition sites for TnpI recombinase. Black triangles denote terminal inverted repeats (IRs). *Orf1* and *orf2* have unknown function.
C. Example of the number of diverse transposable elements associated with the crystal toxin gene in *B. thuringiensis* strains. The example given is present on a large plasmid of *B. thuringiensis* HD-2 (see Mahillon et al., 1994). Double vertical lines indicate that only the genetic elements and not the inter-element regions are drawn to scale.

Frequently, *Bacillus* strains are used as donors of these elements to bacteria that are not easily genetically manipulated (Woolley et al., 1989; Mullany et al., 1994; Roberts et al., 2003).

OTHER MOBILE ELEMENTS

A large number of insertion sequences, non-conjugative transposons and group I and group II introns have been found in the *Bacillus* (Table 1). These have often been found associated with conjugative and mobiliseable elements thus contributing to the spread of genes throughout this genus.

TRANSPOSONS CONTAINING MERCURY RESISTANCE GENES

Mercury is found in the environment as a result of industrial pollution, volcanic and other natural processes and is also present in the amalgam used in restorations of carious lesions in human teeth. Therefore bacteria will be exposed to this toxic metal and will have to develop resistance in order to survive.

In *Bacillus* spp. mercury resistance operons are found both in plasmids and chromosome, and are usually located in different class II transposons, such as Tn*5084* and Tn*MERI*1 (Figure 4A). Tn*5084* is located in plasmid pKLH6 of the *B. cereus* RC607 strain isolated from Boston Harbour. Similar *mer* determinants to that of Tn*5084*, were found in mercury resistant *B. firmus*, *B. subtilis*, *B. lentus* and *B. megaterium* isolates from Minamata Bay, Japan, in *Bacillus* isolates from contaminated areas of Russia and England and in other Gram-positive bacteria (Nakamura and Silver, 1994; Bogdanova et al., 1998; Hart et al., 1998; Gupta et al., 1999; Huang et al., 1999; Bogdanova et al., 2001; Narita et al., 2003).

Tn*MERI*1 is a class II transposon found on the chromosome of *B. megaterium* MB1 isolated from Minamata Bay (Huang et al., 1999). This transposon has a transposition module similar to the Tn*21* subgroup of the Tn*3* family. Tn*MERI*1 is flanked by 38-bp inverted repeats (IR), contains transposase (*tnpA*) and resolvase (*tnpR*) encoding sequences, as well as putative *res* sites upstream of *tnpR*, and carries a mercury resistance determinant related to that of *B. cereus* RC607. However it also contains an insertion of a group II intron in a non coding region between the resolvase gene and the *mer* operon (Huang et al., 1999) (Figure 4A).

Tn*5085* from the Gram-positive *Exiguobacterium* sp. TC38-2b is practically identical to Tn*MERI*1 except that it does not contain the group II intron. Tn*5085* is also nearly identical to Tn*5084* with the exception of the *res* region and the *tnpR* gene where the two transposons differ by 10 % (Bogdanova et al., 1998; Huang et al., 1999; Bogdanova et al., 2001). The different *mer* transposons display a mosaic structure suggesting that recombination events played an important role in the evolution of these elements in *Bacillus*.

CRYPTIC TRANSPOSONS

Two cryptic Tn*3*-like class II transposons have been reported in *B. thuringiensis* strains, Tn*4430* and Tn*5401* (Figue 4B). Tn*4430* is 4149 bp and contains two ORFs encoding a recombinase (TnpI) and a transposase (TnpA), delineated by 38-bp IRs. The resolvase of Tn*4430* is homologous to the phage integrase family of site specific recombinase, differing in that way from other Tn*3* transposons, which belong to the resolvase-invertase family of site-specific recombinases. During transposition TnpA mediates the formation of cointegrates between donor and target replicons (with duplication of Tn*4430*) which are then resolved by TnpI (Mahillon and Lereclus, 1988). The replicative transposition of Tn*4430* occurs with the formation of a 5 bp duplication at the target site. Tn*4430* is frequently found within *B. thuringiensis* conjugative and mobilisable plasmids thereby having an active role in the transfer of plasmids by the formation of cointegrates (Reddy et al., 1987; Green et al., 1989; Mahillon et al., 1994; Mahillon and Chandler, 1998; Andrup et al., 2003). Tn*4430* can also be found in the chromosome and homologous sequences have also been detected in strains of *B. cereus* (Mahillon et al., 1994).

Tn*5401* is 4837 bp in size and contains two large co-transcribed ORFs orientated in the same direction and coding for an integrase-like recombinase, TnpI, and a Tn*3*-like transposase, TnpA. It also has two small upstream ORFs of unknown function which are divergently transcribed from *tnpI-tnpA* (Baum, 1994). Tn*5401* is flanked by unusually long IRs of 53bp and transposes with the generation of 5 bp A+T rich duplication at the target site (Baum, 1994). TnpI binds to a conserved sequence element of 12 bp (ATGTCCRCTAAY) and negatively regulates transcription of *orf1-orf2* and *tnpI-tnpA* (Baum, 1994; Baum, 1995; Baum et al., 1999). Four repeats of this sequence element were identified in the intergenic region between *orf1* and *tnpI*, which are needed for TnpI-mediated recombination in vivo (Baum et al., 1999). The TnpI 12 bp-recognition element is also present at the IRs of Tn*5401* (Baum, 1994; Baum, 1995). Binding of TnpI to these sites appears to modulate binding of TnpA to the IRs and therefore to regulate transposition of Tn*5401* (Baum et al., 1999). Tn*5401* is highly divergent from Tn*4430* at the nucleotide and amino acid level but they nevertheless exhibit a similar structural organization (Figure 4B).

IS

IS*231* is the best studied IS in *Bacillus*, is a member of the IS*4* family, and was first described in large plasmids of *B. thuringiensis*. Different iso-IS*231* elements, and variants corresponding to two different 1.6 kb and 1.9 kb size classes, have been described in *B. thuringiensis* (Lereclus et al., 1984; Mahillon et al., 1987). IS*231*A is flanked by 20-bp IRs, which are reasonably well conserved, and which transpose with duplication of 10 to 12 bp of the target sites (Mahillon et al., 1987; Mahillon et al., 1994). IS*231* elements appear to have a highly preferred insertion site, which at least for IS*231*A is not only dictated by the DNA sequence but also by its conformation (Hallet et al., 1994). IS*231*A inserts preferentially in the IRs of Tn*4430* without disrupting the structure or mobility of the complex (IS*231*::Tn*4430*::IS*231*) (Mahillon et al., 1987; Hallet et al., 1991; Mahillon et al., 1994). Transposition of IS*231* elements appears to occur by a conservative, non-replicative pathway (Leonard and Mahillon, 1998).

Like Tn*4430*, IS*231*, IS*232* (IS*21* family), IS*240* (IS*6* family) are frequently associated with the crystal toxin genes of the *B. thuringiensis* large plasmids (Lereclus et al., 1984; Mahillon et al., 1987; Menou et al., 1990; Rosso and Delecluse, 1997) (Figure 4C). Several variants of iso-IS*232* and iso-IS*240* elements have also been identified (Mahillon et al., 1994; Leonard et al., 1997; Rosso and Delecluse, 1997). The distribution of these elements among strains of *B. thuringiensis*, *B. cereus* and *B. mycoides* is not random. While IS*231* is the most widespread, and IS*240* is also found in the three species, IS*232* seems to be exclusive to *B. thuringiensis* (Leonard et al., 1997; Helgason et al., 1998). IS*231*-like sequences have also been identified in *B. anthracis,* including in its virulence plasmids (Henderson et al., 1995).

The ecology and genetics of these elements have been extensively reviewed by Mahillon et al. (1994).

INTRONS

Several group I introns have been found in *Bacillus* bacteriophages, such as B22, SPO1, SPβ and SPβ-related prophage (Goodrich-Blair et al., 1990; Bechhofer et al., 1994; Lazarevic, 2001). A group I self-splicing intron was also found in the *recA* gene of the *B. anthracis* chromosome, although it lacks the endonuclease essential for intron mobility (Ko et al., 2002). *Bacillus* group II intron sequences were also found on the genome of *B. halodurans* and on the chromosome of the *B. megaterium mer* transposon Tn*MERI*1 (Huang et al., 1999; Takami et al., 2001).

OTHER MECHANISMS OF BACTERIAL GENE TRANSFER: TRANSFORMATION AND TRANSDUCTION

Although conjugation is the most important mechanism for gene transfer between the bacterial genera discussed here, bacterial gene transfer can also occur by transformation (the uptake of free DNA by competent bacteria) and transduction (bacteriophage (bacterial virus) mediated gene transfer).

Several studies have been carried out to investigate survival of free DNA under conditions in the GIT and a few have investigated the ability of this DNA to transform bacteria. DNA becomes resistant to nuclease attack once it attaches to the surface of competent bacterial cells (Romanowski et al., 1993). DNA exposed to rumen fluid at 37°C is extensively degraded within 10 min due to the action of heat-labile nucleases (Ruiz et al., 2000). Not surprisingly, plasmid DNA exposed to rumen fluid for more than 1 minute was unable to transform competent *E. coli* cells (Duggan et al., 2000). Plasmid DNA sequences were still detectable by PCR amplification 24 h after incubation in human saliva (Mercer et al., 1999), although degradation was much more rapid *in vivo* (Duggan et al., 2003). Transformation of the oral bacterium *S. gordonii* occurs readily in the presence of filter sterile saliva (Mercer et al., 1999), and even *in vivo* transformants were detectable for up to 1 min (Mercer et al., 2001). DNA degradation occurs rapidly in the acidic stomach and in the small intestine (Martin-Orue et al., 2002). The conclusions from these experiments are that free DNA survives in the GIT, and that some GIT bacteria are capable of transformation. These two events are most likely to occur together, *in vivo,* in the upper GIT, or oral cavity.

The main limitation to transformation in the GIT may be the number of GIT bacteria capable of becoming competent for DNA uptake. Although a wide range of bacteria are known to have a competence phase in their lifecycle most are pathogens, Gram-negative bacteria or streptococcal relatives (Lorenz and Wackernagel 1994). Transformation has been associated with development of resistance in such bacteria: for example the acquisition of penicillin resistance in *S. pneumoniae* and *Neisseria* species (Dowson et al., 1994). Little is known about the ability of the abundant Gram-positive GIT bacteria to take up exogenous DNA. Some strains of *B. subtilis* are naturally competent, and it has been suggested that DNA transformation is important for bacterial adaptation in soil populations of *B. subtilis* (Graham and Istock, 1978; Graham and Istock, 1979).

Bacteriophages have been implicated both in the spread of antibiotic resistance genes (e.g. Lacey and Kruczenyk, 1986) and bacterial virulence factors (eg. Scotland et al., 1983, Sunagawa et al., 1992) but again most of the detailed research has targeted pathogenic bacteria, in particular enteric pathogens such as *Salmonella* (Rabsch et al., 2002). Phage mediated DNA transfer is the most specific mechanism of bacterial gene transfer, involving recognition of host-specific cell surface receptors. Bacteriophages are classified based on their morphology and lifecycle (see review by Yin and Stotzky, 1997), and many exist indefinitely as temperate phages or lysogens, when they are integrated into the host genome and replicated along with the host chromosomal DNA. Phages only cause host cell death when they enter the lytic lifecycle. At this point they excise from the host genome and become packaged in the phage head, sometimes with adjacent pieces of host DNA, thereby contributing to bacterial gene transfer.

The rumen contains large numbers of bacteriophages (Klieve and Swain, 1993; Swain et al., 1996), and as much as 10% of ruminal bacteria may harbour bacteriophages. Many temperate phages are associated with some of the major ruminal Gram-positive bacteria (Klieve et al., 1989). In contrast little is known about the incidence of bacteriophages in Gram-positive

bacteria from hindgut environments. However since the bacterial populations of the rumen and hindgut ecosystems contain many of the same species or genera (Figure 1), it follows that the bacteriophage population may also be common.

Broad host range generalized transducing bacteriophages have been reported for many *Bacillus* species. Under laboratory conditions, interspecies transduction of the tetracycline resistance plasmid pBC16 by the generalized transducing bacteriophage CP-51 has been demonstrated into and between *B. anthracis*, *B. cereus* and *B. thuringiensis* (Ruhfel et al., 1984). Despite this knowledge of transduction and the potential for transfer of additional DNA, there is no *in vivo* evidence that bacteriophages contribute to gene transfer between Gram-positive spore-forming bacteria.

IN VIVO GENE TRANSFER INVOLVING CLOSTRIDIA AND BACILLUS IN THE GIT

Most gene transfer experiments involving GIT bacteria have been under *in vitro* laboratory conditions, or in gnotobiotic animals using defined donor and recipient bacteria (reviewed in Scott, 2002). It is difficult to look for transfer into the native microflora, mainly due to problems in counter-selecting recipient bacteria, but also because the donor bacterium may not become established against the background flora for a sufficient time for gene transfer to occur. Transfer of the *C. difficile* transposon Tn*5397* was observed in a model of an oral biofilm from an introduced *B. subtilis* donor (Roberts et al., 1999). In this case the native recipient bacteria were identified as *Streptococcus acidominimus*, and the transconjugants persisted for longer than the introduced donor in the biofilm.

INDIRECT EVIDENCE FOR WIDESPREAD BACTERIAL GENE TRANSFER

The presence of identical genes in diverse bacterial species from geographically distinct environments has been used to illustrate that inter-species gene transfer is a common event. Such evidence is compounded when codon usage frequencies and DNA % G+C content is also considered. In general a bacterial coding sequence changes slowly at a rate of 1 % per 1 million years (Ochman and Wilson, 1987), so genes that are 99 % identical may have diverged into different hosts as long ago as one million years. The coding sequence of many antibiotic resistance genes has been used to identify the original Gram-positive host bacterium for genes first identified in Gram-negative bacteria (Courvalin, 1994).

Tetracyclines have been used widely and indiscriminately for many years, consequently there has been an extreme selection pressure to spread these genes among bacterial species. *tet*(W) genes from various Gram-positive anaerobic bacterial species from different GIT ecosystems have remarkable sequence conservation (between 99.6 % to 99.94%; Barbosa et al., 1999; Scott et al., 2000), strongly inferring rapid, recent gene transfer events. The *tet*(W) gene from *B. fibrisolvens* has a different codon usage pattern and a higher % G+C content compared to other *B. fibrisolvens* genes (Barbosa et al., 1999). These facts indicate that the gene originated in a higher DNA % G+C bacterium and was acquired by *B. fibrisolvens* recently.

While sequences of *tet*(O) and *tet*(W) genes appear to be strongly conserved (reviewed in Scott, 2002), the sequences of *tet*(Q) genes from related bacteria from human or animal hosts show a lot more divergence. In fact the sequence variation between different *tet*(Q) genes can be as much as 4% (Nikolich et al., 1994). Sequences of *tet*(M) genes also show much more variability, and often exist as mosaic genes (Oggioni et al., 1996). Such genes presumably result from the co-infection of one host with two slightly different *tet*(M) genes, giving rise to a mosaic gene by homologous recombination. In fact the sequence homologies between tetracycline resistance genes from different classes (usually 60% to 70%, with regions of strong sequence conservation interspersed with hypervariable regions) is such that homologous recombination between different genes may be a common route for the evolution of new tetracycline resistance determinants. *Megasphaera elsdenii* strains contain two different types of mosaic genes, involving recombinations between *tet*(O) and *tet*(W) (Stanton and Humphrey, 2003). A strong identity to each different class of gene is retained within that part of the mosaic gene.

Much of the evidence for antibiotic resistance gene transfer into and between *Bacillus* spp. comes from DNA sequence analysis of resistance determinants in different hosts. *tet*(L) is present in many diverse plasmids of different *Bacillus* species (Imanaka et al., 1981; Eccles and Chopra, 1984; Oskam et al., 1991), some of which are mobilisable. The macrolide-lincosamide-streptogramin B (MLS) resistance genes found on different conjugative transposons of *Bacteroides* isolates were virtually identical (99.6%) to *ermG* previously found in a *Bacillus sphaericus* plasmid (Monod et al., 1987; Cooper et al., 1996; Shoemaker et al., 2001; Wang et al., 2003). The *erm*(C) gene residing on the small *B. subtilis* plasmid, pIM13 is nearly identical to that on the *S. aureus* plasmids pE194 and pE5 (Monod et al., 1986). A *vanA* gene cluster with substantial homology (93 - 96% nucleotide sequence identity) to the enterococcal Tn*1546 vanA* gene cluster has been found in the chromosome of a glycopeptide resistant isolate of *B. circulans* (Fontana et al., 1997; Ligozzi et al., 1998). However, the Tn*1546* transposase and resolvase were absent indicating association with a different genetic element.

The advent of genome sequencing has enabled complete bacterial genomes to be analysed for regions of foreign DNA. The genome sequence of the vancomycin resistant *E. faecalis* isolate V583 indicated that at least 25% of the genome was foreign or mobile DNA (Paulsen et al., 2003). This is a much higher proportion than other sequenced bacteria and may help to explain the rapid emergence and spread of drug resistant phenotypes of enterococci (Paulsen et al., 2003). Sequencing of *Thermotoga maritima* revealed a high proportion of genes (almost 25%) that had strong similarities to genes from archaebacteria (Nelson et al., 1999). These genes tended to be clustered together in different regions of the chromosome, which strongly implies that bacterial gene transfer events can occur across kingdoms. The genome of *Clostridium acetobutylicum* also contains operons with greatest similarity to those in archeabacteria (Nolling et al., 2001).

The genome sequence of a limited number of commensal GIT bacteria is now known. *Bacteroides thetaiotaomicron* contains 63 genes encoding transposases, again indicating a role for mobile DNA in the evolution of this organism (Xu et al., 2003). Six regions of the *Bifidobacterium longum* genome apparently consist of recently acquired DNA, and encode restriction modification systems, xylan utilisation genes and, importantly, genes involved in the production of exopolysaccharides (Schell et al., 2002). The indication that bacteria transfer genes important for host-microbe interactions as well as substrate utilisation and antibiotic resistance illustrates the importance of gene transfer events in bacterial evolution and adaptation to environmental niches.

Four *Bacillus* species genome sequences are available: *B. subtilis*, *B. halodurans*, *B. anthracis* and *B. cereus* (Kunst et al., 1997; Takami et al., 2000; Ivanova et al., 2003; Read et al., 2003). Various putative mobile elements (prophages, transposases, insertion (IS) elements, introns, and plasmids) have been identified on these genomes, again providing information about the extent and importance of horizontal gene transfer for microbial evolution and adaptation. The presence of multiple prophages in the genomes of the different *Bacillus* species suggests that transduction might have played an important role in the evolution of these genomes through horizontal gene transfer (Kunst et al., 1997). An exception to this is the genome of *B. halodurans* C-125 where no prophage sequences were found, although two prophage related proteins were identified (Takami et al., 2000). In contrast the *B. halodurans* genome revealed a surprising number of different IS elements belonging to previously described and new IS families (Takami et al., 2000; Takami et al., 2001). This observation contrasts with the genome of *B. subtilis* 168 where no IS sequences were identified (Kunst et al., 1997).

STABILITY OF TRANSFERRED DNA

Despite the range of gene transfer mechanisms available to bacteria and the frequency with which they appear to occur, bacteria still maintain the distinction between different species. This is due to the instability and limited expression of incoming genes. Gram-negative bacteria are often able to express genes originating in Gram-positive bacteria but the converse is not true. This has been shown to be due to translational specificity, at least partly involving recognition of the ribosome binding site (Roberts and Rabinowitz, 1989). The consensus sequence for the ribosome-binding site in Gram-positive bacteria is GGAGG, which is directly complementary to the CCUCC sequence found on the 16S rRNA subunit. Thus the free energy associated with ribosome binding is much higher for translation in Gram-positive bacteria than in *E. coli* where anything between one and four base complementarity of these two regions is sufficient. The possession of restriction modification systems is another mechanism by which bacteria avoid becoming swamped by foreign DNA. Additionally rumen bacteria possess many restriction systems (reviewed in Flint and Scott, 2000). Most restriction systems are inactive against single-stranded DNA, and since DNA transferred by conjugation and often also by transformation enters a cell in a single-stranded form, this protection is not absolute.

The stability of the incoming sequence depends on it conferring a selective advantage, otherwise it will be diluted out as the bacteria replicate. If the incoming DNA does confer a selective advantage (as is the case with antibiotic and heavy metal resistance genes in situations where there is corresponding environmental pressure) the recombinant bacterium will soon take over the population. Once a new trait has become established in a bacterial population removing the selective pressure does not always lead to loss of the acquired gene as the bacteria adapt to the presence of the new DNA. Thus it is much easier to try and prevent the spread of antibiotic resistance by limiting the selective pressures exerted in the first place, than to eliminate an established resistance. Since we effectively only 'see' gene transfer events where a selective advantage is conferred bacterial gene transfer may occur on a much greater scale than we currently perceive, a fact which is being borne out following the analysis of complete genome sequences.

CONCLUDING REMARKS

Bacterial gene transfer undoubtedly contributes to the spread of bacterial antibiotic resistance. This chapter has illustrated the various ways that DNA can transfer from one bacterium to another or from the environment to a bacterium. It can also be seen that the many of the mobile elements that can transfer between different bacteria are modular in their architecture. These modules may be exchanged between different mobile elements leading to the formation of new elements which may be more suited to a new environment than their precursors. It has also been demonstrated that different genes (e.g. *tet*(M)) posses a mosaic structure, due to recombination between related genes, again leading to novel genes and genetic structures. These mechanisms of mobility and genetic variation have produced a myriad of mobile genetic elements in the spore forming bacteria which will, due to the spore forming ability, probably persist in the environment even if the selection pressure is removed.

ACKNOWLEDGEMENTS
Work in the laboratory at ITQB, Oeiras, Portugal, is supported by a grant from the European Union 5th Framework (QLK5-CT-2001-01729). We thank Cláudia R. Serra for help with literature search. Work at the Eastman Dental Institute was supported by the Wellcome Trust and the BBSRC.

REFERENCES
Abraham LJ, Wales AJ, Rood JI. (1985). Worldwide distribution of the conjugative *Clostridium perfringens* tetracycline resistance plasmid, pCW3. Plasmid. *14*, 37-46.

Adams V, Lyras D, Farrow KA, Rood JI. (2002) The clostridial mobilisable transposons. Cell Mol Life Sci. *59*, 2033-43.

Agerso, Y., Jensen, L.B., Givskov, M., and Roberts, M.C. 2002. The identification of a tetracycline resistance gene *tet*(M), on a Tn*916*-like transposon, in the *Bacillus cereus* group. FEMS Microbiol. Lett. *214*, 251-256.

Andrup, L., Damgaard, J., and Wassermann, K. 1993. Mobilization of small plasmids in *Bacillus thuringiensis* subsp. *israelensis* is accompanied by specific aggregation. J. Bacteriol. *175*, 6530-6536.

Andrup, L., Jensen, G.B., Wilcks, A., Smidt, L., Hoflack, L., and Mahillon, J. 2003 The patchwork nature of rolling-circle plasmids: comparison of six plasmids from two distinct *Bacillus thuringiensis* serotypes. Plasmid. *49*, 205-232.

Andrup, L., Jorgensen, O., Wilcks, A., Smidt, L., and Jensen, G.B. 1996. Mobilization of "nonmobilizable" plasmids by the aggregation-mediated conjugation system of *Bacillus thuringiensis*. Plasmid. *36*, 75-85.

Ash, C., Farrow, J.A., Dorsch, M., Stackebrandt, E., and Collins, M.D. 1991. Comparative analysis of *Bacillus anthracis*, *Bacillus cereus*, and related species on the basis of reverse transcriptase sequencing of 16S rRNA. Int. J. Syst. Bacteriol. *41*, 343-346.

Awad MM, Rood JI. (1997) Isolation of alpha-toxin, theta-toxin and kappa-toxin mutants of *Clostridium perfringens* by Tn*916* mutagenesis. Microb Pathog. *22*, 275-84.

Babb BL, Collett HJ, Reid SJ, Woods DR. (1993) Transposon mutagenesis of *Clostridium acetobutylicum* P262, isolation and characterization of solvent deficient and metronidazole resistant mutants. FEMS Microbiol Lett. *114*, 343-8.

Barbosa T.M., Scott K.P.,and Flint H.J. 1999. Evidence for recent intergeneric transfer of a new tetracycline resistance gene, *tet*(W), isolated from *Butyrivibrio fibrisolvens*, and the occurence of *tet*(O) in ruminal bacteria. Environ. Microbiol. 1: 53-64.

Battisti, L., Green, B.D., and Thorne, C.B. 1985. Mating system for transfer of plasmids among *Bacillus anthracis*, *Bacillus cereus*, and *Bacillus thuringiensis*. J. Bacteriol. *162*, 543-550.

Baum, J.A. 1994. Tn*5401*, a new class II transposable element from *Bacillus thuringiensis*. J. Bacteriol. *176*, 2835-2845.

Baum, J.A. 1995. TnpI recombinase: identification of sites within Tn*5401* required for TnpI binding and site-specific recombination. J. Bacteriol. *177*, 4036-4042.

Baum, J.A., Gilmer, A.J., and Light Mettus, A.M. 1999. Multiple roles for TnpI recombinase in regulation of Tn*5401* transposition in *Bacillus thuringiensis*. J. Bacteriol. *181*, 6271-6277.

Bechhofer, D.H., Hue, K.K., and Shub, D.A. 1994. An intron in the thymidylate synthase gene of *Bacillus* bacteriophage beta *22*, evidence for independent evolution of a gene, its group I intron, and the intron open reading frame. Proc. Natl. Acad. Sci. U S A. *91*, 11669-11673.

Bernhard, K., Schrempf, H., and Goebel, W. 1978. Bacteriocin and antibiotic resistance plasmids in *Bacillus cereus* and *Bacillus subtilis*. J. Bacteriol. *133*, 897-903.

Bertram, J., Stratz, M., and Durre, P. 1991. Natural transfer of conjugative transposon Tn*916* between gram-positive and gram-negative bacteria. J. Bacteriol. *173*, 443-448.

Bogdanova, E.S., Bass, I.A., Minakhin, L.S., Petrova, M.A., Mindlin, S.Z., Volodin, A.A., Kalyaeva, E.S., Tiedje, J.M., Hobman, J.L., Brown, N.L., and Nikiforov, V.G. 1998. Horizontal spread of *mer* operons among gram-positive bacteria in natural environments. Microbiology. *144*, 609-620.

Bogdanova, E., Minakhin, L., Bass, I., Volodin, A., Hobman, J.L., and Nikiforov, V. 2001. Class II broad-spectrum mercury resistance transposons in Gram-positive bacteria from natural environments. Res. Microbiol. *152*, 503-514.

Bringel, F., Van Alstine, G.L., and Scott, J.R. 1991. A host factor absent from *Lactococcus lactis* subspecies *lactis* MG1363 is required for conjugative transposition. Mol. Microbiol. 5: 2983-2993.

Briolat V, and Reysset G. (2002) Identification of the *Clostridium perfringens* genes involved in the adaptive response to oxidative stress. J Bacteriol. *184*, 2333-43.

Brynestad S, Sarker MR, McClane BA, Granum PE, Rood JI. (2001). Enterotoxin plasmid from *Clostridium perfringens* is conjugative. Infect Immun. *69*, 3483-7.

Celli J, and Trieu-Cuot P. (1998) Circularization of Tn*916* is required for expression of the transposon-encoded transfer functions: characterization of long tetracycline-inducible transcripts reading through the attachment site. Mol Microbiol. *28*, 103-17.

Chopra, I., Bennett, P.M., and Lacey, R.W. 1973. A variety of Staphylococcal plasmids present as multiple copies. J. Gen. Microbiol. *79*, 343-345.

Christie, P.J., Korman, R.Z., Zahler, S.A., Adsit, J.C., and Dunny, G.M. 1987. Two conjugation systems associated with *Streptococcus faecalis* plasmid pCF10, identification of a conjugative transposon that transfers between *S. faecalis* and *Bacillus subtilis*. J. Bacteriol. *169*, 2529-2536.

Cooper, A.J., Shoemaker, N.B., and Salyers, A.A. 1996. The erythromycin resistance gene from the Bacteroides conjugal transposon Tc[r] Em[r] 7853 is nearly identical to *erm*G from *Bacillus sphaericus*. Antimicrob. Agents Chemother. *40*, 506-508.

Courvalin P. 1994. Transfer of antibiotic resistance genes between gram-positive and gram-negative bacteria. Antimicrob. Agents Chemother. *38*, 1447-1451.

Delecluse, A., Bourgouin, C., Klier, A., Rapoport, G. 1989 Nucleotide sequence and characterization of a new insertion element, IS*240*, from *Bacillus thuringiensis israelensis*. Plasmid. *21*, 71-78.

Dowson, C.G., Coffey, T.J., and Spratt, B.G. 1994. Origin and molecular epidemiology of penicillin-binding-protein-mediated resistance to beta-lactam antibiotics. Trends Microbiol. 2: 361-366.

Duggan, P.S., Chambers, P.A., Heritage, J. and Forbes, J.M. 2000. Survival of free DNA encoding antibiotic resistance from transgenic maize and the transformation activity of DNA in ovine saliva, ovine rumen fluid and silage effluent. FEMS Microbiology Letters *191*, 71-77.

Duggan, P.S., Chambers, P.A., Heritage, J., and Forbes, M.J. 2003. Fate of genetically modified maize DNA in the oral cavity and rumen of sheep. Br. J. Nutr. *89*, 159-166.

Eccles, S.J., and Chopra, I. 1984. Biochemical and genetic characterization of the *tet* determinant of *Bacillus* plasmid pAB124. J. Bacteriol. *158*, 134-140.

Flannagan SE, Zitzow LA, Su YA, Clewell DB. (1994) Nucleotide sequence of the 18-kb conjugative transposon Tn916 from *Enterococcus faecalis*. Plasmid *32*, 350-4.

Flint, H.J. and Scott, K.P. 2000. Genetics of rumen microorganisms: gene transfer, genetic analysis and strain manipulation. In: Ruminant Physiology: Digestion, Metabolism, Growth and Reproduction. P.B. Cronjé, ed. CAB International.

Fontana, R., Ligozzi, M., Pedrotti, C., Padovani, E.M., and Cornaglia, G. 1997. Vancomycin-resistant *Bacillus circulans* carrying the *vanA* gene responsible for vancomycin resistance in enterococci. Eur. J. Clin. Microbiol. Infect. Dis. *16*, 473-474.

Franke AE, and Clewell DB.(1981) Evidence for a chromosome-borne resistance transposon (Tn916) in *Streptococcus faecalis* that is capable of "conjugal" transfer in the absence of a conjugative plasmid. J Bacteriol *145*, 494-502.

Gawron-Burke C, and Clewell DB (1984). Regeneration of insertionally inactivated streptococcal DNA fragments after excision of transposon Tn916 in *Escherichia coli*: strategy for targeting and cloning of genes from gram-positive bacteria. J Bacteriol. *159*, 214-21.

Ghosh, A.C. 1978. Prevalence of *Bacillus cereus* in the faeces of healthy adults. J. Hyg. (Lond.) *80*, 233-236.

Goodrich-Blair, H., Scarlato, V., Gott, J.M., Xu, M.Q., and Shub, D.A. 1990. A self-splicing group I intron in the DNA polymerase gene of *Bacillus subtilis* bacteriophage SPO1. Cell. *63*, 417-424.

Graham, J.B., and Istock, C.A. 1978. Genetic exchange in *Bacillus subtilis* in soil. Mol. Gen. Genet. *166*, 287-290.

Graham, J.P., and Istock, C.A. 1979. Gene exchange and natural selection cause *Bacillus subtilis* to evolve in soil culture. Science. *204*, 637-639.

Green, B.D., Battisti, L., and Thorne, C.B. 1989. Involvement of Tn4430 in transfer of *Bacillus anthracis* plasmids mediated by *Bacillus thuringiensis* plasmid pXO12. J. Bacteriol. *171*, 104-113.

Gryczan, T.J., Contente, S., and Dubnau, D. 1978. Characterization of *Staphylococcus aureus* plasmids introduced by transformation into *Bacillus subtilis*. J. Bacteriol. *134*, 318-329.

Guffanti, A.A., Quirk, P.G., and Krulwich, T.A. 1991. Transfer of Tn925 and plasmids between *Bacillus subtilis* and alkaliphilic *Bacillus firmus* OF4 during Tn925-mediated conjugation. J. Bacteriol. *173*, 1686-1689.

Gupta, A., Phung, L.T., Chakravarty, L., and Silver, S. 1999. Mercury resistance in *Bacillus cereus* RC607, transcriptional organization and two new open reading frames. J. Bacteriol. *181*, 7080-7086.

Hallet, B., Rezsohazy, R., and Delcour, J. 1991. IS231A from *Bacillus thuringiensis* is functional in *Escherichia coli*: transposition and insertion specificity. J. Bacteriol. *173*, 4526-4529.

Hallet, B., Rezsohazy, R., Mahillon, J., and Delcour, J. 1994. IS231A insertion specificity: consensus sequence and DNA bending at the target site. Mol. Microbiol. *14*, 131-139.

Haraldsen JD, and Sonenshein AL. (2003) Efficient sporulation in *Clostridium difficile* requires disruption of the sigmaK gene. Mol Microbiol. *48*, 811-21.

Hart, M.C., Elliot, G.C., Osborn, A.M., Ritchie, D.A., and Strike, P. 1998. Diversity amongst *Bacillus merA* genes amplified from mercury resistant isolates and directly from mercury polluted soil. FEMS Microbiol. Ecol. *27*, 73-84.

Helgason, E., Caugant, D.A., Lecadet, M.M., Chen, Y., Mahillon, J., Lovgren, A., Hegna, I., Kvaloy, K., and Kolsto, A.B. 1998. Genetic diversity of *Bacillus cereus*/*B. thuringiensis* isolates from natural sources. Curr. Microbiol. 37, 80-87.

Henderson, I., Yu,D., and Turnbull, P.C. 1995. Differentiation of *Bacillus anthracis* and other 'Bacillus cereus group' bacteria using IS231-derived sequences. FEMS Microbiol. Lett. *128*, 113-118.

Hendrick, C.A., Johnson, L.K., Tomes, N.J., Smiley, B.K., and Price, J.P. 1991. Insertion of Tn916 into *Bacillus pumilus* plasmid pMGD302 and evidence for plasmid transfer by conjugation. Plasmid. *26*, 1-9.

Herzog-Velikonja, B., Podlesek, Z., and Grabnar, M. 1994. Conjugal transfer of transposon Tn916 from *Enterococcus faecalis* to *Bacillus licheniformis*. Plasmid. *31*, 201-206.

Hold, G.L., Pryde, S.E., Russell, V.J., Furrie, E., and Flint, H.J. 2002. Assessment of microbial diversity in human colonic samples by 16S rDNA sequence analysis. FEMS Microbiol. Ecol. *39*, 33-39.

Huang, C.C., Narita, M., Yamagata, T., Itoh, Y., and Endo, G. 1999. Structure analysis of a class II transposon encoding the mercury resistance of the Gram-positive Bacterium *Bacillus megaterium* MB1, a strain isolated from minamata bay, Japan. Gene. *234*, 361-369.

Hyronimus, B., Le Marrec, C., Urdaci, M.C. 1998. Coagulin, a bacteriocin-like inhibitory substance produced by *Bacillus coagulans* I4. J. Appl. Microbiol. *85*, 42-50.

Imanaka, T., Fujii, M., and Aiba, S. 1981. Isolation and characterization of antibiotic resistance plasmids from thermophilic bacilli and construction of deletion plasmids. J. Bacteriol. *146*, 1091-1097.

Ivanova, N., Sorokin, A., Anderson, I., Galleron, N., Candelon, B., Kapatral, V., et al. 2003. Genome sequence of *Bacillus cereus* and comparative analysis with *Bacillus anthracis*. Nature. *423*, 87-91.

Ivins, B.E., Welkos, S.L., Knudson, G.B., and Leblanc, D.J. 1988. Transposon Tn916 mutagenesis in *Bacillus anthracis*. Infect. Immun. *56*, , 176-181.

Jarrett, P., and Stephenson, M. 1990. Plasmid transfer between strains of *Bacillus thuringiensis* infecting *Galleria mellonella* and *Spodoptera littoralis*. Appl. Environ. Microbiol. *56*, 1608-1614.

Jaworski DD and Clewell DB.(1995) A functional origin of transfer (oriT) on the conjugative transposon Tn916. J Bacteriol. *177*, 6644-51.

Jennert KC, Tardif C, Young DI, Young M. (2000) Gene transfer to *Clostridium cellulolyticum* ATCC 35319. Microbiology. *146*, 3071-80.

Jensen, G.B., Andrup, L., Wilcks, A., Smidt, L., and Poulsen, O.M. 1996. The aggregation-mediated conjugation system of *Bacillus thuringiensis* subsp. *israelensis*: host range and kinetics of transfer. Curr. Microbiol. *33*, 228-234.

Jensen, G.B., Wilcks, A., Petersen, S.S., Damgaard, J., Baum, J.A., and Andrup, L. 1995. The genetic basis of the aggregation system in *Bacillus thuringiensis* subsp. *israelensis* is located on the large conjugative plasmid pXO16. J. Bacteriol. *177*, 2914-2917.

Klieve, A.V., Hudman, J.F., and Bauchop, T. 1989. Inducible bacteriophages from ruminal bacteria. Appl. Environ. Microbiol. *55*, 1630-1634.

Klieve, A.V., and Swain, R.A. 1993. Estimation of ruminal bacteriophage numbers by pulsed-field gel electrophoresis and laser densitometry. Appl. Environ. Microbiol. *59*, 2299-2303.

Ko, M., Choi, H., and Park, C. 2002. Group I self-splicing intron in the *recA* gene of *Bacillus anthracis*. J. Bacteriol. *184*, 3917-3922.

Koehler, T.M. 2000. *Bacillus anthracis*. In: Gram-Positive Pathogens. Fischetti V.A., Novick, R.P., Ferretti, J.J., Portnoy, D.A. and Rood, J.I. ed. ASM Press, Washington. p. 519-528.

Koehler, T.M., and Thorne, C.B. 1987. *Bacillus subtilis* (*natto*) plasmid pLS20 mediates interspecies plasmid transfer. J. Bacteriol. *169*, 5271-5278.

Kunst, F., Ogasawara, N., Moszer, I., Albertini, A.M., Alloni, G., Azevedo, V., Bertero, and et al. 1997. The complete genome sequence of the gram-positive bacterium *Bacillus subtilis*. Nature. *390*, 249-256.

Lacey R.W. and Kruczenyk S.C. 1986 Epidemiology of antibiotic resistance in *Staphylococcus aureus*. J. Antimicrob. Chemother. *18*, 207-14.

Lazarevic, V. 2001. Ribonucleotide reductase genes of *Bacillus* prophages: a refuge to introns and intein coding sequences. Nucleic Acids Res. *29*, 3212-3218.

Leonard, C., Chen, Y., and Mahillon, J. 1997. Diversity and differential distribution of IS231, IS232 and IS240 among *Bacillus cereus*, *Bacillus thuringiensis* and *Bacillus mycoides*. Microbiology. *143*, 2537-2547.

Leonard, C., and Mahillon, J. 1998. IS231A transposition: conservative versus replicative pathway. Res. Microbiol. *149*, 549-555.

Lereclus, D., Ribier, J., Klier, A., Menou, G., and Lecadet, M.M. 1984. A transposon-like structure related to the delta-endotoxin gene of *Bacillus thuringiensis*. Embo J. 3: 2561-2567.

Ligozzi, M., Lo Cascio, G., and Fontana, R. 1998. *vanA* gene cluster in a vancomycin-resistant clinical isolate of *Bacillus circulans*. Antimicrob. Agents Chemother. *42*, 2055-2059.

Lin WJ, Johnson EA. (1995) Genome analysis of *Clostridium botulinum* type A by pulsed-field gel electrophoresis. Appl Environ Microbiol. *61*, 4441-7.

Lorenz, M.G., and Wackernagel, W. 1994. Bacterial gene transfer by natural genetic transformation in the environment. Microbiol. Rev. *58,* 563-602.

Lovett, P.S., Duvall, E.J., and Keggins, K.M. 1976. *Bacillus pumilus* plasmid pPL*10,* properties and insertion into *Bacillus subtilis* 168 by transformation. J. Bacteriol. *127,* 817-828.

Louis, P., Duncan, S.H., McCrae, S., Millar, J., Jackson, M., and Flint, H.J. Restricted distribution of the butyrate kinase pathway among butyrate-producing bacteria from the human colon [J. Bact, submitted 2003]

Mahillon, J., and Chandler, M. 1998. Insertion sequences. Microbiol. Mol. Biol. Rev. *62,* 725-774.

Mahillon, J., and Lereclus, D. 1988. Structural and functional analysis of Tn*4430,* identification of an integrase-like protein involved in the co-integrate-resolution process. EMBO J. 7: 1515-1526.

Mahillon, J., Rezsohazy, R., Hallet, B., and Delcour, J. 1994. IS*231* and other *Bacillus thuringiensis* transposable elements: a review. Genetica. 93, 13-26.

Mahillon, J., and Seurinck, J. 1988. Complete nucleotide sequence of pGI2, a *Bacillus thuringiensis* plasmid containing Tn*4430.* Nucleic Acids Res. *16,* 11827-11828.

Mahillon, J., Seurinck, J., Delcour, J., and Zabeau, M. 1987. Cloning and nucleotide sequence of different iso-IS*231* elements and their structural association with the Tn*4430* transposon in *Bacillus thuringiensis.* Gene. 51, 187-196.

Mahler, I., and Halvorson, H.O. 1980. Two erythromycin-resistance plasmids of diverse origin and their effect on sporulation in *Bacillus subtilis.* J. Gen. Microbiol. *120,* 259-263.

Marra, D., Pethel, B., Churchward, G.G., and Scott, J.R. 1999. The frequency of conjugative transposition of Tn*916* is not determined by the frequency of excision. J. Bacteriol. *181,* 5414-5418.

Marra, D., and Scott, J.R. 1999. Regulation of excision of the conjugative transposon Tn*916.* Mol. Microbiol. *31,* 609-621.

Martin-Orue S.M., O'Donnell A.G., Arino J., Netherwood T., Gilbert H.J., and Mathers J.C. 2002. Degradation of transgenic DNA from genetically modified soya and maize in human intestinal simulations. Br. J. Nutr. *87,* 533-542.

Mattsson DM, and Rogers P. (1994) Analysis of Tn916-induced mutants of *Clostridium acetobutylicum* altered in solventogenesis and sporulation. J Ind Microbiol.: 258-68.

Melville C.M., Scott K.P., Mercer D.K., and Flint H.J. 2001. Novel tetracycline resistance gene, *tet*(32), in the Clostridium-related human colonic anaerobe K10 and its transmission in vitro to the rumen anaerobe *Butyrivibrio fibrisolvens.* Antimicrob. Agents Chemother. *45,* 3246-3249.

Melville, C.M., Brunel, R., Flint, H.J. and Scott, K.P. The *Butyrivibrio fibrisolvens tet*(W) gene is carried on the novel conjugative transposon Tn*B1230,* which contains duplicated nitroreductase coding sequences. [J. Bact. submitted].

Menou, G., Mahillon, J., Lecadet, M.M., and Lereclus, D. 1990. Structural and genetic organization of IS*232,* a new insertion sequence of *Bacillus thuringiensis.* J. Bacteriol. *172,* 6689-6696.

Mercer, D.K., Melville, C.M., Scott, K.P. and Flint, H.J. 1999. Natural genetic transformation in the rumen bacterium *Streptococcus bovis* JB1. FEMS Microbiology Letters. *179,* 485-490.

Mercer, D.K., Scott, K.P., Melville, C.M., Glover, L.A., and Flint, H.J. 2001. Transformation of an oral bacterium via chromosomal integration of free DNA in the presence of human saliva. FEMS Microbiol. Lett. *200,* 163-167.

Monod, M., Denoya, C., and Dubnau, D. 1986. Sequence and properties of pIM13, a macrolide-lincosamide-streptogramin B resistance plasmid from *Bacillus subtilis.* J. Bacteriol. *167,* 138-147.

Monod, M., Mohan, S., and Dubnau, D. 1987. Cloning and analysis of *ermG*, a new macrolide-lincosamide-streptogramin B resistance element from *Bacillus sphaericus.* J. Bacteriol. *169,* 340-350.

Mullany, P. (2000) Gene transfer in the GI and oral cavity. Microbial Ecology in Health and Disease Supplement p 73-80.

Mullany, P. (2002) Introduction to the multi-author review on conjugative transposons. Mol. Cell. Life Sci. *59,* 2015-6.

Mullany, P Roberts AP and Wang H (2002) Mechanism of integration and excision in conjugative transposons. Mol. Cell. Life Sci. *59,* 2017-22.

Mullany, P., Wilks, M., Lamb, I., Clayton, C., Wren, B., and Tabaqchali, S. 1990. Genetic analysis of a tetracycline resistance element from *Clostridium difficile* and its conjugal transfer to and from *Bacillus subtilis.* J. Gen. Microbiol. *136,* 1343-1349.

Mullany, P. Wilks, M. Puckey L. and Tabaqchali S. (1994) Gene cloning in *Clostridium difficile* using Tn*916* as a shuttle conjugative transposon. Plasmid **31** 320-323.

Mullany, P., Wilks, M., and Tabaqchali, S. 1991. Transfer of Tn*916* and Tn*916* delta E into *Clostridium difficile*: demonstration of a hot-spot for these elements in the *C. difficile* genome. FEMS Microbiol. Lett. *63,* 191-194.

Mullany, P., Wilks, M., and Tabaqchali, S. 1995. Transfer of macrolide-lincosamide-streptogramin B (MLS) resistance in *Clostridium difficile* is linked to a gene homologous with toxin A and is mediated by a conjugative transposon, Tn*5398.* J. Antimicrob. Chemother. *35,* 305-315.

Nagai, T., Phan Tran, L.S., Inatsu, Y., and Itoh, Y. 2000. A new IS*4* family insertion sequence, IS*4Bsu1*, responsible for genetic instability of poly-gamma-glutamic acid production in *Bacillus subtilis.* J. Bacteriol. *182,* 2387-2392.

Naglich, J.G., and Andrews, R.E., Jr. 1988. Tn*916*-dependent conjugal transfer of pC194 and pUB110 from *Bacillus subtilis* into *Bacillus thuringiensis* subsp. *israelensis.* Plasmid. *20,* 113-126.

Naglich, J.G., and Andrews, R.E., Jr. 1988a. Introduction of the *Streptococcus faecalis* transposon Tn*916* into *Bacillus thuringiensis* subsp. *israelensis.* Plasmid. *19,* 84-93.

Nakamura, K., and Silver, S. 1994. Molecular analysis of mercury-resistant *Bacillus* isolates from sediment of Minamata Bay, Japan. Appl. Environ. Microbiol. *60,* 4596-4599.

Narita, M., Chiba, K., Nishizawa, H., Ishii, H., Huang, C.C., Kawabata, Z., Silver, S., and Endo, G. 2003. Diversity of mercury resistance determinants among *Bacillus* strains isolated from sediment of Minamata Bay. FEMS Microbiol. Lett. *223,* 73-82.

Natarajan, M.R., and Oriel, P. 1991. Conjugal transfer of recombinant transposon Tn*916* from *Escherichia coli* to *Bacillus stearothermophilus.* Plasmid. *26,* 67-73.

Nelson, K.E., Clayton, R.A., Gill, S.R., Gwinn, M.L., Dodson, R.J., Haft, D.H., Hickey, E.K., Peterson, J.D., Nelson, W.C., Ketchum, K.A., McDonald, L., Utterback, T.R., Malek, J.A., Linher, K.D., Garrett, M.M., Stewart, A.M., Cotton, M.D., Pratt, M.S., Phillips, C.A., Richardson, D., Heidelberg, J., Sutton, G.G., Fleischmann, R.D., Eisen, .JA., Fraser, C.M., et al. 1999. Evidence for lateral gene transfer between Archaea and bacteria from genome sequence of *Thermotoga maritima.* Nature. *399,* 323-329.

Nikolich, M.P., Hong, G., Shoemaker, N.J., and Salyers, A.A. 1994. Evidence for natural horizontal transfer of *tet*(Q) between bacteria that normally colonise humans and bacteria that normally colonise livestock. Appl. Environ. Microbiol. *19,* 3255-3260.

Nolling J, Breton G, Omelchenko MV, Makarova KS, Zeng Q, Gibson R, Lee HM, Dubois J, Qiu D, Hitti J, Wolf YI, Tatusov RL, Sabathe F, Doucette-Stamm L, Soucaille P, Daly MJ, Bennett GN, Koonin EV, Smith DR. 2001. Genome sequence and comparative analysis of the solvent-producing bacterium *Clostridium acetobutylicum.* J. Bacteriol. *183,* 4823-4838.

Norgren, M., and Scott, J.R. 1991. The presence of conjugative transposon Tn*916* in the recipient strain does not impede transfer of a second copy of the element. J. Bacteriol. *173,* 319-324.

Ochman, H. and Wilson A.C. 1987. Evolution in bacteria: evidence for a universal substitution rate in cellular genomes. J. Mol. Evol. *26,* 74-86

Oggioni, M.R., Dowson, C.G., Maynard-Smith, J., Provvedi, R. and Pozzi, G. 1996. The tetracycline resistance gene *tet*(M) exhibits mosaic structure. Plasmid *35,* 156-163.

O'Keeffe T, Hill C, Ross RP.(1999) In situ inversion of the conjugative transposon Tn916 in Enterococcus faecium DPC3675. FEMS Microbiol Lett. *173,* 265-71.

Okinaka, R.T., Cloud, K., Hampton, O., Hoffmaster, A.R., Hill, K.K., Keim, P., et al. 1999. Sequence and organization of pXO1, the large *Bacillus anthracis* plasmid harboring the anthrax toxin genes. J. Bacteriol. 181, 6509-6515.

Oskam, L., Hillenga, D.J., Venema, G., and Bron, S. 1991. The large *Bacillus* plasmid pTB19 contains two integrated rolling-circle plasmids carrying mobilization functions. Plasmid. *26,* 30-39.

Pannucci, J., Okinaka, R.T., Williams, E., Sabin, R., Ticknor, L.O., and Kuske, C.R. 2002. DNA sequence conservation between the *Bacillus* anthracis pXO2 plasmid and genomic sequence from closely related bacteria. BMC Genomics. 3: 34.

Paulsen, I.T., Banerjei, L., Myers, G.S., Nelson, K.E., Seshadri, R., Read, T.D., Fouts, D.E., Eisen, J.A., Gill, S.R., Heidelberg, J.F., Tettelin, H., Dodson, R.J., Umayam, L., Brinkac, L., Beanan, M., Daugherty, S., DeBoy, R.T., Durkin, S., Kolonay, J., Madupu, R., Nelson, W., Vamathevan, J., Tran, B., Upton, J., Hansen, T., Shetty, J., Khouri, H., Utterback, T., Radune, D., Ketchum, K.A.,

Dougherty, B.A., Fraser, C.M. 2003. Role of mobile DNA in the evolution of vancomycin-resistant *Enterococcus faecalis*. Science. *299*, 2071-4.

Rabsch, W., Mirold, S., Hardt, W.D., and Tschape, H. 2002. The dual role of wild phages for horizontal gene transfer among Salmonella strains. Berl. Munch. Tierarztl. Wochenschr. *115*, 355-359.

Read, T.D., Peterson, S.N., Tourasse, N., Baillie, L.W., Paulsen, I.T., Nelson, K.E., and et al. 2003. The genome sequence of *Bacillus anthracis* Ames and comparison to closely related bacteria. Nature. *423*, 81-86.

Reddy, A., Battisti, L., and Thorne, C.B. 1987. Identification of self-transmissible plasmids in four *Bacillus thuringiensis* subspecies. J. Bacteriol. 169, 5263-5270.

Roberts, A. P., Braun, V.. von-Eichel Streiber C and Mullany P. (2001a). Demonstration that the group II intron from the clostridial conjugative transposon Tn*5397* undergoes splicing *in vivo* Journal of Bacteriology *183*, 1296-1299.

Roberts, A. P., Hennequin, C Elmore, M Collignon, A Karjalainen, T Minton, Nl Mullany P (2003) Development of an integrative vector for the expression of antisense RNA in *Clostridium difficile* Journal of Microbiological Methods. 55 (3). 617-24

Roberts, A. P., Johanesen P, Lyras, D. Mullany P. and Rood. J. I. (2001) Comparison of Tn*5397* from *Clostridium difficile*, Tn*916* from *Enterococcus faecalis* and the CW459*tet*(M) element from *Clostridium perfringens* shows that they have similar conjugation regions but different insertion and excision modules. Microbiology. *147*, 1243-1251.

Roberts, A.P., Pratten, J., Wilson, M., and Mullany, P. 1999. Transfer of a conjugative transposon, Tn*5397* in a model oral biofilm. FEMS Microbiol. Lett. *177*, 63-66.

Roberts, M.W., and Rabinowitz, J.C. 1989. The effect of *Escherichia coli* ribosomal protein S1 on the translational specificity of bacterial ribosomes. J. Biol. Chem. *264*, 2228-2235.

Romanowski G., Lorenz M.G., and Wackernagel W. 1993. Plasmid DNA in a groundwater aquifer microcosm--adsorption, DNAase resistance and natural genetic transformation of *Bacillus subtilis*. Mol Ecol. 2: 171-181.

Rosso, M.L., and Delecluse, A. 1997. Distribution of the insertion element IS*240* among *Bacillus thuringiensis* strains. Curr. Microbiol. *34*, 348-353.

Ruhfel, R.E., Robillard, N.J., and Thorne, C.B. 1984. Interspecies transduction of plasmids among *Bacillus anthracis*, *B. cereus*, and *B. thuringiensis*. J. Bacteriol. *157*, 708-711.

Ruiz, T.R., Andrews, S., and Smith, G.B. 2000. Identification and characterization of nuclease activities in anaerobic environmental samples. Can. J. Microbiol. *46*, 736-740.

Schell, M.A., Karmirantzou, M., Snel, B., Vilanova, D., Berger, B., Pessi, G., Zwahlen, M.C., Desiere, F., Bork, P., Delley, M., Pridmore, R.D., and Arigoni, F. 2002. The genome sequence of *Bifidobacterium longum* reflects its adaptation to the human gastrointestinal tract. Proc. Natl. Acad. Sci. USA. *99*, 14422-14427.

Schnepf, E., Crickmore, N., Van Rie, J., Lereclus, D., Baum, J., Feitelson, J., Zeigler, D. R., and Dean, D.H. 1998. *Bacillus thuringiensis* and its pesticidal crystal proteins. Microbiol. Mol. Biol. Rev. *62*, 775-806.

Scotland, S.M., Smith, H.R., Willshaw, G.A., and Rowe, B. 1983. Vero cytotoxin production in strain of *Escherichia coli* is determined by genes carried on bacteriophage. Lancet. 2: 216.

Scott, J.R., Bringel, F., Marra, D., Van Alstine, G., and Rudy, C.K. 1994. Conjugative transposition of Tn*916*, preferred targets and evidence for conjugative transfer of a single strand and for a double-stranded circular intermediate. Mol. Microbiol. *11*, 1099-1108.

Scott, J.R., Kirchman, P.A., and Caparon, M.G. 1988. An intermediate in transposition of the conjugative transposon Tn*916*. Proc. Natl. Acad. Sci. U S A. *85*, 4809-4813.

Scott K.P., Barbosa T.M., Forbes K.J., and Flint H.J. 1997. High-frequency transfer of a naturally occurring chromosomal tetracycline resistance element in the ruminal anaerobe *Butyrivibrio fibrisolvens*. Appl. Environ. Microbiol. *63*, 3405-3411.

Scott, K.P., Melville, C.M., Barbosa, T.M., and Flint, H.J. 2000. Occurrence of the new tetracycline resistance gene *tet*(W) in bacteria from the human gut. Antimicrob. Agents Chemother. *44*, 775-777.

Scott K.P. 2002. The role of conjugative transposons in spreading antibiotic resistance between bacteria that inhabit the gastrointestinal tract. Cell Mol. Life Sci. *59*, 2071-2082.

Selinger, L.B., McGregor, N.F., Khachatourians, G.G., and Hynes, M.F. 1990. Mobilization of closely related plasmids pUB110 and pBC16 by *Bacillus* plasmid pXO503 requires *trans*-acting open reading frame beta. J. Bacteriol. *172*, 3290-3297.

Sen, S., and Oriel, P. 1990. Transfer of transposon Tn*916* from *Bacillus subtilis* to *Thermus aquaticus*. FEMS Microbiol. Lett. *55*, 131-134.

Senghas E, Jones JM, Yamamoto M, Gawron-Burke C, Clewell DB. (1988) Genetic organization of the bacterial conjugative transposon Tn*916*. J Bacteriol. *170*, 245-9.

Shoemaker, N.B., Vlamakis, H., Hayes, K., and Salyers, A.A. 2001. Evidence for extensive resistance gene transfer among *Bacteroides* spp. and among *Bacteroides* and other genera in the human colon. Appl. Environ. Microbiol. *67*, 561-568.

Showsh, S.A., and Andrews, R.E., Jr. 1992. Tetracycline enhances Tn*916*-mediated conjugal transfer. Plasmid. *28*, 213-224.

Showsh, S.A., and Andrews, R.E., Jr. 1996. Functional comparison of conjugative transposons Tn*916* and Tn*925*. Plasmid. *35*, 164-173.

Showsh, S.A., and Andrews, R.E., Jr. 1999. Analysis of the requirement for a pUB110 mob region during Tn*916*-dependent mobilization. Plasmid. *41*, 179-186.

Stackebrandt E and Rainey F.A. (1997) Phylogenetic relationships. In: The clostridia: Molecular biology and Pathogenesis, pp.3-20, Rood J.I., McClane, B.A., Songer, J.G., and Titball R.W. (eds) Academic Press, San Diego, CA.

Stahl, S.R. 1991. Plasmids in *Bacillus stearothermophilus* coding for bacteriocinogeny and temperature resistance. Plasmid. *26*, 94-107.

Stanton, T.B. and Humphrey, S.B. 2003. Isolation of tetracycline-resistant *Megasphaera elsdenii* strains with novel mosaic gene combinations of *tet*(O) and *tet*(W) from swine. Appl. Environ. Microbiol. *69*, 3874-3882.

Stevens, A.M., Shoemaker, N.B. and Salyers, A.A. 1990. Genes on a *Bacteroides* conjugal tetracycline resistance element which mediate production of plasmid-like forms from unlinked chromosomal DNA may be involved in transfer of the resistance element. J. Bact. *172*, 4271-4279.

Storrs, M.J., Poyart-Salmeron, C., Trieu-Cuot, P., and Courvalin, P. 1991. Conjugative transposition of Tn*916* requires the excisive and integrative activities of the transposon-encoded integrase. J. Bacteriol. 173, 4347-4352.

Su YA, He P, Clewell DB.(1992). Characterization of the tet(M) determinant of Tn*916*, evidence for regulation by transcription attenuation. Antimicrob Agents Chemother. *36*, 769-78.

Suau, A., Bonnet, R., Sutren, M., Godon, J-J., Gibson, G.R., Collins, M.D., and Doré, J. 1999. Direct analysis of genes encoding 16S rRNA from complex communities reveals many molecular species within the human gut. Appl. Environ. Microbiol. *65*, 4799-4807.

Sunagawa H, Ohyama T, Watanabe T, Inoue K.(1992) The complete amino acid sequence of the *Clostridium botulinum* type D neurotoxin, deduced by nucleotide sequence analysis of the encoding phage d-16 phi genome. J Vet Med Sci. *54*, 905-13.

Swain, R.A., Nolan, J.V., and Klieve, A.V. 1996. Natural variability and diurnal fluctuations within the bacteriophage population of the rumen. Appl. Environ. Microbiol. *62*, 994-997.

Tajima, K., Aminov, R.I., Nagamine, T., Ogata, K., Nakamura, M., Matsui, H. and Benno, Y. 1999. Rumen bacterial diversity as determined by sequence analysis of 16S rDNA libraries. FEMS Microbiol. Ecol. *29*, 159-169.

Takami, H., Han, C.G., Takaki, Y., and Ohtsubo, E. 2001. Identification and distribution of new insertion sequences in the genome of alkaliphilic *Bacillus halodurans* C-125. J. Bacteriol. *183*, 4345-4356.

Takami, H., Nakasone, K., Takaki, Y., Maeno, G., Sasaki, R., Masui, N., Fuji, F., and et al. 2000. Complete genome sequence of the alkaliphilic bacterium *Bacillus halodurans* and genomic sequence comparison with *Bacillus subtilis*. Nucleic Acids Res. *28*, 4317-4331.

Thomas, D.J., Morgan, J.A., Whipps, J.M., and Saunders, J.R. 2000. Plasmid transfer between the *Bacillus thuringiensis* subspecies *kurstaki* and *tenebrionis* in laboratory culture and soil and in lepidopteran and coleopteran larvae. Appl. Environ. Microbiol. *66*, 118-124.

Thomas, D.J., Morgan, J.A., Whipps, J.M., and Saunders, J.R. 2001. Plasmid transfer between *Bacillus thuringiensis* subsp. *israelensis* strains in laboratory culture, river water, and dipteran larvae. Appl. Environ. Microbiol. *67*, 330-338.

Torres, O.R., Korman, R.Z., Zahler, S.A., and Dunny, G.M. 1991. The conjugative transposon Tn*925*, enhancement of conjugal transfer by tetracycline in *Enterococcus faecalis* and mobilization of chromosomal genes in *Bacillus subtilis* and *E. faecalis*. Mol. Gen. Genet. *225*, 395-400.

Turnbull, P.C., and Kramer, J.M. 1985. Intestinal carriage of *Bacillus cereus*: faecal isolation studies in three population groups. J. Hyg. (Lond) *95*, 629-638.

Vilas-Boas, L.A., Vilas-Boas, G.F., Saridakis, H.O., Lemos, M.V., Lereclus, D., and Arantes, O.M. 2000. Survival and conjugation of *Bacillus thuringiensis* in a soil microcosm. FEMS Microbiol. Ecol. *31,* 255-259.

Volk, W.A., Bizzini, B., Jones, K.R., Macrina, F.L. (1988) Inter- and intrageneric transfer of Tn916 between *Streptococcus faecalis* and *Clostridium tetani.* Plasmid *19,* 255-9

Wang, H. and Mullany, P. (2000) The large resolvase TndX is required and sufficient for the integration and excision of derivatives of the novel conjugative transposon Tn*5397* J. Bacteriol. *182,* 6577-83.

Wang, H. Roberts, A., and Mullany, P. (2000) DNA sequence of the insertional hot-spot of Tn916 in the *Clostridium difficile* genome and discovery of a Tn916-like element in an environmental isolate integrated in the same hot spot. FEMS Microbiol. Lett. *192,* 15-20

Wang, H., Roberts, A., Lyras, D., Rood, J., Wilks, M., and Mullany, P. (2000a) Characterisation of the ends and target sites of the novel conjugative transposon Tn*5397* from *Clostridium difficile* : excision and circularisation is mediated by the large resolvase TndX. Journal of Bacteriology. 182 p3775-3783.

Wang, Y., Wang, G.R., Shelby, A., Shoemaker, N.B., and Salyers, A.A. (2003). A newly discovered bacteroides conjugative transposon, CTn*GERM1*, contains genes also found in gram-positive bacteria. Appl. Environ. Microbiol. *69,* 4595-4603.

Wilcks, A., Jayaswal, N., Lereclus, D., and Andrup, L. 1998. Characterization of plasmid pAW63, a second self-transmissible plasmid in *Bacillus thuringiensis* subsp. *kurstaki* HD73. Microbiology. *144,* 1263-1270.

Wilcks, A., Smidt, L., Okstad, O.A., Kolsto, A.B., Mahillon, J., Andrup, L. 1999 Replication mechanism and sequence analysis of the replicon of pAW63, a conjugative plasmid from *Bacillus thuringiensis.* J. Bacteriol. *181,* 3191-3200.

Woolley, R.C., Pennock, A., Ashton, R.J., Davies, A., and Young, M. 1989. Transfer of Tn*1545* and Tn*916* to *Clostridium acetobutylicum.* Plasmid. 22, 169-174.

Xu, J., Bjursell, M.K., Himrod, J., Deng, S., Carmichael, L.K., Chiang, H.C., Hooper, L.V., and Gordon, J.I. 2003. A genomic view of the human-*Bacteroides thetaiotaomicron* symbiosis. Science. *299,* 2074-2076.

Xu, K., He, Z.Q., Mao, Y.M., Sheng, R.Q. and Sheng, Z.J. 1993 On two transposable elements from *Bacillus stearothermophilus.* Plasmid. 29: 1-9.

Yin, X. and Stotzky, G. 1997. Gene transfer among bacteria in natural environments. Adv. in Appl. Microbiol. 45: 153-212.

Chapter 11

Bacillus Spores as Probiotic Products for Human Use

Sonia Senesi

SUMMARY

Increasingly, published scientific literature on probiotics supports the concept that consumption of exogenously supplied live bacteria can positively influence human health. The worldwide interest in probiotics has opened new perspectives in the application of microorganisms as biotherapeutic tools and favoured the expanding marketing of probiotic-based products. Lactic acid-producing bacteria belonging to the genus *Lactobacillus* and *Bifidobacterium* are the most studied probiotic organisms and the most frequently marketed for human use as commercial products. Spores of the genus *Bacillus* are used as probiotics especially in Europe and in Southeast Asia, where most of them are currently marketed to treat and prevent a broad variety of gastrointestinal disorders in man. Studies addressing the mechanism by which spores of *Bacillus* spp. exert probiotic activities are so far limited, although an increasing number of studies are proving encouraging. This chapter will focus on *Bacillus*-based probiotics marketed for human use particularly the taxonomic position, characterisation, efficacy and safety of the *Bacillus* strains present in the commercial formulations.

INTRODUCTION

Live microbial feed supplements (probiotics) have been used for as long as people have eaten fermented food. The notion that consumption of fermented products, in particular fermented milk products, promotes human health has a historical tradition empirically adopted since the Middle Ages. However, the principle of using harmless bacteria for combating pathogens was formally hypothesised only at the beginning of last century by Elie Metchnikoff (Metchnikoff, 1907). He first advanced the hypothesis that organisms present in fermented milk products (e.g., lactobacilli) positively influenced the intestinal microbial balance contributing to colonisation resistance. Since then, an increasing number of studies describing various beneficial effects exerted by different microbial feed supplements on both animals and humans have been published. One of the earliest reports on humans was published in 1915 by Newman, who described a successful application of "Lactobacillus-containing foods" for the treatment of humans cystitis (Newman, 1915). In the 1960s, several microbial feed supplements became commercially available and some were specifically marketed for human use. Early published "human trials" (Beck and Necheles, 1961; Settel, 1962; Schwartz and Smith, 1963) described several beneficial effects exerted on human well-being by an early probiotic product ("Bacid", commercialised in N.Y., USA). The beneficial effects claimed at that time by scientists and traders lacked scientific validity since no control groups were included in these "clinical trials" and the properties of the probiotic strains were completely unknown.

More substantial support to the belief that consumption of exogenously supplied live bacteria could positively influence human health, arose from numerous *in vivo* and *in vitro* studies, showing that the normal intestinal flora behaves as an extremely effective barrier against infections caused by pathogenic and opportunistic microorganisms (Fuller, 1991). The higher susceptibility of germ-free animals to infection when compared to conventional counterparts (Hentges, 1992) was attributed, at least in part, to the lack of colonisation resistance (competition of normal bacterial flora with invading microorganisms) (Vollard and Clasener, 1994), but mainly to the impoverished or absent intestinal flora, that had been proven to play a key role in the development of a balanced immune phenotype (Cross, 2002 and references herein). These considerations favoured the idea that oral administration of microorganisms could promote beneficial effects in the human host when perturbation of normal intestinal flora due to specific factors (such as antibiotics, medications, diet or surgery) had adversely affected the functionality of the intestinal barrier. Consequently, several probiotic products were developed, licensed for human use, and marketed for the prevention and treatment of intestinal disorders *prior* to their purported benefits being demonstrated. The majority of probiotic preparations for human consumption contain lactic acid-bacteria, especially species allotted to the genus *Lactobacillus*, although several *Bacillus*-based products, developed since at least the 1960's, are currently used as probiotics and most marketed for human use.

More recently, probiotic organisms and the related commercial products have been more extensively studied *in vitro* and *in vivo*. However, most studies have been addressed to lactobacilli, while spore-based probiotics have received scant attention. Indeed, while searches in the Medline Library cited 22 published articles on probiotic bacteria up to 1990, 1470 reports were published between 2000 and 2002, with 866 papers dated 2001 and 2002. These numbers clearly indicate an increasing interest in this research area and, most likely, this may not yet have reached its threshold, since nearly 400 reports have already been quoted in Medline during the first 9 months of 2003. Despite the remarkable increase in interest in probiotic bacteria, the total number of papers specifically addressed to *Bacillus* probiotics is only 72; it should be emphasised, however, that nearly half of these are dated 2002 and 2003. This recent increase in reports dealing with *Bacillus*-based probiotics might hopefully be considered as a renewal of scientific interest to explain the mechanisms by which *Bacillus* spores, once ingested, beneficially influence the balance of normal intestinal flora.

Since most of our knowledge on the benefits exerted by probiotic bacteria is from experimental studies performed with *Lactobacillus*, these studies should be briefly mentioned to give an indication of the potential applications of biotherapeutical agents to human health. Probiotics have known benefits in the prevention and treatment of many gastrointestinal disorders (e.g., inflammatory bowel disease, infectious and antibiotic-related diarrhoeas, post-resection disorders, such as pouchitis), in reducing the risk of infection in patients not subjected to prophylactic antibiotic treatment, in the treatment of rheumatoid arthritis, in lowering cholesterol and improving lactose intolerance (Dunne et al., 1999; Gorbach, 2000; Marteau, 2001; Bonn, 2002; Gorbach, 2002). Interesting results have demonstrated that probiotic bacteria generate immune responses that mediate protection against microbial pathogens (Cross, 2002). In this perspective, a relevant finding is that the DNA released by, or purified from, probiotic bacteria plays a major role in host immunomodulation of allergic (Giochetti et al., 2000) and inflammatory diseases (Hooper et al., 2001) by interacting with Toll-like receptors and dendritic cells in the gastrointestinal tract (Rachmilewitz et al., 2002). This considerable and compelling evidence has also encouraged the exploitation of probiotics in the field of intestinal carcinogenesis (given their influence in gut flora), in the enhancement of oral vaccine administration, in balancing impaired-immune responsiveness in allergic and autoimmune diseases, and possibly, in microbial interference therapy, addressed to reduce the use of antibiotics in the face of increasing development of antimicrobial-resistant bacteria. Future areas for probiotic applications arise from the possibility of introducing appropriate genes in probiotic microorganisms, thus, making them able to produce and release pharmacological substrates into the intestinal lumen and possibly in other organs and tissues of the human body (Gorbach, 2002). However, much work still remains and serious criticisms have been raised by many physicians and scientists about the role played by probiotics in human well-being, since: (i) information on the stability/survival of probiotic microbes in both the products and in the human intestinal tract is often lacking; (ii) evidence for their mechanism of action is still minimal, and controversial; and (iii) controlled clinical trials to demonstrate the safety other than the claimed health benefits of probiotics are limited or absent.

DEFINITION OF PROBIOTIC BACTERIA FOR HUMAN USE

The definition of probiotics has been modified several times (FAO/WHO, 2001; FAO/WHO 2002). Most changes resulted from the improved knowledge of functional properties of microbes selected for their use as probiotics; however, until now, a formal definition establishing ideal characteristics of probiotics for human use was lacking (Mecfarlane, 2002).

Lilly and Stillwell (1965) first used the term "probiotics" to indicate "substances produced by microorganisms which promote the growth of other microorganisms". Later, Parker (1974) substantially modified this definition by using the term "probiotics" to describe "organisms which contribute to the intestinal microbial balance". The definition of probiotics was further modified by Fuller (1989), who described probiotics as "live microbial feed supplements which beneficially affect the animal host by improving its intestinal microbial balance". This definition stresses the need for probiotics to contain live microbes and implies that ingested organisms may actively interact with the host. Therefore, such a definition implies that probiotic bacteria may potentially exert diverse effects on the host, depending on both the targeted probiotic organism and on the host *status*. More recently, many other definitions of probiotic bacteria have been proposed, which, in general highlight that probiotics may be either mono- or mixed-microbial cultures; each microbial component of probiotic formulations must be identified and fully characterised; the probiotic organisms, when administered to humans in adequate amounts, must confer reproducibly demonstrable health benefits; and, essentially, the safety of each probiotic strain must be properly assayed under controlled experimental conditions, including animal models and clinical trials (FAO/WHO, 2002).

Notably, none of the definitions adopted to delineate features of probiotic bacteria contain any guidelines as to what bacteria constitute optimum probiotic species and are more able than others to exert specific beneficial effects on human health. To date, no scientific data has provided comparative studies of potential "probiotic" bacteria. Rather, strain-specific effects are more frequently described in diverse assay procedures, animal models, and clinical trials. Even though direct comparison between strains has rarely been performed, different bacterial strains, even belonging to diverse genera, have been reported to produce similar effects. Therefore, until the mechanisms by which probiotic bacteria exert their activities are better understood and assessment criteria to compare different probiotic microorganisms are established, it is more appropriate that the definition of microorganisms as probiotics for human use should include: (i) accurate characterisation of individual strains selected as probiotics; and (ii) unambiguous demonstration of the strain-specific health effects, by means of properly designed biological assays and controlled clinical trials (FAO/WHO, 2002).

OVERVIEW OF SPORE-BASED PROBIOTICS FOR HUMAN USE

A substantial number of *Bacillus*-based probiotic products have been developed by several industrial companies in Europe and in Southeast Asia. *Bacillus*-spores have been licensed for human use since the 1960s and employed in bacterioprophylaxis and bacteriotherapy because of their purported ability to prevent and treat a broad variety of gastrointestinal disorders, including infectious and antibiotic-related diarrhoeas (Mazza, 1994).

The position of *Bacillus*-based probiotics is unique among all other probiotic products since they are delivered only in the form of live spores instead of actively growing cells. Obviously, this feature has great industrial relevance since spore-based probiotic formulations: (i) have a long, if not indefinite, shelf life; (ii) do not require complicated conditions for storage and transport; and, eventually, (iii) the producers may guarantee to the consumers that the spores making up their commercial products are not expected to undergo any changes in both stability and viability. However, the use of bacterial spores as probiotics addresses the important and paradoxical question of whether these dormant cells can possess the required probiotic properties that are possessed by other probioitcs that exist in the vegetative state and do not form spores. A listing of desirable properties has been identified, driven by experimental studies performed with lactic acid-producing bacteria, in particular *Lactobacillus* and *Bifidobacterium* spp. These microorganisms are the most commonly used bacteria as probiotics and all of them are delivered exclusively in the form of metabolically active vegetative cells. The "catalogue" of requirements for lactobacilli to be effective probiotics includes the ability to:

(i)   adhere to epithelial cells;
(ii) persist and multiply in the intestinal tract;
(iii) exclude or reduce adherence of pathogens to intestinal mucosa;
(iv) produce substrates (metabolites, peptides, bacteriocins) antagonistic to pathogen growth; and
(v)   be safe and therefore non-invasive, non-carcinogenic, and non-pathogenic.

A robust body of evidence has documented that lactobacilli are components of the normal intestinal flora, essentially harmless, resist gastric acid, bile salts and pancreatic enzymes, adhere to intestinal mucosa, readily colonise the intestinal tract and inhibit the *in vitro* growth of many enteric pathogens including *Salmonella typhimurium*, *Escherichia coli*, *Clostridium perfringens*, and *Clostridium difficile*. In addition, the beneficial effects exerted by lactobacilli on human health, debated for

decades, is now increasingly supported by hundreds of convincing published reports through *in vitro* assay procedures, experimental challenge in animals models, and controlled clinical trials in humans (Reid, 1999; Gorbach, 2002; Sullivan and Nord, 2002; Macfarlane and Cummings, 2002).

Studies focused on the characterisation of *Bacillus* spores as candidate probiotics are too few to allow speculation as to which features they should posses to exert beneficial effects on the human host, although some of the properties regarded as probiotic requirements for lactobacilli have been explored. The main challenges that must be addressed with *Bacillus* spores as valid probiotics are to:

(i)   identify *Bacillus* species already marketed as probiotics for human use;
(ii) characterise substances produced *in vitro* that can exert inhibitory activities towards pathogen growth; and
(iii) explore which mechanism may account for the purported ability of *Bacillus*-based probiotics to play a role in enhancing human well-being.

In addition, *Bacillus* spores could also be applied in the following research areas: (i) microbial delivery of immune-protective antigens by probiotic bacterial spores in order to elicit a mucosal and systemic immune stimulation (see Chapter 18); (ii) insertion of heterologous genes in probiotic spores thus making the corresponding vegetative cells able to produce and release pharmacological substances into the intestinal lumen and possibly, in other human body districts.

Table 1. Status of *Bacillus* species present in commercial probiotic products.

Probiotic Product	Manufacturer	Bacteria listed
Bactisubtil	Marion Merrell S.A., Bourgoin-Jallieu, France	*B. subtilis*
Biosubtyl	Biophar Co. Ltd., Nha Trang, Vietnam	*B. subtilis*
Biosubtyl "Dalat"	National Institute of Vaccines and Biological Substances, Da Lat, Vietnam	*B. subtilis*
Biosporin	D. K. Zabolotny Institute of Microbiology and Virology, Kyiv, Ukraine	*B. subtilis 3* and *B. licheniformis*
Domuvar	Bioprogress, Anagni-FR, Italy	*B. subtilis*
Enterogermina[1]	Sanofi-Synthelabo O.T.C., Milan, Italy	*B. clausii*
Enterum	Angelini SpA, Rome, Italy	*Bacillus subtilis*
Flora Source	MBA Company, Sewickley, PA	16 species including "*Lactobacillus sporogenes*", lactobacilli and bifidobacteria
Lacbon, Lacris[2]	Uni-Sankyo	"*Lactobacillus sporogenes*"
Lactipan *plus*	Istituto Biochimico Italiano, Milan, Italy	"*Lactobacillus sporogenes*"
Lactospore	Sabinsa Corp., Piscataway, NJ	"*Lactobacillus sporogenes*"
Lactopure	Pharmed Medicare	"*Lactobacillus sporogenes*"
Medilac[3]	Hanmi Pharmaceutical Co., Ltd., Korea and Beijing, China	*B. subtilis* and *Enterococcus fecium*
Nature's First Food	Nature's First Law, San Diego, CA	42 species including *B. subtilis*, *B. pumilus*, *B. polymyxa*,and lactobacilli
NaturaFlora	Natura Health, Hayes, UK	"*Lactobacillus sporogenes*"
Primal Defense	Garden of Life, West Palm Beach, FL	14 species including *B. subtilis*, *B. licheniformis*, lactobacilli, and bifidobacteria
Subalin	D. K. Zabolotny Institute of Microbiology and Virology, Kyiv, Ukraine	Recombinant strain of *B. subtilis 3*
Subtyl	Pharmaceutical Factory 24, Ho Chi Minh City, Vietnam	*B. subtilis*

[1] The older label, stating *B. subtilis*, has been amended since 2000.
[2] The label stating "*Lactobacillus sporogenes*" has not been amended, although it is an invalid name.
[3] Suggested as a new species

MOLECULAR IDENTIFICATION OF *BACILLUS* SPECIES MARKETED AS PROBIOTICS FOR HUMAN USE

The genus *Bacillus* comprises 77 different recognised species according to the International Code of Nomenclature of Bacteria (ICNB; www.bacterio.cict.fr). Among these, only a limited number has been used or considered for human use. As shown in Table 1, a common problem of the *Bacillus* strains contained in commercial products is their miss-classification. This is not surprising since the taxonomic position of members allotted to the genus *Bacillus* was essentially based on differences in phenotypic traits. Although such traits should not be underestimated, their discriminatory power is very poor. Sequencing of ribosomal 16S RNA and DNA/DNA reassosciation experiments have shown that the genus *Bacillus* comprises a wide variety of species whose taxonomic position in a single genus is unsatisfactory. This is the case of the members previously encompassed in the *Bacillus brevis* group, which has been recognised as a new genus, named *Brevibacillus*, encompassing 11 different species (Logan et al., 2002). In contrast, the members of the *B. cereus* sub-group, which comprises *B. thuringiensis* and *B. anthracis*, are indistinguishable by 16S rRNA sequence analysis, while they differ significantly for several phenotypic traits including the expression of virulence factors encoded by plasmid-carried genes (Mock and Fouet, 2001). Combination of phenotypic characterisation and genome typing led to the unambiguous identification of some of the *Bacillus* strains contained as live spores in the corresponding probiotic products (Table 1). The revised taxonomic position of some probiotic *Bacillus* strains will hopefully lead to revised labelling of the corresponding product formulations since proper product labelling of all products for human use whether as a novel food or for clinical use is subject to EU directives. Product labelling should state the real name assigned to the bacterium and the identified strain should be deposited in an internationally recognised culture collection. To date, only the bacterial strains constituting the product "Enterogermina" have been relabelled. "Enterogermina" (Table 2) is made up of four distinct *Bacillus* strains, each of them previously allotted to *B. subtilis*, as for most probiotic *Bacillus* species listed in Table 1. A revision of the taxonomic position of one of the strains contained in the probiotic formulation "Enterogermina" showed that such a strain was

more closely related to members of the subgroup *B. alcalophilus* than to *B. subtilis* (Green et al., 1999). Later, a further revision of all "Enterogermina" strains was addressed to verify the original taxonomic position of the four "Enterogermina" strains isolated from spore mixtures found in original commercial preparations dated 1975, 1984, 2000 (Senesi et al., 2001). The analysis of physiological and biochemical traits, complete 16S tRNA gene sequences, and DNA-DNA reassociation of the four "Enterogermina" strains (O/C, N/R, T, and SIN) demonstrated that the "Enterogermina" strains belong to a unique genospecies, which was unequivocally identified as the alkalitolerant species *B. clausii*. In particular, the chromosomal DNA extracted from the "Enterogermina" strains O/C, N/R, T, and SIN, showed 75% reassociation (DT_m 3.5°C) with chromosomal DNA extracted from *B. clausii* type strain DSM8716 and the sequencing of the 16S rRNA of each strain (Accession number AJ297491 to AJ297494, respectively) revealed 100% identity among the strains O/C, N/R, T, and SIN (Senesi et al., 2001). Moreover, other molecular evidence was produced (tRNA intergenic spacer length polymorphism, single-strand conformation polymorphism of PCR-amplified spacer regions of tRNA genes, and randomly amplified polymorphic DNA), that demonstrated that the strains *B. clausii* O/C, N/R, T, and SIN, in contrast to several reference strains of *B. clausii*, were characterised by a notable low level of intraspecific genome diversity and that each strain has remained unchanged for at least 25 years (Senesi et al., 2001).

Some products still use the invalid name "*Lactobacillus sporogenes*" to designate the probiotic bacterium contained in their commercial probiotic products (Table 2 and Appendix 1). "Lactospore" (Sabinsa Corp., Piscataway, NJ), "Lacbon" (Uni-Sankyo), "Lacris" (Uni-Sankyo), "Lactopure" (Pharmed Medicare), "Lactipan *plus*" (Istituto Biochimico Italiano, Milan, Italy), and "NaturaFlora" (Natura Health, Hayes, UK) are currently labelled as "*Lactobacillus sporogenes*". Since this strain was originally described as a spore-forming bacterium (Horowitz-Wlassova and Nowotelnow, 1932), it could not be considered a member allotted to the genus *Lactobacillus* and, therefore, the label "*Lactobacillus sporogenes*" should be amended. "*Lactobacillus sporogenes*" is an invalid name since it is not cited in either *Bergey's Manual* or in the *Approved List of Bacterial Names*. The valid name to designate "*Lactobacillus sporogenes*" is *B. coagulans*, as recently identified (Sanders

Table 2. Commercial *Bacillus*-based probiotic products.

Probiotic Product	Manufacturer	*Bacillus* species in the commercial product	
		Older Label	Identified
Bactisubtil	Marion Merrell S.A., Bourgoin-Jallieu, France	*B. subtilis*	*B. cereus*
Biosubtyl	Biophar Co. Ltd., Nha Trang, Vietnam	*B. subtilis*	*B. pumilus*
Biosubtyl "Dalat"	National Institute of Vaccines and Biological Substances Substances, Da Lat, Vietnam	*B. subtilis*	*B. cereus*
Enterogermina	Sanofi-Synthelabo O.T.C., Milan, Italy	*B. subtilis*	*B. clausii*[1]
Domuvar	Bioprogress, Anagni-FR, Italy	*B. subtilis*	*B. clausii*
Lactipan *plus*	Istituto Biochimico Italiano, Milan, Italy	"*Lactobacillus sporogenes*"	*B. subtilis*[2]
Lactospore	Sabinsa Corp. Piscateway, N J	"*Lactobacillus sporogenes*"	*B. coagulans*[2]
Subtyl	Pharmaceutical Factory 24, Ho Chi Minh City, Vietnam	*B. subtilis*	*B. vietnami*[3]

[1] Approved by the Italian Ministry of Health
[2] Approved by the Japanese Ministry of Health and Welfare
[3] Approved by the Chinese State Drug Authority

et al., 2001). However, not all the probiotic products labelled as "*Lactobacillus sporogenes*" have been identified as *B. coagulans*. An example is the probiotic product "Lactipan plus" (Istituto Biochimico Italiano, Milan, Italy), which is labeled as "*Lactobacillus sporogenes*" yet contains an aerobic spore-bearing bacterium now identified as *B. subtilis* (Hoa et al., 2000). As shown in Table 1, the revision of the taxonomic position of several species of probiotic *Bacillus* strains led to the demonstration that *B. subtilis* is not the predominant *Bacillus* species employed as probiotic for human use. Rather, the species of *Bacillus* most commonly found in commercialised probiotic products (Table 2) comprise the following: *coagulans, clausii, cereus, laterosporus, pumilus, licheniformis, mesentericus,* and *polymyxa*. For the *Bacillus* strain contained in "Subtyl" (Pharmaceutical Factory 24, Ho Chi Minh City, Vietnam) the name *B. vietnami* has been proposed (Green et al., 1999; Hoa et al., 2000), since the strain did not show any significant 16S rRNA homology with any known *Bacillus* species. A notable and intriguing observation is the finding that both "Bactisubtil" and "Biosubtyl Dalat", labelled as *B. subtilis,* contained *B. cereus* as the probiotic bacterium (Hoa et al., 2000). This was based on the similarity of the two strains with *B. thuringiensis*, based on 16S rRNA sequencing analysis. However, the strains carried an exosporium, which is peculiar to *B. cereus* strains and they did not produce parasporal crystal toxins, which is a phenotypic trait peculiar to *B. thuringiensis*. So, while 16S rRNA sequencing analysis predicted *B. thuringiensis* as an unambiguous species designation could not be made and is in agreement with Ash et al. who have reported the problems in designating species to members of the "*Bacillus cereus* sub-group" (Ash et al., 1991). However, aside from the exact taxonomic position of the *Bacillus* strains contained in "Bactisubtil" and "Biosubtyl Dalat" within the "*Bacillus cereus* sub-group", it has been shown that some strains belonging to *B. cereus* and *B. thuringiensis* are responsible for food-borne intestinal infections, the severity of which is dependent on the expression and secretion of a large variety of enterotoxins by the actively growing cells. The *in vitro* production of hemolysins and the ability to grow under anaerobic conditions have been reported for the probiotic strains contained in "Bactisubtil" and "Biosubtyl Dalat" (Hoa et al., 2000): these features indicate that the spores contained in the commercialised products can germinate and grow in the intestinal tract, thus potentially releasing toxins into the intestinal lumen as recently demonstrated by Kniehl et al. (2003). Nevertheless, probiotic products containing strains of *B. cereus* have been, until recently, used in the agricultural industry (see Chapter 13). For example, "Pacifor" a probiotic preparation of *B. cereus* used for piglets and sows was recently withdrawn in 2000 due to the potential risk for transmission of toxinogenic microorganism to humans (http://europa.eu.int.comm/food/fs/sc/scan/out41.pdf) (see Appendix II).

COMMERCIAL PROBIOTIC PRODUCTS CONTAINING *BACILLUS* SPORES

As already mentioned, probiotic products containing *Bacillus* spores have been commercialised for a considerable time, especially in Europe and in South East Asian Countries.

As an example, Enterogermina *Bacillus* spores have been licensed for human use and marketed to prevent and treat gastrointestinal disorders as early as 1958. A list of *Bacillus*-based probiotic products available in Europe is shown in Table 2 as well further information provided in the appendices. Some commercial products carry non-spore-forming species in addition to the spore forming species (e. g., "Flora Source", "Nature's First Food", "Primal Defense" and "Medilac") (Table 2 and Appendix 1), while others carry more than one *Bacillus* species, for example, "Biosporin" which contains *B. subtilis* and *B. licheniformis* and "Enterogermina" that carries four closely related strains belonging to the same species (*B. clausii*). "Subalin" is a peculiar probiotic product since it is comprised of a recombinant *B. subtilis* strain (*B. subtilis* 2335/ 105 carrying the human interferon gene). This probiotic has been licensed for veterinary use in Russia (but not in Europe; see Chapter 13) and has been proposed as a new generation of probiotic for medicinal use since the recombinant strain has been reported to retain the antibacterial activity of the host strain, and in addition, to display anti-viral and immunomodulatory properties associated with the secreted interferon (Sorokulova, 1998). Many benefits have been claimed by the producers for the probiotic products they market. Because of this, it should be mentioned that some probiotic products (Table 2) have obtained authorisation for human use in different formulations (e.g., aqueous spore suspensions or dried spore powders) by the Ministry of Health in the country where the probiotics are produced. Unfortunately, most literature supporting and dealing with the licensing procedure are not available.

Today, the probiotic industry is flourishing and the desire by consumers to use natural methods for health maintenance, linked to their expectation that novel foods are a source of enhanced well-being has encouraged the belief that the probiotic market will continue to expand. Much of this growth depends on the reliability of claims being made and whether these will indeed be proven founded. A real interest in establishing scientific credibility has now become important for many companies and scientists. Unlike *bona fide* drugs that must be exposed to substantial and extensive scientific scrutiny before use the situation regarding the current use of probiotics is unusual and possibly contentious since only now are studies being made to support the claims being made.

On the other hand, the legislator needs to provide clear rules and regulations, based on scientific evidence, allowing measurable biomarkers and criteria for carefully selected strains for human use. Interestingly, as shown in Chapter 13, such controls are already in place for the use of probiotics for animal use but they do not yet exist for use in humans. Clear definition of what constitutes a probiotic and what criteria are necessary for making 'beneficial' claims will be important (if not essential) for defining probiotic preparations as either a food, a novel food supplement, or as pharmaceutical preparations. In addition, extra vigilance regarding the detection of symptomatic infection due to probiotic microbial strains should be taken, and isolates should be sent to reference centres for molecular characterisation and confirmation. The first attempts at defining rules for what constitutes a probiotic bacterium, safety assessments and product labelling have been taken and the reader should consult the following references for further information (FAO/WHO, 2002; Borriello, 2003; Sanders, 2003).

QUESTIONING THE ROLE OF *BACILLUS* SPORES AS PROBIOTICS FOR HUMAN USE

Bacillus-based probiotics have long been used as biotherapeutic agents and are currently administered (as prescription and non-prescription products) by the oral route in the form of a dried spore powder or as aqueous spore suspensions. The main beneficial effect that has been claimed for probiotic *Bacillus* spores is their effectiveness in the maintenance/restoration of intestinal microbial balance (Coppi et al., 1985; Mazza, 1994). How this is achieved by ingested spores is not intuitive; rather, it is generally assumed that restoration of gastrointestinal health by consumption of probiotics is brought about mainly or exclusively by actively growing bacteria that reside in the intestinal tract until such time that the normal flora can be re-established (Rolfe, 2000). In this way, the beneficial effect exerted by probiotics could be reached through colonisation of intestinal mucosa and the release of metabolically active substances by growing cells during their residence in the intestinal tract. This mechanism of action may be exerted by probiotic lactobacilli, which are consumed only as vegetative cells, but not by bacterial spores, unless the ingested spores undergo germination followed, subsequently, by active replication of vegetative cells.

Only recently, the role of bacterial spores as probiotics has been investigated by examining:

(i) the ability of ingested spores to adhere to epithelial cells and undergo germination, giving rise to populations of actively growing cells able to colonise, even transiently, the gastrointestinal tract;
(ii) the residence time of spores and/or vegetative cells in the intestinal tract in animal models and in humans;
(iii) *in vitro* production of metabolites, peptides, bacteriocins exerting inhibitory activities towards pathogen growth;
(iv) competitive *in vitro* and *in vivo* inhibition for bacterial growth;
(v) ability of vegetative cells to release vitamins and convert genotoxic compounds into unreactive products; and essentially
(vi) safety in animal models of infection.

All these aspects, have contributed to a catalogue of "probiotic" features shared by *Bacillus* spores, although the mechanism by which spores may influence human well-being through the activities they exert during their time in the intestinal tract requires further investigation both *in vitro* and *in vivo*.

ADHESION OF *BACILLUS* SPORES TO EPITHELIAL CELLS

The ability to adhere to the intestinal mucosa is one of the main selection criteria for potential probiotics as it prolongs their persistence in the intestinal tract and allows the probiotic to exert its effect for longer (Ouwehand, 1999). Moreover, adhesion of probiotic bacteria to intestinal mucosa may block

adhesion sites for pathogens, if the probiotic strain and the pathogen share the same adhesion receptor molecules. Since competitive inhibition (sometimes referred to as competitive exclusion) for bacterial adhesion is regarded as a mechanism of action for probiotics, some probiotic strains have been chosen for their ability to adhere to epithelial cells *in vitro*. Good adhesion to both epithelial cells and intestinal mucus has been shown for several strains of *Lactobacillus*, while this trait has been studied for a limited number of *Bacillus* spores and/or vegetative cells. A comparative study on the ability exhibited by vegetative cells of several *Bacillus* and *Lactobacillus* strains to adhere to erythrocytes showed that *Lactobacillus* generally possess a higher degree of adhesion than *Bacillus* strains (Smirnov and Kosiuk, 1997). However, a higher degree of adhesion to Hep-2 and Caco-2 cells has been reported for the *B. clausii* spores "Enterogermina" in comparison to the corresponding vegetative cells (Angioi et al., 1995); the authors show by scanning electron micrograph analysis that *B. clausii* "Enterogermina" spores (named as *B. subtilis* by the authors) adhere to Caco-2 cells and undergo germination under *in vitro* conditions. However, although adhesion may account for a prolonged residence of probiotic bacteria in the intestinal tract, such a property is also considered a potential virulence factor involved in the translocation of pathogens across the mucosa surface (Finlay and Falcow, 1997). In this view, probiotics should not be administered to any patients suffering from severe immunodeficiency (FAO/WHO, 2002), since these individuals are more susceptible to endogenously borne severe bloodstream infections. There is an interesting *in vitro* study, performed with *Lactobacillus* strains, in which isolates from bloodstream infection and strains of dairy origin are compared *in vitro* for adhesion ability (Apostolou et al., 2001): although the authors showed a trend for blood isolates to bind to intestinal mucus in higher numbers than strains from dairy origin, other factors were suggested to be involved in the development of systemic episodes, mainly the host *status*, since several clinical *Lactobacillus* isolates were found to bind at low levels to mucus *in vitro*. Therefore, a high degree of adhesion established *in vitro* for probiotic bacteria does not imply a higher ability to translocate across the mucosal surface *in vivo*, while it may reflect the propensity to prolong their residence in the intestinal tract.

GERMINATION OF *BACILLUS* SPORES IN THE INTESTINAL TRACT

This trait is of major interest in order to evaluate the probiotic activity of spores. It has recently been studied in animal models and in healthy individuals. Indeed, if spores can germinate in the intestinal tract, the mechanism by which ingested *Bacillus* spores exert their probiotic activity may be similar to that proposed for *Lactobacillus* probiotics, which are administered only in vegetative forms. Otherwise, a different mode of action should be hypothesised to explain the recent demonstration that oral administration of spores competitively exclude colonisation by pathogens (La Ragione et al., 2001). The validity of spores as probiotics has been recently demonstrated, at least in animals, by showing that oral inoculation of 1-day-old chicks with 2.5×10^8

B. subtilis spores suppressed all aspects of infection when chicks were challenged with *Escherichia coli* 078:K80 (La Ragione et al., 2001). Although an older observation (Hisanga, 1980) showed that *Bacillus* spores may germinate in ligated ileum loops in rabbits, only recently has the fate and dissemination of ingested spores been explored in animal models of infection. Spinosa et al., using spores of *B. subtilis* and *B. clausii* in a murine model (Spinosa et al., 2000), found that only spores could be detected in the intestinal tract up to 1 week following oral dosing and concluded that no probiotic activity should be claimed for *Bacillus* spores. However, in this study the presence of Bacilli was only examined at defined times following oral dosing. Hoa et al., using *B. subtilis* spores given orally to mice (Hoa et al., 2001), examined the total number of Bacilli that were excreted from mice orally dosed with a preparation of spores. They reported that the total number of spores rescued from faeces of the majority of mice examined was higher than the size of original inoculum, and suggested that a portion of ingested spores underwent germination. However, since germinated spores and/or vegetative cells were not unequivocally confirmed to be present in mouse faeces, these authors suggested that spore germination must have been followed by a round of sporulation within the intestinal tract, most likely induced by the anoxic conditions of the gut environment (Hoa et al., 2001). These results were thought to support the view that, at least in mice: (i) *B. subtilis* spores may have transient residency in the gut; and (ii) the oral dosing of *B. subtilis* spores does not lead to dissemination of spores since no spores or vegetative cells were recovered from other organs or tissues. A further insight into spore germination in the intestinal tract was given by Casula and Cutting using a *B. subtilis* strain carrying a genetically chimeric gene (*ftsH-lacZ*), which was expected to be expressed only in vegetative cells (Casula and Cutting, 2002). The oral administration of this engineered *B. subtilis* strain to mice led to the demonstration that spores do germinate in the intestinal tract (jejunum and ileum), thus potentially colonising the small intestine, albeit briefly. Despite the occurrence of vegetative forms, derived from spores, implied from the detection of *ftsH-lacZ* gene expression, no evidence was given supporting the presence of actively growing cells of *Bacillus* in the intestinal tract. The high sensitivity of the growing *Bacillus* cell to bile salts was considered a possible explanation for the lack of vegetative forms in the intestinal tract: indeed, it has been shown for several *Bacillus* species (13 strains tested) that the vegetative forms are killed *in vitro* by bile salts, with the exception of *B. racemilacticus* and *B. coagulans,* which were found to tolerate oxagall above 0.3% (Hyronimus et al., 2000). Moreover, in addition to the susceptibility of vegetative forms to bile salts, it should be also considered that the ingested spores, once it has germinated in the intestinal tract, may give rise a population of vegetative cells in a hostile environment, where anaerobic conditions do not favour growth of aerobic *Bacillus* species. These two factors could account for the absence of a recovery of vegetative forms from the intestinal tract of animals to which *B. subtilis* spores had been orally administered, despite the expression of *ftsH-lacZ* gene by vegetative cells (Casula and Cutting, 2002). As was the case for mice, in humans no vegetative Bacilli were detected in stool samples collected from healthy volunteers (19 individuals) to whom spores of *B. clausii* (6.0×10^9 spores comprising strains O/C, N/R, T, and SIN of *B. clausii* "Enterogermina") were orally administered in a single dose (unpublished, this laboratory). Only spores were recovered from stool samples, since no differences in bacterial counts were observed between heated (80°C *per* 10 min) and non-heated samples seeded on strain selective plates (Senesi et al., 2001). Analysis of the transit of spores through the gut showed that the majority of ingested *B. clausii* spores (nearly 60% of the ingested spores) were recovered within 2 days post ingestion, with exponential washout kinetics occurring in all volunteers within the following 15 days (data not shown). The observation that: (i) the transit time of the majority of ingested spores through the gut occurred at day 2 post ingestion is in agreement with the experimentally calculated mean transit time of a solid marker through the gut in humans (Graff et al., 2001); and (ii) the 15 days washout kinetics of ingested spores is in accordance with the requirement that probiotic bacteria should reside in the intestinal tract for at least 2 weeks for exerting the suggested health-promoting properties (Rolfe, 2000). Moreover, the ratio between the recovered and ingested spores was lower than 1.0 in nearly 15% of the volunteers. These data were explained by assuming that a number of ingested spores underwent germination and the germinated spores gave rise vegetative forms which, most likely, sustained some rounds of growth. Thereafter, *B. clausii* vegetative cells, which are susceptible to bile salts (Spinosa et al., 2000) and are aerobic (Nielsen et al., 1995), may be killed in the intestinal tract or, alternatively, may differentiate into spores in the hostile gut environment, as suggested in order to account for the increased number of *B. subtilis* spores recovered in the faeces from mice (Hoa et al., 2001; Casula and Cutting, 2002). These results suggest that the intestinal tract of both mice and humans may be colonised, briefly and not extensively, by vegetative cells derived from ingested spores. Indeed, as proposed by Casula and Cutting for mice treated with *B. subtilis* spores as probiotics (Casula and Cutting, 2002), the potential benefits of *B. clausii* spores as probiotics for human use appears to rely on the following:

(i) *B. clausii* spores can survive transit across the stomach intact, since they are resistant to gastric juice *in vitro* (Ciffo et al., 1987);
(ii) 100% of them may be potentially delivered into the gastrointestinal tract; and
(iii) a portion of ingested spores may germinate and give rise, although briefly, to a population of vegetative cells in the human intestinal environment.

Therefore, the *in vitro* production of metabolites, peptides, and bacteriocins as well as the ability of vegetative cells to release vitamins and convert genotoxic compounds into unreactive products (see below) could also occur *in vivo* thus contributing to the purported probiotic activities of *Bacillus*-based probiotics.

IN VITRO PRODUCTION OF ACTIVE SUBSTANCES AND MICROBIAL INTERFERENCE THERAPY WITH *BACILLUS* PROBIOTICS

The use of probiotics to restore intestinal homeostasis is considered an attractive alternative to conventional antibiotic therapy. The World Health Organization has advocated where possible a policy of microbial interference therapy in order to reduce antibiotic administration (O'Sullivan, 2001). The validity of microbial interference therapy relies on the demonstration that probiotic bacteria may effectively protect the host from intestinal disorders and enteric infection. Some reports with *Bacillus* probiotics has been produced in such a perspective; but much work remains to be done in order to classify the mechanism of action of particular probiotics against particular pathogens since the same probiotic bacterium may inhibit different pathogens by different mechanisms.

Many reports describing the use of *Bacillus* probiotics deal with the properties exerted by Biosporin (*B. subtilis* 3 *plus B. licheniformis*) and by its recombinant derivative Subalin. Some articles describe that oral administration of these probiotics to mice may be used for microbial interference therapy since they: (i) prevent the development of dysbacteriosis induced in animals by antibiotics such as polymyxin and kanamycin; and (ii) efficiently cure mouse dysbacteriosis in a week of oral treatment (Furzicova et al., 2000). Biosporin has been administered to more than 800 patients suffering from acute intestinal disorders, dysbacteriosis of different aetiology, and non-specific ulcerative colitis: the application of this probiotic was found to restore the normal microflora in all patients, who, during the course of the treatment, did not suffer from any adverse reactions (Smyrnov et al., 1994). Biosporin as well as Subalin are described to inhibit the *in vitro* growth of *Helicobacter pylori* and that a single dose administration of these probiotics exerts a protective effect (100% protection) in mice infected with *H. pylori* at LD_{50} (Sorokuloza et al., 1997). The anti-*H. pylori* activity exerted *in vitro* by *B. subtilis* 3 was demonstrated to be due to antibiotics (at least two different compounds), one of them identified as amicoumacin A, which is actively secreted by the probiotic strain *B. subtilis* 3 (Pinchuk et al., 2001 and see also Chapter 15). A newly identified bacteriocin (polyfermenticin), produced by *Bacillus polyfermenticus*, having a narrow spectrum of activity against both Gram-positive and Gram-negative bacteria, has been proposed for use in the treatment of long term intestinal disorders (Lee et al., 2001). Eventually, it was demonstrated that the daily administration of *B. coagulans* to piglets allows this bacterium to be integrated into the enteric microflora, where it is transient; this finding also suggested that this probiotic *Bacillus* may represent an alternative to antibiotic therapy (Adami and Cavazzoni, 1999).

Vitamins

Among probiotic bacteria, organisms able to produce and actively secrete B-group vitamins are more attractive for being used for human consumption to treat or to prevent intestinal disorders. Indeed, active B-group vitamin secretion may compensate for the host vitamin deficiency often resulting from an imbalance in the intestinal flora and/or a prolonged oral antibiotic therapy.

Some probiotic bacteria (Owvehand et al., 1999) behave as net consumers of vitamins, particularly the B-complex, and may be potentially responsible for a decrease in vitamin bioavailability in hosts. Several probiotic *Bacillus* and *Lactobacillus* strains have been compared for their ability to produce and actively secrete riboflavin (VB_2), by the use of a VB_2-auxotrophic mutant of *B. cereus*, generated by mini-Tn*10* insertion in the *ribG* gene (Salvetti et al., 2003). It was demonstrated that *B. cereus* IP5832 (Bactisubtil) and *Lactobacillus ramnosus* GG (Dicoflor 30) do not release any VB_2, while *B. clausii* 17A1 (Domuvar), *B. subtilis* (Enterum), and the *B. clausii* strains O/C, N/R, T, and SIN (Enterogermina) are able to secrete riboflavin with the highest amounts of secreted riboflavin being detected in strain SIN and T (Salvetti et al., 2003). Taking into consideration the fact that *Bacillus clausii* spores are resistant to gastric acid and may germinate in the human gut this supports the hypothesis that these bacteria may actively release VB_2 into the intestinal lumen.

Genotoxicity inhibition

In a recent study, the effect exerted by several *Bacillus* spp., including strains contained in some probiotics on the genotoxicity induced by the mutagen 4-nitroquinoline-1-oxide (4-NQO) has been examined (Caldini et al., 2002). All probiotic strains were shown to have the potential for deactivating 4-NQO (0.1 mM): genotoxicity inhibition, determined by the use of short-term bacterial assay SOS-chromotest with *Escherichia coli* PQ37 as the test organism, ranged from 92.9 to 100% and the activity of 0.1 mM 4-NQO was significantly reduced after coincubation with *Bacillus* cell suspensions (10^8 CFU/ml) for 150 min at 37°C. The antigenotoxicity activity was associated with modification of 4-NQO, which was shown to undergo reductive bioconversion at its nitro group moJety, thus generating 4-amino-quinolone, that is inactive (Takahashi et al., 1987). This observation, although *in vitro*, supports the hypothesis that these bacteria may be potentially involved in the bioconversion of genotoxic compounds to unreactive products in the intestinal tract, which could be brought about by *Bacillus* cells, when they are produced in the gut environment.

BACILLUS PROBIOTICS AND IMMUNOMODULATION

An additional benefit that ingestion of spores may promote, once ingested, relies on the interaction of *Bacillus* spores with the host immune system. Although many studies have been centred on *B. anthracis*, there are many other spore forming *Bacillus* spp. that possess immunostimulating and immunomodulating activities which, however, are devoid of pathogenicity. It has long been described that *B. subtilis* "Enterogermina" (now identified as *B. clausii)* spores, once ingested, stimulate expression of activation markers on lymphocytes (Caruso et al., 1993), the synthesis of membrane bound and secreted IgA (Fiorini, 1985), restore interferon production in aged mice (Grasso et al.,1994). More recently, it has been shown that, *in vitro*, *B. subtilis* spores determine an increase in the percentage of monocytes synthesising TNF-α or IL-1β, while other non-spore-forming Gram-positive bacteria do not stimulate cytokine production (Oggioni et al., 2003).

SAFETY CONSIDERATIONS OF *BACILLUS* PROBIOTICS AS ALLOCHTHONOUS MICROBES TO HUMAN INTESTINAL TRACT

Unless inadvertently ingested with contaminated foods or deliberately consumed, as in the case of "natto" (a traditional Japenese fermented legumes), *Bacillus* species are not considered normal members of the mammalian intestinal tract, rather they are ubiquitously spread in the environment and frequently isolated from water, air, and from soil (see Chapter 1). The extensive use of *Bacillus* probiotics documents an excellent safety record in humans when taken orally; however, just because *Bacillus* species are allochthonous to the human intestinal tract, their use addresses the question of whether they are sufficiently safe to be deliberately consumed in large amounts. The general belief that *Bacillus* species are non-pathogenic (with the notable exception of *B. anthracis,* and some strains of *B. cereus*) has been formalised by the Federal Drug Administration (FDA) as *Bacillus* species are included among bacteria Generally Recognised as Safe (GRAS). Nevertheless, increasing evidence supports the view that exogenous live microbes, including environmental isolates, may potentially behave as opportunistic pathogens in severely ill and/or immunocompromised hosts, including hospitalised patients who frequently suffer from severe opportunistic endogenously born infections. Therefore, the use of probiotics, both *Bacillus-* and *Lactobacillus*-based probiotic formulations, should not be administered to severely compromised hosts, as recommended by the Joint FAO/WHO (2002); indeed, the few reports describing infections due to administration of probiotic *Bacillus* and *Lactobacillus* strains document the occurrence of symptomatic episodes in severely immunocompromised patients (Oggioni et al., 1998; Rautio et al., 1999; Mackay et al., 1999). On the other hand, the interest in reducing the prophylactic use of antibiotics in patients undergoing major surgical interventions, such as transplants (Bonn 2002), should lead to a preferential use of probiotic bacteria for delivering a microbial interference therapy in patients temporarily treated with immuno-suppressing drugs; this application of probiotics, therefore, should be supported by assessment of their safety in immunocompromised animals (FAO/WHO, 2002). For a more detailed review of the safety of *Bacillus* spp. see Chapter 8.

There is the assumption that an ideal probiotic bacterium targeted towards functionality in the intestinal tract should behave or resemble existing members of intestinal microflora, presumably ensuring prolonged colonisation. This principle is largely theoretic, since: (i) the mechanisms by which probiotics influence and alter the intestinal microbial balance is still not clearly understood; (ii) the 400-500 bacterial species of indigenous normal microflora of intestinal tract are still not characterised; and (iii) the mechanisms by which indigenous species interact with each other and with the human body has not been fully elucidated. Moreover, there are clear examples showing that probiotic benefits may be exerted by microbes that are allochthonous to the human intestinal tract, such as in the case of *Saccharomyces boulardii*. This yeast is used for preventing relapses in patients suffering from recurrent pseudomembranous colitis due to *Clostridium difficile*. The probiotic activity exerted by *Saccharomyces boulardii* consists of the proteolytic cleavage of the protein receptor that specifically binds the *C. difficile* toxin A to the surface of intestinal epithelial cells (Czerucka and Rampal, 2002). Specific concerns over safety should be addressed when a genitcally engineered strain is to be used as is the case with the probiotic *Bacillus* strain, Subalin. An *in vivo* evaluation of Subalin was determined by oral administration to calves, chickens, and mice, bacteriological analysis and PCR analysis. These studies showed that oral administration of Subalin produced no disturbances in the microbial ecology of intestinal tract and did not lead to the appearance of spontaneous transformants (Beliavskaia et al., 2001); moreover, when injected intravenously and intraperitoneally into mice in doses of 5×10^9 Subalin cells did not produce any lethal or toxic effects (Osipova et al., 1998). Subalin has been used in healthy volunteers and its oral administration was proven to be safe and without any side-effects (Sorokulova, 1998). A moderate anti-tumor effect exerted by Subalin alone, administered *per os* in mice receiving tumor transplants, has been described together with the increase in the antitumor activity exerted by cyclophosphamide when administered in combination with Subalin (Cherdyntseva et al., 1997).

ANTIBIOTIC RESISTANCE OF *BACILLUS* PROBIOTICS

Resistance to a wide variety of antibiotics has been reported for most *Bacillus* strains present in probiotic products. This feature, often regarded as a positive trait in the view of probiotic administration in combination with antimicrobial drugs, merits proper study though. First, it is imperative that no transmission of antibiotic resistance from probiotic bacteria to human commensal colonisers of the intestinal tract or other body districts should occur (see Chapter 10). In contrast, the few published reports that describe a resistant-susceptible phenotype rarely give information about the characterisation of genes responsible for the drug-resistance. Among the described patterns of antibiotic resistance exhibited by *Bacillus* probiotics there are: *B. clausii* (Enterogermina) strains O/C, N/R, T, and SIN (Ciffo 1984), *B. subtilis* 3 (Biosporin and Subalin) (Smirnov et al., 1994), *B. clausii* 17A1 (Domuvar), *B. subtilis* (Lactipan *plus*), *B. cereus* (Biosubtil "Dalat"), and *B. subtilis* (Subtil), *B. cereus* (Bactisubtil) (Hoa et al., 2000; Green et al., 1999). In one study, a gene, called *aadD2*, identified in all *B. clausii* (Enterogermina) strains, which encodes a aminoglycoside 4'-nucleotidyltransferase, been fully characterised (Accession number AF539790) (Bozdogan et al., 2003). The gene product was responsible for nucleotidylation of kanamicin, tobramycin, and amicacin. The *aadD2* gene was not transferable by conjugation. The *aadD2* gene was located on the chromosome in all *B. clausii* strains as well as in the reference type strain *B. clausii* DSM8716, and was described as specific to this species. The characterisation of the *B. clausii* *aadD2* gene may be regarded as an example of how insights on the drug-resistance patterns described for probiotic strains could contribute to a better understanding of the basis of genomic background shared by probiotic *Bacillus* strains exhibiting a drug-resistance phenotype.

CONCLUSIONS

Although the health value of *Bacillus*-based probiotics remains controversial, there is sufficient evidence, principally from European studies, that supports the application of *Bacillus* spores as biotherapeutic agents for human use. The basic contributions supporting the use of spores as candidate probiotics rely on the demonstration that spores can survive transit across the stomach intact, are delivered into the intestinal environment without any loss of viability, and a portion of ingested spores may germinate giving rise, although briefly, to a population of vegetative cells enabling colonisation of human gastrointestinal tract; moreover, the 15 days washout kinetics of ingested spores is in accordance with the requirement that probiotic bacteria should reside in the intestinal tract for at least two weeks to exert their purported health-promoting properties. The ability of *Bacillus* spores to germinate in the human gut makes it possible that the *in vitro* production of metabolites, peptides, and bacteriocins as well as the release of VB_2 and the conversion of genotoxic compounds into unreactive products could also occur *in vivo*. Additional benefits contributing to the purported probiotic activities of *Bacillus*-based probiotics rely on the interaction of spores with the host immune system: *Bacillus* spores have been proven to stimulate expression of activation markers on lymphocytes, synthesis of membrane bound and secreted IgA, restoration of interferon production, and increase in the percentage of monocytes synthesising TNF-α or IL-1β. All these aspects have contributed to a catalogue of "probiotic" features shared by *Bacillus* spores, although the mechanism by which spores may influence human well-being through the activities they exert during their residence time in the intestinal tract requires further investigation both *in vitro* and *in vivo*.

REFERENCES

Adami, A., and Cavazzoni, V. 1999. Occurrence of selected bacterial groups in the faeces of piglets fed with *Bacillus coagulans* as probiotics. J. Basic Microbiol. *39*, 3-9.

Angioi, A., Zanetti, S., Sanna, A., Delogu, G., and Fadda, G. 1995. Adhesiveness of *Bacillus subtilis* to epithelial cells cultured *in vitro*. Microbial Ecol. Health Dis. *8*, 71-77.

Apostolou, E., Kirjavainen, P.V., Saxelin, M., Rautelin, H., Valtonen, V., Salminen, S.J., Ouwehand, A.C. 2001. Good adhesion properties of probiotics: a potential risk for bacteriemia? FEMS Immunol. Med. Microbiol. *31*, 35-39.

Ash, C., Farrow, J.A.E., Wallbanks, S., and Collins, M.D. 1991. Phylogenetic heterogeneity of the genus *Bacillus* revealed by comparative analysis of small-subunit-ribosomal RNA sequence. Lett. Appl. Microbiol. *13*, 202-206.

Beck, C., and Neckeles, H. 1961. Beneficial effects of administration of *Lactobacillus acidophilus* in diarrheal and other intestinal disorders. Am. J. Gastroenterol. *35*, 522-530.

Beliavskaia, V.A., Igna'ev, G.G., Cherdyntseva, N.V., and Litviakov, N.V. 2001. Adjuvant properties of subalin, a recombinant-interferon producing probiotic. Zh. Mikrobiol. Epidemiol. Immunobiolol. *6*, 77-82.

Bonn, P. 2002. Probiotics reduce the risk of gut infections. Lancet *2*, 16.

Boriello, S.P., Hammes, W.P., Holzapfel, W., Marteau, P., Schrezenmeir, J., Vaara, M., and Valtonen, V. 2003. Safety of probiotics that contain lactobacilli and bifidobacteria. Clin. Infect. Dis. *15*, 775-780.

Bozdogan, B., Galopin, S., Gerbaud, G., Courvalin, P., and Leclercq, R. 2003. Chromosomal *aadD2* encodes an aminoglycoside nucleotidyltransferase in *Bacillus clausii*. Antimicrob. Agents Chemother. *47*, 1343-1346.

Caldini, G., Trotta, F., and Cenci, G. 2002. Inhibition of 4-nitroquinoline-1-oxide genotoxicity by *Bacillus* strains. Res. Microbiol. *153*, 165-171.

Caruso, A., flaminio, G., Fulghera, S., Foresti, I., Balsari, A., and Turano, A. 1993. Expression of activation markers on pheripheral blood lymphocytes following oral administration of *Bacillus subtilis* spores. Int. J. Immunopharmacol. *15*, 87-92.

Casula, G., and Cutting, S.M. 2002. *Bacillus* probiotics: spore germination in the gastrointestinal tract. Appl. Env. Microbiol. *68*, 2344-2352.

Cherdyntseva, N.V., Litviakov, N.V., Smol'ianicov, E.S., Beliavskaia, V.A., and Masycheva, V.I. 1997. Modulation of the antitumor effect of cyclophosphamide by the recombinant probiotic Subalin. Vopr. Onkol. *43*, 313-315.

Ciffo, F. 1984. Determination of the spectrum of antibiotic resistance of the *Bacillus subtilis* strains of Enterogermina. Chemioterapia *1*, 45-52.

Ciffo, F., Decarro, C., Giovannetti, M., and Mazza, P.G. 1987. Gastric resistance of *Bacillus subtilis* spores used in oral bacteriotherapy: *in vitro* studies. Farmaci Terapia, *4*, 163-169.

Coppi, F., Rappuolo, M., Mandressi, A., Bellorofonte, C., and Trinchieri, A. 1985. Results of treatment with *Bacillus subtilis* (Enterogermina) after antibiotic therapy in 95 patients with infection calculosis. Chemioterapia, *4*, 467-470.

Cross, M.L. 2002. Microbes versus microbes: immune signals generated by probiotic lactobacilli and their role in the protection against microbial pathogens. FEMS Immunol. Microbiol. Letters, *34*, 345-353.

Czerucka, D. and Rampal, P. 2002. Experimental effect of *Saccharomyces boulardii* on diarrheal pathogens. Microbes Infect. *4*, 733-739.

Dunne, C., Murphy, L., Flynn, S., O'Mahony, L., O'Halloran, S., Feeney, M., Morrisey, D., Thornton, G., Daly, C. et al., 1999. Probiotics: from myth to reality. Demonstration of functionality in animal models of disease and in human clinical trials. Antonie von Leeuwenhoek, *76*, 279.292.

[FAO/WHO] Food and Agriculture Organization/World Health Organization (2001). Report on a Joint FAO/WHO Expert Consultation on evaluation of health and nutritional properties of probiotics in food including powder milk with live lactic acid bacteria (http: //www.fao.org/es/ESN/food/foodandfood_probio_en.stm)

[FAO/WHO] Food and Agriculture Organization/World Health Organization (2002). Report on a Joint FAO/WHO Working Group on Drafting Guidelines for the evaluation of probiotics in food (ftp: //fp.faohttp: //www.fao.org/es/ESN/Probio/probio.htm)

Finlay, B.B., and Falcow, S. 1997. Common themes in microbial pathogenicity revisited. Microbial. Mol. Biol. Rev. *6*, 136-169.

Fiorini, G., Cimminiello, C., Chianese, R. Visconti, G.P., Cova, D., Uberti, T., and Gibelli, A. 1985. *Bacillus subtilis* selectively stimulates the synthesis of membrane bound and secreted IgA. Chemioterapia, *4*, 310-312.

Fuller, R. 1989. Probiotic in man and animals. J. Appl. Bacteriol. *66*, 365-378.

Fuller, R. 1991. Probiotics in human medicine. Gut, *32*, 439-442.

Furzicova, T.M., Sorokulova, I.B., Serhiichuk, M.H., Sichkar, S.V., and Smirnov, V.V. 2000. The effect of antibiotic preparation and their combination with probiotics on the intestinal microflora of mice. Mickrobiol. J. *62*, 26-35.

Gionchetti, P., Rizzello, F., Venturi, A., Brigidi, P., Matteuzzi, D., Bazzocchi, G., Poggioli, G., Miglioli, M., and Campieri, M. 2000. Oral bacteriotherapy as maintenance treatment in patients with chronic pouchitis: a double-blind, placebo-controlled trial. Gastroenterology, *119*, 305-309.

Green, D.H., Wakely, P.R., Page, A., Barnes, A., Baggicalupi, L., Ricca, E., and Cutting, S. 1999. Characterization of two *Bacillus* probiotics. Appl. Environ. Microbiol. *65*, 4288-4291.

Gorbach, S.L. 2000. Probiotics and gastrointestinal tract. Am. J. Gastroenterol. *95*, 2S-4S.

Gorbach, S.L. 2002. Probiotics in the third millennium. Dig Liver Dis. *34*, S2-7.

Graff, J., Brinch, K., and Madsen, J.L. 2001. Gastrointestinal mean transit times in young and middle-aged healthy subjects. Clin. Physiol. *21*, 253-259.

Grasso, G., Migliaccio, P., Tanganelli, C., Brugo, A.M., and Muscettola, M. 1994. Restorative effect of *Bacillus subtilis* spores on interferon production in aged mice. Ann. N.Y. Acad. Sci. *717*, 198-208.

Green, D.H., Wakely, P.R., Page, A., Barnes, A., Baggicalupi, L., Ricca, E., and Cutting, S. 1999. Characterization of two *Bacillus* probiotics. Appl. Environ. Microbiol. *65*, 4288-4291.

Hentges, D.J. 1992. Gut flora and disease resistance. In Probiotics: The Scientific Basis, R. Fuller, ed. (Chapman and Hall, London), pp. 87-109.

Hisanga, S. 1980. Studies on the germination of genus *Bacillus* in the rabbit and canine intestine. J. Nagoya City Med. Assoc. *30*, 465-469.

Hoa, T.T., Baggicalupi, L., Huxham, A., Smertenko, A., Van, P.H., Ammendola, S., Ricca, E., and Cutting, S. 2000. Characterization of *Bacillus* species used for bacteriotherapy and bacterioprophylaxis of gastrointestinal disorders. Appl. Environ. Microbiol. *66,* 5241-5247.

Hoa, T.N., Duc, L.H., Isticato, R., Baggicalupi, L., Ricca, E., Van, P., and Cutting, S. 2001. Fate and dissemination of *Bacillus subtilis* spores in a murine model. Appl. Env. Microbiol. *67,* 3819-3823.

Hooper, L.V., Wong, M.H., Thelin, A., Hansson, L., Falk, P.G., and Gordon, J.I. 2001. Commensal host-bacterial relationships in the gut. Science, *292,* 1115-1118.

Howrowitz-Wlassova, L.M. and Nowotelnow, N.W. 1932. Ober eine sporegene Milchsaurebacterienart, bacillus sporogenes n. sp. Cent F. Bak, II Abt., *87,* 311.

Hyronimus, B., Le Marrec, C., Hadj Sassi, A., and Deshamps, A. 2000. Acid and bile tolerance of spore forming lactic acid bacteria. Int. J. Food Microbiol. *61,* 193-197.

Isticato, R., Cangiano, G., Tran, H.T., Ciabattini, A., Medaglini, D., Oggioni, M.R., De Felice, M., Pozzi, G. and E. Ricca. 2001. Surface display of recombinant proteins on *Bacillus subtilis* spores. J. Bacteriol. *183,* 6294-6301.

Kniehl, E., Becker, A., and Forster, D.H. 2003. Pseudo-outbreak of toxigenic *Bacillus cereus* isolated from stools of three patients with diarrhoea after oral administration of a probiotic medication. J. Hosp. Infect. *55,* 33-38.

La Ragione, R., Casula, G., Cutting, S.M., and Woodward, M.J. 2001. *Bacillus* spores competitively exclude *Escherichia coli* O78, K80 in poultry. Vet. Microbiol. *79,* 133-142.

Lee, K.H., Jun, K.D., Kim, W-S., and Paik, H.D. 2001. Partial characterization of polyfermenticin SCD, a newly identified bacteriocin of *Bacillus polyfermenticus*. Lett. Appl. Microbiol. *32,* 146-151.

Lilly, D.M., and Stillwell, R.H. 1965. Probiotics: growth-promoting factors produced by microorganisms. Science, *147,* 747-748.

Logan, N.A., Turnbull, P.C.B. 1999. *Bacillus* and recently derived genera. In Murray P.R. Baron E.J. Pfaller MA Tenover FC Volken RH ed. Manual of Clinical Microbioloiy, 7th edition, Wshington DC: ASM Press pp: 357-369.

Karagiannidis, A-C., Kritas, S.K., Boscos, C., Georgoulakis, I.E., and Kyriakis, S.C. 2001. Field evaluation of a bioregulator containing live *Bacillus cereus* spores on health and performance of sows and their litters. J. Vet. Med. Physiol. Pathol. Clin. Med. *48,* 137- 145.

Kuipers, O.P. 1999. Genomics for food biotechnology: prospects of the use of high-throughput technologies for the improvement of food microorganisms. Curr. Opin. Biothec. *10,* 511-516.

Macfarlane, G.T. and Cummings, J.H. 2002. Probiotics, infection and immunity. Curr. Opin. Infect. Dis. *15,* 501-506.

MacKay, A.D., Taylor, M.B., Kibbler, C.C., and Hamilton-Miller, J.M.T. 1999. *Lactobacillus* endocarditis caused by a probiotic organism. Clin. Microbiol. Infect. *5,* 290-292.

Marteau, P.R., Vrese, M.D., Cellier, C.J., and Schrenzenmeir, J. 2001. Protection from gastrointestinal diseases with the use of probiotics. Am. J. Clin. Nutr. *73,* 430S-436S.

Mazza, G. 1994. The use of *Bacillus subtilis* as an antidiarrhoeal microorganism. Boll. Chim. Farm. *113,* 3-18.

Metchnikoff, E. 1907. The prolongation of life. Heinemann, London, UK.

Mock, M. and Fouet, A. 2001. Anthrax .Annu. Rev. Microbiol. *55,* 647-571.

Nielsen, F., Fritze, D., and Priest, G. 1995. Phenetic diversity of alkalophilic *Bacillus* strains: proposal for nine new species. Microbiol. *147,* 1745-1761.

Newman, R. 1915. The treatment of cystitis by intravescical injection of lactic bacillus cultures. Lancet, *14,* 330-332.

Oggioni, M., Ciabattini, A., Cuppone, A.M., and Pozzi, G. 2003. *Bacillus* spores for vaccine delivery. Vaccine *21,* S2/96-S2/101.

Oggioni, M.R., Pozzi, G., Valensin, P.E., Galieni, P., and Bigazzi, C. 1998. Recurrent septicemia in an immunocompromised patient due to probiotic strains of *Bacillus subtilis*. J. Clin. Microbiol. *36,* 325-326.

Opisova, I.G., Sorokulova,I.B., Tereshkina, N.V., and Grigor'eva, L.V. 1998. Safety of bacteria of the genus *Bacillus*, forming the base of some probiotics, Zh. Mikrobiol. Epidemiol. Immunobiol. *6,* 68-70.

Ouwehand, A.C., Kirjavainen, P.V., Shortt, C. and Salminen, S. 1999. Probiotics: mechanisms and established effects. Int. Dairy. J. *9,* 43-52.

O'Sullivan G.C. 2001. Probiotics. Brit. J. Surg. *88,* 161-162.

Parker, R.B. 1974. Probiotics, the other half of the antibiotics story. Animal *29,* 4-8.

Pinchuk, I.V., Bressollier, P., Verneuil, B., Fenet, B., Sorokulova, I.B., Megraud, F., and Urdaci, M. 2001. *In vitro* anti-*Helicobacter pylori* activity of the probiotic strain *Bacillus subtilis* 3 is due to secretion of antibiotics. Antimicrob. Agents Chemother. *45,* 3156-3161.

Rachmilewitz, D., Karmeli, F., Takabayashi, K., Hayashi, T., Leider-Trejo, L., Lee, J., Leoni, L.M., and Raz, E. 2002. Immunostimulatory DNA ameliorates experimental and spontaneous murine colitis. Gastroenterology, *122,* 1428-1441.

Rautio, M., Jousimies-Somer, H., Kauma, H., Pietarinen, I., Saxelin, M., Tynkkynen, S., and Koskela, M. 1999. Liver abscess due to *Lactobacillus rhamnosus* strain indistinguishable from *L. rhamnosus* strain GG. Clin. Infect. Dis. *28,* 1159-1160.

Reid, G. 1999. The scientific basis for probiotic strains of *Lactobacillus*. Appl. Env. Microbiol. *65,* 3763-3766.

Rolfe, R.D. 2000. The role of probiotic cultures in the control of gastrointestinal health. J. Nutr. *130,* 396S-402S.

Salvetti, S., Celandroni, F., Ghelardi, E., Baggiani, A., and Senesi, S. 2003. Rapid determination of vitamin B2 secretion by bacteria growing on solid media, J. Appl. Microbiol. *(in press)*

Sanders, M.E., Morelli, L., Bush, S. 2001. "Lactobacillus sporogenes" is not a *Lactobacillus* probiotic. AMS News, *67,* 385-6.

Sanders, M.E. 2003. Probiotics: consideration for human health. Nutr. Rev. *61,* 91-99.

Schartz, E.D. and Smith, J.J. 1963. Preliminary investigation of viable *Lactobacillus acidophilus* instillation in vaginal pruritus. Clin. Med. *70,* 1120-1122.

Senesi, S., Celandroni, F., Tavanti, A., and Ghelardi, E. 2001. Molecular chararterization and identification of *Bacillus clausii* strains marketed for use in oral bacteriotherapy. Appl. Env. Microbiol. *67,* 834-839.

Settel, E. 1962. *Lactobacillus acidophilus* in the treatment of functional gastrointestinal disorders. Clin. Med. *69,* 700-704.

Smirnov, V.V. and Kosiuk, I.V. 1997. The adhesive properties of bacteria in the genus *Bacillus* – the components of a probiotic. Mikrobiol. Z. *59,* 36-43.

Smirnov, V.V., Reznyk, S.R., and Sorokulova, I.V. 1994. The highly effective biological preparation biosporin. Lik Sprava, 5-6, 133-138.

Smirnov, V.V., Rudenko, A.V., Samgorodskaia, N.V., Sorokulova, I.B., Rezn, S.R., and Sergeichuk, T.M. 1994. Susceptibility to antimicrobial drugs of strain of bacilli used as a basis for various probiotics. Antibiot. Khimioter. *39,* 23-28.

Sorokulova, I.B. 1998. The safety and reactogenicity of the new probiotic subalin for volunteers. Mikrobiol. Z. *60,* 43-46.

Sorokulova, I.G., Kirik, D.L., and Pinchuk, I.V. 1997. Probiotics against *Campylobacter* pathogens. J. Travel. Med. *4,* 167-170.

Spinosa, M.R., Braccini, T., Ricca, E., De Felice, M., Morelli, L., Pozzi, G., and Oggioni, M. 2000. On the fate of ingested spores. Res. Microbiol. *151,* 361-368.

Sullivan, A. and Nord, C.E. 2002. The place of probiotics in human intestinal infections. Antimicrob. Agents. *20:* 313-319.

Takahashi, K., Kaiya, T., and Kawazoe, Y. 1987. Structure-mutagenicity relationship among aminoquinolines, aza-analogues of naphthylamine, and their N-acetyl derivatives. Mutat. Res. *187:* 191-197.

Vollard, E.J. and Clasener, H.A.L. 1994. Colonisation resistance. Atimicrobial Agents Chemother. 38: 409-414.

Chapter 12

Production and Probiotic Effects of Natto

Tomohiro Hosoi and Kan Kiuchi

SUMMARY

This review summarizes the production of natto and probiotic effects of *Bacillus subtilis* (*natto*). Natto is a popular food made by fermenting cooked soybeans with *Bacillus subtilis* (*natto*) in Japan. Natto has a characteristic aroma and stickiness and its quality is affected by the type of soybean and *B. subtilis* (*natto*) strains. Processing methods such as the soaking and steaming of soybeans, inoculation of *B. subtilis* (*natto*) spores, packaging, and fermentation are all important for the highest quality natto. Natto contains numerous nutrients originating from both soybeans as well as from intact cells and metabolites of *B. subtilis* (*natto*) and many of these have physiological activity. *B. subtilis*, *B. subtilis* (*natto*), and natto are considered to have the potential as probiotics. Although *B. subtilis* is not a predominant bacterium in the human intestine, characteristics of the cells such as spore formation, resistance of the spore to oxygen and acidity seem to be desirable for a probiotic. In addition, spores are thought to be able to germinate in the intestine under certain conditions (see Chapter 7). Ingestion of intact *B. subtilis* cells and natto probably increase the *Lactobacillus* spp. and *Bifidobacterium* spp. in the intestine. *B. subtilis* cells induce cytokine responses of human intestinal epithelial-like Caco-2 cells with less cytotoxicity than nonpathogenic *Escherichia coli*, pathogenic *Salmonella enteritidis*, and *Pseudomonas aeruginosa*. In addition, a serine protease, subtilisin, produced by *B. subtilis* degrades soybean allergens and shows fibrinolytic activity. Ingestion of vitamin K_2 (menaquinone-7) helps coagulant activity and prevent osteoporosis. Natto contains phytoestrogens (isoflavones) that originate in soybeans and they seem to have preventive effects on breast and prostate cancer, osteoporosis, menopausal symptoms, and heart disease.

FERMENTED SOYBEAN FOODS IN ASIA

Fermented soybean foods made with *Bacillus subtilis* cells are produced in China, where they are called 'dou chi' (or dauchi). They consist of salted, sweet, and non-salted types. The salted type of dou chi (xian-dou chi) contains 10-20% salt to inhibit their putrefaction by contaminating bacteria. The most typical, sweet, dou chi is called 'tian-dou chi' where it is used as a seasoning for Beijing duck. Nonsalted dou chi has been developed into various kinds of natto (Ito et al., 1996) called 'itohiki-natto' (hereafter abbreviated to natto) in Japan, 'kinema' in Nepal and Myanmar, 'tua nao' in Thailand, and 'chungkuk-jang' in Korea. Natto is produced only with *B. subtilis* (*natto*) (formerly called *B. natto*, see section '*Bacillus subtilis* (*natto*) spores' below).

These fermented soybeans are consumed in a variety of forms. For instance, 'tua nao' is used as a raw ingredient in salads while 'Chungkuk-jang', which contains cayenne peppers and garlic, is used as an ingredient for Korean soup called chige. In Japan, natto is mixed with soy sauce, sliced Welsh onion (similar to stone leeks), mustard, dried seaweed, and/or a raw egg. Seasoned natto are usually eaten with rice in Japan. A looser natto (hikiwari-natto) and a dried type of natto are also used. Before the Second World War, the transportation infrastructure was not well developed in Japan. Hence, dried, hard natto, having a long shelf life, was usually manufactured and consumed. Of the three types of natto, itohiki-natto is the most popular at present in Japan. Hikiwari-natto is also used for preparing sushi. Consumption of natto has increased compared to other soybean foods during the 1990's, generating sales of 160 billion yen in 1996 (Statistics Bureau, Ministry of Public Management, Home Affairs, Post and Telecommunications, Japan, 2001).

INGREDIENTS OF NATTO

The raw materials required for natto production are *B. subtilis* (*natto*) spores called 'natto bacilli', soybeans, and water. *B. subtilis* (*natto*) strains are used to initiate fermentation of steamed whole soybeans. The quality of natto is affected by the quality of the soybeans (see below) and the *B. subtilis* (*natto*) strains but not the water. Normal tap water is sufficient for the production of natto.

Bacillus subtilis (*natto*) spores

In 1913, Dr. S. Sawamura of Imperial University of Tokyo isolated natto bacillus and named it *Bacillus natto* (Sawamura, 1913). The first commercially available natto using a pure culture of *B. natto* was sold through the efforts of Professor J. Hanzawa of Hokkaido Imperial University in 1928 (Federation of Japan Natto Manufactures Cooperative Society, 1971). In 1957, *B. natto* strains were included in and documented as *B. subtilis* strains in the 7th edition of Bergey's Manual of Determinative Bacteriology (Smith and Gordon, 1957). However, only *B. natto* and no other *B. subtilis* strains are used for natto production. This is because natto with the desired characteristics can be produced only by *B. natto* strains (Itaya and Matsui, 1999; Qiu et al., 2003). Using *B. natto* strains gives natto a distinctive aroma, a wrinkly bacterial layer on the surface of soybeans, and the desirable degree of stickiness. When natto of high quality is picked up, it tends to elongate into strings of soybeans and not to lump. In addition, biotin greatly enhances the growth of *B. natto* (Ohsugi and Imanishi, 1985). Japanese scientists, therefore, usually refer to the natto bacillus as *B. subtilis* (*natto*) to distinguish it from other strains of *B. subtilis*. Lyophilized spores of *B. subtilis* (*natto*) or spore suspensions in water are sold as a starter for natto by three companies in Japan. Some natto manufactures have utilized their own *B. subtilis* (*natto*) strains in order to produce natto of the characteristic quality. Contamination of natto products by other microorganism must be avoided, as it affects the quality of natto and may cause food poisoning.

Soybeans

The total volume of soybeans consumed in Japan has been approximately 5 million metric tons over the past ten years. Total consumption of soybeans for food production was 1.01 million metric tons in 2000. Japan imports the majority of soybeans (more than 80%) required for domestic food production. There was a 13 % increase in soybean consumption for natto products during 1991 and 2000, and consumption reached 122 thousand metric tonnes in 2000. Domestic soybeans are considered to be superior to foreign ones for natto production by Japanese natto manufacturers, although the supply of Japanese-grown soybean is insufficient to meet the demands for natto production. Since they are grown in limited domestic regions and production is small, Japanese soybeans command a high price. Considering the low rate of self-sufficiency in soybean, and in order to increase their quality using domestic soybeans, a national initiative has been launched to increase the production of domestic cultivars of soybeans in Japan. To date, some new cultivars of soybeans have been bred in many districts of Japan and have been used in the production of natto (Sohma and Matsukawa, 2000). The names of these soybean cultivars and the locations where they were bred have been compiled and published by the Japanese Ministry of Agriculture, Forestry, and Fisheries.

Natto manufacturers prefer certain kinds of soybean cultivars and their preferences can also vary from region to region because of consumer preferences. Numerous processing tests have been conducted in an attempt to elucidate which cultivars of soybeans are most appropriate for making high-quality natto. Popular soybean varieties for natto production are 'suzuhime' and 'suzumaru,' which are grown in Hokkaido; 'kosuzu,' in Iwate, Miyagi and Akita Prefectures; and 'natto-shoryu,' in Ibaraki Prefecture.

Desirable qualities of soybeans for natto are generally as follows:
1. Extra small or small size
2. Easily washable
3. Yellow surfaces and hila
4. A suitable degree of stickiness when made into natto
5. Sweet taste
6. Slight changes in constituents during storage

Color

Soybeans with a brilliant light yellow or yellow coloured surface and with a light yellow hilum are favored. Natto made from brown or black soybeans has traditionally not been sold well in Japan due to its unappetizing appearance and lack of characteristic aroma. However, natto using black soybeans has recently been sold, and its high content of polyphenols has been emphasized.

Size

Soybean size is classified by diameter into four groups in Japan: extra small refers to less than 5.5 mm in diameter, small ranges from 5.5 mm to no more than 7.3 mm, medium ranges from 7.3 mm to no more than 7.9 mm, and large constitutes a diameter of 7.9 mm or more. Consumers in Kanto and more northern regions of Japan regard the extra small and small soybeans as the most suitable ones for natto. On the other hand, those in Kansai region prefer larger soybeans. Natto processed from small and extra small soybeans tends to have a distinctive natto flavor and a strong taste due to excess fermentation. It was reported that activities of alkaline protease (subtilisin) were higher in natto produced from small soybeans than that from large soybeans (Takahashi et al., 1996; Pero and Sloma, 1993). However, the activities of neutral proteases (metalloprotease) did not differ with the size of soybeans used (Takahashi et al., 1996). This result suggests that degradation of proteins in natto produced from small soybeans progresses more strongly than that from large soybeans during the latter half of the fermentation period when the pH value of the product is increased by metabolites from the *B. subtilis* (*natto*) cells. This may cause the taste and smell of natto from small soybeans to be stronger than those from large soybeans. Natto made from large soybeans has a weak smell of ammonia and shows a low degradation rate of proteins, although its nattto-like taste is still perceivable (Takahashi et al., 1996).

Protein content

B. subtilis (*natto*) cells utilize the proteins, peptides, and amino acids in soybeans for their growth. The kinds and quantities of peptides and amino acids produced by the activities of *B. subtilis* (*natto*) during fermentation affects the flavor appeal of natto. Hence, soybeans with high protein content are also preferred. Using these criteria, 'suzuhime' and 'zizuka' cultivars, with high protein contents, bright color, and polished appearance, are highly regarded for natto production.

Sugar content

In order to produce a natto with a good quality and flavor, it is important that available carbohydrates are supplied to *B. subtilis* (*natto*) and that the hydrolysis of proteins proceeds appropriately during fermentation. Because extra small soybeans tend to have higher sugar content, they are regarded as superior for the production of natto. Since the storage life of natto is determined by the ammonia flavor, the content of ammonia is considered to be the most important quality control characteristic in natto production (Taira et al., 1987). However, it has also been reported that soybeans with high sugar content are not necessarily best; free sugar content is more important than total sugar content for natto processing (Taira, 1989 and 1990). This is because *B. subtilis* (*natto*) can utilize certain saccharides such as sucrose, raffinose, and stachyose but not starch for growth (Kanno et al., 1982).

Small soybeans tend to become softer than medium or large ones, when steamed under identical conditions (Taira et al., 1987). A positive correlation between the firmness of the steamed soybeans and the ammonia nitrogen levels of natto products has been reported (Taira et al., 1987). In this report, it was suggested that in hard, steamed soybeans, hydrolysis by *B. subtilis* (*natto*) cells occurs just under the surface of the soybeans. Degradation and usage of the constituents inside the soybeans by the bacteria will not occur quickly. This means that *B. subtilis* (*natto*) cells can utilize sugars near the surface in a relatively short time, and then protein degradation begins in the early stages of fermentation. This induces a strong ammonia flavor. Hence, it is important to select soybeans with a high sugar content and to steam them until they become soft. A typical steam treatment is 1.5 kg/cm^2 steam pressure for 20 min.

Washing and storage methods

Natto is produced year round and therefore harvested soybeans must be stored before they are processed. Soybeans contaminated with soil are cleaned, washed, and then stored for months in a cool room. It is desirable that they are stored in a refrigerated room at a temperature below 15°C with a relative humidity of about 60% (Hase et al., 1980). If soybeans are stored between 25°C and 35°C, the raffinose in soybean increases and the stachyose decreases (Hase et al., 1980). Germination rates of soybeans are often examined to check their quality. A low germination rate suggests that the soybeans have been preserved under undesirable temperature and humidity conditions (Taira, 1983). There is also a negative correlation between germination rates and the solid matter content of soybeans in soaked water (Taira, 1983) and the quality of natto made with such soybeans is generally not acceptable. Therefore, soybeans with low germination rates are not suitable for natto production. However, germinated soybeans are not used to produce natto.

NATTO PROCESSING

Natto is processed as shown in Figure 1 and individual steps explained below.

Washing and soaking of soybeans

Sieves are used to separate small or extra-small soybeans at the beginning of the process. Contaminant and foreign substances such as plant stalks and leaves, soil and sand are removed. Soybeans are weighed, washed in a screw-wash press, rubbed in a maelstrom flow, polished with burhstone, and finally washed in clean tap water. Soybeans must be softened in water before natto processing, because natto produced from unsoftened soybeans is too hard to eat. Soaking should be performed until the weight of soybeans increases approximately 2.2-2.7 fold (Taira et al., 1983; Taira et al., 1987). To achieve this, soybeans are soaked in tap water at 10°C for about 20 h. It is important to clean the soaking tanks regularly, since these are prone to contamination with lactic acid-producing bacteria that can inhibit the growth of *B. subtilis* (*natto*) and impair fermentation. Some processors soak soybeans overnight by running tap water through the tanks for cooling. However, too much changing of water during soaking or soaking for too long a time will reduce the taste of the natto.

Steaming of soybeans

After soaking, soybeans must be steamed (using steam at 1.5 kg/cm^2 pressure for 20 min) in order to soften the beans further and denature undesirable soybean proteins such as hematoglutenin and trypsin inhibitor (Chang et al., 1987). At the same time, contaminating bacteria are killed. Steaming vats that can hold 60-120 kg of raw soybeans at a time are usually used.

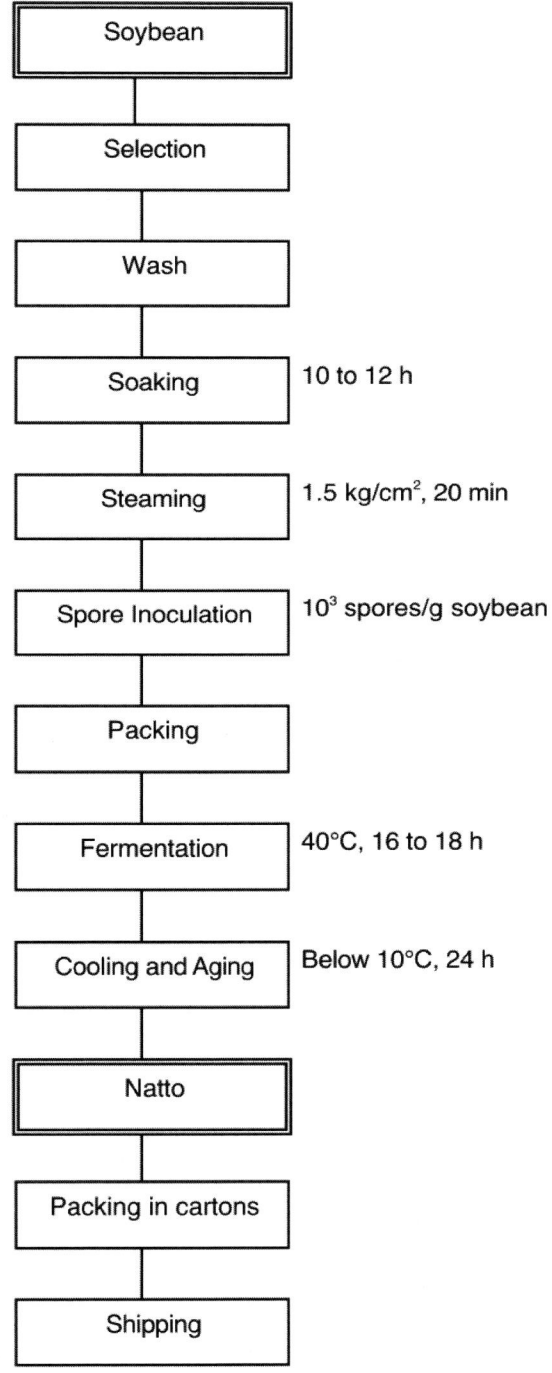

Figure 1. Procedure involved in processing natto.

Inoculation of *Bacillus subtilis* (*natto*) spores

Bacillus subtilis (*natto*) exists both as a vegetative cell and spore form. Spores are more suitable for storage so spore suspensions for natto production are used. Immediately after steaming, while the soybeans are still hot (for example 85°C), the soybeans are tipped from the vat, and sprayed with a *B. subtilis* (*natto*) spore suspension. The concentration of *B. subtilis* (*natto*) spores in the soybean stock should be approximately 10^3 colony forming units (CFU)/g soybean. Inoculation of *B. subtilis* (*natto*) spores at much higher concentrations inhibits the development of the desired stickiness and appearance (Takahashi et al., 1996).

Packaging

Paper cups or polystyrene paper trays are commonly used to package 25 to 150 g of processed soybeans. The most common size holds 50 g of natto (Figures 2A and 2B). The historical changes in packaging methods are described later (section *Changes in packages*). In recent years, almost all filling processes are mechanized and manual assistance is not required. In general, 60 g of steamed soybeans inoculated with *B. subtilis* (*natto*) spores are packed as a 50 g package by an automatic filler, because moisture loss causes an approximate 20% reduction in weight during fermentation. After filling, soybeans inside the packages are covered with a perforated polyethylene film (Figure 2B). Plastic sachets of soy sauce seasoning and/or other condiments such as mustard and freeze-dried sliced Welsh onion can be put on the films. The lid of the package is fixed in place with a bond, using a heat sealer, or a rubber ring.

Fermentation

Fermentation rooms are equipped with ventilators and air conditioners in order to control the humidity of the fermenting soybeans. Conditions in the fermentation room are controlled automatically by a computer for optimum fermentation. Packages are placed horizontally in a box and stacked in the fermentation room. In order to produce natto of high quality, a sufficient oxygen supply, a fine adjustment of room temperature, humidity, and fermentation period are essential (Muramatsu et al., 1995). Typical fermentation conditions and product status are as follows. The room temperature is set at 40°C, and the initial humidity is controlled between 85-90%. The humidity is reduced to 75% between hours 6 and 16 and to 55% between hours 16 and 24. Under these conditions, after 8 h of fermentation, a viscous substance begins to be produced, creating the stickiness of natto. After 10 h of fermentation, the internal temperature of the product begins to increase, and it reaches 48 to 52°C by hour 14. If this optimum temperature of the product is not reached (i.e., if it is too low or too high), the quality of the final natto tends to be lower. The temperature largely affects the stickiness and taste of natto. Thereafter, the temperature of the product is gradually dropped down to 40°C. After 18-20 hours of fermentation, the temperature of the fermentation room is set so that the product temperature decreases to 10°C at hour 24. At this point, the boxed product is transferred from the fermentation room to a refrigerated room where the temperature is below 10°C and is stored for more than 1 day. It is important to control the condition of the fermentation rooms so that the temperature

of the product changes as described above. The temperature near the ceiling of the fermentation rooms can easily increase more than that of the lower part of the room near the floor. This difference causes different fermentation conditions and qualities of products. Therefore, it is necessary to change the fermentation periods slightly among products at different positions of the room and/or to change regularly the position of the products in the room during the fermentation. The optimum conditions for fermentation also vary depending upon the strains of *B. subtilis* (*natto*) and soybeans used.

Packing for Shipment

Automated shrink packaging of two or three polystyrene paper (PSP) trays or cups is the conventional packing method and an automated carton packer has recently been developed specifically for natto. Cartons are stored in a refrigerated room until they are transported to stores by trucks. Since the odor of natto easily intensifies due to the progressive production of ammonia under warm temperature conditions, natto is treated as a perishable product and its shelf life is determined to be one week if refrigerated below 10°C.

Changes in Packages

The type of packaging affects the fermentation process, but in addition, the following factors are important for natto packaging:

Figure 2. Photographs of natto. (A) Natto in polystyrene paper tray package, (B) Natto covered with polyethylene film with numerous small holes, (C) Natto stirred with chopsticks.

1. Ease of manufacture
2. Ease of distribution
3. Easy to serve
4. Disposable and able to be incinerated

Before the Second World War, China jars and packages made of rice straw or wood shavings were often used to pack natto in Japan. Rice straw is an excellent packaging material for natto because it promotes growth of *B. subtilis* (*natto*) by maintaining a warm internal temperature and absorbing extra moisture emitted from the steamed soybeans during the initial stages of fermentation. Adequate moisture is required for good growth of *B. subtilis* (*natto*) cells. If moisture inside the packages is lost during fermentation, the resulting natto tends to be dry, with sticky and strong strings. However, these traditional packaging methods did not meet the requirements of being easy to store or the trends towards spoil-proof and mass-produced food after the 1960s. The development of fillers has also promoted the standardization of natto packaging. Paper tips replaced the rice straw first and then PSP packages were developed. At present, straw packages are used only for souvenirs or special local products. When straw packages are used, natto is packed in thin polyethylene films first with additional packing in steamed and dried straw.

The initial PSP packages containing 100-120 g natto were considered to be too large, because most consumers prefer only to eat an entire cup of natto at one time. Hence, PSP packages and easily printable paper cups containing 40 or 50 g natto are now mainly being used. Smaller cups containing 20 or 30 g natto are also used in the provision for school meals. However, since PSP packaging and paper cups do not regulate temperature and moisture automatically during fermentation, control of the conditions of the fermentation room becomes even more important. Moreover, steamed soybeans mixed with *B. subtilis* (*natto*) spores should be covered with a perforated inner packaging. The purpose of this inner packaging is to assist airflow and maintain the moisture inside the package. It also allows sachets of seasoning to be included in one package. The corners of the perforated lining are pressed down in order to prevent the steamed soybeans from drying out.

A functional plastic membrane that allows for automatic handling as described above was developed as a packaging material for natto (Ishikawa, 1985). This was called the 'respiramy cup', and it did indeed meet many of the requirements for natto packaging. It functioned to maintain the temperature and allowed for the free movement of oxygen, carbon dioxide, ammonia gas, and other gases, but not of water. This membrane, however, was too expensive to use as a packaging material and is not used at the present time.

ASSESSMENT OF QUALITY

Chemical composition
The chemical composition of natto does not differ greatly from that of soybeans except for the vitamin K content (Table 1) (Resources Council, Science and Technology Agency, Japan, 2000). *B. subtilis* (*natto*) produces vitamin K_2 (menaquinone-7)

(Kudo, 1990; Sumi, 1999). Domestic soybeans, U.S.A. soybeans, Chinese soybeans, and natto contain 21, 39, 39, and 2148 µg/100 g dry weight of vitamin K, respectively. Natto is covered with sticky substances produced by *B. subtilis* (*natto*) during fermentation. These sticky substances are composed of polyglutamic acid and levan (fructan) (Claus and Berkeley, 1986). During fermentation, *B. subtilis* (*natto*) also produces proteases and an amylase (Ferrari et al., 1993). Peptides or amino acids produced during fermentation constitute part of the characteristic natto taste. *B. subtilis* (*natto*) utilizes soybean saccharides and produces the characteristic flavor of natto.

Sensory tests
The quality of natto is determined by the sensory tests described in the book *Methods of Natto Research* (Society for Study of Natto and National Food Research Institute, Ministry of Agriculture, Forestry and Fisheries, Japan, 1989). Natto of high quality has the following properties:

1. Even surface layer- Bacterial layer should be formed entirely on the surface of soybeans.
2. No lysis and no glistening- Uneven spots or glistening of soybeans by bacterial cell lysis should not be observed.
3. No damaged soybeans- There should be few split, crushed, and/or peeled soybeans.

Table 1. Compositions of soybeans and itohiki-natto (/100 g dry).				
	Dried and raw soybean			Itohiki-natto
	Japan	USA	China	
Energy (kcal)	476	490	482	493
Protein (g)	40.3	37.4	37.5	40.7
Lipid (g)	21.7	24.6	22.2	24.7
Carbohydrate (g)	32.2	32.6	35.2	29.9
Ash (g)	5.7	5.4	5.0	4.7
Vitamin A Retinol (µg)	0	0	0	0
Carotene (µg)	7	8	10	0
Retinol equivalents (µg)	1	1	1	0
Vitamin D (µg)	0	0	0	0
Vitamin E (mg)	4.1	3.9	4.9	3.0
Vitamin K (µg)	21	39	39	2148*
Vitamin B_1 (mg)	0.95	1.00	0.96	0.17
Vitamin B_2 (mg)	0.34	0.34	0.34	1.38
Niacin (mg)	2.5	2.4	2.5	2.7
Vitamin B_6 (mg)	0.61	0.52	0.67	0.59
Vitamin B_{12} (µg)	0	0	0	Tr
Folate (µg)	263	249	297	30
Pantothenic acid (mg)	1.74	1.69	1.87	8.89
Ascorbic acid (mg)	Tr	Tr	Tr	Tr

Note: Water contents of Japanese, American, and Chinese dried and raw soybeans, and itohiki-natto are 12.5, 11.7, 12.5, and 59.5 (/100 g wet), respectively. Tr: Trace.
*Including menaquinone-7.
Source: Modified from Resources Council, Science and Technology Agency, Standard tables of food composition in Japan, 5th revised edition., Ministry of Finance, Japan, 2000.

4. Bright color- The color of soybean surfaces should be brown or light brown and should not be dark brown or blackish.
5. Good aroma- The product should have a sweet aroma without an ammonia-like, scorched, undesirable or acidic odor.
6. Proper firmness- The soybeans should be properly soft and have a smooth texture.
7. Good taste- Relish, sweet, not bitter, and slightly astringent. The preferred taste is created by amino acids, peptides, and saccharides.
8. Proper stickiness- When stirred with a pair of chopsticks, the viscosity of the natto should increase to form strong strings (Figure 2C).

Taste panels can evaluate natto according to these criteria. If any foreign substances are found or tyrosine crystals are formed on the surface of soybeans, they are noted and the natto given a low quality score.

Changes in consumers' preferences

Traditionally, extra small and small soybeans have been used to make natto. However, in 1999, a natto made from medium-sized soybeans won the contest held by the Federation of Japan Natto Manufactures Cooperative Society. Nattos made with large soybeans are also preferred now by Japanese consumers (Federation of Japan Natto Manufactures Cooperative Society, 2000). Present-day consumers are tending to buy a natto with markedly weaker odors and strings. Natto with traditional characteristics (distinctive odor and strong strings) may no longer have mass appeal. Some manufactures have used their own *B. subtilis* (*natto*) strains to produce less aromatic products.

HEALTH BENEFITS

Bacillus subtilis (*natto*) cells:
Effects on intestinal microflora and feed efficiency

Lactobacillus spp. and *Bifidobacterium* spp. are mainly used as probiotics for humans and animals. However, other bacteria and fungi can also be used as probiotics. For example, *Bacillus* spp., *Enterococcus* spp., *Streptococcus* spp., and *Saccharomyces cerevisiae* seem to have potential as probiotics (Goldin, 1998). In the screening and selection of certain microbial strains as probiotics, phenotype and genotype stability, carbohydrate and protein utilization patterns, safety, acid and bile stability, adhesion characterization, production of antimicrobial substances, antibiotic resistance patterns, immunogenicity, and viability and properties during processing and storage are considered to be important (Tuomola et al., 2001). *B. subtilis* is an aerobic spore-forming bacterium. *B. subtilis* spores are relatively resistant to oxygen, active oxygen species, acid, drying, and heating compared to other bacteria (Smith and Gordon, 1957; Claus and Berkeley, 1986). *B. subtilis* can also grow under O_2-reduced conditions. These characteristics are desirable for potential probiotics. Unfortunately, however, *B. subtilis* is not strongly resistant to bile acid and is not a predominant bacterium in the human intestine (Spinosa et al., 2000).

Several reports have demonstrated the effects of orally administered *B. subtilis* on the intestinal microflora, body weight gain and increased feed efficiency of animals, birds, and shrimp (Ozawa et al., 1981; Jiraphocakul et al., 1990; Maruta et al., 1996a and 1996b; Hosoi et al., 1999; Dalmin et al., 2001; Rengpipat et al., 2003; Vaseeharan and Ramasamy et al., 2003). These results indicate that ingestion of live *B. subtilis* cells or spores can actually improve the intestinal microflora. When weanling piglets were fed a diet including spores of *B. subtilis* (*natto*), the changes in intestinal microflora varied depending upon the region of the intestine examined (Ozawa et al., 1981). In the jejunum, the numbers of *Streptococcus* spp. and *Bifidobacterium* spp. increased, while no difference was observed in the colon when compared with the control diet group. Turkeys fed *B. subtilis* cultures showed significant increases in body weight and cumulative feed efficiency by 2.5% (Jiraphocakul et al., 1990). In chickens given *B. subtilis* the detection rate of the intestinal pathogen *Campylobacter jejuni* decreased in the laboratory portion of the experiment (Maruta et al., 1996a) as well as a decrease in the number of *Salmonella typhilimurium*. In field trials, feeding a *B. subtilis* strain decreased the cell number and/or detection rates of intestinal *Enterobacteriaceae*, *Clostridium perfringens*, and *Campylobacter* sp. When sows and gilts were fed an experimental diet containing *B. subtilis*, the number and/or detection rates of faecal *Bifidobacterium* spp. and *Lactobacillus* spp. increased, but *Streptococcus* spp., *Enterobacteriaceae*, *Clostridium perfringens*, and *Bacteroidaceae* decreased (Maruta et al., 1996b). The diarrheal rate of the piglets up to 10 days old and mortality rate up to 25 days old also decreased. When mice were intubated with intact and autoclaved *B. subtilis* (*natto*) spores for 8 days, only intact spores changed the fecal microflora, and the patterns of the changes differed depending upon the diets fed (Hosoi et al., 1999). Feeding a diet including egg white decreased fecal *Lactobacillus* spp., although the administration of *B. subtilis* (*natto*) spores inhibited the decrease. On the other hand, feeding a diet including casein and administering *B. subtilis* (*natto*) spores increased only *Bacteroidaceae* but not lactobacilli. Administration of *Bacillus* spp. to shrimp showed the reduction of *Vibrio* spp. and improved the growth and survival rates of black tiger shrimp *Penaeus monodon* and the water quality in pond (Dalmin et al., 2001; Rengpipat et al., 2003; Vaseeharan and Ramasamy et al., 2003).

Ingestion of natto (50 g/day) significantly affected the composition and metabolic activity of the human fecal microflora (Terada et al., 1999). Ingestion of natto specifically increased the number of *B. subtilis* (*natto*) and *Bifidobacterium* spp. (the latter increased from 15% of the total bacterial count before consumption to 39% after 14 days consumption), although it decreased the number and detection rates of lecithinase-positive clostridia including *Clostridium perfringens*. The concentrations of fecal acetic acid, total organic acids, and succinic acid increased, while fecal concentrations of indole, ethylphenol, and skatol decreased. Fecal ammonia, cresol, and fecal pH values also decreased.

Mechanisms of the above effects have not been clarified. However, germination and/or some metabolites from *B. subtilis* cells seem to be necessary in order to explain their effects, because it was shown that in mice, administration of autoclaved spores did not affect the fecal microflora (Hosoi et al., 1999). The possibility of germination of *Bacillus* spp. spores in the intestine has been examined. When *B. subtilis* (*natto*) spores were inoculated in the ligated loops of the ileum of dogs, some spores did germinate, but died after germination (Hisanaga et al., 1978). It has been shown that *B. thuringiensis* spores germinate in the gut fluid of the tobacco horn worm (Wilson and Benoit, 1990) and *B. subtilis* spores in the mouse gut (Hoa et al., 2001). In the murine experiments, the number of spores excreted in the feces of the mice was, in some experiments, larger than spores inoculated. An extensive discussion of the fate of *Bacillus* spores in the gastrointestinal tract is given in Chapter 7. In general, when foods are ingested, the pH value in the stomach sometimes increases to 3-4 and spores of *Bacillus* spp. appear to be resistant to such pH values. Some of the *B. subtilis* spores ingested together with other food may be able to sustain their viability and germinate in the upper intestine once the surrounding pH value is neutralized. This then allows them to produce probiotic activity.

Catalase and subtilisin have been proposed as the active molecules responsible for the effects of *B. subtilus* (*natto*) on intestinal microflora (Hosoi et al., 2000). The growth of three strains of lactobacilli co-cultured aerobically with *B. subtilis* (*natto*) has been examined. Addition of *B. subtilis* (*natto*) to the culture medium *in vitro* resulted in an increase in the number of viable cells of all lactobacilli tested. Both catalase and *B. subtilis* (*natto*) enhanced the growth of *Lactobacillus reuteri*, whereas *B. subtilis* (*natto*), but not catalase, enhanced the growth of *Lactobacillus acidophilus*. In a medium containing 0.1 mM hydrogen peroxide, its toxic effect on *L. reuteri* was abolished by catalase or *B. subtilis* (*natto*). Catalase has been reported to exhibit a growth-promoting effect on lactobacilli (Kono and Fridovich, 1983). The viability of lactobacilli readily decreases in the presence of active oxygen species and is partly attributable to the fact that lactobacilli do not generally produce a defense molecule against active oxygen species. However, aerobic bacteria including *B. subtilis* (*natto*) can produce catalase. Vegetative cells of *B. subtilis* primarily produce catalase-1 in the logarithmic phase of growth, and additionally produce catalase-2 and -3 as growth progresses (Loewen and Switala, 1987a, 1987b, and 1988). Intact *B. subtilis* spores contain only catalase-2 in the spore coat. Some other anaerobic bacteria in the intestine such as *Escherichia coli*, *Bacteroides* spp., and *Eubacterium* spp. also produce catalase (Claiborne and Fridovich, 1979; Claiborne et al., 1979; Wilkins et al., 1978; Takamine and Imamura, 1995). It may be important for these bacteria to scavenge hydrogen peroxide in order to colonize the intestine where active oxygen species are produced. The addition of a serine protease, subtilisin, from *Bacillus licheniformis* to the culture medium improved the growth and viability of *L. reuteri* and *L. acidophilus* in the absence of hydrogen peroxide (Hosoi et al., 2000). *B. subtilis* (*natto*) secretes two serine proteases, subtilisin NAT with an isoelectric point (pI) of 8.7 and a 90-kDa serine proteinase (pI 3.9) (Ichishima et al., 1986; Kamata et al., 1989; Nakamura et al., 1992; Kato et al., 1992). Taken

together these results indicate that *B. subtilis* (*natto*) can enhance the growth and/or viability of lactobacilli possibly through production of catalase and subtilisin.

Effects on the immune system

The mucosal immune system is the first line of defense against foreign antigens. The system consists of many kinds of cells, including epithelium, macrophage, dendritic cells, intra-epithelial T lymphocytes, B lymphocytes, and neutrophils (Nagler-Anderson, 2001). The effects of *B. subtilis* cells on some of these intestinal cells have been examined in order to determine whether *B. subtilis* cells possess immunostimulating effects. A recent study questioned whether human intestinal epithelium-like Caco-2 cells can produce cytokines to *B. subtilis* (*natto*) strains, in addition to other pathogenic and non-pathogenic bacteria (Hosoi et al., 2003). It is not clear whether epithelial cells can respond to non-pathogenic strains of bacterial cells. Live cells of non-pathogenic *B. subtilis* or *B. subtilis* (*natto*) strains, as well as non-pathogenic *E. coli*, pathogenic *Salmonella enteritidis*, and *Pseudomonas aeruginosa*, all induced secretion of interleukin 6 (IL-6) and/or IL-8 but not of IL-7 and IL-15. The amounts of cytokines induced by *B. subtilis* (*natto*) cells were dependent upon the strain used. Cytokine induction of epithelial cells may differ between bacterial species or strains regardless of their pathogenicity. Some nonpathogenic bacteria as well as pathogenic ones seem to be able to induce cytokine secretion from normal intestinal epithelial cells when they are orally ingested. Nitrite formation in the macrophage cell line J774.2 in the presence of heat-killed *B. subtilis* cells has been reported (Kengatharan et al., 1998) and peptidoglycan from *B. subtilis* has been shown to induce nitrite formation in macrophages. Lipoteichoic acid from *Staphylococcus aureus* was more potent than lipoteichoic acid from *B. subtilis*.

Translocation of *Bacillus* spp. spores and/or cells has been examined (Spinosa et al., 2000). When mice were fed spores intragastrically, both spores and vegetative cells were detected in the lymph nodes and spleens. In another report, low numbers of spores were detected in mesenteric lymph nodes, livers, and spleens (Hoa et al., 2001). However, the effect of spore ingestion on the increase of bacterial numbers in these organs was not significant. The authors indicated that spores do not appear to translocate substantially across the mucosal surfaces. The effect of *B. subtilis* (*natto*) on T and B lymphocytes in the chicken spleen has also been examined (Inooka et al., 1986). When chickens were fed 10^7 CFU/g of *B. subtilis* (*natto*) spores, the percentages of T and B lymphocytes in the spleens increased compared to those of control groups. Although *B. subtilis* cells are not predominant in the intestine, these results indicate that ingested bacterial cells can affect the mucosal immune system (see Chapter 9 for a more in depth analysis).

The mechanisms causing these effects of *B. subtilis* cells on the immune system remain unclear, although interactions of bacterial cellular components with Toll-like receptors and/or NOD proteins expressed in the host cells seem to be essential. Many bacterial components, including peptidoglycans, lipoproteins, lipoteichoic acid, flagellin and unmethylated CpG dinucleotides in bacterial DNA are known to bind to

Toll-like receptors and induce cytokine responses (Medzhitov, 2001). Binding of peptidoglycans to intracellular NOD proteins participate in apoptosis and inflammation of cells (Inohara and Nuñez, 2003). *B. subtilis* and *B. subtilis* (*natto*) cells would presumably also contain some of these active substances.

Anti-allergy effect of subtilisin

About 15 soybean proteins have been shown to be recognized by sera of soybean-sensitive patients with atopic dermatitis (Ogawa et al., 1991). Three major allergens were identified and designated as *Gly m* Bd 60K, *Gly m* Bd 30K, and *Gly m* Bd 28K, respectively (Ogawa et al., 2000). *Gly m* Bd 60K is the α-subunit of β-conglycinin. *Gly m* Bd 30K is a soybean oil-body-associated glycoprotein homologous to Der p (or f) 1, a major allergen of house dust mite. *Gly m* Bd 28K is a vicilin-like glycoprotein, which is a minor component fractionated into 7S globulin fraction. *Bacillus subtilis* (*natto*) produces a serine protease of subtilisin NAT during its growth (Ichishima et al., 1986; Kamata et al., 1989; Nakamura et al., 1992) andSubtilisin NAT appears to be able to degrade *Gly m* Bd 28K (Bando et al., 1998). Various non-fermented soybean products such as soybean protein isolate (SPI), tofu, kori-dofu, and yuba contain *Gly m* Bd 28K at high concentrations, although fermented soybean products such as natto, soy sauce, and miso do not (Bando et al., 1998).

Recent studies have implied that the intestinal microbial flora are important to the interrelationship between infection and allergy (Wills-Karp et al., 2001). Endogenous flora of the gut stimulate the immune system. It is likely that *Lactobacillus* spp. administration may have preventive and therapeutic effects on allergic diseases (Pelto et al., 1998; Kalliomaki et al., 2001). It has also been shown that allergic children are less colonized with *Lactobacillus* spp. compared to non-allergenic children (Majamaa and Isolauri, 1997). In addition, *Lactobacillus* spp. differentially modulate expression of cytokines and maturation surface markers in murine dendritic cells dependent upon the strains (Christensen et al., 2002). Dendritic cells, present throughout the gastrointestinal tract, play a pivotal immunoregulatory role in the Th1 and Th2 cell balance (Wills-Karp et al., 2001). Th2 cells produce IL-4, IL-13, and IL-5, which coordinately regulate the allergic response. Unfortunately, it has not been examined whether similar effects can be obtained by the administration of *B. subtilis* strains.

Fibrinolytic activity of subtilisin

Circulating platelets and blood-derived proteins (fibrin) are essential for the formation of blood clots, which prevent bleeding long enough for healing to occur (Brass 2001). However, excess coagulation prevents normal physiologic blood flow, which causes thrombotic disorders. Thrombolytic therapy is the most direct means of restoring blood flow (Becker 1991). *Bacillus* spp. produce serine proteases called subtilisins, which are known to have fibrinolytic activity (Fayek and El-Sayed, 1980; Ichishima et al., 1986; Sumi et al., 1987; Kamata et al., 1989; Sumi et al., 1990; Nakamura et al., 1992; Fujita et al., 1993; Urano et al., 2001). Subtilisin NAT produced by *B. subtilis* (*natto*) (also called nattokinase) is 99.5% homologous to subtilisin E (Nakamura

et al., 1992). Oral administration of subtilisin or natto induced mild and frequent enhancement of fibrinolytic activity in plasma and production of tissue plasminogen activator (Sumi et al., 1990). Euglobulin fibrinolytic activity, degradation products from fibrin/fibrinogen, and the amount of tissue plasminogen activator increased by administration of subtilisin. However, whole blood clot lysis time did not significantly decrease. In addition, ingestion of natto decreased euglobulin lysis time and increased euglobulin fibrinolytic activity. The mechanisms of fibrinolytic activity of subtilisin NAT are not fully understood, although subtilisin NAT appears to digest fibrin directly and cleave and inactivate plasminogen activator inhibitor 1 (PAI-1) (Urano et al., 2001). Elevation of PAI-1 in plasma is found in patients with thrombotic disease (Wiman, 1995). High PAI-1 activity is related to impaired fibrinolysis, and its low activity is associated with bleeding disorders (Hamsten et al., 1987; Schleef et al., 1989; Fay et al., 1992).

Role of vitamin K₂ (Menaquinone-7) in the prevention of osteoporosis

Vitamin K has important roles in blood coagulation and bone metabolism (Vermeer et al., 1995; Shearer, 2000). The most abundant forms of vitamin K are phylloquinone (vitamin K_1) and menaquinone (vitamin K_2). Menaquinone refers to a series of vitamin K homologues with polyunsaturated aliphatic side chains of varying length. These compounds are generally referred to as menaquinone-n (MK-n), where n is the number of isoprenoid residues of which the side chain is composed. A marked deficiency of menaquinone-7 and menaquinone-8 has been demonstrated in patients with osteoporotic fractures (Hodges et al., 1991). An effect of vitamin K_2 treatment on spinal bone mineral density (BMD) in postmenopausal women has been demonstrated (Iwamoto et al., 1999). In addition, low serum and bone vitamin K status in patients with longstanding Crohn's disease has been reported (Schoon et al., 2001). Some bacteria including *B. subtilis* (*natto*) produce vitamin K_2 (menaquinone-7) (see section *Assessment of quality*) (Kudo, 1990; Sumi, 1999). Ingestion of 100 g of natto increases vitamin K levels in serum (Kudo, 1990). There were no significant differences in serum vitamin K_1 levels after intake of natto, although vitamin K_2 levels and total vitamin K levels after 24 hr intake significantly increased from 0.90 to 6.21 and from 1.95 to 7.14 ng/mL. A large geographic difference in serum vitamin K_2 levels in postmenopausal women in Tokyo and Hiroshima in Japan and in England has been reported (Kaneki et al., 2001). A trial using a bacterial strain with a high productivity of vitamin K_2 has been performed where vitamin K_2 production was compared to that of a commercial strain (Tsukamoto et al., 2001). The results showed that the selected bacterial strain was capable of producing twice the concentration of vitamin K_2 than the original commercial strain.

Phytoestrogens: Effects on cancer and osteoporosis

It has been hypothesized that ingestion of phytoestrogens (isoflavones) contained in soybeans, play an important role in the prevention of cancer, osteoporosis, menopausal symptoms, and heart disease (Kurzer and Xu, 1997; Messina, 1999). Many epidemiologic studies have examined the relationship between soybean consumption and cancer risk (reviewed in references by Kurzer and Xu, 1997; Messina, 1999), although no significant conclusion has been reached. Early exposure (during the neonatal or prepubertal period of life) to soybean products is thought to be essential for cancer protection (Brown and Lamartiniere, 1995; Lamartiniere et al., 1995; Murrill et al., 1996; Wu et al., 1996). Some soybean products with a high isoflavone content have been developed and sold in Japan by using selected soybeans varieties and specific strains of *B. subtils* (*natto*).

Isoflavones are a subclass of the more ubiquitous flavonoids. The primary isoflavones in soybeans are genistein and daidzein and their respective β-glycosides, genistin and daidzin. Much lower amounts of glycitein and its glycoside, glycitin, are also present in soybeans (Wang and Murphy, 1994a and 1994b). Isoflavones in non-fermented soybean foods appear mostly as the conjugate, whereas in fermented food such as natto and miso, the aglycones dominate (Fukutake et al., 1996). Hence, levels of genistein in the fermented soybean products are higher than those in soybeans and non-fermented foods. The calculated daily intake levels of genistein and genistin ingested from soybeans and related soybean products by the Japanese are 1.5-4.1 and 6.3-8.3 mg/person, respectively.

In order to clarify the mechanisms of the effects of isoflavones, many studies have been performed. Intestinal microbial cells can convert daidzein into several different products, including the isoflavonoids equol, dihydrodaidzein, and *O*-desmethylangolensin (Joannou et al., 1995). It has been proposed that genistein was metabolized to dihydrogenistein and 6'-hydroxy-*O*-desmethylangolensin in humans (Joannou et al., 1995). Estrogenic activities of isoflavones are quite weak when compared to physiologic estrogens (Farmakalidis et al., 1985). However, consumption of soybean foods increased blood isoflavone concentrations to several orders of magnitude higher than those of physiologic estrogens (Xu et al., 1995). Isoflavones can also have antiestrogenic effects when placed in a high-estrogen environment (Folman and Pope, 1966; Mäkela et al., 1995). Genistein inhibits several enzyme activities involved in signal transduction and DNA topoisomerases I and II (Akiyama et al., 1987; Okura et al., 1988; Markovits et al., 1989; Constantinou et al., 1990; Kiguchi et al., 1990; Linassier et al., 1990; Thorburn and Thorburn, 1994). In addition, genistein increases the *in vitro* concentrations of transforming growth factor β (TGF-β), an inhibitor of epithelial cell growth (Sathyamoorthy et al., 1998). This effect is thought to be an important contributor to the anticancer effect. Isoflavones have been shown to bind the estrogen receptors (ERs). Recently, a new receptor, ERβ, has been discovered, to which genistein can bind but the binding is weaker than for natural estrogen, 17β-estradiol (Kuiper et al., 1996). It is possible that this finding, and the fact that ERβ is expressed in differing amounts depending on the types of cells, support the observation that isoflavones have antiosteoporotic but weakly antiuterotrophic effects (Kuiper et al., 1997; Barnes et al., 2000).

ACKNOWLEDGEMENTS

The authors would like to thank CRC Press for permission to reproduce this chapter from "Handbook of Fermented Functional Foods" (2003) Edited by E. R. Farnworth.

REFERENCES

Akiyama, T., Ishida, J., Nakagawa, S., Ogaware, H., and Watanabe, S., Itoh, N., Shibuya, M., and Fukami, Y. (1987). Genistein, a specific inhibitor of tyrosine-specific protein kinases. J. Biol. Chem. 262, 5592-5595.

Bando, N., Tsuji, H., Hiemori, M., Yoshizumi, K., Yamanishi, R., Kimoto, M., and Ogawa, T. (1998). Quantitative analysis of *Gly m* Bd 28K in soybean products by a sandwich enzyme-linked immunosorbent assay. J. Nut. Sci. Viaminol., 44, 655-664.

Barnes, S., Kim, H., Darley-Usmar, V., Patel, R., Xu, J., Boersma, B., and Luo, M. (2000). Beyond ERα and ERβ: Estrogen receptor binding is only part of the isoflavone story. J. Nutr. 130, 656S-657S.

Becker, R.C. (1991). Seminars in thrombosis, thrombolysis and vascular biology. 4. Fibrinolysis. Cadiology 79, 188-210.

Brass, S. (2001). Platelets and proteases. Nature 413, 26-27.

Brown, N.M. and Lamartiniere, C.A. (1995). Xenoestrogens alter mammary gland differentiation and cell proliferation in the rat. Environ. Health Perspect., 103, 708-713.

Chang, C.J., Tanksley, T.D., Knabe, D.A., and Zebrowska, T. (1987). Effects of different heat treatments during processing on nutrient digestibility of soybean meal in growing swine. J. Anim. Sci. 65, 1273-1282.

Christensen, H.R., Frøkiær, H., and Pestka, J.J. (2002). Lactobacilli differentially modulate expression of cytokines and maturation surface markers in murine dendritic cells. J. Immunol. 168, 171-178.

Claiborne, A. and Fridovich, I. (1979). Purification of the *O*-dianisidine peroxidase from *Escherichia coli* B. J. Biol. Chem. 254, 4245-4252.

Claiborne, A., Malinowski, D.P., and Fridovich, I. (1979). Purification and characterization of hydroperoxidase II of *Escherichia coli* B. J. Biol. Chem. 254, 11664-11668.

Claus, D. and Berkeley, R.C.W. (1986). Genus *Bacillus subtilis* Cohn, 1872, 174. In Bergey's manual of determinative bacteriology, Volume 2, P.H.A. Sneath, N.S. Mair, M.E. Sharpe, and J.G. Holt, eds. (Baltimore: Williams and Willkins), pp. 1105-1139.

Constantinou, A., Kiguchi, K., and Huberman, E. (1990). Induction of differentiation and DNA strand breakage in human HL-60 and K-562 leukemia cells by genistein. Cancer Res. 50, 2618-2624.

Dalmin, G., Kathiresan, K., and Purushothaman, A. (2001). Effect of probiotics on bacterial population and health status of shrimp in culture pond ecosystem. Indian J. Exp. Biol. 39, 939-942.

Farmakalidis, E., Hathcock, J.N., and Murphy, P.A. (1985). Oestrogenic potency of genistin and daidzin in mice. Food Chem. Toxicol. 23, 741-745.

Fay, W.P., Shapiro, A.D., Shih, J.L., Schleef, R.R., and Ginsburg, D. (1992). Brief report: complete deficiency of plasminogen-activator inhibitor type 1 due to frame-shift mutation, N. Eng. J. Med. 327, 1729-1733.

Fayek, K.I. and El-Sayed, S.T. (1980). Fibrinolytic activity of enzyme produced by *Bacillus subtilis*. Z. Ernährungswiss 19, 21-23.

Federation of Japan Natto Manufactures Cooperative Society. (1971). In History of natto (Tokyo, Japan).

Federation of Japan Natto Manufactures Cooperative Society. (2000). In Report on the Promoting Enterprise of Cooperative Domestic Soybean Utilization (Tokyo, Japan).

Ferrari, E., Jarnagin, A.S., and Schmidt, B.F. (1993). Commercial production of extracellular enzymes, In *Bacillus subtilis* and other Gram-positive bacteria. A.L. Sonenshein, J.A. Hoch, and R. Losick, Eds. (Washington, D.C.: American Society for Microbiology), pp. 917-937.

Folman, Y. and Pope, G.S. (1966). The interaction in the immature mouse of potent oestrogens with coumesrol, genstein and other utero-vaginotrophic compounds of low potency. J. Endocrinol. 34, 215-225.

Fujita, M., Nomura, K., Hong, K., Ito, Y., Asada, A., and Nishimuro, S. (1993). Purification and characterization of a strong fibrinolytic enzyme (nattokinase) in the vegetable cheese natto, a popular soybean fermented food in Japan. Biochem. Biophys. Res. Commun. 197, 1340-1347.

Fukutake, M., Takahashi, M., Ishida, K., Kawamura, H., Sugimura, T., and Wakabayashi, K. (1996). Quantification of genistein and genistin in soybeans and soybean products. Food Chem. Toxicol. 34, 457-461.

Goldin, B.R. (1998). Health benefits of probiotics. Br. J. Nutr. 80 (Suppl. 2), S203-S207.

Hamsten, A., de Faire, U., Walldius, G., Dahlen, G., Szamosi A., Landou, C., Blomback, M., and Wiman, B. (1987). Plasminogen activator inhibitor in plasma: risk factor for recurrent myocardial infarction. Lancet 2, 3-9.

Hase, S., Yasui, T., Nagashima, S., and Ohta, T. (1980). Changes in quality of soybeans during storage. Changes in sugars and starch content. Rpt. Natl. Food Res. Inst. (Japan) 36, 7-13.

Hisanaga, S., Mase, M., Mizuno, A., Hayakawa, Y., Ohkubo, T., and Hachisuka, Y. (1978). Germination of Bacillus natto spores in the canine intestine. Nippon Saikingaku Zasshi 33, 689-696.

Hoa, T.T., Duc, L.H., Isticato, R., Baccigalupi, L., Ricca, E., Van, P.H., and Cutting, S.M. (2001). Fate and dissemination of Bacillus subtilis spores in a murine model. Appl. Environ. Mcrobiol. 67, 3819-3823.

Hodges, S.J., Pilkington, M.J., Stamp, T.C., Catterall, A., Shearer, M.J., Bitensky, L., and Chayen, J. (1991). Depressed levels of circulating menaquinones in patients with osteoporotic fractures of the spine and femoral neck. Bone 12, 387-389.

Hosoi, T., Ametani, A., Kiuchi, K., and Kaminogawa, S. (1999). Changes in fecal microflora induced by intubation of mice with Bacillus subtilis (natto) spores are dependent upon dietary components. Can. J. Microbiol. 45, 59-66.

Hosoi, T., Ametani, A., Kiuchi, K., and Kaminogawa, S. (2000). Improved growth and viability of lactobacilli in the presence of B. subtilis (natto), catalase, or subtilisin. Can. J. Microbiol. 46, 892-897.

Hosoi, T., Hirose, R., Saegusa, S., Ametani, A., Kiuchi, K., and Kaminogawa, S. (2003). Cytokine responses of human intestinal epithelial-like Caco-2 cells to the non-pathogenic bacterium Bacillus subtilis (natto). Int. J. Food Microbiol. 82, 255-264.

Ichishima, E., Takada, Y., Taira, K., and Takeuchi, M. (1986). Specificities of extracellular and ribosomal serine proteinases from Bacillus natto, a food microorganism. Biochim. Biophys. Acta. 869, 178-184.

Inohara, N. and Nuñez, G. (2003). NODs: intracellular proteins involved in inflammation and apoptosis. Nat. Rev. Immunol. 3, 371-382.

Inooka, S., Uehara, S., and Kimura, M. (1986). The effect of Bacillus natto on the T and B lymphocytes from spleens of feeding chickens Poult. Sci. 65, 1217-1219.

Ishikawa, T. (1985). Utility Model: S60-150035.

Itaya, M. and Matsui, K. (1999). Conversion of Bacillus subtilis 168: natto producing Bacillus subtilis with mosaic genomes. Biosci. Biotchnol. Biochem. 63, 2034-2037.

Ito, H., Tong, J., and Li, Y. (1996). Chinese dauchi, from itohiki-natto to non-mashed miso, Miso Sci. Technol. 44, 224-250.

Iwamoto, I., Kosha, S., Noguchi, S., Murakami, M., Fujino, T., Douchi, T., and Nagata, Y. (1999). A longitudinal study of the effect of vitamin K2 on bone mineral density in postmenopausal women a comparative study with vitamin D3 and estrogen-progestin therapy. Maturitas 31, 161-164.

Joannou, G.E., Kelly, G.E., Reeder, A.Y., Waring, M., and Nelson, C. (1995). A urinary profile study of dietary phytoestrogens. The identification and mode of metabolism of new isoflavonoids. J. Steroid Biochem. Mol. Biol. 54, 167-184.

Jiraphocakul, S., Sullivan, T.W., and Shahani, K.M. (1990). Influence of a dried Bacillus subtilis culture and antibiotics on performance and intestinal microflora in turkeys. Poult. Sci. 69, 1966-1973.

Kamata, H., Yamagata, Y., Nakamura, T., Nakajima, T., Oda, K., Murano, S., and Ichishima, E. (1989). Characterization of the complex between α2-macroglobulin and a serine proteinase from Bacillus natto. Agric. Biol. Chem. 53, 2695-2702.

Kalliomaki, M., Salminen, S., Arvilommi, H., Kero, P., Koskinen, P., and Isolauri, E. (2001). Probiotics in primary prevention of atopic disease: a randomized placebo controlled trial. Lancet 357, 1076-1079.

Kaneki, M., Hedges, S.J., Hosoi, T., Fujiwara, S., Lyons, A., Crean, S.J., Ishida, N., Nakagawa, M., Takechi, M., Sano, Y., Mizuno, Y., Hoshino, S., Miyao, M., Inoue, S., Horiki, K., Shiraki, M., Ouchi, Y., and Orimo, H. (2001). Japanese fermented soybean food as the major determinant of the large geographic difference in circulating levels of vitamin K2: possible implications for hip-fracture risk. Nutrition 17, 315-321.

Kanno, A., Takamatsu, H., Takano, N., and Akimoto, T. (1982). Change of saccharides in soybeans during manufacturing of natto. Nippon Shokuhin Kogyo Gakkaishi 29, 105-110.

Kato, T., Yamagata, Y., Arai, T., and Ichishima, E. (1992). Purification of a new extracellular 90-kDa serine proteinase with isoelectric point of 3.9 from Bacillus subtilis (natto) and elucidation of its distinct mode of action. Biosci. Biotech. Biochem. 56, 1166-1168.

Kengatharan, K.M., De Kimpe, S., Robson, C., Foster, S.J., and Thiemermann, C. (1998). Mechanism of Gram-positive shock: identification of peptidoglycan and lipoteichoic acid moieties essential in the induction of nitric oxide synthase, shock, and multiple organ failure. J. Exp. Med. 188, 305-315.

Kiguchi, K., Constantinou, A., and Huberman, E. (1990). Genistein induced cell differentiation and protein-linked DNA strand breakage in human melanoma cells. Cancer Commun. 2, 271-278.

Kono, Y. and Fridovich, I. (1983). Functional significance of manganese catalase in Lactobacillus plantarum. J. Bacteriol. 155, 742-746.

Kudo, T. (1990). Warfarin antagonism of natto and increase in serum vitamin K by intake of natto. Artery 17, 189-201.

Kuiper, G.G.J.M., Enmark, E., Pelto-Huikko, M., Nilsson, S., and Gustafsson, J.-Å. (1996). Cloning of a novel receptor expressed in rat prostate and ovary. Proc. Natl. Acad. Sci. U.S.A. 93, 5925-5930.

Kuiper, G.G.J.M., Grandien, K., Enmark, E., Haggblad, J., Nilsson, S., and Gustafsson, J.-Å. (1997). Comparison of the ligand binding specificity and transcript tissue distribution of estrogen receptors alpha and beta. Endocrinology 138, 863-870.

Kurzer, M.S. and Xu, X. (1997). Dietary phytoestrogens. Ann. Rev. Nutr. 17, 353-381.

Lamartiniere, C.A., Moore, J.B., Brown, N.M., Thompson, R., Hardin, M.J., and Barnes, S. (1995). Genistein suppresses mammary cancer in rats. Carcinogenesis 16, 2833-2840.

Linassier, C., Pierre, M., Le Pecq, J.B., and Pierre, J. (1990). Mechanisms of action in NIH-3T3 cells of genistein, an inhibitor of EGF receptor tyrosine kinase activity. Biochem. Pharmacol. 39, 187-193.

Loewen, P.C. and Switala, J. (1987a). Multiple catalases in Bacillus subtilis. J. Bacteriol. 169, 3601-3607.

Loewen, P.C. and Switala, J. (1987b). Purification and characterization of catalase-1 from Bacillus subtilis. Biochem. Cell Biol. 65, 939-947.

Loewen, P.C. and Switala, J. (1988). Purification and characterization of spore-specific catalase-2 from Bacillus subtilis. Biochem. Cell Biol. 66, 707-714.

Majamaa, H. and Isolauri, E. (1997). Probiotics: a novel approach in the management of food allergy. J. Allerg Clin. Immunol. 99, 179-185.

Mäkela, S.I., Pylkkänen, L.H., Santti, R.S.S., and Adlercreutz, H. (1995). Dietary soybean may be antiestrogenic in male mice. J. Nutr. 125, 437-445.

Markovits, J., Linassier, C., Fosse, P., Couprie, J., Jacquemin-Sablon, A., Saucier, J.M., Le Pecq, J.B., and Larsen, A.K. (1989). Inhibitory effects of the tyrosine kinase inhibitor genistein on mammalian DNA topoisomerase II. Cancer Res. 49, 5111-5117.

Maruta, K., Miyazaki, H., Masuda, S., Takahashi, M., Marubashi, T., Tadano, Y., and Takahashi, H. (1996a). Exclusion of intestinal pathogens by continuous feeding with Bacillus subtilis C-3102 and its influence on the intestinal microflora in broilers. Anim. Sci. Technol. 67, 273-280.

Maruta, K., Miyazaki, H., Tadano, Y., Masuda, S., Suzuki, A., Takahashi, H., and Takahashi, M. (1996b). Effects of Bacillus subtilis C-3102 intake on fecal flora of sows and on diarrhea and mortality rate of their piglets. Anim. Sci. Technol. 67, 403-409.

Messina, M. J. (1999). Legumes and soybeans: overview of their nutritional profiles and health effects. Am. J. Clin. Nutr. 70 (Suppl.), 439S-450S.

Medzhitov, R. (2001). Toll-like receptors and innate immunity. Nat. Rev. Immunol. 1, 135-145.

Muramatsu, K., Kanai, S., Kimura, N., Miura, N., Yoshida, K., and Kiuchi, K. (1995). Production of natto with high elastase activity. Nippon Shokuhin Kogyo Gakkaishi 42, 575-582.

Murrill, W.B., Brown, N.M., Zhang, J.X., Manzolillo, P.A., Barnes, S., and Lamartiniere, C.A. (1996). Prepubertal genistein exposure suppresses mammary cancer and enhances gland differentiation in rats. Carcinogenesis 17, 1451-1457.

Nagler-Anderson, C. (2001). Man the barrier! Strategic defences in the intestinal mucosa. Nat. Rev. Immunol. 1, 59-67.

Nakamura, T., Yamagata, Y., and Ichishima, E. (1992). Nucleotide sequence of the subtilisin NAT gene, aprN, of Bacillus subtilis (natto). Biosci. Biotech. Biochem. 56, 1869-1871.

Ogawa, A., Samoto, M., and Takahashi, K. (2000). Soybean allergen and hypoallergenic soybean products. J. Nutr. Sci. Vitaminol. 46, 271-279.

Ogawa, T., Bando, N., Tsuji, H., Okajima, H., Nishikawa, K., and Sasaoka, K. (1991). Investigation of the IgE-binding proteins in soybeans by immunoblotting with the sera of the soybean-sensitive patients with atopic dermatitis. J. Nutr. Sci. Vitaminol. 37, 555-565.

Ohsugi, M. and Imanishi, Y. (1985). Microbiological activity of biotin-bitamers. J. Nutr. Sci. Vitaminol. *31*, 563-572.

Okura, A., Arakawa, H., Oka, H., Yoshinari, T., and Monden, Y. (1988). Effect of genistein on topoisomerase activity and on the growth of [Val 12]Ha-ras-transformed NIH 3T3 cells. Biochem. Biophys. Res. Commun. *157*, 183-189.

Ozawa, K., Yokota, H., Kimura, M., and Mitsuoka, T. (1981). Effects of administration of *Bacillus subtilis* strain BN on intestinal flora of weanling piglets. Jpn. J. Vet. Sci. *43*, 771-775.

Pelto, L., Isolauri, E., Lilius, E.M., Nuutila, J., and Salminen, S. (1998). Probiotic bacteria down-regulate the milk-induced inflammatory response in milk-hypersensitive subjects but have an immunostimulatory effect in healthy subjects. Clin. Exp. Allergy *28*, 1474-1479.

Pero, J. and Sloma, A. (1993). Proteases. In *Bacillus subtilis* and other Gram-positive bacteria, A.L. Sonenshein, J.A. Hoch, and R. Losick, Eds. (Washington, D.C.: American Society for Microbiology), pp. 939-952.

Qiu, D., Oshima, H., Ohashi, Y., and Itaya, M. (2003). Construction of physical maps of *Bacillus subtilis* (*natto*) strains. Nucleic Acids Res. Suppl. 207-8.

Rengpipat, S., Tunyanun, A., Fast, A.W., Piyatiratitivorakul, S., and Menasveta P. (2003). Enhanced growth and resistance to *Vibrio* challenge in pond-reared black tiger shrimp *Penaeus monodon* fed a *Bacillus* probiotic. Dis. Aquat. Organ. *55*, 169-173.

Resources Council, Science and Technology Agency, Japan. (2000). Standard tables of food composition in Japan, 5th revised ed. (Tokyo: Printing Bureau, Ministry of Finance).

Sathyamoorthy, N., Gilsdorf, J.S., and Wang, T.T. (1998). Differential effect of genistein on transforming growth factor beta 1 expression in normal and malignant mammary epithelial cells. Anticancer Res. *18*, 2449-2453.

Sawamura, S. (1913). On *Bacillus natto*. Bulletin of Agricultural College, Tokyo *5*, 107-109.

Schleef, R.R., Higgins, D.L., Pillemer, E., and Levitt, L.J. (1989). Bleeding diathesis due to decreased functional activity of type 1 plasminogen activator inhibitor. J. Clin. Invest. *83*, 1747-1752.

Schoon, E.J., Muller, M.C., Vermeer, C., Schurgers, L.J., Brummer, R.J., and Stockbrugger, R.W. (2001). Low serum and bone vitamin K status in patients with longstanding Crohn's disease: another pathogenetic factor of osteoporosis in Crohn's disease? Gut *48*, 473-477.

Shearer, M.J. (2000). Role of vitamin K and Gla proteins in the pathophysiology of osteoporosis and vascular calcification. Curr. Opin. Clin. Nutr. Metab. Care *3*, 433-438.

Smith, N.R. and Gordon, E. (1957). Genus *Bacillus subtilis* Cohn, 1872. In Bergey's manual of determinative bacteriology, 7th ed., R.S. Breed, E.G.D. Murray, and N.R. Smith, Eds. (Baltimore: Williams and Willkins), pp. 620-622.

Society for Study of Natto and National Food Research Institute, Ministry of Agriculture, Forestry and Fisheries. (1989). Methods of Natto Research (Tokyo: Kohrin).

Sohma, S. and Matsukawa, I. (2000). Soybean. In Legume Encyclopedia, T. Watanabe ed. (Tokyo: Saiwai Shobo), pp. 47-59.

Spinosa, M.R., Braccini, T., Ricca, E., De Felice, M., Morelli, L., Pozzi, G., and Oggioni, M.R. (2000). On the fate of ingested *Bacillus* spores. Res. Microbiol. *151*, 361-368.

Statistics Bureau, Ministry of Public Management, Home Affairs, Post and Telecommunications, Japan. (2001). Annual report on the family income and expenditure survey 2000 (Tokyo: Japan Statistical Association).

Sumi, H. (1999). Determination of the vitamin K (menaquinone-7) content in fermented soybean natto and in the plasma of natto-ingesting subjects. J. Home Econ. Jpn. *50*, 309-312.

Sumi, H., Hamada, H., Nakanishi, K., and Hiratani, H. (1990). Enhancement of the fibrinolytic activity in plasma by oral administration of nattokinase. Acta. Haematol. *84*, 139-143.

Sumi, H., Hamada, H., Tsushima, H., Mihara, H., and Muraki, H. (1987). A novel fibrinolytic enzyme (nattokinase) in the vegetable cheese Natto; a typical and popular soybean food in the Japanese diet. Experientia *43*, 1110-1111.

Taira, H. (1983). Quality of soybeans seeds grown in Japan. Part 3. Correlation between physical property, chemical component, and suitability for food processing pairs. Rpt. Natl. Food Res. Inst. (Japan) *42*, 27-39.

Taira, H. (1989). Quality and suitability of domestic soybeans for food processing. Nogyo Gijutsu *44*, 385-391.

Taira, H. (1990). Quality of soybeans for processed foods in Japan. Jpn. Agr. Res. Quart. *24*, 224-230.

Taira, H., Suzuki, N., Tsukamoto, C., Kainuma, Y., Tanaka, H., and Saito, M. (1987). Quality of soybeans seeds grown in Japan. Part 15. Suitability for natto processing of small seed cultivars and quality of the natto. Rpt. Natl. Food Res. Inst. (Japan) *51*, 48-58.

Taira, H., Takahashi, H., Okano, H., and Nagashima, S. (1983). Quality of soybeans seeds grown in Japan. Part 4. Suitability of soybean varieties grown in Ibaraki Prefecture for natto production. Rpt. Natl. Food Res. Inst. (Japan) *43*, 62-71.

Takahashi, J., Furuguchi, K., Miyama, K., and Kikuchi, K. (1996). Effect of inoculation size on quality of natto. Rpt. Food Res. Inst., Tochigi Prefecture *10*, 17-18.

Takamine, F. and Imamura, T. (1995). Isolation and characterization of bile acid 7-dehydroxylating bacteria from human feces. Microbiol. Immunol. *39*, 11-18.

Terada, A., Yamamoto, M., and Yoshimura, E. (1999). Effect of the fermented soybean product "Natto" on the composition and metabolic activity of the human fecal flora. Jpn. J. Food Microbiol. *16*, 221-230.

Thorburn, J. and Thorburn, T. (1994). The tyrosine kinase inhibitor, genistein, prevents α-adrenergic-induced cardiac muscle cell hypertrophy by inhibiting activation of the Ras-MAP kinase signaling pathway. Biochem. Biophys. Res. Commun. *202*, 1586-1591.

Tsukamoto, Y., Kasai, M., and Kakuda, H. (2001). Construction of a *Bacillus subtilis* (*natto*) with high productivity of vitamin K_2 (menaquinone-7) by analog resistance. Biosci. Biotechnol. Biochem. *65*, 2007-2015.

Tuomola, E., Crittenden, R., Playne, M., Isolauri, E., and Salminen, S. (2001). Quality assurance criteria for probiotic bacteria. Am. J. Clin. Nutr. *73* (Suppl.), 393S-398S.

Urano, T., Ihara, H, Umemura, K., Suzuki, Y., Oike, M., Akita, S., Tsukamoto, Y., Suzuki, I., and Takada, A. (2001). The profibrinolytic enzyme subtilisin NAT purified from *Bacillus subtilis* cleaves and inactivates plasminogen activator inhibitor type 1. J. Biol. Chem. *276*, 24690-24696.

Vaseeharan, B. and Ramasamy, P. (2003). Control of pathogenic *Vibrio* spp. by *Bacillus subtilis* BT23, a possible probiotic treatment for black tiger shrimp *Penaeus monodon*. Lett. Appl. Microbiol. *36*, 83-87.

Vermeer, C., Jie, K. -S.G., and Knapen, M.H.J. (1995). Role of vitamin K in bone metabolism. Ann. Rev. Nutr. *15*, 1-22.

Wang, H.-J. and Murphy, P.A. (1994a). Isoflavone composition of American and Japanese soybeans in Iowa: effects of variety, crop year, and location. J. Agric. Food Chem. *42*, 1674-1677.

Wang, H.-J. and Murphy, P.A. (1994b). Isoflavone content in commercial soybean foods. J. Agric. Food Chem. *42*, 1666-1673.

Wilkins, T.D., Wagner, D.L., Vertri, B.J., Jr., and Gregory, E.M. (1978). Factors affecting production of catalase by *Bacteroides*. J. Clin. Microbiol. *8*, 553-557.

Wills-Karp, M., Santeliz, J., and Karp, C.L. (2001). The germless theory of allergic disease: revisiting the hygiene hypothesis. Nat. Rev. Immunol. *1*, 69-75.

Wilson, G.R. and Benoit, T.G. (1990). Activation and germination of *Bacillus thuringiensis* spores in *Manduca sexta* larval gut fluid. J. Invertebr. Pathol. *56*, 233-236.

Wiman, B. (1995). Plasminogen activator inhibitor 1 (PAI-1) in plasma: its role in thrombotic disease. Thromb. Haemost. *74*, 71-76.

Wu, A.H., Ziegler, R.G., Horn-Ross, P.L., Nomura, A.M., West, D.W., Kolonel, L.N., Rosenthal, J.F., and Hoover, R.N. (1996). Tofu and risk of breast cancer in Asian-Americans. Cancer Epidemiol. Biomerkers Prev. *5*, 901-906.

Xu, X., Harris, K.S., Wang, H.-J., Murphy, P.A., and Hendrich, S. (1995). Bioavailability of soybean isoflavones depends upon gut microflora in women. J. Nutr. *125*, 2307-2315.

Chapter 13

Spore Probiotics as Animal Feed Supplements

Stephen T. Cartman and Roberto M. La Ragione

SUMMARY

Concern over the emergence of bacterial pathogens with multiple antibiotic resistance has led the European Commission to restrict the use of antibiotic feed supplements on European farms, with a complete ban proposed for 2006. Therefore, alternative approaches are required to maximise livestock performance and reduce morbidity and mortality rates through controlling infectious enteric disease. Probiotics are a promising alternative and currently there are six bacterial spore-based products available for veterinary application. However, only two of these are approved for use within the European Union. The remaining products are deemed to be unsafe due to the bacterial strains concerned having either donor potential in antibiotic resistance gene transfer, or potential for producing toxins. Recent studies have demonstrated benefits to both animal health and performance following supplementation of feed with bacterial spores. Furthermore, the intrinsic resistance of bacterial spores to adverse environmental conditions means they are ideally suited to long term storage and distribution as a commercial product. However, as is the case for probiosis in general, evidence explaining spore probiotic functional mechanisms of action is extremely limited at present. Ongoing research aims to identify these as well as desirable traits of effective probiotic bacteria so that spore-based feed supplements may be optimised for veterinary use and become an effective alternative to antibiotic feed supplements.

INTRODUCTION

Domestic animals may benefit from feed supplements through two broad means; increased resistance to infectious disease and enhanced performance characteristics. Since the 1940's, antibiotic feed supplements have been used extensively. However, the recent emergence of clinically important bacteria harbouring multiple antibiotic resistance, in both human and veterinary medicine, has increased pressure to reduce widespread and liberal antibiotic usage.

Within Europe, in 1997, farming was the second largest consumer of antibiotics after the medical profession, with approximately one-third of all antibiotics used being in the form of feed supplements (Anon., 1997). For ethical reasons, therapeutic antibiotic applications (accounting for the remaining two-thirds of antibiotic usage on European farms) are unlikely to be subject to strict regulation (although greater vigilance may be exercised when diagnosing disease and prescribing antibiotics). Therefore, reducing the use of antibiotics as feed supplements has been a primary target for control. Following a 1999 assessment (SCAN, 1999a), four antibiotics (bacitracin,

spiramycin, tylosin and virginiamycin) were banned for use as feed supplements in the European Union (EU) due to their direct use, or relatedness to antibiotics used in human medicine. Consequently, only four antibiotics remain licensed for use as feed supplements in Europe (bambermycin, avilamycin, salinomycin and monensin), none of which have been used therapeutically in humans to date. However, a complete ban on the use of antibiotic feed supplements is due to take effect in 2006.

The benefits conferred by antibiotic feed supplements diminish as biosecurity levels improve. Therefore, in the post-antibiotic feed supplement era, effective farm management will be paramount, requiring strict policies on cleaning and disinfection procedures, animal housing quality and animal husbandry. The importance of vaccination programmes in protecting against infectious disease may also become more pronounced. However, it is unrealistic to rely on these measures alone to maintain cost efficiency and public health safety standards of domestic livestock operations. Therefore, research for alternative feed supplements is intensifying. Approaches under investigation include prebiotics, probiotics and synbiotics. Prebiotics are "non-digestible food ingredients which beneficially affect the host by selectively stimulating the growth, activity, or both, of one or a limited number of bacterial species present in the colon" (Gibson and Roberfroid, 1995). Probiotics are "live microbial feed supplements which beneficially affect the host by improving intestinal balance" (Fuller, 1989). Lastly, synbiotics combine pre- and probiotic application in a single product.

Although the term "spore probiotics" potentially includes bacterial preparations of *Bacillus* spp., *Sporolactobacillus* spp., *Brevibacillus* spp. and *Clostridium* spp., as well as fungal spore preparations, the vast majority of work conducted in this area is concerned with the application of non-pathogenic *Bacillus* spp. as livestock probiotic feed supplements. Accordingly, the discussion in this chapter will focus on *Bacillus* spp. and the term "spore probiotics" refers to *Bacillus* spore preparations. However, many of the principles discussed are relevant to probiotic supplementation of animal feed in general.

CURRENT PERSPECTIVE
Enhancing resistance to infectious disease

Milner and Shaffer (1952) were the first to document an increased natural resistance to *Salmonella* infection with age in poultry. Nurmi and Rantala (1973) formalised this observation by proposing the widely accepted "competitive exclusion" (CE) concept, which attributes this increased resistance against infectious enteric disease to the native gastrointestinal (GI) microflora. Accordingly, establishing this protective effect requires oral intake of viable bacteria, especially anaerobes (Fuller, 1999; Schneitz and Mead, 2000). Proposed modes of action for CE agents include competition for host mucosal receptor sites, secretion of bacteriocin-like compounds, production of inhibitory fermentative by-products such as volatile fatty acids, competition for essential nutrients and stimulation of host immune functions. However, scientific evidence in support of one or more of these hypotheses is scarce. In reality, it is likely that a combination of factors are important to CE and these may subtly differ depending upon the bacterial and host interactions concerned.

Interestingly, except for immune stimulation, none of the hypotheses proposed as CE mechanisms of action require direct stimulation of host functions. However studies have reported enhanced animal health or performance due to probiotic feed supplementation (Cavazzoni et al., 1998; Santoso et al., 1995). This implies that positive stimulation of host functions is a feature and indeed, a desirable objective of successful probiotic application.

Enhancing performance characteristics

Probiotic feed supplements may potentially enhance feed conversion efficiency, weight gain rate, milk quality, milk yield, egg quality and egg yield. However, there are few studies which have investigated possible mechanisms of action. The current lack of understanding regarding the multitude and complexity of factors involved in probiosis may explain why some studies have noted positive effects on animal performance as a result of probiotic feed supplementation (Cavazzoni et al., 1998; Santoso et al., 1995) while others have not (Jiraphocakul et al., 1989; Garza-Cazares et al., 2001). This illustrates the need for further research so that effective probiotics may be developed and administered such that animal performance can be consistently and reproducibly enhanced.

WHY USE *BACILLUS* BASED PROBIOTIC PREPARATIONS AS ANIMAL FEED SUPPLEMENTS?

To date, the CE concept has been developed through poultry studies in the main. The fundamental concept of the indigenous gut microflora protecting against infectious enteric disease has been demonstrated on a number of occasions. Newly hatched chicks have been dosed orally with microbial preparations originating from adult bird GI contents or faecal material, and subsequently challenged with *Salmonella* or other enteric foodborne pathogens (reviewed by Mead, 2000). However, despite this efficacy in protecting against infectious enteric disease, undefined microbial preparations such as these are unsuitable for commercial application due to the potential for unintentional pathogen transfer. Such an incident may cause animal morbidity and mortality, and also presents a public health risk because zoonotic pathogens may be transferred into the human food chain.

Mixed cultures of defined microorganisms, which have included *Bacillus* spp., have been shown to be beneficial to animals (Fukushima and Nakano, 1996; Endo et al., 1999;

Fukushima et al., 1999). Use of multi-strain probiotics can be justified by the claim that they are broad spectrum preparations, effective in several different target species, promoting both health and growth rates in animals (Fuller, 1999). Yet, from a commercial perspective, it is likely that cost-effective manufacture of such preparations, consistent product quality and consistent field performance will be harder objectives to achieve than for defined monoculture probiotics. However, despite successes, defined cultures have not been as effective CE agents as undefined preparations (Starvic and D'Aoust, 1993).

Nevertheless, studies have clearly demonstrated the ability of *Bacillus* spp. to enhance host resistance against enteric infections and promote animal health (Maruta et al., 1996; Kyriakis et al., 1999; La Ragione et al., 2001; La Ragione and Woodward, 2003). Furthermore, the resistance of bacterial spores to adverse environmental conditions means that spore probiotics have longevity and will retain viability, even without refrigeration, which are essential aspects of shelf-life for a live microbial product. This intrinsic property of bacterial spores also favours cost-effective distribution, negating the need for transport and storage under closely controlled conditions. Therefore, although the mechanisms of action are still to be elucidated before spore probiotic application can be optimised, such products are an attractive commercial prospect.

At present, a degree of caution may be required however. Although the Scientific Committee on Animal Nutrition (SCAN; a group of independent scientists appointed by the European Commission) have recommended that "...*provided toxin production cannot be detected, products based on species of Bacillus, other than those from the B. cereus taxonomic group, should be accepted* [as safe for use in animal nutrition]" (SCAN, 2000), public confidence must be established over the safety of using not only *Bacillus* spp. (particularly toxigenic strains), but microorganisms in general, for dietary applications in both animals and humans (see Chapters 9 and 10 for a more detailed discussion on the safety of *Bacillus* spp.).

SPORE PROBIOTIC PRODUCTS FOR VETERINARY APPLICATION

Table 1 details seven veterinary spore probiotic products, of which only two (Bioplus 2B® and Toyocerin®) are approved for use in the EU (see Appendix 2 for a detailed discussion on EU regulations relating to Bacilliary probiotics for animal feeds). Three of the products listed (Esporafeed Plus®, Pacifor C10® and Toyocerin®) are *B. cereus* preparations which can potentially produce toxins associated with emetic and diarrhoeic food poisoning. One of these, Paciflor C10®, is no longer manufactured. The *B. cereus* strain which was used in this product is able to produce at least two major enterotoxins known to be associated with food poisoning, therefore, SCAN concluded that it posed a risk to human health (SCAN, 2001b). Similarly, inadequate evidence to demonstrate that the *B. cereus* strain in Esporafeed Plus® does not produce enterotoxins led SCAN to conclude that this product poses a risk to human health (SCAN, 1999b). However, *B. cereus* var. *toyoi*, comprising the Toyocerin® product, has been shown to not produce toxins and

Table 1. Commercial veterinary spore probiotic products.

Product	Active *Bacillus* spp.	Culture collection accession numbers	Target species	Supplier	Recommended dose (cfu / kg feed)	Additional information
AlCare™	*B. licheniformis*	NCTC 13123	Swine	Alpharma Inc. Melbourne, Australia	$1 \times 10^9 - 1 \times 10^{10}$	A non-bacitracin producing strain.
Biogrow®	*B. subtilis* and *B. licheniformis*	-	Calves, Poultry and Swine	Provita Eurotech Ltd. County Tyrone, N. Ireland	$4.8 \times 10^8 - 3.2 \times 10^9$ (dependent on target species)	-
Bioplus 2B®	*B. subtilis* and *B. licheniformis*	DSM 5750 DSM 5749	Calves, Poultry and Swine	Chr. Hansen BioSystems A/S. Denmark	$4.8 \times 10^8 - 3.2 \times 10^9$ (dependent on target species)	EU approved.
Esporafeed Plus®	*B. cereus*	CECT 953	Swine	Norel, S. A. Madrid, Spain	1×10^9	-
Neoferm BS 10®	*B. clausii* (2 different strains)	CNCM MA23/3V CNCM MA66/4M	???	Sanofi Santé Nutrition Animale. France	???	-
Paciflor C10®	*B. cereus*	CIP 5832 and ATCC 14893	Calves, Poultry, Rabbits and Swine	Intervet Ltd. International	$2 \times 10^8 - 5 \times 10^9$ (dependent on target species)	Withdrawn from production in 2002.
Toyocerin®	*B. cereus* var. *toyoi*	CNCM I 1012 and NCBI-40112	Calves, Poultry, Rabbits and Swine	ASAHI Vet SA, Spain	$1 \times 10^8 - 1 \times 10^9$ (dependent on target species)	A non-toxin producing strain. EU approved.

Abbreviations: CECT – Spanish Type Culture Collection, Spain; CNCM – Collection Nationale de Cultures de Microorganismes, France; CIP – Collection de L'Institute Pasteur, France; DSM – Deutsche Sammlung von Mikroorganismen, Germany; NCBI – National Center for Biotechnology Information, USA; NCTC – National Collection of Type Cultures, United Kingdom; ??? – Information unknown. (To the best of our knowledge, all information presented is correct at the time of going to press.)

is one of the two spore probiotics currently approved for use as an animal feed supplement in the EU (SCAN, 2001a).

Concern over bacterial pathogens acquiring antibiotic resistance is the major driving force for research into probiotics and other alternatives to antibiotic feed supplements. Considering this, another issue which contributed to SCAN concluding that Esporafeed Plus® poses a risk to human health, was the presence of the tetracycline resistance gene *tetB* (typically located on a transposon) in the genome of this *B. cereus* strain (SCAN., 1999b). Similar reasoning also led SCAN to declare that Neoferm BS 10® was unsafe for use as a feed additive (SCAN, 1999c). This product is derived from a probiotic registered for use in humans which is intended for combined administration with antibiotics. Therefore, the two *B. clausii* strains concerned were deliberately selected for their resistance against a number of clinically important antibiotics.

These examples illustrate the need for strict regulation of animal probiotic feed supplements. Widespread use of probiotics harbouring antibiotic resistance genes risks transfer of resistance to bacterial pathogens of clinical importance. Furthermore, use of probiotics which consist of undefined microbial preparations, risks introducing zoonotic pathogens into the human food chain. Therefore, in the interests of public safety, complete characterisation of livestock probiotic products is essential. Somewhat alarmingly however, characterisation studies relating to seven different human spore probiotic preparations found only one to be labelled with the correct *Bacillus* spp. (Green et al., 1999; Hoa et al., 2000; Senesi et al., 2001). This highlights the need for an effective regulatory compliance system.

Much of the information on probiotic use in farm animals is derived from trials performed by manufacturers or suppliers (Fuller, 1999). Due to commercial interest, such studies frequently focus on the benefits livestock receive as a result of

probiotic feed supplementation. Conversely, academic studies have investigated both the benefits animals can receive and the mechanisms of action through which probiotics function. Therefore, in an attempt to identify bacterial traits which are important for successful spore mediated probiosis, investigations have often employed well-characterised laboratory isolates, not necessarily commercial strains. The assumption is that spore probiotics share common modes of action, and as such, specific variations between different *Bacillus* spp. have not been a focus of investigations to date. The following sections discuss spore probiotic studies in poultry, monogastrics and ruminants.

SPORE PROBIOTICS AND POULTRY

The use of hatcheries in large scale farming operations results in chicks emerging into a controlled environment, completely deprived of contact with adult birds. As a result, intensively farmed chicks take a number of weeks to develop a protective GI microflora post-hatch. In a natural scenario, chicks would rapidly acquire a protective microflora within days, through foraging amongst eggshells and adult faeces. However, the delay in microflora acquisition, brought about by modern farming practice, means that chicks are susceptible to rapid and persistent colonisation by both commensal and pathogenic bacteria in the first weeks of life (Barrow et al., 1988). This increased susceptibility of young birds to colonisation by enteric bacterial pathogens is further exacerbated because newly hatched chicks are immunologically naïve (Jeurissen et al., 1989; Bar-Shira et al., 2003). Indeed, the GI microflora is implicated in host immune system development (Isolauri et al., 2001).

Antibiotic feed supplements have been an effective control measure to date, but in light of current concern and EU policy, alternatives are required to ensure poultry farming remains cost-effective and continues to meet public health safety standards.

Strict hygiene practices will certainly be necessary. The use of vaccination programmes however, is likely to be limited. Mass vaccination programmes are often expensive and logistically complicated to execute. Unlike the broad-spectrum methods of antibiotic and probiotic feed supplements, vaccinations target specific pathogens. Therefore it would be unfeasible to attempt vaccinating poultry against a range of zoonotic and veterinary pathogens. Furthermore, the value of vaccinating immunologically naïve birds is questionable (Mast and Goddeeris, 1999) and may even be detrimental to the avian immune system (Holt et al., 1999). It is for reasons such as these that poultry have been a focus for probiosis development.

La Ragione et al. (2001) demonstrated that one-day old chicks dosed orally by gavage with spores of laboratory strain *B. subtilis* PY79 displayed greater resistance to an avian pathogenic *Escherichia coli* (APEC) challenge twenty-four hours later. *E. coli* O78:K80 colonisation of the spleen, liver and ceaca was significantly reduced, as was the level of this bacterium shed in faeces. Interestingly, the suppressive effect of spores was apparently abolished by seven-days of age. A similar but separate study demonstrated that *B. subtilis* PY79 also suppresses colonisation and persistence of *Salmonella* Enteritidis and *Clostridium perfringens* in chicks (La Ragione and Woodward, 2003). Faecal shedding of both these pathogens was reduced, particularly *Salmonella* Enteritidis. These experiments lasted 35 and 36 days respectively, during which *B. subtilis* PY79 persisted in the avian intestine, although in decreasing numbers. A number of explanations may account for this observation: *B. subtilis* may not colonise, or only briefly colonise, the chicken gut, resulting in dwindling numbers and this organism eventually being cleared. Another possibility is that *B. subtilis* may die in this environment following germination due to the anoxic conditions in the gut. However, *B. subtilis* can grow under anaerobic conditions when suitable nutrients are available (Hoffmann et al., 1995; Nakano and Zuber, 1998), suggesting that a different explanation may be more likely. Intestinal fluids present in the gut may simply prohibit cell proliferation and ultimately kill vegetative cells. Alternatively it is possible that an aspect of host mucosal immunity is lethal to this non-pathogenic bacterium. Indeed, if probiotic cell death is a feature of *Bacillus* mediated probiosis, this explanation is favoured as an important mechanism of action for CE, because other hypotheses proposed (see section titled 'current perspectives') require a probiotic culture to be metabolically active in the host GI tract.

Health benefits and enhanced performance have also been documented in chickens as a direct result of supplementing feed with *Bacillus* cultures. Santoso et al. (1995) reported that the inclusion of *B. subtilis* in the diet of broilers improved feed conversion efficiency, reduced abdominal fat, reduced the concentrations of triacylglycerol and cholesterol in the liver and carcass, and the serum triacylglycerol concentration. Similarly, when broiler feed was supplemented with *B. coagulans*, Cavazzoni et al. (1998) reported enhanced feed conversion efficiency and bird growth which were comparable to birds fed a diet supplemented with the antibiotic virginiamycin (now banned for use as a feed supplement in the EU). However,

neither of these studies investigated potential mechanisms through which these *Bacillus* isolates may have conferred these benefits.

Samanya and Yamauchi (2002) noted a positive effect on intestinal morphologies in chickens fed dried *B. subtilis* var. *natto* (the fermentative organism used in production of the Japanese health food Natto). Increased villus height, enterocyte cell area and cell mitosis were among the benefits observed which interestingly, correlated with a reduced blood ammonia concentration. Reduced ammonia concentrations have also been noted in the caeca of probiotic fed chickens (Endo et al., 1999) and the small intestine of pigs fed with *B. cereus* (Kirchgessner et al., 1993). Ammonia has been shown to stimulate *B. cereus* spore germination (Preston and Douthit, 1984) and may serve as a germination signal in addition to the acid shock experienced by spores passing through the avian ventriculus (stomach). Intestinal ammonia originates from the process of ureolysis in gut. It has been shown to be detrimental to villus histology in rats (Warzecha et al., 2000) and may also enter the blood stream (Yeo and Kim, 1997). The implication is that reduced GI ammonia concentration may be the result of a probiotic effect on the intestinal microflora. Subsequently, reduced levels of toxic ammonia allows increased villus height resulting in a greater gut lumen surface area which in turn, enables increased nutrient absorption. This may go towards explaining observations of increased feed conversion efficiency as a result of probiotic administration. However, further morphometric and nutritional studies are required to substantiate this idea.

SPORE PROBIOTICS AND MONOGASTRIC ANIMALS

Spore probiotic products are available for oral prophylaxis in humans (Sanders et al., 2003) and have documented application in dogs (Hisanga, 1980; Biourge et al., 1998) and pigs (Kirchgessnser et al., 1993; Kyriakis et al., 1999). From a commercial perspective, pigs are the most important animals in this category because they are farmed. During weaning, piglets are subject to many stresses. The end of the suckling period marks the end of maternally derived immunity (acquired from the colostrum) and brings about separation from the sow. Mixing with other piglets and changes in environment and diet also add to stress. These factors may lead to an intestinal microflora imbalance, leaving piglets more susceptible to enteric pathogens and post-weaning diarrhoea syndrome (PWDS). Enterotoxigenic *E. coli* (ETEC) are of particular concern, frequently being a major contributing factor. Cases of mortality are usually only a small percentage of those affected unless complications occur. Nevertheless, in addition to morbidity cases, financial loss is still an issue for the farmer.

A study conducted by Kyriakis et al. (1999) found that a *B. licheniformis* spore preparation from the company Alpharma (but not specified as AlCare™ – see Table 1) was effective in reducing the incidence and severity of PWDS in piglets. This result was consistent amongst piglets in the positive control group which received feed supplemented with Toyocerin®. Animals receiving either probiotic also displayed enhanced weight gain, enhanced feed conversion efficiency and, in contrast to pigs in the negative control group, none were shedding ETEC strains 22 days after weaning. However, modes of action were not investigated in this study.

A later study of Toyocerin® (Jadamus et al., 2001) suggested that *B. cereus* var. *toyoi* germinates rapidly in the upper intestine of piglets and chickens and this is subsequently followed by sporulation further down the gastrointestinal tract. This finding is consistent with those of Hisanga et al. (1980), Hoa et al. (2001) and Casula and Cutting (2002). Spore germination and outgrowth following ingestion, may be important in spore probiosis and is indeed essential if metabolic activity is required to exert a probiotic effect (see Chapter 5 for a discussion on the fate of ingested spores).

The idea that animal performance may be enhanced through probiotic mediated modification of intestinal barrier functions is further supported by the work of Baum et al. (2002). In agreement with findings in poultry fed *B. subtilis* var. *natto* (Samanya and Yamauchi, 2002), pigs receiving *B. cereus* var. *toyoi* displayed taller villi in the small intestine. Furthermore, a thicker mucosa and more mature mucins were observed in this region of the pig gut. In contrast, a thinner mucosa and more goblet cells, secreting less mature mucins were seen in the large intestine. These results are further evidence that probiosis extends beyond the CE concept and that cross-talk between host and probiotic microorganisms is important for improving animal health and performance. Interestingly, in the same study, the yeast *Saccharomyces boulardii* conferred similar benefits to pigs when administered as a probiotic which suggests that different probiotics may stimulate host functions through common mechanisms.

SPORE PROBIOTICS AND RUMINANTS

The digestive tract of ruminants is more complex than that of monogastric animals. Following ingestion and mastication, a food bolus is swallowed and arrives in the rumen (forestomach) where a complex commensal microflora facilitates feed fermentation. Ingesta can flow freely between here and the reticulum (the second forestomach), which is essentially an extension of the rumen. Mechanical breakdown accompanies the feed fermentation process with the regurgitation of food and further mastication occurring during rumination (chewing the cud). Partially fermented food then passes into the omasum where water and soluble nutrients are absorbed. The remaining ingesta passes into the abomasum (true stomach), then through to the small intestine and down the remainder of the GI tract which is anatomically similar to that of monogastric animals. Like other animals, ruminants must acquire a GI microflora after birth and microbial colonisation occurs rapidly during the suckling period. In these first days of life, the rumen is not fully functional. Milk bypasses the rumen via the oesophageal groove into the abomasum. As a calf begins to consume solid feed, the microbial population of the rumen increases and diversifies to resemble that of the adult, so that by the time of weaning, the rumen is fully developed.

Probiotic application may be used to rapidly establish a functional rumen microflora. This is desirable for a swift and successful transition from liquid to solid feed which reduces the likelihood of GI problems. However, the full potential of supplementing ruminant feed with spore probiotics is yet to be firmly established as this area has received little attention to date. Due to the costs and logistics involved in large animal studies, this is likely to remain the case until further knowledge is acquired regarding desirable traits of probiotic microorganisms and the mechanisms of action by which they work. Such studies can then be reasonably justified both financially and ethically. Nevertheless, Jenny et al. (1991) have documented small but positive effects on both feed conversion efficiency and weight gain in calves fed *B. subtilis* in the first month post-weaning. Although, no improvement in animal performance was documented in a more recent study where bull feed was supplemented with *B. cereus* spores (Garza-Cazares et al., 2001).

Scouring is also a problem for calves at the time of weaning and, as is the case for pigs, ETEC strains are a common cause of diarrhoea. Consequently, it is likely that spore probiotics will prove to be most effective as CE agents in young ruminants reducing enteric infections which cause morbidity and mortality in weaning calves.

CONCLUSIONS AND FUTURE DIRECTION

Due to concern regarding the acquisition of multiple antibiotic resistance by bacteria of medical concern, attempts have been made to reduce antibiotic usage and this is reflected in EU policy on antibiotics as animal feed supplements. As a result research interests have focused on probiotics as an alternative to antibiotic feed supplements. However, probiotics are still in the early stages of development with much of the evidence for their beneficial effects being anecdotal.

To date, studies of *Bacillus* based probiotics have reported effects on the host intestinal microflora (Ozawa et al., 1981; Maruta et al., 1996; Kyriakis et al., 1999) and suppression of pathogens (La Ragione et al., 2001; La Ragione and Woodward, 2003). However, other studies have found that spore probiotics have not conferred an effect (Jiraphocakul et al., 1990; Garza-Cazares et al., 2001). Host factors may partially explain this discrepancy, although other influences including diet composition (Hosoi et al., 1998) and administered medications (Jiraphocakul et al., 1990) are also likely to be important.

Ongoing research aims to identify desirable traits of microorganisms for probiosis, elucidate the mechanisms through which they work as probiotics and optimise administration regimes with regard to dosing quantities and time of application. Live microbial feed supplements may then be administered effectively and ethically to improve animal health and performance. Subsequently, the field of spore probiotics may then be able to effectively embrace pioneering work which makes use of *Bacillus* spores as oral vaccine vehicles (Isticato et al., 2001; Oggioni et al., 2003; Duc et al., 2003a; Duc et al., 2003b). This would produce highly effective feed supplements which promote animal health by stimulating lasting immunity against specific pathogens, offer general protection against enteric infection and also enhance animal performance by optimising feed conversion efficiency and weight gain.

REFERENCES

Anonymous. 1997. Report for the International Federation for Animal Health (IFAH): Estimated annual usage of antimicrobials in humans and animals in the EU. http://fedesa.be/europe/topics/antibio/kit3.htm

Barrow, P.A., Hassan, J.O., Lovell, M.A. 1988. Intestinal colonisation in the chicken by food poisoning *Salmonella* serotypes microbial characteristics associated with faecal excretion. Avian Pathol. 17: 571-588.

Bar-Shira, E. Sklan, D. and Friedman, A. 2003. Establishment of immune competence in the avian GALT during the immediate post-hatch period. Dev Comp Immunol. 27: 147-157.

Baum, B. Liebler-Tenorio, E.M., Enß, M-L., Pohlenz, J.F. and Breves, G. 2002. *Saccharomyces Boulardii* and *Bacillus cereus* var. *toyoi* influence the morphology and mucins of the intestine of pigs. Z Gastroenterol. 40: 277-284.

Biourge, V., Vallet, C. Levesque, A., Sergheraert, R., Chevalier, S. and Robertson, J-L. 1998. The use of probiotics in the diet of dogs. J. Nutr. 128(12 Suppl): 2730S-2732S.

Casula, G. and Cutting, S. M. 2002. *Bacillus* probiotics: Spore germination in the gastrointestinal tract. Appl Environ Microbiol. 68: 2344-2352.

Cavazzoni, V., Adami, A. and Castrovilli, C. 1998. Performance of broiler chickens supplemented with *Bacillus coagulans* as probiotic. Br Poult Sci. 39: 526-529.

Duc, L.H., Hong, H.A. and Cutting, S.M. 2003a. Germination in the spore gastrointestinal tract provides a novel route for heterologous antigen delivery. Vaccine. 21: 4215-4224.

Duc, L.H., Hong, H.A., Fairweather, N., Ricca, E. and Cutting, S.M. 2003b. *Bacillus* spores as vaccine vehicles. Infect Immun. 71: 2810-2818.

Endo, T., Nakano, M., Simizu, S., Fukushima, M. and Miyoshi, S. 1999. Effects of a probiotic on the lipid metabolism of cocks fed on a cholesterol-enriched diet. Biosci Biotechnol Biochem. 63: 1569-1575.

Fukushima, M. and Nakano, M. 1996. Effects of a mixture of live organisms, *Lactobacillus acidophilus* or *Streptococcus faecalis* on cholesterol metabolism in the livers of rats fed a fat- and a cholesterol-enriched diet. Br J Nutr. 76: 857-867.

Fukushima, M., Yamada, A., Endo, T. and Nakano, M. 1999. Effects of a mixture of live organisms, *Lactobacillus acidophilus* or *Streptococcus faecalis* on delts6-desaturase activity in the livers of rats fed a fat- and a cholesterol-enriched diet. Nutrition. 15:373-378.

Fuller, R. 1989. Probiotics in man and animals. J Appl Bacteriol. 66: 365-378.

Fuller, R. 1999. Probiotics for farm animals. In: Probiotics: A Critical Review. G.W. Tannock, ed. Horizon Scientific Press, Wymondham. p. 15-22.

Garza-Cazares, F. Daenickle, R and Flachowsky, G. 2001. Research note: Effect of *Bacillus cereus* on performances of growing bulls. Arch Tierernahr. 55: 161-165.

Gibson, G.R. and Roberfroid, M.B. 1995. Dietry modulation of the human colonic microbiota: introducing the concept of prebiotics. J Nutr. 125: 1401.

Green, D.H., Wakeley, P.R., Page, A., Barnes, A., Baccigalupi, L. Ricca, E and Cutting, S. 1999. Characterization of two *Bacillus* probiotics. Appl Environ Microbiol. 65:4288-4291.

Hisanga, S. 1980. Studies on the germination of *Bacillus* spores in rabbit and canine intestines. Journal of Nagoya City Medical Association. 30: 456-469.

Hoa, N.T., Baccigalupi, L., Huxham, A., Smertenko, A., Van, P.H., Ammendola, S., Ricca, E. and Cutting, S. 2000. Characterization of *Bacillus* species used in oral bacteriotherapy and bacterioprophylaxis of gastrointestinal disorders. Appl Environ Microbiol. 66:5241-5247.

Hoffman, T., Troup, B., Szabo, A. Hungerer, C. and Jahn, D. 1995. The anaerobic life of *Bacillus subtilis*: Cloning of the genes encoding the respiratory nitrate reductase system. FEMS Microbiol Lett. 131: 219-225.

Holt, P.K., Gast, R.K., Porter, R.E. and Stone, H.D. 1999. Hyperesponsiveness of the systemic and mucosal humoral immune systems in chickens infected with *Salmonella enterica* serovar Enteritidis at one day of age. Poult Sci. 78:1510-1517.

Hosoi, T. Ametani, A. Kiuchi, K. and Kaminogawa, S. 1998. Changes in fecal microflora induced by intubation of mice with *Bacillus subtilis* (*natto*) spores are dependent upon dietary components. Can J Microbiol. 45: 59-66.

Isolauri, E., Sütas, Y., Kankaanpää, Arvilommi, H. and Salminen, S. 2001 Probiotics: effects on immunity. Am J Clin Nutr. 73(suppl): 444S-450S.

Isticato, R., Cangiano, G., Tran, H.T., Ciabattini., A., Medaglini, D., Oggioni, M.R., De Felice, M., Pozzi, G. and Ricca, E. 2001. Surface display of recombinant proteins on *Bacillus subtilis* spores. J Bacteriol. 183: 6294-6301.

Jadamus, A. Vahjen, W. and Simon, O. 2001. Growth and behaviour of a spore forming probiotic strain in the gastrointestinal tract of broiler chicken and piglets. Arch Anim Nutr. 54: 1-17.

Jenny, B.F., Vandijk, H.J. and Collins, J.A. 1991. Performance and fecal flora of calves fed a *Bacillus subtilis* concentrate. J Dairy Sci. 74: 1968-1973.

Jeurissen, S.H.M., Janse, E.M., Kock, G. and DeBoer, G.F. 1989. Postnatal development of mucosa-associated lymphoid tissue in chickens. Cell Tissue Res. 258: 119-124.

Jiraphocakul, S., Sullivan, T.W. and Shahani, K.M. 1990. Influence of a dried *Bacillus subtilis* culture and antibiotics on performance and intestinal microflora in turkeys. Poult Sci. 69: 1966-1973.

Kirchgessner, M., Roth, F.X., Eidelsburger, U. and Gedek, B. 1993. The nutritive efficiency of *Bacillus cereus* as a probiotic in the raising of piglets. 1. Effect on the growth parameters and gastrointestinal environment. Arch Tierernaher. 44: 111-121. In German

Kyriakis, S.C., Tsiloyiannis, V.K., Vlemmas, J., Sarris, K., Tsinas, A.C., Alexopoulos, C. and Jansegers, L. 1999. The effect of probiotic LSP 122 on the control of post-weaning diarrhoea syndrome of piglets. Res Vet Sci. 67: 223-228.

La Ragione, R.M. and Woodward, M.J. 2003. Competitive exclusion by *Bacillus subtilis* spores of *Salmonella enterica* serotype Enteritidis and *Clostridium perfringens* in young chickens. Vet Micro. 94: 245-256.

La Ragione, R.M., Casula, G. Cutting, S.M. and Woodward, M.J. 2001. *Bacillus subtilis* spores competitively exclude *Escherichia coli* O78:K80 in poultry. Vet Micro. 79: 133-142.

Maruta, K., Miyazaki, H., Masuda, S., Takahashi, M., Marubashi, T., Tadano, Y. and Takahashi, M. 1996. Exclusion of intestinal pathogens by continuous feeding with *Bacillus subtilis* C-3102 and its influence on the intestinal microflora in broilers. Anim Sci Technol. 67: 273-280.

Mast, J. and Goddeeris, B.M. 1999. Development of immunocompetence of broiler chickens. Vet Immunol Imunopathol. 70: 245-256.

Mead, G.C. 2000. Prospects for 'competitive exclusion treatment to control Salmonellas and other foodborne pathogens in poultry. Veterinary Journal. 159: 111-123.

Milner, K.C. and Shaffer, M.F. 1952. Bacteriologic studies of experimental *Salmonella* infections in chick. J Infect Dis. 90: 81.

Nakano, M.M. and Zuber, P. 1998. Anaerobic growth of a "strict aerobe" (*Bacillus subtilis*). Annu Rev Microbiol. 52: 165-190.

Nurmi, E.V. and Rantala, M. 1973. New aspects of *Salmonella* infection in broiler production. Nature. 241: 210.

Oggioni, M.R., Ciabattini, A. Cuppone, A.M. and Pozzi, G. 2003. *Bacillus* spores for vaccine delivery. Vaccine. 21(Suppl 2): S96-S101.

Ozawa, K., Yokota, H., Kimura, M. and Mitsuoka, T. 1981. Effects of administration of *Bacillus subtilis* strain BN on intestinal flora of weaning piglets. Jpn J Vet Sci. 43: 771-775.

Preston, R.A. and Douthit, H.A. 1984. Stimulation of germination of unactivated *Bacillus cereus* spores by ammonia. J Gen Microbiol. 130: 1041-1050.

Samanya, M. and Yamauchi, K-e. 2002. Histological alterations of intestinal villi in chickens fed dried *Bacillus subtilis* var. *natto*. Comp Biochem Physiol A. 133: 95-104.

Sanders, M.E. Morelli, L. and Tompkins, T. 2003. Sporeformers as human probiotics: *Bacillus*, *Sporolactobacillus* and *Brevibacillus*. Comprehensive Reviews in Food Science and Food Safety. 2: 101-110.

Santoso, U., Tanaka, K. and Ohtani, S. 1995. Effect of dried *Bacillus subtilis* on growth, body composition and hepatic lipogenic enzyme activity in female broiler chicks. Br J Nutr. 74: 523-529.

SCAN. 1999a. Opinion of the Scientific Steering Committee on antimicrobial resistance. http://europa.eu.int/comm/food/fs/sc/ssc/out50_en.pdf

SCAN. 1999b. Assessment by the Scientific Committee on Animal Nutrition (SCAN) of a micro-organisms product: Esporafeed Plus®. http://europa.eu.int/comm/food/fs/sc/scan/out39_en.pdf

SCAN. 1999c. Assessment by the Scientific Committee on Animal Nutrition (SCAN) of a micro-organisms product: Neoferm BS10®. http://europa.eu.int/comm/food/fs/scan/out28_en.pdf

SCAN. 2000. Opinion of the Scientific Committee on Animal Nutrition on the safety of use of *Bacillus* species in animal nutrition. http://europa.eu.int/comm/food/fs/sc/scan/out41_en.pdf

SCAN. 2001a. Report of the Scientific Committee on Animal Nutrition on product Toyocerin® for use as feed additive. http://europa.eu.int/comm/food/fs/sc/scan/out72_en.pdf

SCAN. 2001b. Assessment by the Scientific Committee on animal nutrition of the safety of product Paciflor® for use as feed additive. http://europa.eu.int/comm/food/fs/sc/scan/out62_en.pdf

Schneitz, C. and Mead, G. 2000. Competitive exclusion. In: *Salmonella* in domestic animals. C. Wray and A. Wray, eds. CABI Publishing, UK. p. 301-322.

Senesi, S. Celandroni, F., Tavanti, A. and Ghelardi, E. 2001. Molecular characterization and identification of *Bacillus clausii* strains marketed for use in oral bacteriotherapy. Appl Environ Microbiol. 67:834-839.

Starvic, S. and D'Aoust, J.Y. (1993) Undefined and defined bacterial preparations for the competitive exclusion of *Salmonella* in poultry – a review. J Food Prot. 56: 173-180.

Warzecha, Z. Dembinski, A., Brzozowski, T., Ceranowicz, P., Pajdo, R., Niemiec, J., Drozdowicz, D., Mitis-Musiol, M. and Konturek, S.J. 2000. Gastroprotective effect of histamine and acid secretion on ammonia-induced gastric lesions in rats. Scand J Gastroenterol. 35: 916-924.

Yeo, J. and Kim, K.I. 1997. Effect of feeding diets containing an antibiotic, a probiotic or yucca extract on growth and intestinal urease activity in broiler chicks. Poult Sci. 76: 381-385.

Chapter 14

The Potential Use of *Bacillus* spp. in Sports Turf Management

Alan C. Gange and Karen J. Hagley

SUMMARY

Sports turf is a long-term crop, heavily reliant on water, fertilizers and pesticides to maintain it in perfect condition. Recent studies have shown that microbial populations in turf soils, including numbers of *Bacillus* spp., are low compared with natural grasslands. The main reason for this is carbon limitation, but compaction, pesticides and fertilizers may also be responsible. It has been argued that *Bacillus* spp. have great potential for use in turf management, because many of the problems (poor root growth, pest and disease attack and weed control) could be ameliorated by these bacteria. There is now research on all of these topics and the promise of *Bacillus* as plant protectants has been shown by their incorporation into new biopesticides. However, a cause for concern is that a number of bacterial products are available for use on turf and, while the theory behind their use is clear, their efficacy and mode of action is largely untested.

INTRODUCTION

Sports turf is a remarkable ecosystem in many respects. It is a perennial crop, often remaining undisturbed for tens or even hundreds of years. It is species-poor, the ideal grass composition being between one and three species. It must display continuous cover (and ideally some growth) throughout a calendar year, even in temperate climates. The nature of the different games played upon it and the demands of televised sport means that the surface must be in immaculate condition, whatever the weather.

These requirements would be arduous in most situations, but sports turf has other demands placed upon it. As a general rule, the aim of the groundsman or greenkeeper is to reduce above-ground biomass, while simultaneously trying to maximise root production. However, the very fact that a sport is played upon grass means that there are considerable mechanical stresses also placed upon the grass plants, in the form of trampling or wear, caused by players and machinery. In the light of all these demands it may seem remarkable that turf is produced to the high standards that we see, but over the years there has been considerable research into the construction and maintenance of sports turf areas. An excellent review of the former is provided by Stewart (1994), while Adams and Gibbs (1994) provide a comprehensive overview of the entire subject.

Many different sports take place on grass and the nature of the grass species in the sward is determined by the game or activity that takes place upon it. Thus, in games such as golf or bowls, where a short, smooth and even playing surface is required, the desirable grasses are species of bents (*Agrostis* spp.) and fescues (*Festuca* spp.). Meanwhile, for football, rugby pitches or horse racing tracks, perennial ryegrass (*Lolium perenne*) is the species of choice, due to its capacity to resist wear. Adams and Gibbs (1994) describe the grass species required for different sports in more detail.

The study of turfgrass microbiology is still in its infancy, and virtually all of the research that has been conducted pertains to golf putting greens or football pitches. The current situation is a peculiar one, in that there are various commercial products available for use on sports turf that contain *Bacillus* spp., yet there is an astonishing lack of research on their efficacy or mode of action. Much of the relevant research has either been conducted with non-turf grasses or has taken place in controlled environments, rather than field situations. Hence, much of this review is aimed at showing the potential of *Bacillus* in turf management, since there are many good reasons why future turf managers may come to rely on microbial products. This chapter will focus on golf and football turf in temperate climates, because this represents the majority of research, as well as the majority of turf grass areas, on a worldwide scale. However, it should be noted that in tropical climates, the grass species in these turfs will differ and their relations with soil microorganisms will differ also.

PROBLEMS WITH SPORTS TURF PRODUCTION

Here, the problems associated with sports turf and the peculiarity of its construction are outlined. This is to provide a background so that the need and potential uses of *Bacillus* spp. in turf can be appreciated.

The priority of any turf manager is to maintain the one or two desirable grass species in the sward. Broad leaved weeds can be controlled by a wide range of herbicides but by far the most important weed of fine turf is *Poa annua* (annual meadow grass or annual bluegrass) (Gange et al., 1999). While a certain amount of *Poa* does little harm in a football pitch, it is most problematic in golf putting greens. It is a rapid invader of such areas (Hagley et al., 2003) and can quickly dominate the sward. It is a problem because its upright habit disrupts the game, while its shallow rooting means it is subject to drought stress. It is also susceptible to a wide range of diseases that cause disfigurement of the sward (Fermanian et al., 1997). However, it has an ability to set seed at the height to which a golf green is cut (about 4.5 mm) meaning that, unlike the perennial desirable grasses, the seedbank becomes dominated by *Poa* and a green can become a self-regenerating sward of this weed. Various chemical methods, involving selective herbicides or plant growth regulators have been tried for its control (Johnson, 1982; Johnson and Murphy, 1995), but none of them are satisfactory. Microbiological methods involving *Xanthomonas* bacteria (Imaizumi et al., 1997) or arbuscular mycorrhizal fungi (Gange et al., 1999) have also been tried with some degree of success. Many *Bacillus* spp. are considered to be plant growth promoters, rather than plant antagonists (Chanway, 2003 and Appendix III), but if these bacteria could be used indirectly in the control of *Poa* (see below) in a cost effective manner, then almost every golf club in cooler climates would probably buy the product.

The main problem afflicting sports turf in temperate areas is the number of disease-causing fungi that occur (Vargas, 1994). Of these, *Microdochium nivale* (known as microdochium patch, Fusarium patch, pink snow mould and a host of other names) is by far the most damaging and important pathogen in the UK (Mann, 2004). It is a low-temperature pathogen, attacking a very wide range of grasses (Smith, et al., 1989), although it has recently been shown that different varieties seem to occur in fine turf, and some of these are active throughout the year, while others are only so in spring and autumn (Gange and Case, 2003).

Sports turf is attacked by a wide range of insect pests, though due to the reduced nature of the above-ground vegetation, many of these are subterranean, attacking the roots. In the USA, various species of chafers or white grubs (Coleoptera: Scarabaeidae), including the Japanese beetle (*Popillia japonica*), and European chafer (*Rhizotrogus majalis*) are important pests (Fermanian et al., 1997). In the UK, tipulids (Diptera: Tipulidae) are the most numerous of turfgrass pests (Mann, 2004), although the resulting damage is usually done by birds that dig in the turf to eat the larvae, rather than the grubs themselves (Perris and Evans, 1996).

Undoubtedly one of the major problems with sports turf is the degree of trampling that the sward suffers. When one considers that on an average golf course, approximately 70,000 rounds of golf may be played per annum or that in an average game of football, about 250,000 foot imprints per pitch occur (Adams and Gibbs, 1994), it is not surprising that most putting greens and football pitches suffer from compaction of the rootzone. Such compaction leads to poor root growth, poor drainage and waterlogging and ultimately, grass plants that are nutrient-stressed and susceptible to disease. Compaction favours the growth of *P. annua* and is also a major reason as to why microbial levels in turf soils are low (Hagley, 2002) (see below).

To combat the effect of compaction and allow good drainage of the soil, a number of specifications have been developed for the rootzone of putting greens and football pitches. In essence, modern rootzones are sand-dominated, a good example being the specification developed by the United States Golf Association (USGA) for putting greens (USGA Green Section Staff, 1993). A USGA green has a rootzone which is about 95% sand, the remainder being organic matter, consisting of milled peat. Football pitches too have a rootzone dominated by sand (known as the sand carpet construction) in which little organic matter is present at the outset.

A final problem with sports turf is that fact that it is a carbon-limited ecosystem. Golf putting greens are cut to an average height of 4 – 4.5 mm in the summer (higher in winter), while football pitches are cut to approximately 25 mm during the playing season and 50 mm in the closed season (Adams and Gibbs, 1994). It is well known that severe foliage removal of grasses means a lack of photosynthetic area, with reduced amounts of carbon fed to the root system, resulting in reduced root biomass (e.g. Bremer et al., 1998). Furthermore, a reduction in carbon fed to the root system means a reduction in root exudation, thereby resulting in a lower soil microbial population (Grayston et al., 1997).

MICROBIAL POPULATIONS IN SPORTS TURF

Remarkably little is known about the microbial communities present in sports turf soils (Gange, 1998) and only recently has information become available on the community structure present in golf putting greens (Hagley, 2002). The longest-running study appears to be that reported in Elliott et al. (1996, 1999). In a project funded by the United States Golf Association, these workers have studied the bacterial composition (populations and diversity) of USGA putting greens for a period of years following construction.

Three major conclusions have arisen out of this work. The first is that the bacterial composition of newly constructed rootzones in 1996 was dominated by species of *Acidovorax*, *Burkholderia* and *Pseudomonas*, while three years later, the dominant genera were *Arthrobacter*, *Bacillus*, and *Pseudomonas* (Elliott et al., 2000). It is clear from these data that *Bacillus* spp. are rare in new putting greens and populations build up slowly, over a period of years. The second conclusion was that there were distinct seasonal changes in species populations, with *Bacillus* spp. showing higher populations in the warmer times in the year. Thirdly, total abundance of bacteria was lower in the USGA rootzone, compared with areas of natural grassland. Recent work by Skipper (2001) has begun to characterise species of bacteria found in USGA rootzones. A total of four *Bacillus* species have so far been identified in these greens, indicating that there is the potential for augmentation of existing populations, if suitable products are applied.

Slightly more encouraging results were obtained by Bigelow et al. (2002), in which bacterial populations were measured in USGA putting green rootzones during the first two years of establishment. Enumeration of different bacterial groups was performed using selective media and by this method it was found that total abundance (as measured by colony forming units, cfu) was similar in these young greens to that in mature (5 y old) greens. Furthermore, *Bacillus* spp. formed a significant part of the community in the young greens, reaching levels of 2.2×10^5 cfu g^{-1} of dry soil. The dynamics of *Bacillus* populations were particularly interesting, in that a slow but steady increase was observed, in marked contrast to the fluctuations seen in other bacterial groups. It was concluded that the ability of *Bacillus* to produce drought-resistant endospores meant that they were more able to survive periods of adverse environmental conditions, leading to relatively stable populations. Such stability has also been observed in agricultural field soils (Kirchner et al, 1993)

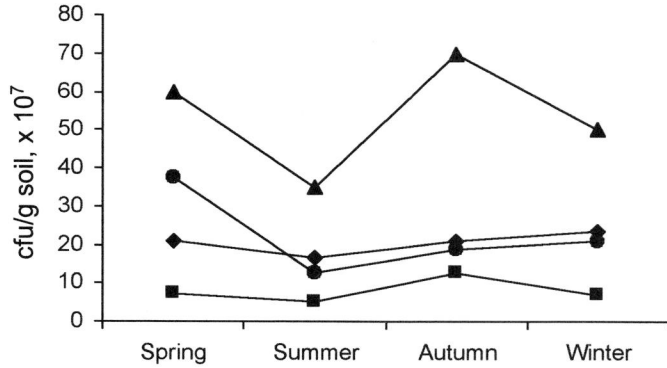

Figure 1. Total colony-forming units of heat-tolerant bacteria in soils from four putting green rootzones, over the course of one year, in southern England. Key to symbols: (▲) neutral loam; (●) alkaline soil; (◆) acidic sand; (■) USGA specification.

and is most encouraging, as it suggests that any inoculants of *Bacillus* applied to sports turf may be successful, because the spore phase will enable survival in times of stress, unlike sensitive species (Karnok, 2000). Indeed, in a study of the effects of shading on populations of *B. megaterium* applied to turf, Giesler et al. (2000) recorded little change in numbers post inoculation, although the time scale over which this was done was short (16 days).

There are two drawbacks with the above-mentioned studies. The first is that they have mainly relied on dilution plating as a method to enumerate populations, yet it has been estimated that as little as 1% of bacteria present in a sample may be culturable on any known medium (Atlas and Bartha, 1997). Indeed, studies have shown that the soil bacterial levels that are recorded in putting greens can depend heavily on the medium that is used for growth (Mancino et al., 1993; Elliott and Des Jardin, 1999a). The second problem is that while USGA greens have received much attention, the fact is that the majority of golf putting greens in the UK are not of such construction, having been in existence for many years. These greens are soil-based, in which the rootzone was constructed from the local topsoil. In the UK

today, older golf courses tend to have soil-based rootzones while courses of about 12 y old or less tend to have sand-dominated rootzones, conforming to USGA specifications.

In an attempt to overcome these problems, Hagley (2002) sampled soil from a range of putting greens in southern England in which the rootzones were of four types: an acidic sand, pH about 5.4, a neutral loam, pH 6.8, a calcareous soil, overlying chalk, pH 8.1 and a USGA specification, pH 6.3. Three types of bacterial community profiling were performed. The first was enumeration using 10% tryptone soya broth agar following heating to 80°C for heat resistant bacteria such as *Bacillus* (following the protocol and for comparison with Elliott et al., 1996). The second method involved total bacterial abundance as measured by the levels of bacterial-specific phospholipid fatty acids (PLFA) (Frostegård et al., 1991) while the third involved use of the BIOLOG method for community physiological profiling (Garland and Mills, 1991).

Dilution plating showed that there were dramatic differences in bacterial abundance between the four rootzone types (Figure 1). Putting greens composed of neutral loam had the highest counts, while those from USGA rootzones had the lowest. In most cases there were seasonal differences too, with numbers peaking in spring and autumn. However, the differences between rootzone types were maintained, despite this seasonality.

When total bacterial abundance was measured by PLFA analysis, there were some interesting differences, showing the value of using different techniques (Figure 2). In this case, the highest bacterial abundance was found in rootzones with a high pH, but there was also a degree of consistency, in that USGA greens still produced the lowest levels.

Diversity of a community is a complex measure, not just of the abundance of individuals, but involving also the spread of numbers across species (Magurran, 1988). When diversity (recorded as the Shannon-Weiner index *H*) of resource use on BIOLOG plates was measured, there were again differences between rootzone types (Figure 3). USGA rootzones had a considerably less diverse community of bacteria within them, i.e., one that was able to utilise a much smaller variety of carbon sources.

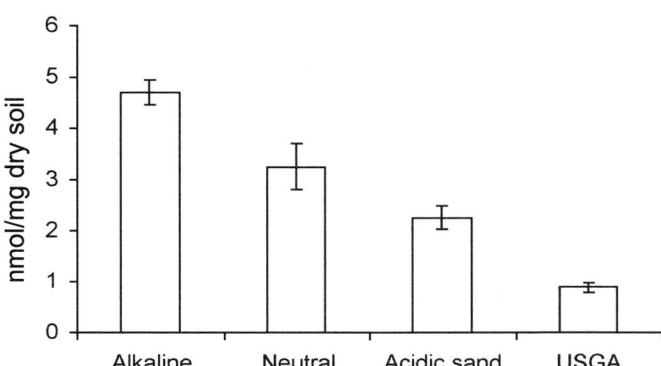

Figure 2. Total bacterial abundance, as measured by total phospholipid fatty acids, in soils from four putting green rootzones in southern England. Bars represent means with one standard error.

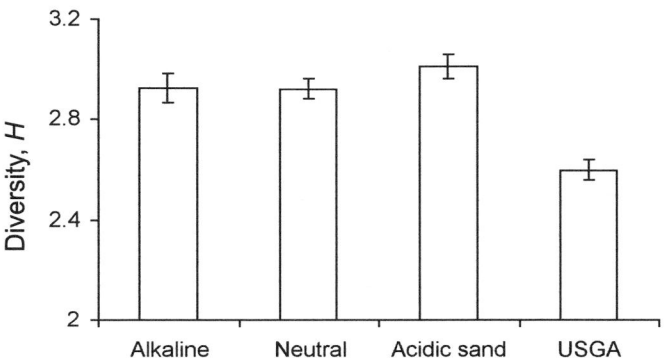

Figure 3. Bacterial phenotypic diversity (Shannon Weiner, *H*), as measured by carbon source utilization, from four putting green rootzones in southern England. Bars represent means with one standard error.

The results of Hagley (2002) are important because they show that the bacterial community differs greatly between different types of golf putting green, both in terms of total abundance and the diversity of species. How these differences relate to the use of commercial microbial inoculant products that contain *Bacillus* species is unknown. However, they do suggest that the efficacy of such products is likely to differ between different sports turf surfaces, due to variation in the indigenous fauna within them.

POTENTIAL USES OF *BACILLUS* SPP. IN SPORTS TURF

Species of *Bacillus* show a wide variety of positive effects on plant growth several of which have important potential for exploitation in sports turf. A summary of plant growth promoting effects observed and the mechanisms involved is provided by Chanway (2003).

Direct plant growth effects

Many species of *Bacillus* are known to produce plant growth hormones, such as auxins, cytokinins and gibberellins (Timmusk et al., 1999; Chanway 2003). The production of very small amounts of such hormones in the rhizosphere can increase root growth (Benizri et al., 2001), which is of great potential benefit to sports turf. Given that fine turf is regularly mown to a height that can be as low as 4 mm, it is obvious that the reduction in photosynthetic area will mean a reduced amount of carbon translocated to the root system for growth (Bremer et al., 1998). Turf is therefore a carbon-limited ecosystem (Hagley, 2002) and enhancement of root production by microbial agents may be successful (though see below for reasons why this may not always be so). Indeed, Holl et al. (1988) found that *Bacillus polymyxa* (now called *Paenibacillus polymyxa*) addition resulted in increased growth of *Lolium perenne*, which is a major component of turf used for winter games or horse racing in the UK. To date, no published studies have investigated whether *Bacillus* spp. can enhance root production in working sports turf in the UK.

A second direct growth effect that these bacteria may have is through the fixation of nitrogen. Many *Bacillus* strains are known to possess nitrogenase and nitrogen fixation has been detected in the rhizosphere of several grass species (Lindberg and Granhall, 1984). This method of direct growth promotion was discounted by Holl et al. (1988) as a reason for positive effects in *L. perenne*, though other grasses, such as wheat, have been shown to benefit significantly (Rennie and Larson, 1979). In golf putting greens especially, such nitrogen fixation effects are likely to be inconsequential at the present time. This is because the amount fixed will be tiny, compared with the level of nitrogen already in the rootzone, together with that applied each year by inorganic fertilizer (Perris and Evans, 1996). However if bacterial numbers in turf soils could be increased to levels similar to those found in natural grasslands, then N fixation could become more important and thereby reduce the heavy reliance on inorganic fertilizers that currently exists.

A similar argument applies to the management of phosphorus in sports turf soils. Several species of *Bacillus* have been shown to increase phosphate mineralization in soils, either by the secretion of enzymes or of organic acids that lower the pH and thus increase the solubility of inorganic phosphorus compounds (Chanway, 2003). However, soil P levels are high in many forms of sports turf, especially golf putting greens, where levels can easily exceed 100 ppm (Baker et al., 1997). It is unlikely that this method of growth promotion will ever be of significance in older putting greens, though there would be potential for it in new (USGA) greens or other sports turf where less phosphatic fertilizer is applied (Adams and Gibbs, 1994).

In summary, plant growth hormone production is potentially the most important direct plant growth promoting effect that these bacteria may have in sports turf soils. In fact, indirect growth effects are probably of much greater importance and these are dealt with below.

Antagonism of plant pathogens

This is the most important mechanism by which *Bacillus* spp. may have positive effects on sports turf and is certainly the area in which most relevant research has been conducted.

Brown patch disease (causative agent *Rhizoctonia solani*) is common in the United States and the incidence of it has increased in the UK in recent years (Perris and Evans, 1996; Mann, 2004). Marten et al. (1999) found that a strain of *B. subtilis* (strain B2g) was effective against this fungus while Ryder et al. (1999) tested a number of strains of *B. subtilis* and *B. cereus* for activity. It was found that *B. subtilis* strain B931 and *B. cereus* strains A47 and M22 were highly effective against *R. solani* in seedling wheat. Large growth promotion effects were found and in some cases these were the equal to those obtained with a standard biocontrol isolate of *Pseudomonas corrugata*. In the same study, strains B931, A47 and M22 were also found to protect plants against 'take-all' disease (causative organism *Gaeumannomyces graminis* var. *tritici*).

Take-all is another relatively new disease in the UK in sports turf (where the causative organism is *G. graminis* var. *avenae*) and it is most prevalent on young USGA specification putting greens (Mann, 2004). It shows a unique disease cycle of being virulent for a few years, followed thereafter by a stage known as 'take-all decline'. The decline stage is thought to be due to the build up of fungal and bacterial antagonists in the soil (Smith et al., 1989), although exactly which species are most important in turf soils has yet to be identified. One reason for this may be the strain specific effects that exist between *Bacillus* spp. and the disease. Thus while Ryder et al. (1999) found antagonism by *B. subtilis* strain B908 and *B. cereus* strains A47-2 and A47-3 against *G. graminis* var. *tritici*, Landschoot et al. (1993) tested *B. cereus* isolate UW-85, *B. megaterium* isolate B-153-2-2 and *B. subtilis* isolate LT-40 against *G. graminis* var. *avenae* on creeping bentgrass (*A. palustris*) and found no effect of any of the isolates on the pathogen. These results show that strain identity is a major factor in disease suppression by *Bacillus* and indicates that any successful commercial inoculant may have to be very fined-tuned.

Bacillus spp. have been shown to provide plant growth promotion effects through antagonism to a number of other important turf diseases. Stier and Tetrault (1999) found that

B. licheniformis showed some activity against Dollar spot (causative organism *Sclerotinia homeocarpa*) and this bacterium has now been incorporated into a commercial fungicide product, registered by Novozyme (M. Ward, pers comm.). Meanwhile Raikes (1997) reported that microdochium patch disease could be reduced in severity on rye grass by several species of *Bacillus*, while Viji and Uddin (2001) found that *B. lentimorbus* was effective against a number of diseases, including microdochium patch, anthracnose (causative organism *Colletotrichum graminicola*) and leaf spot or 'melting out' (causative organism *Drechslera poae*). *B. subtilis* has been shown to reduce the incidence of summer patch disease (causative organism *Magnaporthe poae*) in both laboratory and field trials, but as with take-all, the effect was dependent upon the strain of bacterium used (Thompson et al., 1996). *B. subtilis* strain QST713 has also been incorporated into a commercial fungicide (Serenade™) for use in America (Yuan and Flanagan, 2000) and strain N4 has been incorporated into compost that has a highly suppressive effect on *R. solani* (Nakasaki et al., 1998). Taken together, these studies show that there is great potential for the use of *Bacillus* as antagonists of plant pathogens and this approach is likely to become an important part of turf grass management in the next few years. The main reason for this is the fact that many of the fungicides currently approved for use on turf will be withdrawn (Wood, 2001) meaning that alternative, biological control strategies will have to be sought.

Control of insect pests

Entomopathogenic bacteria are of great importance in pest control and a review of their efficacy and mode of action is provided by Lacey et al. (2001). Perhaps the best known is *B. thuringiensis* but in turf grass situations, the most widely used is *Paenibacillus popilliae* which is active against white grubs or chafers (Coleoptera: Scarabaeidae) (Klein and Kaya, 1995). The insecticidal activity of these bacteria is associated with the production of proteinaceous toxins, known as Cry proteins which are produced at the time of sporulation. They can be highly specific insect gut toxins and have the added advantage of low toxicity to mammals (Siegel, 2001).

P. popilliae causes milky disease in chafer grubs and has been targeted frequently against the Japanese beetle, *P. japonica* in the USA (Koppenhöffer et al., 2000). However, the efficacy of this bacterium has been limited because of problems with spore production *in vitro* (Redmond and Potter, 1995) and lower levels of infection than expected having been obtained in field trials (Klein and Kaya, 1995). Nevertheless, several commercial products containing this bacterium are currently available in the USA.

B. thuringiensis (Bt) occupies the largest share of the biopesticide market, with most recent estimates of annual sales being about $125 million (Lacey et al., 2001). However, it has not been widely used in the turf grass market, although some studies indicate great potential. Alm et al. (1997) showed that *B. thuringiensis* serovar *japonensis* strain Buibui toxin caused a large reduction in Japanese beetle larval populations in turfgrass. This was achieved through two formulations, one in which the bacterium was arrested in late sporulation phase and the other

in which the toxin was expressed in recombinant *Pseudomonas fluorescens*.

B. thuringiensis subsp. *israelensis* (Bti) is used around the world for the control of many medically important insects, such as blackflies and mosquitoes (Skovmand et al., 2000). Interestingly, Damgaard et al. (1998) found 32 different isolates of Bt occurring on the phylloplane of ryegrass foliage in a pasture in the Netherlands, of which 75% were Bti. Furthermore, several of the Bti isolates were found to have activity against *Tipula oleracea* (Diptera: Tipulidae). Tipulid flies are the most important insect pests of turf in the UK (Mann, 2004) and therefore there may be potential for the use of Bt or Bti in turf, when the currently approved insecticides are withdrawn.

Interactions with arbuscular mycorrhizal fungi

A final indirect benefit that *Bacillus* spp. may have in sports turf is through the growth of arbuscular mycorrhizal (AM) fungi. These fungi form symbiotic associations with the roots of many vascular plants (Smith and Read, 1997). However, in certain conditions, the association may sometimes act in favour of the fungus, meaning that the mycorrhiza becomes parasitic on the plant (Gange and Ayres, 1999). One such situation is with *Poa annua*, the major weed of sports turf, where it has been shown that AM fungi can reduce the growth of this grass, while at the same time promoting growth of the desirable grasses in the sward (Gange, 1998; Gange et al., 1999). However, AM abundance in turf rootzones is very low and addition of fungal inocula does not always increase mycorrhizal colonisation of roots (Gange et al., 1999).

Experiments have shown that *B. subtilis* can act as a mycorrhiza helper bacterium, enhancing the colonisation of roots by indigenous AM fungi or promoting the establishment of colonisation by introduced species (Toro et al., 1997; Vosatka and Gryndler, 2000). A recent experiment (Gange, unpublished) has shown how this interaction may be of benefit in sports turf, but that the result will be determined by the availability of carbon in the system. Replica putting greens containing *Agrostis stolonifera* were created in flower pots using a USGA rootzone mix and maintained at a height of 5 mm. Treatments were established of bacterial addition (the product Organica, Headland Amenity, containing species of *Bacillus*, *Pseudomonas* and *Streptomyces* applied at the recommended rate equivalent to 150g of product in 30 l of water for a 500 m² square area), mycorrhizal addition (Vaminoc-G, MicroBio Ltd, containing four species of *Glomus*, applied at the rate of 19g m⁻²) and carbon addition (the product Fulcrum Blade, Cargill Plc at the rate of 75 l in 400 l water ha⁻¹). Mycorrhizal colonisation results for the treatments in which the fungi were applied are given in Figure 4. It can be seen that addition of the bacterial mixture alone did not affect AM colonisation levels, but that when carbon was added to the system, bacteria did enhance the abundance of the fungi. This simple experiment shows that bacterial inoculants do have the potential to increase AM abundance in replica turf systems, but that their efficacy is limited by the availability of carbon. Therefore, bacterial inoculants in field situations would need to be applied together with a carbon source for the best results to be achieved.

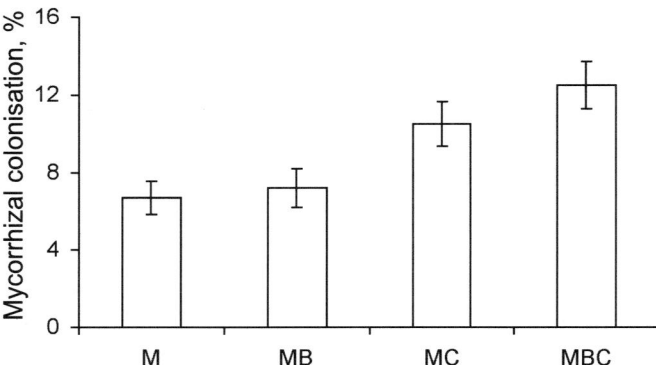

Figure 4. Mycorrhizal colonization of *Agrostis stolonifera* in replica golf greens with addition of mycorrhizal inoculum (M), bacteria (B) and carbon (C).

WHAT FACTORS MAY CONSTRAIN THE EFFICACY OF *BACILLUS* SPP IN WORKING SPORTS TURF?

It has already been shown that turf is a carbon-limited ecosystem and its availability is likely to be the major factor that affects the use of *Bacillus* spp. if they are applied in inoculants to turf. Carbon limitation is caused by the extreme mowing regime that is applied and one example of the deleterious effect of this on an insect pathogen was shown by Rothwell and Smitley (1999). These authors found that the infection rate of larvae of the chafer *Ataenius spretulus* by *P. popilliae* was 68% in the rough areas of a golf course, compared to only 31% in a fairway. The conclusion was that mowing reduced the efficacy of the pathogen, through its removal along with leaf clippings.

Another important factor is the degree of shading of the turf. Most golf putting greens are in relatively open situations, exposed to the sun, but shading is a major feature of football stadia. The design of modern stadia is such that large areas of the pitch, particularly along each long side, are heavily shaded (Newell, 2000). Several studies have examined the effects of shading on populations of *Bacillus* added to turf. In all cases it has been found that shading helped to promote population establishment (Giesler and Yuen, 1995; 2001; Giesler et al., 2000). This is because in shaded areas, leaf wetness duration and relative humidity are increased, while canopy air and foliage temperatures are decreased. This suggests that bacterial inoculants could have a very important role to play in football pitches, as grass growth is known to be reduced in the shaded areas (Anon., 2000). If microbial populations could be established that have plant growth promoting effects, then more even grass growth may be obtained, without recourse to artificial lighting. Many bacteria species are harmed by excessive exposure to ultra violet radiation (Atlas and Bartha, 1997) and this may be another reason why populations establish less easily, if they are applied to areas of turf in full sun. However, virtually all microbial inoculant products now available are in liquid form, meaning that the spores within them will be delivered to a rootzone relatively quickly and will not remain on the surface. The fact that the bacterial species are in spore form is extremely important for their application, since the spore is much more resistant to environmental stresses than is the vegetative cell (Wolken et al., 2003).

Turf grass receives an array of pesticides and some of the fungicides used are known to have bactericidal effects (Yang et al., 2000). However, it is also true that many bacterial species are able to use organic pesticides as a carbon source, causing their degradation (Mercadier et al., 1997; Armbrust, 2001). Furthermore, bacterial inoculants using spore formulations may be compatible for tank mixing with certain pesticides, due to the resistance of the spore to these chemicals (Wolken et al., 2003). Two chemicals that are applied commonly to turf (chlorothalonil and iprodione) have both been shown to reduce the level of spore viability of *P. popilliae* and hence its effectiveness (Dingman, 1994). Iprodione is the most used turf fungicide in the UK yet Dingman's study showed that it was toxic to *P. popilliae* at concentrations less than the recommended rate in the field. While it is accepted that far fewer chemicals for use on turf will be available in the future (Wood, 2001), this does not mean that they will be eliminated. Future studies of bacterial inoculants for use on turf must include work on their compatibility with the chemicals in use or under current development.

Sports turf also receives large amounts of fertilizer each year (Adams and Gibbs, 1994). The addition of nitrogen or phosphorus will clearly alter nutrient dynamics in the rhizosphere, as well as affecting other factors such as pH. It is unknown whether fertilizer application is detrimental to soil bacterial populations in UK turf soils, but one study in the USA suggested that this may not be a significant problem. Elliott and Des Jardin (1999b) could find no effects of a range of organic and inorganic fertilizers over a two year period on various members of the soil microbial community.

PRODUCT AVAILABILITY IN THE UK

Despite the fact that there is a paucity of published research on the bacterial communities associated with sports turf in the UK, there are now a number of products available to the turf manager that are mixtures of bacterial species. In most cases, the species composition of these products is a closely guarded secret, though it is likely that all of them contain at least one species of *Bacillus* (most probably *subtilis*). It would be pointless and divisive to list all the products here, because the market is constantly changing and the information would rapidly become incorrect, as products come and go.

All companies have a web site and some of these are informative and even show results from research that has been conducted with a particular product. A search using Google employing the key words 'bacteria product turf' and stating the domain as .uk will provide a number of web sites which list the products available. If the reader contacts any of these companies, they will find the response to be highly variable. In preparation of this manuscript, we sent the same e-mail to all the companies purporting to offer a bacterial product for use on turf, asking for details. One company responded the next day with considerable information, some refused to provide any details, while others failed to respond. It is therefore easy to understand the degree of scepticism that exists amongst turf managers to these products, when so many claims are made about them and yet so little information is forthcoming (Karnok, 2000). This is unfortunate, because as this review has hopefully shown, such products have great potential benefit and should become an accepted part of integrated turf management in the future.

ACKNOWLEDGEMENTS

We are grateful to the Natural Environment Research Council for funding the experimental studies on microbial community structure.

REFERENCES

Adams, W. A., and Gibbs, R. J. (1994). Natural Turf for Sport and Amenity: Science and Practice (Wallingford, CABI).

Alm, S. R., Villani, M. G., Yeh, T., and Shutter, R. (1997). *Bacillus thuringiensis* serovar *japonensis* Buibui strain for control of Japanese and oriental beetle larvae (Coleoptera: Scarabaeidae). Appl. Entomol. Zool. *32*, 477-484.

Anonymous (2000). Stadium shade still a major problem. Groundsman *54*, 14.

Armbrust, K. L. (2001). Chlorothalonil and chlorpyrifos degradation products in golf course leachate. Pest Manage. Sci. *57*, 797-802.

Atlas, R. M., and Bartha, R. (1997). Microbial Ecology, Fundamentals and Applications (Menlo Park, Benjamin Cummings Science Publishing).

Baker, S. W., Binns, D. J., and Cook, A. (1997). Performance of sand-dominated golf greens in relation to rootzone characteristics. J. Turf Sci. *73*, 43-57.

Benizri, E., Baudoin, E., and Guckert, A. (2001). Root colonization by plant-growth-promoting rhizobacteria. Biocont. Sci. Technol. *11*, 557-574.

Bigelow, C. A., Bowman, D. C., and Wollum, A. G. (2002). Characterization of soil microbial population dynamics in newly constructed sand-based rootzones. Crop Sci. *42*, 1611-1614.

Bremer, D., Ham, J., Owensby, C., and Knapp, A. (1998). Responses of soil respiration to clipping and grazing in a tallgrass prairie. J. Env. Qual. *27*, 1539-1548.

Chanway, C. P. (2003). Plant growth promotion by *Bacillus* and relatives. In Applications and Systematics of *Bacillus* and relatives, R. Berkeley, M. Heyndrickx, N. Logan, and P. De Vos, eds. (Oxford, Blackwell Publishing), pp. 219-235.

Damgaard, P. H., Abdel-Ameed, A., Eilenberg, J., and Smits, P. H. (1998). Natural occurrence of *Bacillus thuringiensis* on grass foliage. World J. Microbiol. Biotechnol. *14*, 239-242.

Dingman, D. W. (1994). Inhibitory effects of turf pesticides on *Bacillus popilliae* and the prevalence of milky disease. Appl. Environ. Microbiol. *60*, 2343-2349.

Elliott, M. L., and Des Jardin, E. A. (1999a). Comparison of media and dilutents for enumeration of aerobic bacteria from bermudagrass golf course putting greens. J. Microbiol. Methods *34*, 193-202.

Elliott, M. L., and Des Jardin, E. A. (1999b). Effect of organic nitrogen fertilizers on microbial populations associated with bermudagrass putting greens. Biol. Fertil. Soils *28*, 431-435.

Elliott, M. L., Guertal, E., and Skipper, H. D. (1996). Bacterial populations and diversity within new USGA putting greens. USGA Turfgrass and Environmental Research Summary, 94-95.

Elliott, M. L., Guertal, E., and Skipper, H. D. (1999). Bacterial populations and diversity within new USGA putting greens. USGA Turfgrass and Environmental Research Summary, 8-9.

Elliott, M. L., Skipper, H. D., Kim, J. H., Miller, L. C., Mazur, A. R., Guertal, E. A., McInroy, J. A., and Kloepper, J. W. (2000). Dynamics of rhizobacteria in new USGA greens. USGA Annual Meeting Abstracts *2000*, 164.

Fermanian, T. W., Shurtleff, M. C., Randall, R., Wilkinson, H. T., and Nixon, P. L. (1997). Controlling Turfgrass Pests (Upper Saddle River, Prentice Hall International).

Frostegård, A., Tunlid, A., and Bååth, E. (1991). Microbial biomass measured as total lipid phosphate in soils of different organic content. J. Microbiol. Methods *14*, 151-163.

Gange, A. C. (1998). Turf needs bacteria. Groundsman, 41.

Gange, A. C., and Ayres, R. L. (1999). On the relation between arbuscular mycorrhizal colonization and plant 'benefit'. *87*, 615-621.

Gange, A. C., and Case, S. J. (2003). Incidence of microdochium patch disease in golf putting greens and a relationship with arbuscular mycorrhizal fungi. Grass For. Sci. *58*, 58-62.

Gange, A. C., Lindsay, D. E., and Ellis, L. S. (1999). Can arbuscular mycorrhizal fungi be used to control the undesirable grass *Poa annua* on golf courses? J. Appl. Ecol. *36*, 909-919.

Garland, J. L., and Mills, A. L. (1991). Classification and characterisation of heterotrophic microbial communities on the basis of patterns of community level sole-carbon-source utilisation. Appl. Environ. Microbiol. *57*, 2351-2359.

Giesler, L. J., and Yuen, G. Y. (1995). Effects of shading on short term colonization by bacterial biological control agents in turfgrass. Phytopathology *85*, 1182.

Giesler, L. J., and Yuen, G. Y. (2001). Biological control may work better in the shade. Golf Course Manage. *69*, 49-52.

Giesler, L. J., Yuen, G. Y., and Horst, G. L. (2000). Canopy microenvironments and applied bacteria population dynamics in shaded tall fescue. Crop Sci. *40*, 1325-1332.

Grayston, S. J., Vaughan, D., and Jones, D. (1997). Rhizosphere carbon flow in trees, in comparison with annual plants: the importance of root exudation and its impact on microbial activity and nutrient availability. Appl. Soil Ecol. *5*, 29-56.

Hagley, K. J. (2002) Microbial Community Structure in Sports Turf Soils, PhD, London.

Hagley, K. J., Miller, A. R., and Gange, A. C. (2003). Variation in life history characteristics of *Poa annua* in golf putting greens. J. Turf. Sports Surf. Sci. *78*, 16-24.

Holl, F. B., Chanway, C. P., Turkington, R., and Radley, R. (1988). Growth response of crested wheatgrass (*Agropyron cristatum* L.), white clover (*Trifolium repens* L.), and perennial ryegrass (*Lolium perenne* L.) to inoculation with *Bacillus polymyxa*. Soil Biol. Biochem. *20*, 19-24.

Imaizumi, S., Nishino, T., Miyabe, K., Fujimori, T., and Yamada, M. (1997). Biological control of annual bluegrass (*Poa annua* L.) with a Japanese isolate of *Xanthomonas campestris* pv. *poae* (JT-P482). Biol. Cont. *8*, 7-14.

Johnson, B. J. (1982). Frequency of herbicide treatments for summer and winter weed control in turfgrass. Weed Sci. *30*, 116-124.

Johnson, B. J., and Murphy, T. R. (1995). Effect of paclobutrazol and flurprimidol on suppression of *Poa annua* spp. *reptans* in creeping bentgrass (*Agrostis stolonifera*) greens. Weed Technol. *9*, 182-186.

Karnok, K. J. (2000). Promises, promises: can biostimulants deliver? Golf Course Manage. *68*, 67-71.

Kirchner, M. J., Wollum, A. G., and King, L. D. (1993). Soil microbial populations and activities in reduced chemical input agroecosystems. Soil Sci. Am. J. *57*, 1289-1295.

Klein, M. G., and Kaya, H. K. (1995). *Bacillus* and *Serratia* species for scarab control. Mem. Inst. Oswaldo Cruz *90*, 87-95.

Koppenhöffer, A. M., Wilson, M., Brown, I., Kaya, H. K., and Gaugler, R. (2000). Biological control agents for white grubs (Coleoptera: Scarabaeidae) in anticipation of the establishment of the Japanese beetle in California. J. Econ. Entomol. *93*, 71-80.

Lacey, L. A., Frutos, R., Kaya, H. K., and Vail, P. (2001). Insect pathogens as biological control agents: Do they have a future? Biol. Cont. *21*, 230-248.

Landschoot, P. J., Osburn, R. M., and Hoyland, B. F. (1993). Influence of selected bacteria on take-all patch of creeping bentgrass. Biological and Cultural Tests for Control of Plant Diseases *8*, 116.

Lindberg, T., and Granhall, U. (1984). Isolation and characterization of dinitrogen-fixing bacteria from the rhizosphere of temperate cereals and forage grasses. Appl. Environ. Microbiol. *48*, 683-689.

Magurran, A. E. (1988). Ecological diversity and its measurements (London, Chapman and Hall).

Mancino, C. F., Barakat, M., and Maricic, A. (1993). Soil and thatch microbial populations in an 80% sand: 20% peat bentgrass putting green. Hortscience *28*, 189-191.

Mann, R. L. (2004). A survey to determine the spread and severity of pests and diseases on golf greens in the UK and Ireland. J. Turf. Sports Surf. Sci. *in press*.

Marten, P., Bruckner, S., and Luth, P. (1999). Plant growth promotion of different cultivated plants and biological control of soil-borne phytopathogenic fungi by *Bacillus subtilis* strain B2g. Z. Pflanz. Pflanz. *106*, 74-81.

Mercadier, C., Vega, D., and Bastide, J. (1997). Iprodione degradation by isolated soil microorganisms. FEMS Microbiol. Ecol. *23*, 207-215.

Nakasaki, K., Hiraoka, S., and Nagata, H. (1998). A new operation for producing disease-suppressive compost from grass clippings. Appl. Environ. Microbiol. *64*, 4015-4020.

Newell, A. V. (2000). Stadia design: where the sun doesn't shine! Int. Turf. Bull. *210*, 21-22.

Perris, J., and Evans, R. D. C. (1996). The Care of the Golf Course (Bingley, Sports Turf Research Institute).

Raikes, C. (1997). Fusarium patch disease on winter sports turf in the UK: A potential biocontrol candidate. J. Turf. Manage. *2*, 35-50.

Redmond, C. T., and Potter, D. A. (1995). Lack of efficacy of *in-vivo*-produced and putatively *in-vitro*-produced *Bacillus popilliae* against field populations of Japanese beetle (Coleoptera: Scarabaeidae) grubs in Kentucky. J. Econ. Entomol. *88*, 846-854.

Rennie, R. J., and Larson, R. I. (1979). Nitrogen fixation associated with disomic chromosome substitution lines of spring wheat. Can. J. Bot. *57*, 2771-2775.

Rothwell, N. L., and Smitley, D. R. (1999). Impact of golf course mowing practices on *Ataenius spretulus* (Coleoptera: Scarabaeidae) and its natural enemies. Environ. Entomol. *28*, 358-366.

Ryder, M. H., Yan, Z. N., Terrace, T. E., Rovira, A. D., Tang, W. H., and Correll, R. L. (1999). Use of strains of *Bacillus* isolated in China to suppress take-all and rhizoctonia root rot, and promote seedling growth of glasshouse-grown wheat in Australian soils. Soil Biol. Biochem. *31*, 19-29.

Siegel, J. P. (2001). The mammalian safety of *Bacillus thuringiensis*-based insecticides. J. Invert. Pathol. *77*, 13-21.

Skipper, H. D. (2001). Identification and metabolic diversity of rhizobacteria from bentgrass and bermudagrass greens. USGA Turfgrass and Environmental Research Summary, 13.

Skovmand, O., Kerwin, J., and Lacey, L. A. (2000). Microbial control of mosquitoes and black flies. In Field Manual of Techniques in Invertebrate Pathology: Application and Evaluation of Pathogens for Control of Insects and Other Invertebrate Pests, L. A. Lacey, and H. K. Kaya, eds. (Dordrecht, Kluwer Academic), pp. 767-785.

Smith, J. D., Jackson, N., and Woolhouse, A. R. (1989). Fungal Diseases of Amenity Turfgrasses (London, E. and F.N. Spon Ltd).

Smith, S. E., and Read, D. J. (1997). Mycorrhizal Symbiosis (San Diego, Academic Press).

Stewart, V. I. (1994). Sports Turf Science, construction and maintenance (London, E. and F.N. Spon).

Stier, J., and Tetrault, D. (1999). Use of bacteria and biostimulants to ameliorate dollar spot damage in creeping bentgrass. Wisconsin Turf Res. *17*, 49-55.

Thompson, D. C., Clarke, B. B., and Kobayashi, D. Y. (1996). Evaluation of bacterial antagonists for reduction of summer patch symptoms in Kentucky bluegrass. Pl. Dis. *80*, 856-862.

Timmusk, S., Nicander, B., Granhall, U., and Tilberg, E. (1999). Cytokinin production by *Paenibacillus polymyxa*. Soil Biol. Biochem. *31*, 1847-1852.

Toro, M., Azcon, R., and Barea, J. M. (1997). Improvement of arbuscular mycorrhiza development by inoculation of soil with phosphate-solubilizing rhizobacteria to improve rock phosphate bioavailability (P-32) and nutrient cycling. Appl. Environ. Microbiol. *63*, 4408-4412.

USGA Green Section Staff (1993). Specification for a method of putting green construction (Far Hills, N.J., United States Golf Association).

Vargas, J. M. (1994). Management of Turfgrass Diseases (Boca Raton, CRC Press).

Viji, G., and Uddin, W. (2001). Biocontrol of gray leaf spot (blast) of perennial ryegrass by *Bacillus lentimorbus*. Phytopathology Suppl. *91*, S92.

Vosatka, M., and Gryndler, M. (2000). Response of micropropagated potatoes transplanted to peat media to post-vitro inoculation with arbuscular mycorrhizal fungi and soil bacteria. Appl. Soil Ecol. *15*, 145-152.

Wolken, W. A. M., Tramper, J., and van der Werf, M. J. (2003). What can spores do for us? Trends Biotechnol. *21*, 338-345.

Wood, J. (2001). What will be left to spray? HDC News *74*, 14-15.

Yang, Y. H., Yao, J., Hu, S., and Qi, Y. (2000). Effects of agricultural chemicals on DNA sequencing diversity of a soil microbial community: a study with RAPD markers. Microb. Ecol. *39*, 72-79.

Yuan, C., and Flanagan, S. R. (2000). The mechanism underlining the efficacy of a novel biofungicide Serenade™ WP in disease control. Phytopathology Suppl. *90*, S87.

Chapter 15

Antimicrobial Activity of *Bacillus* Probiotics

Maria C. Urdaci and Irina Pinchuk

SUMMARY

The production of antimicrobials is one of the mechanisms by which probiotic bacteria are believed to exert their beneficial effect on the host, presumably by inhibiting the growth and colonization of the gastrointestinal tract by pathogenic bacteria. This chapter summarises what is now known about the production of bacteriocins, bacteriocin-like inhibitory substances (BLIS) and antibiotics in *Bacillus* species. An attempt is also made to demonstrate the potential that these antimicrobial substances may have in specific commercial formulations.

INTRODUCTION

During the last three decades probiotics have been successfully introduced in the international market for human or animal use. Many of these probiotic bacteria are sold in fermented foods, and their activity is often attributed to the beneficial effects of functional dairy products. The ability of *Bacillus* to form heat-resistant spores is attractive for commercial exploitation. Currently, although limited, some documentation appears to support the beneficial physiological effects of *Bacillus* probiotics in humans and, particularly, in animals (Hattori et al., 1965; Mazza, 1994; Sorokulova et al., 1997; Kiers et al., 2003). However, data describing the functional mechanisms of *Bacillus* probiosis remains very limited.

One century ago E. Metchnikoff, who proposed the scientific basis for the use of probiotics, hypothesized that "vital" microbial competition played one of the major roles in protection of the host against pathogenic microorganisms (Metchnikoff, 1905). In fact, bacteria interact with each other as they attempt to establish themselves in, and then, dominate their natural environment or host. Depending upon the particular microorganism, they interact synergistically or antagonistically as well as competing for their own ecological niche. These competitive interactions take place naturally on the surfaces of the skin as well as on the mucus surface. Most probably, these microbial interactions play a major role in the maintenance of the normal microflora of skin and mucous (especially in the gastro-intestinal and vaginal mucous surfaces) and contribute to the prevention of infectious disease (Wannamaker, 1980; Brook, 1999).

Bacterial interactions operate through a number of mechanisms, including the production of antimicrobial substances, changes in the bacterial microenvironment and competition for nutritional substrates. The mediators of bacterial interference vary and include the production of bacteriophages, complex antimicrobial proteic/peptidic molecules (e.g., certain enzymes and bacteriocins), antibiotics, and less complex molecules such as hydrogen peroxide, organic acids, and ammonia (Wannamaker, 1980; Smith, 1995).

Based on the theory of bacterial interference, it has been proposed that production of antimicrobial substances by probiotics is one of the principle mechanisms by which they protect the host against pathogenic microorganisms. These compounds may not only reduce the number of viable cells but they may also affect bacterial metabolism or toxin production (Fuller, 1989; Dunne and Shanahan 2002). The majority of currently published scientific reports and reviews concern the studies of antimicrobial activity and its mediators in lactic acid probiotic bacteria (Vescovo et al., 1993; Bernet-Camard et al., 1997; Boris et al., 2001; Eijsink et al., 2002). However, analysis of those involved in *Bacillus* probiosis is significantly less understood. Therefore, in this chapter we will examine what is known about the antimicrobial activity of the *Bacillus* spp., especially in those strains that have been reported to be used as probiotics. Specifically, we will review the production of antimicrobial substances with a specific mode of action (bacteriocins and antibiotics) by these probiotic *Bacilli* bacteria and their potential health-promoting effects.

BACILLUS SPP. AS ANTIMICROBIAL SUBSTANCES PRODUCERS

The potential of *Bacillus* spp. to synthesize a wide variety of biologically active metabolites, in particularly those with antibacterial and/or antifungal activity has been intensively exploited in medicine and industry. Moreover, the ability of these bacteria to produce antimicrobial substances is essential for their ability to control plant diseases when they are using as a biological control agents (Raaijmakers et al., 2002 and Appendix III).

As mentioned above the antagonistic activity of a particular bacterial strain depends upon their ability to produce different substances, including compounds with very specific spectrums and modes of action such as bacteriolytic enzymes, bacteriocins, and antibiotics. All of these substances are produced by *Bacillus*.

ROLE OF LYTIC ENZYME PRODUCTION IN THE ANTAGONISTIC ACTIVITY OF *BACILLUS* SPP.

The saprophytic bacteria of the *Bacillus* genus are known to produce many hydrolytic enzymes with different substrate specificity. Based on their antimicrobial activity these lytic enzymes can be divided in two essential groups, those with ability to cause lysis of bacteria and those with anti-fungal activity (Biziulevieius and Sukaite, 2002; Kim et al., 1999).

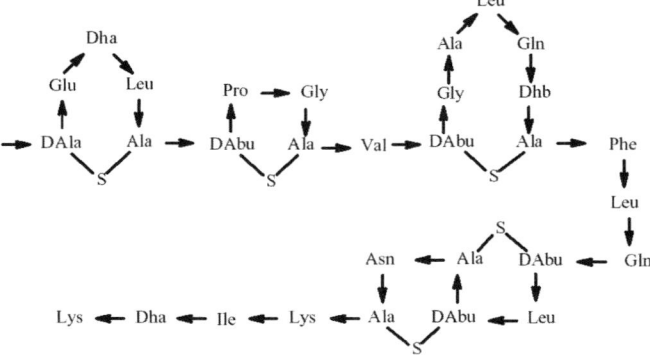

Figure 1. Structural formula of subtilin (Gross and Kiltz, 1973). Dha = dehydroalanine; Dhb = dehydrobutyrin (-methyldehydroalanine), DAbu = aminobutyric acid, Ala-S-Ala = lanthionine; DAbu-S-Ala = -methyllanthionine.

These enzymes are of considerable interest as potentially useful chemotherapeutics (particularly in the veterinary field), food preservatives and biocontrol agents (Biziulevieius and Sukaite, 2001). Many of these hydrolases are autolysins involved in sporulation and can lyse both live and heat-treated bacterial cells (Work, 1959). This is the case for murein hydrolases isolated from many species of this genus, which are reported to have an activity against some Gram-positive bacteria, in particularly against *Micrococcus* and *Streptococcus* spp. Some of the extracellular hydrolases, that are not involved in the process of autolysis, such as elastases, chitinases, laminarinases, and cellulases, also have been reported to have high lytic activity against bacteria and fungi. It has been reported that elastase of *B. subtilis* 6 caused lysis of freshly grown cells of Gram-negative bacteria, such as *Proteus vulgaris*, *Klebsiella pneumoniae*, *Salmonella typhi* and *Pseudomonas aeruginosa* (Gupta et al., 1992). A strain identified as *B. licheniformis* YS-1005 reported to secrete two endopeptidases with highly lytic activity against cariogenic *Streptococcus mutans* strains (Kim et al., 1999). A complex endopeptidase preparation known as Lysosubtilin (Biosinteze, Lithuania) has been reported to have broad antimicrobial spectrum due to the synergistic action of enzymes included in the preparation. Moreover, this enzymatic complex caused the lysis of microorganisms resistant to lysozyme (Biziulevieius and Sukaite, 2002). The main function of *Bacillus* extracellular hydrolases is the release of nutrients from different substrates. In addition, excreted enzymes together with antimicrobial compounds may be used by bacteria to compete with other microbial species and maintain an ecological niche (Helisto et al., 2001).

BACTERIOCINS AND BACTERIOCIN-LIKE INHIBITORY SUBSTANCES (BLIS) PRODUCED BY *BACILLUS* SPP.

Bacteriocins are one of the most abundant and diverse classes of antibacterial molecules having been detected in all major lineages of Eubacteria. They were originally defined as proteinaceous compounds possessing bactericidal activity other than to the producing strain. Unlike traditional antibiotics,

some have a broad spectrum, whereas others possess a narrow spectrum of activity and inhibit only bacteria of the same or closely related species. Bacteriocins are ribosomally synthesized peptides with great variation in size and primary sequence but tend to be small cationic peptides of 20 to 60 amino acids with amphipathic characteristics and high isoelectric points. The first bacteriocins to be studied in depth were those produced by the *Enterobacteriaceae*, which are commonly referred to as colicins (Braun et al., 1994). More recently, it has been discovered that many Gram-positive bacteria produce such antimicrobial peptides. In particular, bacteriocins produced by lactic acid bacteria have attracted considerable interest because of their potential application as natural food preservatives against pathogenic bacteria.

The bacteriocins have been grouped into four distinct classes defined by Klaenhammer (Klaenhammer, 1988). Bacteriocins belonging to class I (lantibiotics) are small (<5KDa) peptides characterized by the presence of unusual constituents like non-proteinogenic didehydroamino acids and lanthionines, that form intramolecular rings and are generated through post-translational modifications. These peptides generally have a broad spectrum of activity and form unstable pores (McAuliffe et al., 2001). Out of the about 26 known lantibiotics, the nisins (A and Z) are the best-studied members which are also of commercial value. Class II bacteriocins represent small (<10 kDa), heat-stable, membrane-active peptides (Nes and Holo, 2000). This class represents the largest group of bacteriocins, and is divided into 3 subgroups. Class IIa include pediocin-like peptides, and this subgroup has attracted much of the attention due to their anti-*Listeria* activity. Unlike the lantibiotics, class II bacteriocins do not undergo extensive post-translational modification. Pediocin PA-1 have fairly broad inhibitory spectra and can inhibit some less closely related Gram-positive bacteria such as *S. aureus*, *Clostridium* spp and *Bacillus*. Bacteriocins belonging to class III consist of large (>30 kDa), heat-labile proteins, while class IV represents complex bacteriocins that contain essential lipid or carbohydrate moieties in addition to a protein component. Moreover, there are bacteriocins that are not biochemically characterized and certain bacteriocin-like compounds that do not correspond to the Klaenhammer classification, which are named as bacteriocin-like inhibitory substances (BLIS).

Bacillus species including *B. subtilis*, *B. licheniformis* and *B. cereus* have been reported to produce bacteriocins and BLIS. Subtilin (Figure 1) has been the most widely characterized bacteriocin: and is a lantibiotic of 32-amino acids, produced by *B. subtilis* ATCC 6633, whose structure and mode of action are similar to those of nisin, exhibiting 59% of sequence homology in the nisin pre-peptides (Gross and Kiltz, 1973 ; Entian and de Vos, 1996). Subtilin is highly stable at high temperature and low pH values. Both subtilin and nisin show similar destructive tendencies toward Gram-positive bacteria. A variant of subtilin (subtilin B) was found to have reduced antibiotic activity due to posttranslational succinylation of the amino group of the N-terminal tryptophan residue (Chan et al., 1993). Sublancin (Figure 2A) from *B. subtilis* 168 is quite different and contains a single lanthionine linkage and two disulfide bridges (Paik et al., 1998). This strain also produces the subtilosin (Figure 2B),

A

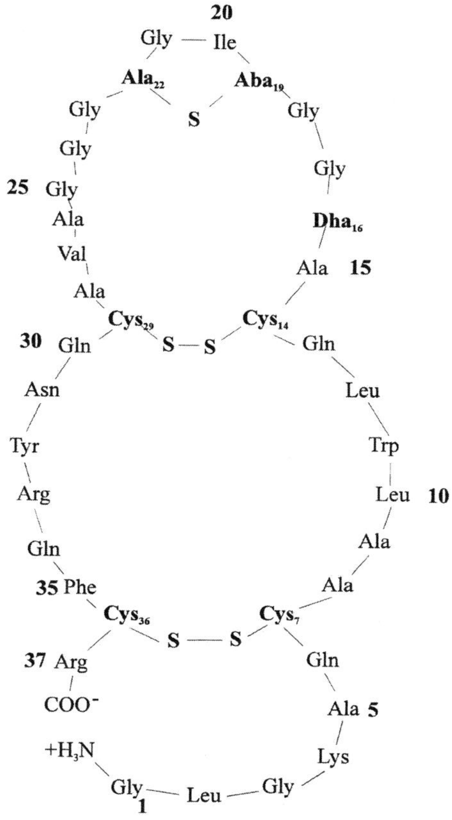

B

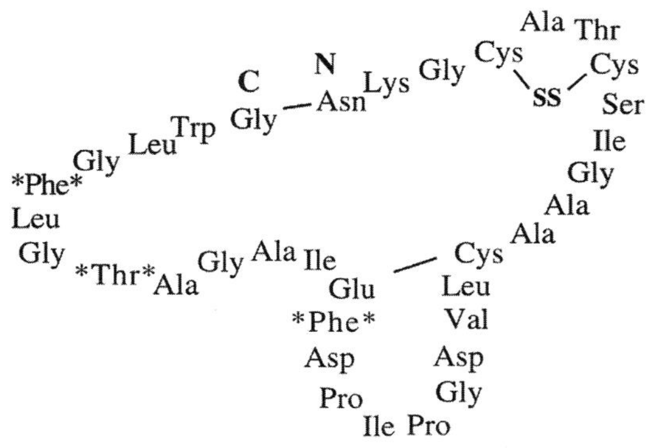

Figure 2. (A) Proposed structure of sublancin (Zheng et al., 1999). The locations of thioether and disulfide bridges are indicated. The representation of the sublancin structure as *three open circles* and a salt-bridge between the N-terminal amino group and the C-terminal carboxyl group is arbitrary, since no information about the secondary structure of sublancin is available. (B) Proposed structure of subtilosin (Babasaki et al., 1985). The proposed link between Glu31 and Cys21 is shown. The asterisks on either side of an amino acid indicate residues that have likely undergone chemical modification. SS, disulfide link between Cys12 and Cys15.

a 35 amino acids cyclic peptide that possess 32 common amino acids and some unusual residues (Babasaki et al., 1985; Zheng et al., 1999). Subtilosin has been reported to be active against *L. monocytogenes* and *B. cereus*. Stöver and Driks (1999) reported that *B. subtilis* 168 produced a 31-kDa sporulation protein with a broad spectrum of antibacterial activity, called TasA. Tas A is secreted into the culture medium early in sporulation and is also incorporated into spores. This latter property may confer a competitive advantage to the spores of these bacteria. The small lantibiotic mersacidin (20 a.a.) has been reported to be produced by *B. subtilis* (Chatterjee et al., 1992). Mersacidin does not form pores in the cytoplasmic membrane but rather inhibits cell wall biosynthesis. This bactericidal effect is based on complex formation with the cell wall precursor lipid II which blocks the transglycosylation reaction of cell wall biosynthesis (Brötz et al., 1998). Its activity *in vivo* against methicillin-resistant *Staphylococcus aureus* strains compares with that of the glycopeptide antibiotic vancomycin. In contrast to vancomycin, the activity of mersacidin was not antagonized by the tripeptide diacetyl-L-Lys-D-Ala-D-Ala, indicating that at the molecular level its mode of action differs from those of glycopeptide antibiotics. The biosynthetic gene cluster of the lantibiotic mersacidin has a size of 12.3 kb and is located at the 348° region of the chromosome as compared to the genome of *Bacillus subtilis* strain 1A1, Marburg 168 (Altena et al., 2000). Moreover, two new distinct lantibiotic peptides ericin S (3,442 Da) and ericin A (2,986 Da) were recently isolated from *B. subtilis* strain A1/3. Ericin S is a subtilin-like lanthionine. Due to only four amino acid exchanges, ericin S and subtilin

showed similar antibiotic activities as well as similar properties in response to heat and protease treatment. Two C-terminal rings of ericin A differed from the subtilin lanthionine pattern. However, this difference is not very important, since compared to the highly active subtilin, ericin A possesses only minor antimicrobial activity (Stein et al., 2002).

B. licheniformis 26 L-10/3RA, isolated from the rumen of buffalo, has been described to produce Lichenin, a BLIS that possesses an bactericidal activity against *S. bovis* and *Eubacterium ruminantium* (Pattnaik et al., 2001). Interestingly, the lichenin was produced only under anaerobic conditions and exposure of the culture supernatant to aerobic conditions resulted in the loss of its inhibitory activity. Another BLIS of low molecular mass (2 kDa), bacillocin 490, is produced by a *B. licheniformis* thermophilic strain. This bacteriocin was active against closely related *Bacillus* spp. Moreover, bactericidal activity of bacillocin 490 was remarkably stable in a wide pH range (Martirani et al., 2002).

The capacity to produce BLIS was also observed for members of the *Bacillus cereus* phylogenetic group (*B. cereus*, *B. thuringiensis*, *B. mycoides* and the psychrotolerant *B. weihenstephanensis*). Cerein produced by *B. cereus* Gn105 (Nacleiro et al., 1993) with a molecular mass of 9 kDa and displays antimicrobial activity against a wide range of *B. cereus* strains without significant inhibitory effect on the growth of other tested Gram-positive bacteria. In contrast, cerein 7 (3.94 kDa), produced by *B. cereus* Bc7 (Oscariz et al., 1999) has been described to inhibit a wide range of Gram-positive bacteria and was classified as a class II bacteriocin. More recently, a

Table 1. Principal bacteriocins and BLIS produced by *Bacillus* spp.

Bacteriocins and BLIS	Type	Activity spectra	Properties	Producer species	Reference
Subtilin	Lantibiotic (32 aa, 3317 Da)	Gram-positive bacteria*	Membrane pore formers	*B. subtilis*	Gross and Kiltz, 1973
Sublancin	Lantibiotic (37 aa, 3877 Da)	Gram-positive bacteria	-	*B. subtilis*	Paik et al., 1998
Ericins A, S	Lantibiotic (2,986 and 3442 Da)	Gram-positive bacteria	-	*B. subtilis*	Stein et al., 2002
Mersacidin	Lantibiotic (20 aa, 1825 Da)	Gram-positive bacteria	Cell wall biosynthesis inhibition	*B. subtilis*	Chatterjee et al., 1992
Coagulin	Pediocin-like (44 aa, 4612 Da)	Gram-positive bacteria	Membrane target	*B. coagulans*	Le Marrec et al., 2000
Lichenin	BLIS (12 aa, appr. 1400 Da)	Gram-positive bacteria	Sensitive to oxygen	*B. licheniformis*	Pattnaik et al., 2001
Bacillocin	BLIS (appr. 2000 Da)	Gram-positive bacteria	-	*B. licheniformis*	Martirani et al., 2002
Subtilosin	Cyclic peptide (35 aa, 3399 Da)	Gram-positive bacteria	New class of BLIS	*B. subtilis*	Babasaki et al., 1985; Zheng et al., 1999

* Generally active against some of the Gram-positive bacteria.

BLIS produced by a *B. thuringiensis* soil isolate which has been named thuricin 439, thermo- and pH- stable bacteriocin, found to exhibit a broad range of inhibitory activity (Ahern et al., 2003). Further examples are thuricin, thuricin 7 and tochicin isolated from different strains of *B. thuringiensis* (Favret and Yousten, 1989; Cherif et al., 2001). Table 1 represents a summary of principal bacteriocins and BLIS produced by *Bacillus* species currently used as probiotics

The last three decades of research on bacteriocins, has led to the discovery and characterization of a broad range of these antimicrobials from *Bacillus* spp. However, direct commercial exploitation in the veterinary and food industries of these antimicrobials is very limited. Major drawbacks to the widespread use of bacteriocins is due primarily to their instability in different food systems, resulting from suboptimal pH, insolubility and sensitivity to proteolytic enzymes. However, characteristics of bacteriocins suggest, that they could play an important role in bacterial competition, particularly among closely related species. Therefore, selection of bacteriocin-producing strains for probiotic design with evaluation of their safety should lead to optimal probiotic species.

BACILLUS SPP. AS ANTIBIOTIC PRODUCERS

In 1974 Berdy (Berdy 1974) reported 167 antibiotics produced by members of the *Bacillus* genus, including 66 different peptides antibiotics from *B. subtilis* and 23 from *B. brevis*. Since then significant numbers of new antibiotics have been isolated from strains of the *Bacillus* genus and many of them have applications in pharmacology, veterinary and food industry.

Antibiotics are secondary metabolites mostly produced at a time when bacterial cells begin to differentiate (i.e., sporulate). The maintenance of a substantial component of the bacterial genome devoted solely to the synthesis of an antibiotic suggests that this molecule is important, if not essential, to the survival of these organisms in their natural habitat. The best-accepted theory is that non-ribosomal antibiotics may play a role in competition with other microorganisms during sporulation, dormancy or the germination process, acting as some sort of hormone or signalling molecule (Mendoza et al., 1993; Zuber et al., 1993).

Bacillus antibiotics share a full range of antimicrobial activity. For example: bacitracin, laterosporin, gramicidin and tyrocidin are active against Gram-positive bacteria; polymyxin possesses activity against Gram-negative bacteria; difficidin is a broad spectrum antibiotic; and mycobacillin and zwittermicin have anti-fungal effects.

Among the dozens of different antibiotics produced by *Bacillus*, the peptide and lipopeptides are the most numerous and *B. subtilis* is the principal producer. *B. subtilis* 168 is the best-studied strain in the genus *Bacillus*, whose genome was completely sequenced in 1997. Strain 168 is known to produce at least two non-ribosomal antibiotics, surfactin (Vollenbroich et al., 1997a) and bacilysin (Ozcengiz et al., 1990). The production of other antibiotics by strain 168 has also been predicted on the basis of genome sequence analysis, such as case for plipastatin (Tsuge et al., 1999). Recently, another antibiotic produced by this *strain* has been identified and named bacilysocin (Tamehiro et al., 2002). It is a novel phospholipid antibiotic (lysophosph atidylglycerol) that may be derived from phosphatidylglycerol through acyl ester hydrolysis. Bacilysocin possess an antifungal activity and described to accumulate within (or associated with) the cells of *Bacillus subtilis* 168.

LIPOPEPTIDE ANTIBIOTICS

Many *Bacillus* strains produce a small peptide(s) with a long fatty moiety, the so-called lipopeptide antibiotics and have specific activities against fungi, bacteria, erythrocytes and different yeasts. The lipopeptides antibiotics are synthesized by multienzyme complexes. The peptide grows in a defined sequence by moving on the template of the multifunctional peptide synthetases (PS). PS are composed of repetitive units (modules), each capable of incorporating one amino acid constituent into the peptide chain (Nakano and Zuber, 1990). A module is defined as the unit that catalyzes the incorporation of a specific amino acid into the peptide product. The arrangement of the modules of a peptide synthetase is usually co-linear with the amino acid sequence of the peptide. The modules can be further subdivided into different domains that are characterized by a set of short conserved sequence motifs. The core of each module is an amino acid adenylation domain that recognizes and activates a specific amino acid. The thiolation domain, located C terminally of the adenylation domain, contains an invariant serine residue essential for the binding of a 4'-phosphopantetheine cofactor. An N-terminal condensation domain is required for the coupling of two consecutively bound amino acids. In addition, modules can be supplemented

		1	2	3	4	5	6	7	
Surfactin	R-CH-CH_2CO-	L-Glu-	L-Leu-	D-Leu-	L-Val-	L-Asp-	D-Leu-	L-Leu	- O
Iturin A	R-CH-CH_2CO-	L-Asn-	D-Tyr-	D-Asn-	L-Gln-	L-Pro-	D-Asn-	L-Ser	- NH
Iturin C	R-CH-CH_2CO-	L-Asp-	D-Tyr-	D-Asn-	L-Gln-	L-Pro-	D-Asn-	L-Ser	- NH
Mycosubtilin	R-CH-CH_2CO-	L-Asn-	D-Tyr-	D-Asn-	L-Gln-	L-Pro-	D-Ser-	L-Asn	- NH
Bacillomycin D	R-CH-CH_2CO-	L-Asn-	D-Tyr-	D-Asn-	L-Pro-	L-Glu-	D-Ser-	L-Thr	- NH
Bacillomycin F	R-CH-CH_2CO-	L-Asn-	D-Tyr-	D-Asn-	L-Gln-	L-Pro-	D-Asn-	L-Thr	- NH
Bacillomycin L	R-CH-CH_2CO-	L-Asp-	D-Tyr-	D-Asn-	L-Ser-	L-Gln-	D-Ser-	L-Thr	- NH

		1	2	3	4	5	6	7	8	9	10
Fengicins	R-CHOH-CH_2CO-	L-Glu-	D-Orn-	D-Try-	D-AThr-	L-Glu-	X	- L-Pro-	L-Gln-	L-Tyr-	L-Ile
Plipastatins	R-CHOH-CH_2CO-	L-Glu-	D-Orn-	L-Try-	D-AThr-	L-Glu-	X	- L-Pro-	L-Gln-	D-Tyr-	L-Ile

Figure 3. Schematic structures of the major lipopeptides antibiotics produced by *Bacillus* strains. R indicates an alkyl moiety (generally C_{14} to C_{17}). AThr, Allo-threonine; Orn, ornithine; X = D-Ala and D-Val for fengycins and plipastatins A and B isoforms. The horizontal square brackets represent the cyclization sites of these cyclopeptides.

with domains that catalyze the modification of the activated amino acid, such as N-methylation, and epimerization from the L-configuration to the D-configuration (Marahiel, 1997; Marahiel et al., 1997).

All these antibiotics are either cyclopeptides (iturinics) or macrolactones (surfactins, fengycins and plipastatins) characterized by the presence of L and D amino acids and variable hydrophobic tails (Figure 3) (Hourdou et al., 1989). Surfactin and iturinics are the lipoheptapeptides wheras fengycin and plipastatins are the lipodecapeptides that also posses the unusual amino acids such as ornithine and allo-threonine. Moreover, surfactin and fengycin contain a β-hydroxy fatty acid, whereas members of the iturinic family, such as mycosubtilin, bacillomycin, and iturin, carry a β-amino fatty acid modification. Frequently, surfactin producer strains also produce iturinic antibiotics (Maget-Dana and Peypoux, 1994). However, it has been observed, that modification of culture conditions (nitrogen source, pH) could lead to the production of additional antibiotics, such as fengycins (Vanittanakom et al., 1986) or plipastatins (Umezawa et al., 1986). These features suggest a high degree of *Bacillus* adaptability, by modulation of the genetic expression in the region of the *Bacillus* genome involved in the synthesis of these two families of compounds (Duitman et al., 1999).

The Iturinic family included the mycosubtilin (Peypoux et al., 1986), bacillomycin D, F, L, (Peypoux et al., 1984), bacillopeptin (Kajimura et al., 1995) and iturin A, C, D, E (Besson et al., 1978; Besson and Michel, 1987). The amino acid compositions of iturin A and mycosubtilin are almost identical, except that the sixth and seventh amino acids are inverted as shown in Figure 3. Iturins share a common mechanism of action, inserting into the cytoplasmic membrane with their hydrophobic tail and auto-aggregating to form a pore, which causes cellular leakage (Maget-Dana and Peypoux 1994). They are described as potent antifungal agents that can be used as biopesticides for plant protection. Moreover, it has been demonstrated that the biocontrol activity of *B. subtilis* RB14, which has a suppressive effect against several phytopathogens, can be attributed mainly to production of iturin A (Asaka and Shoda, 1996).

Surfactins are reportedly produced by several *Bacillus* species (*subtilis, licheniformis, natto, pumilus* and show potent antiviral and antimycoplasma activities (Vollenbroich et al., 1997a, 1997b). Viruses that are covered by a lipoprotein membrane, such as herpes- and retroviruses, are efficiently inactivated by these biosurfactants.

Fengycins are antifungal lipopeptide complexes produced by *B. subtilis* and *B. amyloliquefaciens*. These antibiotics possess high activity against filamentous fungi but have no effect against yeast and bacteria. The lipid moiety of the unsaturated fatty acids is variable and has been identified as anteiso-pentadecanoic acid (ai-C15), iso-hexadecanoic acid (i-C16), n-hexadecanoic acid (n-C16) (Vanittanakom et al., 1986).

Plipastatins, which were isolated from *B. subtilis* and *B. cereus*, have very similar structures when compared to fengycins. Plipastatins are biosynthesised as a mixture of isoforms characterized by variations in both: the nature of the hydrophobic tail and the amino acid composition (Nishikiori et al., 1986). The hydrophobic tail of these antibiotics is composed from 3(*R*)-hydroxy hexadecanoic acid (plipastatins A1 and B1) or a 14(*S*)-methyl-3(*R*)-hydroxy hexadecanoic acid (plipastatins A2 and B2) while the amino acid sequences differ in position 6 with a D-Ala (plipastatin A1 and A2) substituted by a D-Val (plipastatin B1 at B2) (Volpon et al., 2000). Their antimicrobial spectrum remains mostly unknown. However, they were shown to inhibit the phospholipase A_2 (PLA_2), an enzyme involved in a number of physiologically important eukaryotic host cellular processes, such as inflammation, acute hypersensitivity and blood platelet aggregation.

POLYPEPTIDE ANTIBIOTICS

Bacillus species produce polypeptide antibiotics, e.g. gramicidin, tyrocidine, and bacitracin. As well as in the synthesis of the lipopeptide antibiotics, the peptide chain of polypeptide antibiotics grow in a defined sequence using the template of multifunctional peptide synthetases (Nakano and Zuber, 1990). Gramicidins and tyrocidine, that are produced by *B. brevis* strains (Katz and Demain, 1977) and loloatin A, produced by a strain of *B. laterosporus*, are small cyclic decapeptides (Krachkovskii et al., 2002). Traditionally, gramicidins have been considered to be active only against Gram-positive bacteria. However, according to recent data gramicidin S also possesses excellent activity against Gram-negative bacteria and *Candida albicans* (Kondejewski et al., 1996). Polymyxins, branched, cyclic decapeptides linked to a fatty acid residue were isolated from *B. polymyxa* and possess activity against Gram-negative bacteria. They are reported to affect the cell membrane by a detergent-like action (Katz and Demain, 1977). At least 11 different antibiotics of the polymyxin group have been identified, each designated by a letter. Colistins and the circulins are structurally similar to the polymyxins have also been described to be produced by different *Bacillus* strains (Paulus, 1967).

Dodecapeptide bacitracin was isolated from different strains of *B. licheniformis* and *B. subtilis* and possess potent antibiotic activity against Gram-positive bacteria (Johnson et al., 1945; Azevedo et al., 1993). This antibiotic essentially acts though the inhibition of peptidoglycan biosynthesis, binding to C55-isoprenyl pyrophosphate (IPP) and inhibiting the conversion of IPP to C55-isoprenyl phosphate, which is catalyzed by a membrane-associated pyrophosphatase (Toscano and Storm, 1982). Interestingly, producer strains possess some kind of an immunity system against this antibiotic, harboring an ABC transporter system, which has been proposed to pump bacitracin out of the cell for self-protection. Cyclic tridecapeptide antibiotic mycobacillin has been described to be produced by different *B. subtilis* strains and possess a wide-spectrum antibacterial and antifungal activity (Majumdar and Bose, 1958; Katz and Demain, 1977). The mycobacillin target appears to be a lipid-protein site on the membrane of sensitive *Aspergillus niger* (Chowdhury et al., 1998). Moreover, two new peptide antibiotics, produced by *Bacillus polymyxa*, have been described and named as the gavaserin (cyclic octapeptide) and saltavalin (not cyclic polypetide) (Pichard et al., 1995).

Finally, some of the simplest peptide antibiotics have been reported to be produced by *Bacillus* strains. For example, bacilysin (Figure 4) is one of the simplest known peptide antibiotics, and produced by certain strains of *B. subtilis*. It is a dipeptide containing L-alanine and the unusual amino acid L-anticapsine (Walker and Abraham 1970). Another simple peptide antibiotic produced by *B. subtilis* is the rhizocticin (a phosphono-oligopeptide), which has been reported to possess antifungal activity (Kugler et al., 1990) and the tupuseleiamides (new acyldipeptides) produced by a *B. laterosporus* strain (Barsby et al., 2002). Galvez et al. (1993) purified an antifungal and antibacterial peptide (active on Gram-positive bacteria) produced by *B. licheniformis* A12. It is a 0.77-kDa hydrophilic peptide containing two residues of Glu and one of Arg, Ala, Pro, Tyr and Orn. No fatty acids, phosphorus or carbohydrates were detected.

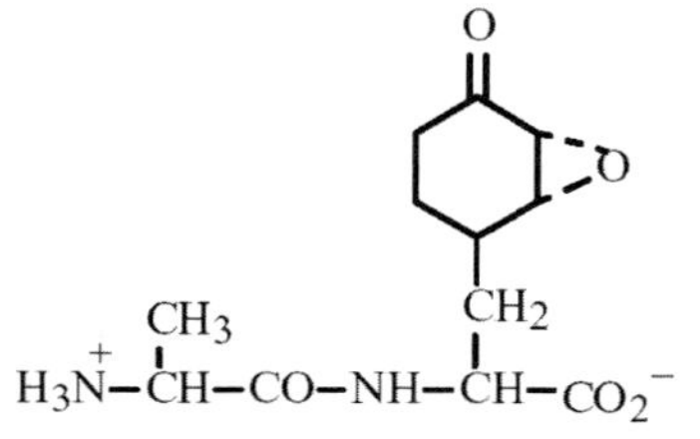

Figure 4. Structural formula of bacilysin a dipeptide containing L-alanine and the unusual amino acid L-anticapsine (Walker and Abraham, 1970).

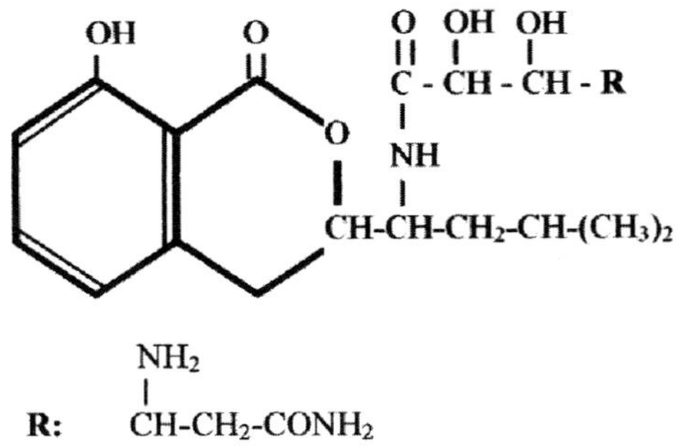

Figure 5. Structural formula of Amicoumacin A (Pinchuk et al., 2002).

Finally, two new peptide antibiotics, produced by *Bacillus polymyxa*, have been described and named as the gavaserin (cyclic octapeptide) and saltavalin (not cyclic polypetide) (Pichard et al., 1995).

ANTIBIOTICS WITH VARIOUS BIOCHEMICAL STRUCTURES

Bacillus strains also produced numerous antibiotics with various biochemical structures. For example, the isocoumarin derivatives baciphelacin produced by *Bacillus thiaminolyticus* have antibacterial and anti-viral effects (Okazaki et al., 1975), kristenin and Y-05460M-A have been described to be produced by *Bacillus* spp. and possess antibacterial and anti-tumor activities (Weisenborn et al., 1984; Sato et al., 1992) and amicoumacin A and B (Figure 5) which have antibacterial, anti-inflammatory and anti-ulcer activity produced by some strains of *B. subtilis* and *B. pumilus* (Itoh et al., 1982; Pinchuk et al., 2002). Pinchuk et al., (2002) screened fifty-one strains of *B. subtilis* for the production of amicoumacin A and B and demonstrated that the production of this antibiotic is not a particular characteristic of one *B. subtilis* strain but it is a more general property for this species with at least 11 of the 51 *B. subtilis* isolates producing amicoumacin A.

The aminosugar antibacterial antibiotic 3,3'-neo trehalosadiamine was produced by a mutant of *B. subtilis* 168 and also has been isolated from a *B. pumilus* strain (Tsuno et al., 1986). Finally, *B. cereus* strain UW85 produces two antibiotics with a wide spectrum of antimicrobial activity kanosamine, an aminoglycoside containing a disaccharide, and zwittermicin A, an aminopolyol with molecular mass of 396 Da. Both antibiotics prevented disease of alfalfa seedlings caused by *P. medicaginis* (Silo-Suh et al., 1994). Table 2 presents a summary of principal antibiotics produced by *Bacillus* species utilised as probiotics.

ANTIMICROBIAL SUBSTANCE PRODUCTION AND HEALTH PROMOTING EFFECT OF *BACILLUS* PROBIOTICS.

Currently, there are 77 species of the genus *Bacillus* officially recognized (www.bacterio.cict.fr). However, only a few of them are used as probiotics for human and animal use: *subtilis*, *licheniformis*, *clausii*, *coagulans*, *cereus*, *pumilus*, *laterosporus* and also some not valid species named as *toyoi*, *natto* and *polyfermenticus*.

Table 2. Principal antibiotics produced by *Bacillus* spp.

Antibiotics	Type	Activity spectra	Properties	Producer species	Reference
Surfactin	Lipoheptapeptide[§]	Antiviral and antimycoplasma	Surfactant	*B. subtilis, B. licheniformis, B. natto, B. pumilus*	Vollenbroich et al., 1977a; 1977b.
Iturins (A, C, D, E)	Lipoheptapeptide	Antifungal	Membrane pore formers	*B. subtilis*	Besson et al., 1978; Besson and Michel, 1987
Mycosubtilin	Lipoheptapeptide	Antifungal	Membrane pore formers	*B. subtilis*	Peypoux et al., 1986
Bacillomycins (D, F, L)	Lipoheptapeptide	Antifungal	Membrane pore formers	*B. subtilis*	Peypoux et al., 1984
Bacillopeptin	Lipoheptapeptide	Antifungal	Membrane pore formers	*B. subtilis*	Kajimura et al., 1995
Fengycin	Lipodecapeptide	Antifungal	-	*B. subtilis* *B. amyloliquefaciens*	Vanittanakom et al., 1986
Plipastatins (A1, A2)	Lipodecapeptide	Antifungal	Phospholipase A2 inhibitors	*B. subtilis* *B. cereus*	Umezawa et al., 1986
Mycobacillin	Cyclic tridecapeptide (1528 Da)	Antibacterial Antifungal	Membrane target	*B. subtilis*	Majundar and Bose, 1958; Katz and Demain, 1977
Bacilysin	Dipeptide (271.1 Da)	Antibacterial Anti-Candida	Cell wall synthesis inhibition	*B. subtilis*	Walter and Abraham, 1970
Bacitracin	Dodecapeptide (1421.7 Da)	Gram-positive bacteria	Peptidoglycan synthese inhibition	*B. subtilis* *B. licheniformis*	Johnson et al., 1945
Loloatin A	Cyclic decapeptide	Antibacterial	-	*B. laterosporus*	Krachkovskii et al., 2002
Gramicidins Tyrocidine	Cyclic decapeptide	Antibacterial	Membrane lipids interactions	*B. brevis*	Katz and Demain, 1977
Rhizocticin	Phosphono-oligopeptide	Antifungal	Threonine synthase inhibition	*B. subtilis*	Kugler et al., 1990
Tupuseleiamide	acyldipeptide	Antifungal	-	*B. laterosporus*	Barsby et al., 2002
Amicoumacin A	Isocoumarins (424 Da)	Antibacterial	-	*B. subtilis* *B. pumilus*	Pinchuk et al.2001; Itoh et al., 1982

* Generally some of Gram-positive bacteria.
§ Molecular mass of lipopetides is not mentioned because the variability of lipid moiety and analogs.

Despite the fact that *Bacillus* spp. are used in human pharmaceutical preparations or in preparations for animal feeds, their mode of action remains poorly understood, particularly the role of antimicrobial production. Very limited information is available about the spectrum of their antimicrobial activity and only few reports describe the identification of antimicrobial substances responsible for this activity. However it has been suggested that the anti-diarrheal effect of the probiotic 'Alpharma' containing *B. licheniformis* LPS122 strain could result from the production of antimicrobials (Kyriakis et al., 1999).

The pharmaceutical probiotic preparation Biosporin®, contains two *Bacillus* strains (*B. subtilis* 3 and *B. licheniformis* 31) and was reported to be active against enteropathogenic *Campylobacter* strains *in vitro* and *in vivo* in a mouse model. Both strains included in this probiotic possess an important spectrum of antibacterial activity (Sorokulova et al., 1997). Moreover, we have reported that *B. subtilis* 3 produces *in vitro* at least two distinct heat-stable, protease-resistant antimicrobial substances that inhibit the growth of many bacteria including *Helicobacter* and *Campylobacter* spp. (Pinchuk et al., 2001). One of these compounds, purified from the cell-free supernatant, was identified as amicoumacin A, an isocoumarin antibiotic with anti-inflammatory properties (Figure 5). This antibiotic was very active against selected strains of *S. aureus, E. faecium, Shigella flexneri*, and *C. jejuni,* but had no significant activity

against the strains of the lactic acid bacteria, *B. fragilis*, and *E. coli*. Interestingly, isolated antimicrobial substances possess an additive inhibitory effect against *Helicobacter pylori in vitro*. Recently we showed that *B. subtilis* 3 can produce another four antibiotics of lipopeptide nature: surfactin, fengicin, iturin N and mycosubtilin (unpublished data). The production of these compounds was dependant upon the carbon source, as well as the composition of microelements in the culture medium used for the antibiotic production. We, therefore hypothesized, that production of antimicrobial substances by *B. subtilis* 3 included in Biosporin® may be one of the mechanisms responsible for the probiotic effects of Biosporin® especially pathogen inhibition (observed *in vivo*).

Another *Bacillus* probiotic Enterogermina®, which includes four strains of *Bacillus* recently reclassified from *subtilis* to *clausii*, has been reported to exert beneficial clinical effects, notably in the treatment of diarrhea and in the prevention of infectious diseases (Mazza, 1994). Analysis of antimicrobial activity of these strains demonstrates that all of them release antimicrobial substances in the medium (Urdaci et al., 2004). Moreover, it was been demonstrated that antimicrobial substances released during stationary growth phase coincided with the onset of sporulation. These substances were active against Gram-positive bacteria, in particular against *S. aureus, E. faecium* and *Clostridium difficile*. No inhibitory effect was observed against Gram-negative bacteria and fungi. The antimicrobial substances were relatively thermostable, since their activity remained after 30 min of incubation at 85°C and they were resistant to subtilisin, proteinase K and chymotrypsin treatment, while sensitive to pronase treatment. Accordingly, they were hypothesized to be bacteriocins. This *in vitro* observed anti-*C. difficile* effect opens new perspectives for the therapeutic use of Enterogermina® in the treatment of *Cl. difficile*-associated diarrhea.

The commercial probiotic *B. polyfermenticus* SCD®, composed of a *B. polyfermenticus* strain (commonly known as Bispan strains), have been successfully used for the treatment of long-term intestinal disorders. *B. polyfermenticus* SCD has been described to produce polyfermenticin, a proteinase K-sensitive and heat labile bacteriocin with a narrow spectrum of activity against some Gram-positive bacteria, which primarily inhibited other *Bacillus* spp. Its production followed typical kinetics of primary metabolite synthesis. Direct detection of polyfermenticin SCD activity on SDS-PAGE suggested that it had an apparent molecular mass of about 14.3 kDa. Polyfermenticin SCD seemed to be very stable throughout the pH range of 2.0 to 9.0 (Lee et al., 2001). Strains of "*B. polyfermenticus*" are very close to *B. subtilis* strains in terms of their morphological and biochemical properties (note that the species name, *polyfermenticus*, name is not officially recognized, see Chapter 8). However, they are still distinct from *B. subtilis*, based on their capacity to metabolize lactose and to produce a larger amount of acetic and lactic acid from glucose and lactose.

We have also recently demonstrated that *B. subtilis* strain included in the probiotic BioPlus 2B® (Chr. Hansen BioSystems A/S, Denmark) possess antagonistic activity against some Gram-positive bacteria and fungi, such as *Listeria inocua* and *Rhizopus* sp. (unpublished data). Analysis of antimicrobial activity of four *B. subtilis* strains, which compose the probiotic PROMARINE® (Sino-Aqua company), demonstrate that at least three of these four strains produced antibacterial substances that inhibit *S. aureus* and *L. inocua* growth, but were inactive against tested Gram-negative bacteria. Moreover, one of these strains possesses a strong antifungal activity against *Fusarium oxysporum* and *Rhizopus* sp. (unpublished data). Some information about the antimicrobial activity of *Bacillus* probiotics is available on the web sites of commercial probiotic companies. For example, Garden of Life® Primal Defense™ is composed of 14 specific soil microorganisms claim to include *B. subtilis* and *B. licheniformis* strains that produces surfactants that supposedly inactivate lipid from the virus envelopes of HIV, CMV, herpes and also possess anti-mycoplasmas activity. It is indicated in this web report (http://www.thehomeostasisprotocol.com/mall/Primal-defense/article3.htm), that some studies concerning Primal Defense™ have been conducted by the University of Berlin's Max-Volmer Institute in Germany.

Bacillus laterosporus (*Brevibacillus laterosporus*) is present in various probiotic preparations (e.g., in Lacbon®, Lactospore® and Flora Balance®). In the case of Flora Balance this contains the strain BOD that is claimed good at destroying pathogenic bacteria and fungi in the digestive tract. *In vitro* inhibition tests demonstrated that this strain is active against

Candida albicans, Salmonella, E. coli and *Streptococcus* (http://www.flora-balance.com/symptom_relief.htm), however any antimicrobial substance responsible for this activity have not yet been characterized. Moreover, we were unable to find any scientific report supporting these antimicrobial activities.

Probiotic strains of *B. coagulans*, (misclassified *Lactobacillus sporogenes*, see Chapter 8) are frequently listed in various dietary probiotic supplements (Lactolase®, Lactospore R®, Lacbon®, Sporlac®, FloraMax®). Recently *B. coagulans* has been successfully used for the treatment of antibiotic-associated diarrhea (La Rosa et al., 2003). The clinically observed *B. coagulans* probiotic protective effect could be due to the production of antimicrobials compounds, since Hyronimus et al. (1998) identified an inhibitory substance, coagulin, produced by the potentially probiotic strain *B. coagulans* I4. This BLIS was plasmid-linked, heat-stable, protease sensitive, and exhibited antimicrobial action against Gram-positive bacteria including *Listeria* spp. Previously, this BLIS has been characterized by N-terminal sequencing of the purified 44-amino acids peptide and found to be quite similar to pediocins produced by *Pediococcus acidilactici*, differing by only a single amino acid. The genetic determinants of coagulin presented a high homology with the genes encoding the pediocin from *Pediococcus* (Le Marrec et al., 2000).

Furthermore, foods and food ingredients that can enhance health, the so-called "functional food" or "novel foods" are without doubt one of the leading trends in the food industry today. Moreover, fermentation technology constitutes an essential part of this market. Many of the traditional fermented foods of Southeast Asia and Africa are receiving new attention for their health promoting or disease preventing/curing effects. Scientific evidence for their physiological functions are accumulating and the technologies enhancing the beneficial effects are developing rapidly by using modern biotechnological techniques. Several fermented products, in particularly alkaline-fermented foods, rely on the participation of various probiotic bacteria of *Bacillus* species, including *B. natto* and *B. subtilis*. They can be made from different raw ingredients, e.g., Japanese natto, Thai thua-nao, and kinema are made from cooked soybeans, soumbala and dawadawa from African locust beans, ogiri from melon seeds, ugba from African oil beans, kawal from fresh legale leaves, owoh from cotton seeds, and pidan from fresh poultry eggs.

Most alkaline fermentations are achieved spontaneously by mixed bacteria cultures, principally dominated by *B. subtilis*. In other cases, pure cultures can be used. For example, Japanese natto is inoculated with a pure culture of *B. subtilis* var *natto*. Natto has a history of use of more than one thousand years, and has been used in traditional medicine and known to have antithrombotic and anti-cancer effects (see Chapter 12 for a complete review of Natto). The antagonistic activities of *B. natto* has been well studied. Ozawa et al. (1979) demonstrated that *B. natto* can inhibit growth of *C. albicans in vitro* and suggested that *B. natto* may also inhibit the growth of *C. albicans* in the intestinal tract. Nagal et al. (1996) characterized a surfactin B2 produced by the *B. natto* which was at least partially responsible for the observed anti-yeast effect. Moreover, the detoxification capacity of *B. natto* fermented soybean food has been recently demonstrated by Osawa et al. (1997). In these experiments cooked rice contaminated with staphylococcal enterotoxin A (SEA) was mixed with 'natto', and the mixture was incubated at 37°C for 1 h. Reversed passive latex agglutination (RPLA) tests performed on the mixture revealed that the RPLA titer against SEA was significantly reduced after incubation. Subsequent analytical tests showed that the SEA protein molecule was fragmented to small peptides by an extracellular protease, subtilisin, produced by *B. natto*. This proteolytic activity of *B. natto* was also found to be effective against other types of staphylococcal enterotoxins.

Bacillus spp. are the main microorganisms responsible for the alkaline fermentation of the African locust bean called Soumbala in Burkina Faso. The predominant compound of this fermented product is *B. subtilis*. However, strains of *B. pumilus*, *Bacillus licheniformis* and *Bacillus megaterium* has also been described to participate in the fermentation process of this food. These strains of *B. subtilis* and *B. pumilus* have been described to possess various lipolytic (Ouoba et al., 2003) and antimicrobial activities (Ouoba, unpublished information).

Microserin® is a composite fermentation consisting of soluble compounds from *Lactobacillus acidophilus* and *Bacillus* organisms (*B. subtilis*, *B. licheniformis* and *B. coagulans*) dried onto a carrier of corn germ meal. It has been hypothesized that the production of peptide antibiotics ("special metabolites") by Microserin® strains improve health and reduce *E. coli* and *Salmonella* (http://www.vi-cor.com/html/microserin/mic-about.html).

A fermented maize product Pozol is consumed by the indigenous Mayan peoples of southeastern Mexico. The Mayans also used Pozol as a medicine to control diarrhea and to cure intestinal infections. Ray et al. (2000) found that a well represented *Bacillus* strain from this product had a broad-spectrum of antagonistic activity Gram-positive and Gram-negative bacteria, yeasts, and molds. The antimicrobial compound(s) was bactericidal and bacteriolytic against *Escherichia coli* V517, bacteriostatic against *Micrococcus luteus*, and fungistatic against *Saccharomyces cerevisiae*. Recently, identification of these compounds has been realized. This strain produce the dipeptides bacilysin and cholotetain and the lipopetide Iturin A (Phister et al., 2004). It will be interesting to see whether these antimicrobials are present in Pozol.

Paenibacillus polymyxa (P13) strain isolated by Piuri et al. (1998) from Argentinean regional fermented sausages and has been reported to secrete a compound that inhibited a wide range of Gram-positive and Gram-negative bacterial species including food-borne pathogens. The antimicrobial substance(s) obtained from the culture supernatant of late stationary growth phase was named polyxin. Polyxin possess BLIS properties such as having a proteinaceous nature (sensitive to proteases), stability to heat (up to 10 min at 90 °C), and acidic pH.

Finally, Zheng and Slavik (1999) isolated from a Chinese fermented soybean seasoning a *B. subtilis* strain producing antibacterial substances against *B. cereus* and *L. monocytogenes*. This inhibitory peptide was characterized first as a possible new BLIS, but subsequently was shown to be identical to subtilosin (Zheng et al., 1999).

CONCLUSIONS AND FUTURE DIRECTIONS

Consumption of certain *Bacillus* probiotic has been shown to have a beneficial impact on both man and animals but reports concerning clinical applicability and the mechanisms of action of these sporeforming probiotics are few. Mazza (1994) in a comprehensive review summarizes the use of a commercial *Bacillus* preparation, Enterogermina®, as an antidiarrheal. Sorokulova et al., (1997) tested the probiotic Biosporin® in the mouse model of *Campylobacter* infection and concluded that the product caused a protective effect. Recently, we demonstrated that Enterogermina® and Biosporin® included *Bacillus* strains able to produce antimicrobial compounds *in vitro* (Urdaci et al., 2004; Pinchuk et al., 2001).

Since the initial probiotic concept was proposed a century ago, it has been hypothesized, that production of antimicrobial compounds by these bacteria play one of the central roles in the probiotic protective mechanism. Since then many scientific reports have demonstrated the production of antimicrobial compounds having a specific mode of action by probiotic bacteria *in vitro*, whereas a certain number of these originate from *Bacillus* spp. Surprisingly, we were unable to find even one report concerning the production of these compounds by probiotic microorganisms *in vivo*. Therefore, continued progress in probiotic research will be required to determine the nature and role of antimicrobial production by *Bacillus* probiotic spp. and to understand their role in probiotic modulation of microbial synergetic/antagonistic interactions with the host. Finally, to improve the probiotic and pharmacological parameters it is crucial to understand the influence of several abiotic and biotic factors, that can affect antibiotic synthesis by the bacteria *in vivo*.

Although, it has been clearly demonstrated that the majority of *Bacillus* probiotics produced many antimicrobial substances *in vitro*, it is still unclear what impact these metabolites will have on representatives of normal human and animal microbiota. For example, it has been demonstrated that the antibiotic amicoumacin A produced by *B. subtilis* 3 doesn't affect the *in vitro* growth of lactic acid bacterial isolates (Pinchuk et al., 2001). Moreover, Hosoi et al. (2000) recently demonstrated that catalase and subtilisin produced by *Bacillus* spp. promote *Lactobacillus* growth *in vitro*. Therefore, for example, application of biofilm formation technology could be very attractive in order to understand the complexities of interaction between pathogenic and non-pathogenic microbiota versus probiotics.

Recent progress in the identification of factors affecting *in vivo* the success of antagonistic bacteria used for the biological control of plant pathogens (Raaijmakers, 2002) has provided significant insights into the development of strategies and methods to investigate those processes in animals and man. For example, it has been demonstrated that inactivation of antibiotic production, in many cases, reduces the ability to control plant pathogens *in vivo*. Mutagenesis has been successfully used to demonstrate antibiotic production by *Bacillus* spp., *Pseudomonas* spp. and some other biocontrol microorganisms that play a major role in the control of plant disease. In several studies evidence of antibiotics in microbial biocontrol has been provided by enhancement of antibiotic production in the producing wild-type strains via introduction/ modification of antibiotic biosynthetic/regulatory genes (Pal et al., 2000; Raaijmakers, 2002). Moreover, it has also been demonstrated in *Pseudomonas fluorescens* that introduction of antibiotic biosynthetic genes in heterologous, non-producing strains lead to significant increases in the protective capacity of recipient strains against *Pythium ultimum* infection of sugar beet (Fenton et al., 1992; Timms-Wilson et al., 2000).

Another recently proposed molecular biology approach involves use of reporter gene systems (Leveau and Lindow, 2002). Since, important numbers of gene clusters involved in synthesis of antimicrobial compounds, in particularly those responsible for the bacteriocin synthesis, produced by *Bacillus* spp. has been identified, application of the strategies described above seems to be promising.

Bioanalytical techniques like high-pressure liquid chromatography (HPLC) analysis have also been proposed for *in situ* quantification of antimicrobials compound produced by biocontrol agents (Raaijmakers, 2002). HPLC followed by mass spectrometry analysis has been successfully introduced in toxicology for the analysis of different micro-compounds in human biological fluids. Presumably application of these technologies could be helpful for monitoring the kinetics of antimicrobial synthesis *in vivo*.

The fact that many probiotic strains, including *Bacillus*, produce bacteriocins and antibiotics raises questions about the safety of their use. Clearly, substantial alterations of the normal mucosal microflora resulting from the application of antimicrobials is something that needs to be addressed. Careful analysis of toxicity as well as the potential development of resistance to these antimicrobial substances by pathogenic microorganisms needs to be examined in the design and application of new probiotic species.

REFERENCES

Ahern, M., Verschueren, S., van Sinderen, D. 2003. Isolation and characterisation of a novel bacteriocin produced by *Bacillus thuringiensis* strain B439. FEMS Microbiol. Lett. *220*, 127-131.

Altena, K., Guder, A., Cramer, C., Bierbaum, G. 2000. Biosynthesis of the lantibiotic mersacidin: organization of a type B lantibiotic gene cluster. Appl. Environ. Microbiol. *66*, 2565-2571.

Asaka, O., Shoda, M. 1996. Biocontrol of *Rhizoctonia solani* damping-off of tomato with *Bacillus subtilis* RB14. Appl. Environ. Microbiol. *62*, 4081-4085.

Azevedo, EC., Rios, EM., Fukushima, K., Campos-Takaki, GM. 1993. Bacitracin production by a new strain of *Bacillus subtilis*. Extraction, purification and characterization. Appl. Biochem. Biotechnol. *42*,1-7.

Babasaki, K., Takao, T., Shimonishi, Y., Kurahashi, K. 1985. Subtilosin A, a new antibiotic peptide produced by *Bacillus subtilis 168*, isolation, structural analysis, and biogenesis. J. Biochem. (Tokyo) *98*,585-603.

Barsby, T., Kelly, MT., Andersen, RJ. 2002. Tupuseleiamides and basiliskamides, new acyldipeptides and antifungal polyketides produced in culture by a *Bacillus laterosporus* isolate obtained from a tropical marine habitat. J. Nat. Prod. *65*, 1447-1451.

Berdy, J. 1974. Recent developments of antibiotic research and classification of antibiotics according to chemical structure. Adv. Appl. Microbiol. *18*, 309-406.

Bernet-Camard, MF., Lievin, V., Brassart D., Neeser, JR., Servin, AL., Hudault, S. 1997. The human *Lactobacillus acidophilus* strain LA1 secretes a nonbacteriocin antibacterial substance(s) active in vitro and in vivo. Appl. Environ. Microbiol. *63*, 2747-2753.

Besson, F., Peypoux, F., Michel, G., Delcambe, L. 1978. Identification of antibiotics of iturin group in various strains of *Bacillus subtilis*. J. Antibiot (Tokyo) *31*, 284-288.

Besson, F., Michel, G. 1987. Isolation and characterization of new iturins: iturin D and iturin E. J. Antibiot (Tokyo) *40*, 437-442.

Biziulevicius, GA., Zukaite, V. 2001. Comparative studies on Polyferm and Fermosorb, two oral (ferment + sorbent) - type preparations designed for therapy/prophylaxis of intestinal infections in animal neonates. J. Vet. Pharmacol. Ther. *24*, 433-438.

Biziulevieius, GA., Sukaite, V. 2002. Comparative antimicrobial activity of lysosubtilin and its acid-resistant derivative, Fermosorb. Int. J. Antimicrob. Agents. *20*, 65-68.

Boris, S., Jimenez-Diaz, R., Caso, JL., Barbes, C. 2001. Partial characterization of a bacteriocin produced by *Lactobacillus delbrueckii* subsp. *lactis* UO004, an intestinal isolate with probiotic potential. J. Appl. Microbiol. *91*, 328-333.

Braun, V., Pilsl, H., Gross, P. 1994. Colicins: structures, modes of action, transfer through membranes, and evolution. Arch. Microbiol. *161*,199-206.

Brook, I. 1999. Bacterial interference. Crit. Rev. Microbiol. *25*, 155-172.

Brötz, H., Bierbaum, G., Leopold, K., Reynolds, PE., Sahl, HG. 1998. The lantibiotic Mersacidin inhibits peptidoglycan synthesis by targeting lipid II. Antimicrob Agents Chemother. *42*, 154-160.

Chan, WC., Bycroft, BW., Leyland, ML., Lian, LY., Roberts, GC. 1993. A novel post-translational modification of the peptide antibiotic subtilin: isolation and characterization of a natural variant from *Bacillus subtilis* A.T.C.C. 6633. Biochem J. *291*, 23-27.

Chatterjee, S., Chatterjee, DK., Jani, RH., Blumbach, J., Ganguli, BN., Klesel, N., Limbert, M., Seibert, G. 1992. Mersacidin, a new antibiotic from *Bacillus. In vitro* and *in vivo* antibacterial activity. J. Antibiot. *45*, 839-845.

Cherif, A., Ouzari, H., Daffonchio, D., Cherif, H., Ben Slama, K., Hassen, A., Jaoua, S. 2001. Thuricin 7: A novel bacteriocin produced by *B. thuringiensis* BMG1.7, a new strain isolated from soil. Lett. Appl. Microbiol. *32*, 243-247.

Chowdhury, B., Das, SK., Bose, SK. 1998. Use of resistant mutants to characterize the target of mycobacillin in *Aspergillus niger* membranes. Microbiology *144*, 1123-1130.

Duitman, EH., Hamoen, LW., Rembold M, Venema, G., Seitz, H., Saenger, W., Bernhard, F., Reinhardt, R., Schmidt, M., Ullrich, C., Stein, T., Leenders, F., Vater J. 1999. The mycosubtilin synthetase of *Bacillus subtilis* ATCC6633, a multifunctional hybrid between a peptide synthetase, an amino transferase, and a fatty acid synthase. Proc. Natl. Acad. Sci. USA *96*, 13294-13299.

Dunne, C., Shanahan, F. 2002. The role of probiotics in the treatment of intestinal infections and inflammation. Curr. Opin. Gastroenterol. *18*, 40-45.

Eijsink, VG., Axelsson, L., Diep, DB., Havarstein, LS., Holo, H., Nes, IF. 2002. Production of class II bacteriocins by lactic acid bacteria; an example of biological warfare and communication. Antonie Van Leeuwenhoek. *81*, 639-654.

Entian, KD., de Vos WM. 1996. Genetics of subtilin and nisin biosyntheses: biosynthesis of lantibiotics. Antonie Leeuwenhoek *69*,109-117.

Favret, M., Yousten, A. 1989. Thuricin: the bacteriocin produced by *B. thuringiensis*. J. Invertebr. Pathol. *53*, 206-216.

Fenton, AM., Stephens, PM., Crowley, J., O'Callaghan, M., O'Gara, F. 1992. Exploitation of gene(s) involved in 2,4-diacetylphloroglucinol biosynthesis to confer a new biocontrol capability to a *Pseudomonas* strain. Appl. Environ. Microbiol. *58*, 3873-3878.

Fuller, R. 1989. Probiotics in man and animals. J. Appl. Bacteriol. *66*, 365-378.

Galvez, A., Maqueda, M., Martinez-Bueno, M., Lebbadi, M., Valdivia, E. 1993. Isolation and physico-chemical characterization of an antifungal and antibacterial peptide produced by *Bacillus licheniformis* A12. Appl. Microbiol. Biotechnol. *39*, 438-42.

Gross, E., Kiltz, H. 1973. The number and nature of a,p-unsaturated amino acids in subtilin. Biochem. Biophys. Res. Commun. *50*, 559-565.

Gupta, M., Kaur, M., Gupta, KG. 1992. Lytic effect of *Bacillus subtilis* elastase on gram-positive and negative bacteria. Indian J. Exp. Biol. *30*, 380-383.

Hattori, Z., Misawa, H., Igarashi, I., Sugiya, Y. 1965. Effects of lactic acid-forming bacteria on *Vibrio coma* inoculated into intestinal segments of rabbits. J. Bacteriol. *90*, 541-545.

Helisto, P., Aktuganov, G., Galimzianova, N., Melentjev, A., Korpela, T. 2001. Lytic enzyme complex of an antagonistic *Bacillus* sp. X-b: isolation and purification of components. J. Chromatogr. B Biomed. Sci. Appl. *758*, 197-205.

Hosoi, T., Ametani, A., Kiuchi, K., Kaminogawa, S. 2000. Improved growth and viability of lactobacilli in the presence of *Bacillus subtilis* (natto), catalase, or subtilisin. Can. J. Microbiol. *46*, 892-897.

Hourdou, ML., F. Besson, I. Tenoux, G. Michel. 1989. Fatty acid and beta-amino acid syntheses in strains of *Bacillus subtilis* producing iturinic antibiotics. Lipids *24*, 940-944.

Hyronimus, B., Le Marrec, C., Urdaci, M.C. 1998. Coagulin, a bacteriocin-like inhibitory substance produced by *Bacillus coagulans* I4. J Appl Microbiol. *85*, 42-50.

Itoh, J., Shomura, T., Omoto, S., Miyado, S., Yuda, Y., Shibata, U., Inouye, S. 1982. Isolation, physicochemical properties and biological activities of amicoumacins produced by *Bacillus pumilus*. Agric. Biol. Chem. *46*, 1255-1259.

Johnson, BA., Anker, H., Meleney, FL. 1945. Bacitracin: a new antibiotic produced by a member of the *B. subtilis* group. Science *102*,376-377.

Kajimura, Y., Sugiyama, M., Kaneda, M. 1995. Bacillopeptins, new cyclic lipopeptide antibiotics from *Bacillus subtilis* FR-2. J. Antibiot (Tokyo) *48*, 1095-1103.

Katz, E., Demain, AL. 1977. The peptide antibiotics of *Bacillus*: chemistry, biogenesis, and possible functions. Bacteriol. Rev. *41*, 449-74.

Kiers, JL., Meijer, JC., Nout, MJ., Rombouts, FM., Nabuurs, MJ., van der Meulen, J. 2003. Effect of fermented soya beans on diarrhoea and feed efficiency in weaned piglets. J. Appl. Microbiol. *95*, 545-552.

Kim, SY., Ohk, SH., Bai, DH., Yu, JH. 1999. Purification and properties of bacteriolytic enzymes from *Bacillus licheniformis* YS-1005 against *Streptococcus mutans*. Biosci. Biotechnol. Biochem. *63*, 73-77.

Klaenhammer, TR. 1988. Bacteriocins of lactic acid bacteria. Biochimie *70*,337-49.

Kondejewski, LH., Farmer, SW., Wishart, DS., Hancock, RE., Hodges, RS. 1996. Gramicidin S is active against both gram-positive and gram-negative bacteria. Int. J. Pept. Protein Res. *47*, 460-466.

Krachkovskii, SA., Sobol', AG., Ovchinnikova, TV., Tagaev, AA., Iakimenko, ZA., Azizbekian, RR., Kuznetsova, NI., Shamshina, TN., Arsen'ev, AS. 2002. Isolation, biological properties, and spatial structure of an antibiotic loloatin A. Bioorg Khim. *28*, 298-302.

Kugler, M., Loeffler, W., Rapp, C., Kern, A., Jung, G. 1990. Rhizocticin A, an antifungal phosphono-oligopeptide of *Bacillus subtilis* ATCC *6633*, biological properties. Arch. Microbiol. *153*, 276-281.

Kyriakis, SC., Tsiloyiannis, VK., Vlemmas, J., Sarris, K., Tsinas, AC., Alexopoulos, C., Jansegers, L. 1999. The effect of probiotic LSP 122 on the control of post-weaning diarrhoea syndrome of piglets. Res. Vet. Sci. *67*, 223-228.

La Rosa, M., Bottaro, G., Gulino, N., Gambuzza, F., Di Forti, F., Ini, G, Tornambe, E. 2003. Prevention of antibiotic-associated diarrhea with *Lactobacillus sporogens* and fructo-oligosaccharides in children. A multicentric double-blind vs placebo study. Minerva Pediatr. *55*, 447-452.

Lee, KH., Jun, KD., Kim, WS., Paik, HD. 2001. Partial characterization of polyfermenticin SCD, a newly identified bacteriocin of *Bacillus polyfermenticus*. Lett. Appl. Microbiol. *32*, 146-151.

Le Marrec, C., Hyronimus, B., Bressollier, P., Verneuil, B., Urdaci, MC. 2000. Biochemical and genetic characterization of coagulin, a new antilisterial bacteriocin in the pediocin family of bacteriocins, produced by *Bacillus coagulans* I(4). Appl. Environ. Microbiol. *66*, 5213-5220.

Leveau, JH., Lindow, SE. 2002. Bioreporters in microbial ecology. Curr. Opin. Microbiol. 5: 259-265.

Maget-Dana, R., Peypoux, F. 1994. Iturins, a special class of pore-forming lipopeptides: biological and physicochemical properties. Toxicology *87*, 151-174.

Majumdar, SK., Bose, SK. 1958. Mycobacillin, a new antifungal antibiotic produced by *Bacillus subtilis*. Nature *181*, 134-135.

Marahiel, MA. 1997. Protein templates for the biosynthesis of peptide antibiotics. Chem Biol. 4: 561-567.

Marahiel, MA., Stachelhaus, T., Mootz, HD. 1997. Modular peptide synthetases involved in nonribosomal peptide synthesis. Chem. Rev. *97*, 2651-2673.

Martirani, L., Varcamonti, M., Naclerio, G., De Felice, M. 2002. Purification and partial characterization of bacillocin 490, a novel bacteriocin produced by a thermophilic strain of *Bacillus licheniformis*. Microb Cell Fact. 1(1):1.

Mazza, P. 1994. The use of *Bacillus subtilis* as an antidiarrhoeal microorganism. Boll. Chim. Farm. *133*, 3-18.

McAuliffe, O., Ross, RP., Hill, C. 2001. Lantibiotics: structure, biosynthesis and mode of action. FEMS Microbiol. Rev. *25*, 285-308.

Mendoza, D., Grau, R., Cronan Jr., JE. 1993. Biosynthesis and function of membrane lipids, p. 411–421. *In* AL. Sonenshein, JA. Hoch, and R. Losick (ed.), *Bacillus subtilis* and other gram-positive bacteria. American Society for Microbiology, Washington, D.C.

Metchnikoff, E. 1905. Immunity of the skin and mucous membrane. *In* : Immunity in infective diseases. Cambridge at University Press. p. 426-427.

Nacleiro, G., Ricca, E., Sacco, M., De Felice, M. 1993. Anti-microbial of a newly identified bacteriocin of *Bacillus cereus*. Appl. Environ. Microbiol. *59*, 4313-4316.

Nagal, S., Okimura, K., Kaizawa, N., Ohki, K., Kanatomo, S. 1996. Study on surfactin, a cyclic depsipeptide. II. Synthesis of surfactin B2 produced by *Bacillus natto* KMD 2311. Chem. Pharm. Bull. (Tokyo) *44*, 5-10.

Nakano, MM., Zuber, P. 1990. Molecular biology of antibiotic production in *Bacillus*. Crit. Rev. Biotechnol. *10*, 223-240.

Nes, I., Holo, H. 2000. Class II antimicrobial peptides from lactic acid bacteria. Biopolymers *55*, 50-61.

Nishikiori, T., Naganawa, H., Muraoka, Y., Aoyagi, T., Umezawa, H. 1986. Plipastatins: new inhibitors of phospholipase A2, produced by *Bacillus cereus* BMG302-fF67. II. Structure of fatty acid residue and amino acid sequence. J. Antibiot (Tokyo) *39*, 745-754.

Okazaki, H., Kishi, T., Beppu ,T., Arima, K. 1975. Letter: a new antibiotic, baciphelacin. J. Antibiot. *28*, 717-719.

Oscariz, JC., Lasa, I., Pisabarro, AG. 1999. Detection and characterisation of cerein 7, a new bacteriocin produced by *Bacillus cereus* with a broad spectrum of activity. FEMS Microbiol. Lett. *178*, 337-341.

Ouoba, LI., Cantor, MD., Diawara, B., Traore, AS., Jakobsen, M. 2003. Degradation of African locust bean oil by *Bacillus subtilis* and *Bacillus pumilus* isolated from soumbala, a fermented African locust bean condiment. J. Appl. Microbiol. *95*, 868-873.

Osawa, R., Matsumoto, K. 1997. Digestion of staphylococcal enterotoxin by *Bacillus natto*. Antonie Van Leeuwenhoek. *71*, 307-311.

Ozawa, K., Yagu-Uchi, K., Yamanaka, K., Yamashita, Y., Ueba, K., Miwatani, T. 1979. *Bacillus natto* and *Streptococcus faecalis* on growth of *Candida albicans*. Microbiol. Immunol. *23*, 1147-1156.

Ozcengiz, G., Alaeddinoglu, NG., Demain, AL. 1990. Regulation of biosynthesis of bacilysin by *Bacillus subtilis*. J. Ind. Microbiol. 6:91–100.

Paik, SH., Chakicherla, A., Hansen, JN. 1998. Identification and characterization of the structural and transporter genes for, and the chemical and biological properties of sublancin 168, a novel lantibiotic produced by *Bacillus subtilis* 168. J. Biol. Chem. *273*, 23134-23142.

Pal, KK., Tilak, KV., Saxena, AK., Dey, R., Singh, CS. 2000. Antifungal characteristics of a fluorescent *Pseudomonas* strain involved in the biological control of *Rhizoctonia solani*. Microbiol Res. *155*, 233-242.

Pattnaik, P., Kaushik, JK., Grover, S., Batish, VK. 2001. Purification and characterization of a bacteriocin-like compound (Lichenin) produced anaerobically by *Bacillus licheniformis* isolated from water buffalo. J. Appl. Microbiol. *91*, 636-645.

Paulus, H. 1967. Polymyxins, p. 254-267. *In* D. Gottlieb and PD. Shaw (ed), Antibiotics, vol.2. Springer-Verlag, Berlin.

Peypoux, F., Pommier, MT., Das, BC., Besson, F., Delcambe, L., Michel, G. 1984. Structures of bacillomycin D and bacillomycin L peptidolipid antibiotics from *Bacillus subtilis*. J. Antibiot (Tokyo). *37*, 1600-1604.

Peypoux, F., Pommier, MT., Marion, D., Ptak, M., Das, BC., Michel, G. 1986. Revised structure of mycosubtilin, a peptidolipid antibiotic from *Bacillus subtilis*. J. Antibiot (Tokyo) *39*, 636-641.

Phister, TG., O'Sullivan DJ., and McKay, LL. 2004. Identification of bacilysin, chlorotetaine, and iturin a produced by *Bacillus* sp. strain CS93 isolated from pozol, a Mexican fermented maize dough. Appl. Environ. Microbiol. *70*, 631-634.

Pichard, B., Larue, JP., Thouvenot, D. 1995. Gavaserin and saltavalin, new peptide antibiotics produced by *Bacillus polymyxa*. FEMS Microbiol. Lett. *133*, 215-218.

Pinchuk, IV., Bressollier, P., Verneuil, B., Fenet, B., Sorokulova, IB., Megraud, F., Urdaci, MC. 2001. In vitro anti-*Helicobacter pylori* activity of the probiotic strain *Bacillus subtilis* 3 is due to secretion of antibiotics. Antimicrob. Agents Chemother. *45*, 3156-3161.

Pinchuk, IV., Bressollier, P., Sorokulova, IB., Verneuil, B., Urdaci, MC. 2002. Amicoumacin antibiotic production and genetic diversity of *Bacillus subtilis* strains isolated from different habitats. Res. Microbiol. *153*, 269-276.

Piuri, M., Sanchez-Rivas, C., Ruzal, SM. 1998. A novel antimicrobial activity of a *Paenibacillus polymyxa* strain isolated from regional fermented sausages. Letters Appl. Microbiol. *27,* 9-13.

Raaijmakers, JM., Vlami, M., de Souza, JT. 2002. Antibiotic production by bacterial biocontrol agents. Antonie Van Leeuwenhoek. *81,*537-47.

Ray, P., Sanchez, C., O'Sullivan, DJ., McKay, LL. 2000. Classification of a bacterial isolate, from pozol, exhibiting antimicrobial activity against several gram-positive and gram-negative bacteria, yeasts, and molds. J. Food Prot. *63,* 1123-1132.

Sato, T., Nagai, K., Suzuki, K., Morioka, M., Saito, T., Nohara, C., Susaki, K., Takebayashi, Y. 1992. A new isocoumarin antibiotic, Y-05460M-A. J. Antibiot. *45,* 1949-1952.

Silo-Suh, LA., Lethbridge, BJ., Raffel, SJ., He, H., Clardy, J., Handelsman, J. 1994. Biological activities of two fungistatic antibiotics produced by *Bacillus cereus* UW85. Appl. Environ. Microbiol. *60,* 2023-2030.

Smith, H. 1995. The revival of interest in mechanisms of bacterial pathogenicity. Bio. Rev. Camb. Philos. Soc. *70,* 277-316.

Sorokulova, IB., Kirik, DL., Pinchuk, IV. 1997. Probiotics against *Campylobacter* pathogens. J. Travel Med. 4:167-170.

Stein, T., Borchert, S., Conrad, B., Feesche, J., Hofemeister, B., Hofemeister, J., Entian, KD. 2002. Two different lantibiotic-like peptides originate from the ericin gene cluster of *Bacillus subtilis* A1/3. J. Bacteriol. *184,* 1703-1711.

Stover, AG., Driks, A. 1999. Secretion, localization, and antibacterial activity of TasA, a *Bacillus subtilis* spore-associated protein. J. Bacteriol. *181,*1664–1672.

Tamehiro, N., Okamoto-Hosoya, Y., Okamoto, S., Ubukata, M., Hamada, M., Naganawa, H., Ochi, K. 2002. Bacilysocin, a novel phospholipid antibiotic produced by *Bacillus subtilis* 168. Antimicrob Agents Chemother. *46,*315-20.

Timms-Wilson, TM., Ellis, RJ., Renwick, A., Rhodes, DJ., Mavrodi, DV., Weller, DM., Thomashow, LS., Bailey, MJ. 2000. Chromosomal insertion of phenazine-1-carboxylic acid biosynthetic pathway enhances efficacy of damping-off disease control by *Pseudomonas fluorescens.* Mol. Plant. Microbe Interact. *13,* 1293-1300.

Toscano WA Jr, Storm DR. 1982. Bacitracin. Pharmacol. Ther. *16,* 199-210.

Phister, TG., O'Sullivan, DJ., McKay, LL. 2004. Identification of Bacilysin, Chlorotetaine, and Iturin A Produced by *Bacillus* sp. Strain CS93 Isolated from Pozol, a Mexican Fermented Maize Dough. Appl. Environ. Microbiol. *70,* 631-634.

Tsuge, K., Ano T., Hirai, M., Nakamura, Y., Shoda, M. 1999. The genes *degQ, pps,* and *lpa-8 (sfp)* are responsible for conversion of *Bacillus subtilis* 168 to plipastatin production. Antimicrob. Agents Chemother. *43,*2183–2192.

Tsuno, T., Ikeda, C., Numata, K., Tomita, K., Konishi, M., Kawaguchi, H. 1986.

3,3'-Neotrehalosadiamine (BMY-28251), a new aminosugar antibiotic. J. Antibiot (Tokyo) *39,* 1001-1003.

Umezawa, H., Aoyagi, T., Nishikiori, T., Okuyama, A., Yamagishi, Y., Hamada, M., Takeuchi, T. 1986. Plipastatins: new inhibitors of phospholipase A2, produced by *Bacillus cereus* BMG302-fF67. I. Taxonomy, production, isolation and preliminary characterization. J Antibiot (Tokyo) *39,* 737-44.

Urdaci, MC., Bressollier, P., Pinchuk, I. 2004. *Bacillus clausii* probiotic strains: Antimicrobial and immunomodulatory activities. J. Clin. Gastroenterol. *38,* S000-S000 In Press.

Vanittanakom, N., Loeffler, W., Koch, U., Jung, G. 1986. Fengycin: a novel antifungal lipopeptide antibiotic produced by *Bacillus subtilis* F-29-3. J. Antibiot (Tokyo) *39,*888-901.

Vescovo, M., Scolari, GL., Caravaggi, L., Bottazzi, V. 1993. Antimicrobial compounds from *Lactobacillus casei* and *Lactobacillus helveticus.* New Microbiol. *16,*171-5.

Vollenbroich, D., Pauli, G., Ozel, M., Vater, J. 1997a. Antimycoplasma properties and application in cell culture of surfactin, a lipopeptide antibiotic from *Bacillus subtilis.* Appl. Environ. Microbiol. *63,*44-49.

Vollenbroich, D., Ozel, M., Vater, J., Kamp, RM., Pauli, G. 1997b. Mechanism of inactivation of enveloped viruses by the biosurfactant surfactin, from *Bacillus subtilis.* Biologicals *25,*289-297.

Volpon, L., Besson, F., Lancelin, JM. 2000. NMR structure of antibiotics plipastatins A and B from *Bacillus subtilis* inhibitors of phospholipase A$_2$. FEBS Lett. *485,* 76-80.

Walker, JE., Abraham, EP. 1970. The structure of bacilysin and other products of *Bacillus subtilis.* Biochem J. *118,* 563-570.

Wannamaker, LW. 1980. Bacterial interference and competition. Scand. J. Infect. Dis. Suppl. *24,* 82-85.

Weisenborn, FL., Brown, WE., Meyers, E. 1984. Antibiotic kristenin from *Bacillus subtilis* ATCC 31340 useful as against gram positive bacteria, Unlisted-Drug. 36 (1984) 9Q.

Work, E. 1959. The action of a lytic enzyme from spores of a bacillus on various species of bacteria. Ann Inst Pasteur (Paris) *96,* 468-480.

Zheng, G., Slavik, MF. 1999. Isolation, partial purification and characterization of a bacteriocin produced by a newly isolated *Bacillus subtilis* strain. Lett. Appl. Microbiol. *28,* 363-367.

Zheng, G., Yan, LZ., Vederas, JC., Zuber, P. 1999. Genes of the sbo-alb locus of *Bacillus subtilis* are required for production of the antilisterial bacteriocin subtilosin. J. Bacteriol. *181:* 7346-7355.

Zuber, P., Nakano, MM., Marahiel, MA. 1993. Peptide antibiotics, p. 897–916. *In* A. L. Sonenshein, J. A. Hoch, and R. Losick (ed.), *Bacillus subtilis* and other gram-positive bacteria. American Society for Microbiology, Washington, D.C.

Chapter 16

Gut Sporeformers

Teresa M. Barbosa, Cláudia R. Serra and Adriano O. Henriques

SUMMARY

Some intestinal symbionts use modified forms of sporulation for propagation and survival between gastrointestinal tracts, or to produce viviparous multiple offspring. Others, such as *Bacillus cereus* and presumably its close relatives show preferential filamentous growth in the invertebrate intestine. Moreover, spores of *B. subtilis* germinate and the vegetative cells appear to be capable of sporulation in the gut of mice. However, with the exception of *Clostridium*, which is a common and well-described component of the gut anaerobic microflora, relatively little is known about the incidence, nature, and properties of sporeformers in gut ecosystems. Without intending to be exhaustive, we try in this chapter to review some of the present knowledge, including the information that we recently obtained from a study aimed towards the isolation of sporeforming bacteria from the gastrointestinal tract of poultry and other animals.

UNUSUAL SPORULATION IN THE GUT

Metabacterium polyspora is a large, strict anaerobic Gram-positive sporeforming bacterium that is a prominent representative of the microbiota of the guinea pig (Angert et al., 1996; Angert and Losick, 1998). The life cycle of *M. polyspora* is coordinated with the passage of the bacteria through the upper part of the gastrointestinal tract of the host. Mature spores but not cells survive passage through the mouth and stomach, and upon entry into the small intestine, the spores germinate. Binary fission appears to be restricted to the period following spore germination, during which the bacteria are in the ileum. However, because in this organism sporulation has evolved into a form of propagation, binary fission appears often to have been bypassed, and the newly-formed cells enter again into the sporulation pathway. Interestingly, germination of *B. subtilis* spores in the jejunum and ileum of mice has also been reported (Casula and Cutting, 2002). In *B. subtilis*, sporulation is initiated by a shift in the site of division septum assembly from a medial position to an asymmetric position near one of the cell poles (Errington, 2003; see Chapter 5). In reality, the sporulation septum starts to be formed near the two cell poles, but the assembly of one is delayed relative to the other. Following the establishment of compartmentalized gene expression in the prespore, through the activation of σ^F, which directs activation of σ^E in the mother cell, the second septum is aborted, ensuring that only one prespore is formed (reviewed by Errington, 2003; see also Chapter 5 this volume). In mutants which are unable to repress formation of the second polar septum, as in cells deficient in the production of σ^E, the two chromosomes present in the cell at the onset of sporulation end-up in the two polar compartments, and thus have an anucleate mother cell, unable to nurture spore formation. In contrast, in *M. polyspora,* spores are produced by division near the two poles of the cell, and following engulfment of the prespores by the mother cell, the polar prespores often undergo further division, to generate additional prespore protoplasts (Angert and Losick, 1998). Division of the engulfed prespores, appears to be medial. In contrast to other sporeformers, *M. polyspora* has multiple nucleoids, which appear to be a pre-requisite to support the formation of multiple prespores (up to nine), which will mature into the characteristic elongated, cigar-shaped spores. It thus seems that in *M. polyspora* the sporulation pathway has been modified to serve as a form of internal reproduction and dispersion, bypassing a period of growth and symmetric (medial) cell division (Angert and Losick, 1998). The acquisition of the capacity to form multiple spore progeny by *M. polyspora*, which is its main mode of propagation, thus appears to be related to its infrequent use of binary cell division.

From the ileum, bacteria gain access to the cecum, where later stages of sporulation are observed, and leave the host via the colon. Since guinea pigs are coprophagous, the mature spores found in the animal faeces, often still inside the mother-cell, are re-introduced into the host, completing the cycle. Propagation by sporulation appears to protect *M. polyspora* from the harsh conditions found outside the GI tract of the host organism. It has been suggested that sporulation in *M. polyspora* could represent an intermediate step in the evolution of a mode of cellular propagation, in which vegetative progeny is generated by internal reproduction (Angert and Losick, 1998; see below).

M. polyspora is closely related to *Epulopiscium* spp. (Angert et al., 1996). *Epulopiscium* spp., are prominent components of the intestinal microflora in certain surgeonfish species (see Angert and Clements, 2004, and refs. therein). *Epulopiscium* spp. are the largest heterotrophic bacteria known, with the cigar-shaped cells reaching lengths of over 600 μm (Angert et al., 1993). A distinctive feature of *Epulopiscium* spp. is their ability to form multiple, internal offspring, as a form of cellular propagation. A single mother cell normally produces two daughter cells, but the production of up to 7 cells has been observed, often associated with the poles of the mother cell (Angert et al., 1993). The cells grow and elongate filling the mother cell cytoplasm, eventually causing lysis of the mother cell. Assembly of polar septa takes place in large progeny cells, just before their release from the mother cell. The cells contain multiple copies of the genome close to the cell periphery, but in preparation to polar cell division, the DNA is reorganized in a polar pattern, a pattern that is reminiscent of the DNA localization in *M. polyspora* prior to polar division (Angert and Losick, 1998; Angert and Clements, 2004). Part of the pole-associated DNA is trapped inside the small cellular compartments formed at the poles, and at later stages, more of the pole-associated DNA is translocated into the offspring cells (Angert and Clements, 2004). In contrast to chromosome partitioning during sporulation in *B. subtilis* (Errington, 2003; and Chapter 5, this volume) following completion of engulfment in *Epulopiscium* spp., only a fraction of the cellular DNA is partitioned into the newly-formed cells (Angert and Clements, 2004).

In contrast to *M. polyspora*, which produces multiple dormant offspring designed to survive transit within or between GI tracts, *Epulopiscium* cells produce active (not dormant) offspring, and show a permanent endosymbiotic relationship with the host. However, both processes may have originated from sporulation, presumably shaped by the co-evolution with hosts (Angert and Clements, 2004, and refs. therein).

Some *Bacillus* and *Clostridia* are able to produce at least occasionally two prespores per mother cell (Smith, 1970), and in the segmented filamentous bacteria each cell produces a prespore following asymmetric cell division, which then undergoes further division following engulfment completion (Chase and Erlandsen, 1976). Segmented filamentous bacteria have been found in the guts of many arthropods, and have been collectively called *Arthromitus*, but recent studies have identified these bacteria as *B. cereus* (Margulis et al., 1998; Feinberg et al., 1999). Presumably, *B. cereus* and closely related bacteria preferentially adopt a filamentous mode of growth in the intestines of invertebrates and similar habitats (Margulis et al., 1998; reviewed by Jensen et al., 2003). It has been suggested that *B. cereus* and close relatives, may experience two distinct but interconnected life cycles, one in which the bacteria establishes a symbiotic relation in the intestines of an invertebrate host, and a second in which the bacteria can multiply in an infected host, vertebrate or invertebrate (Jensen et al., 2003). In any case, the spore appears to be the particle that allows transit from one host to the other, in part through the soil or contaminated food materials. The details of sporulation in each host are not clear. For example, it is not clear if and how the *Arthromitus* stage influences spore formation by *B. cereus* and related bacteria.

The observation that *B. subtilis* spores can germinate and re-initiate sporulation in the GI tract of mice (Hoa et al., 2001; Casula and Cutting, 2002), together with the observation that other *Bacillus* species may adopt a different mode of growth in the intestines of their hosts, and that in species such as *M. polyspora* and *Epulopisium* spp., sporulation in the gut may assume various contours, raise exciting possibilities. It is possible that even familiar species such as *B. subtilis* may adopt particular modes of growth and sporulation when present in gut ecosystems.

BACILLUS SPP. IN THE GI TRACT

While *Bacillus* spp. have traditionally been described as aerobic saprophytic soil organisms, their ubiquitous nature results in the daily intake of considerable numbers of these bacteria, often in the form of spores. This ingestion occurs frequently through food and water or through the use of contaminated feed (Guinebretiere et al., 2001; Vaerewijck et al., 2001), and there have been a number of studies that have reported the isolation of these organisms from faecal material or from different GI tracts (Ghosh, 1978; Turnbull and Kramer, 1985; Gonzalez et al., 1999). In a recent study aimed towards the isolation and characterization of potential *Bacillus* probiotic candidates (Barbosa et al., submitted for publication), we have isolated a large number of presumptive *Bacillus* spp. strains from organic poultry faecal material.

Some of the selected isolates displayed properties that would be advantageous for survival in the gut or for their use as sporeforming CE (competitive exclusion) agents or probiotics. These include tolerance to simulated GI conditions, robust biofilm formation, or the production of antimicrobial compounds. Interestingly, a number of phenotypic characteristics previously described for a standard laboratory strain of *B. subtilis* (strain MB24) differed significantly from those found for their undomesticated counterparts.

ISOLATION AND CHARACTERIZATION OF POULTRY FAECAL SPOREFORMERS

Sporeforming bacteria were selected by heat and ethanol treatment of organic poultry faecal material. The treated samples were plated on a rich medium that supported efficient germination, and incubated aerobically at 37°C. Phase-contrast microscopy of 237 randomly chosen isolates grown on Difco Sporulation Medium (DSM) (Nicholson and Setlow, 1990) revealed a diverse collection of rod-shaped bacteria producing endospores of various sizes and shapes. All the isolates were catalase-positive, a characteristic that differentiates *Bacillus* from the anaerobic sporeformers *Clostridium* spp..

A subset of isolates was further characterized by sequencing of the 16S rDNA, and by testing for biochemical properties using the API 50 CHB system. This permitted a conditional identification of the isolates, although limitations associated with both these approaches, which have previously been described and can result in the mislabeling of isolates, were also found here. While all our isolates could readily be identified to the genus level as *Bacillus* spp., species determination of some proved to be more difficult. This included several strains that were genetically related to *Bacillus* sp., but were unable to utilize any of the carbohydrates present in the API 50 CHB strip, thus producing a biochemical profile that is not currently recognized by the APILAB Plus software (BioMérieux). Despite these drawbacks, species identified among the poultry gut isolates included *B. subtilis*, *B. pumilus*, *B. licheniformis*, *B. cereus* group, *B. clausii*, *B. megaterium* and *B. firmus*.

The mislabeling of sporeformers remains a significant problem (Hoa et al., 2000; Guinebretiere et al., 2001), and the increased use of these strains for industrial and probiotic applications, means that greater attention must be paid to ensure as comprehensive and accurate a characterization as possible. Such characterization is laborious, but is increasingly necessary. For this reason, we employed additional approaches to assist in strain identification and fingerprint. For example, it was noted that two isolates identified by 16S rDNA sequence analysis as *B. clausii* were able to grow under alkaline conditions and presented an extended lag phase when grown in batch culture, both of which are properties that have previously been reported for *B. clausii* isolates present in some probiotic products (Green et al., 1999; Hoa et al., 2000). The characteristic growth properties of these isolates were therefore taken as an additional indication for their correct classification as *B. clausii*.

Another property investigated was the ability to grow at 50 °C, which was demonstrated by all the *B. licheniformis*, *B. pumilus* and *B. subtilis* isolates. Interestingly, one isolate, which exhibited a distant relationship with the *B. cereus* subgroup based on 16S rDNA sequencing and the distinctive small size of its colonies, also grew well at this temperature, despite the fact that this is not a feature normally associated with this group of *Bacillus* (Sneath, 1986).

Further characterization involved testing the ability of the isolates to grow under anaerobic conditions. All the *B. licheniformis* and *B. cereus* group isolates were capable of anaerobic growth (Sneath, 1986). Moreover, some of the *B. subtilis* isolates could also grow to some extent under these conditions, although there were strain-specific differences in the levels of tolerance. *B. subtilis* has been viewed as a strict aerobe, but recent work has shown that it can also grow anaerobically (Nakano and Zuber, 1998). This property would certainly be regarded as advantageous for survival and establishment in the gut anaerobic ecosystem. Interestingly, it has also been shown that *B. subtilis* does not sporulate efficiently under anaerobic conditions (Hoffmann et al., 1998), suggesting that any capacity to sporulate in the gut (Hoa et al., 2001) would be strongly influenced by the particular niche being occupied.

Interestingly, approximately 60 % of the 31 poultry gut *Bacillus* isolates presented some level of hemolysis on 5 % sheep blood agar. These included the isolates from the *B. cereus* group but also isolates of other *Bacillus* species, such as *B. pumilus* and *B. subtilis*. Recent publications have described the production of enterotoxins by species other than *B. cereus*, suggesting that these traits might be more widespread than previously thought (Damgaard et al., 1996; Rowan et al., 2001; Rowan et al., 2003). These observations are relevant considering the increased use of these strains as animal and human probiotics.

The presence of plasmid DNA has frequently been reported in *Bacillus* isolates, and in some cases it has been associated with antibiotic resistance determinants and the production of toxins and antibiotic compounds (Bernhard et al., 1978; Monod et al., 1986; Okinaka et al., 1999). Of the 31 selected isolates analyzed almost 50 % contained detectable plasmid DNA with some isolates harboring multiple plasmids, some of which were rather large. The nature of these plasmids remains to be determined. Plasmid profiles assisted in the discrimination of particular strains belonging to the same species. For example, *B. licheniformis* and *B. megaterium* isolates could be differentiated on the basis of their distinctive plasmid profiles. Occurrence of plasmid DNA is also frequently seen as an indicator of genetic stability and potential for genetic transfer. This observation is of importance since the dense and diverse microbial communities in gut ecosystems are believed to favour genetic transfer between resident and transient bacteria that might for example be ingested with food (see Chapter 10).

Aspects of macroscopic morphology were also used to assist in the identification of distinctive, strain-specific features. The different *B. subtilis* related isolates for example, exhibited clearly

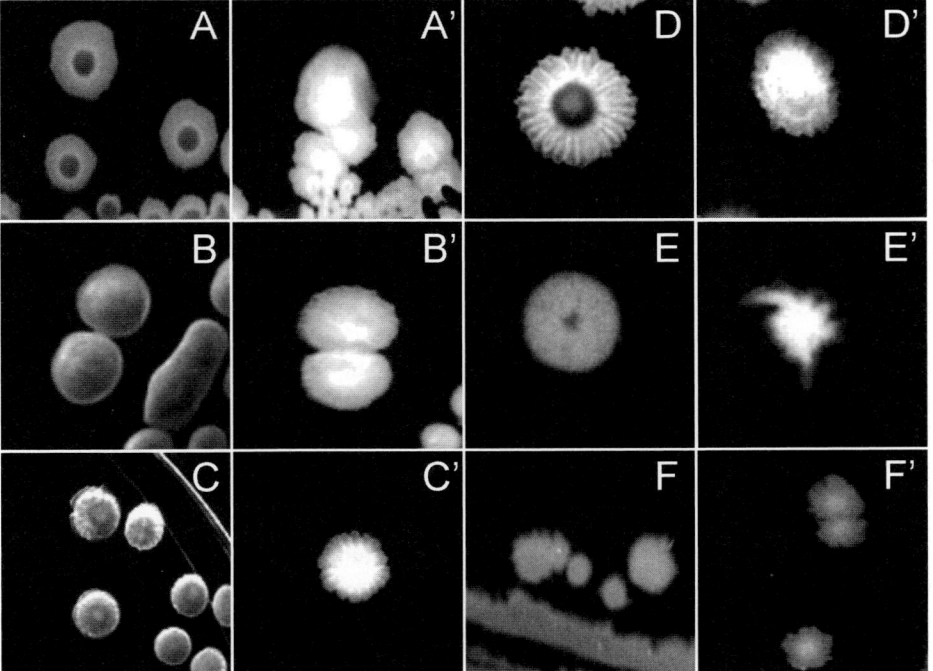

Figure 1. Poultry faecal sporeformers colony morphological diversity. Photographs are at different scales, and thus are merely illustrative. Isolates represented in the different panels were grown in Luria-Bertani, LB (X) and Difco Sporulation Medium, DSM (X') and are as follows: A-A', B-B', C-C', D-D' and E-E', *B. subtilis* related natural isolates; F-F', *B. subtilis* laboratory strain MB24.

different colonial morphologies (Figure 1). Another example is that of an isolate which was easily differentiated from the other *B. cereus* group strains by the release of a dark brown pigment upon growth on both LB and sporulating plates. Accumulation of a dark pigment by sporulating colonies of *B. subtilis* is well documented, and occurs following the synthesis of a laccase that becomes associated with the spore coat (Donovan et al., 1987; Hullo et al., 2001; Martins et al., 2002; see also Chapter 6., this volume). However, pigment formation by this *B. cereus* group isolate was also visible under nutritional conditions that do not support efficient sporulation (e.g. growth on LB), suggesting that its synthesis is not under sporulation control. Hullo et al. (2001) reported on the manganese-dependent, but sporulation-independent production of a pigment on agar media by *B. subtilis* (Hullo et al., 2001). It remains to be determined whether pigment production by the undomesticated *B. cereus* group strain found in our study is linked to the oxidation of manganese.

Microscopical analysis proved to be especially helpful for the diagnosis of some of the isolates. For example, two *B. clausii* isolates mentioned earlier as capable of alkaline growth and showing long lag phase, also had a distinctive microscopical morphology with the formation of long filaments that appeared unseptated by phase-contrast microscopy. However, staining of the cells with the membrane dye FM4-64 revealed that division septa were present. Thus, it appeared that in these isolates of *B. clausii*, physical separation of the sister cells following cytokinesis does not occur, therefore leading to the formation of long chains.

In summary our results suggest that a large number of *Bacillus* isolates belonging to previously classified and some unclassified and possibly new species, constitute part of the gut microflora in poultry. To ascertain whether these bacteria exhibited properties that could be relevant for their survival in the animal gut ecosystem, and by inference relevant to their potential use as probiotics, a group of seven isolates representative of five different species were subjected to a more detailed analysis. The group included three isolates of *B. subtilis*, and one isolate of *B. megaterium*, *B. pumilus*, *B. licheniformis* and *B. clausii*. Below we give an overview of the properties tested and the overall results obtained.

BIOFILMS

While the formation of biofilms by gut microorganisms is well recognised (Probert and Gibson, 2002; Palestrant et al., 2004), its relevance specifically to *Bacillus* strains in the gut has yet to be determined. It is nevertheless conceivable that biofilms on the surface of either the gut or food particles could provide protection for the bacterium against some of the physical and chemical stresses of the gastrointestinal tract, and promote the survival and persistence of some *Bacillus* strains. Alternatively, but not exclusively, the existence of these organism as biofilms could play a role in the protection of the gut epithelium from adhesion by pathogenic agents. Of the poultry gut *Bacillus* isolates tested, the three *B. subtilis* strains, and the *B. licheniformis* isolate produced highly viscous colonies on LB plates (e.g. Figure 1, A, B and C), and also formed strong biofilms. Remarkably for these isolates the film, at the air-medium interface, was strong

enough to sustain the cultures when tubes were inverted. In sharp contrast, the domesticated laboratory strain MB24 grew without forming a ring. This is in agreement with recent findings indicating that robust biofilm formation is an attribute of natural isolates of *B. subtilis* that appears to have become lost in domesticated strains, presumably because of continuous selection for fast planktonic growth (Branda et al., 2001). In their study Branda et al. demonstrated that natural isolates of *B. subtilis* are capable of forming complex biofilms, which produce aerial structures or fruiting bodies. These structures were the preferential sites of sporulation. In keeping with the suggestion that sporulation can occur in the gut (Hoa et al., 2001), one interesting possibility is that biofilms serve as preferable sites for sporulation in this environment.

ANTIBIOTIC SUSCEPTIBILITY

With rare exceptions the group of seven *Bacillus* isolates were susceptible to the collection of antibiotics tested, which included ampicillin, tetracycline, chloramphenicol, gentamycin, ciprofloxacin, erythromycin and lincomycin. Exceptions to this included the *B. clausii* and *B. licheniformis* isolates, which were resistant to erythromycin and lincomycin. These resistance traits were associated with the presence of an MLS$_B$-resistance gene encoding for a rRNA methylase. PCR and Southern blot hybridization identified an *erm*(D) gene on the chromosome of the *B. licheniformis* isolate. Recently, *erm*(34) has been identified in both *B. clausii* probiotic and reference strains (Bozdogan et al., 2004). Sequencing of a PCR product amplified from the genome of the poultry *B. clausii* isolate with *erm*(34) specific primers showed that this determinant is also responsible for MLS$_B$ resistance in the *B. clausii* strain characterized in this study.

ANTIMICROBIAL ACTIVITY

Spores of a laboratory strain of *B. subtilis* have been shown to inhibit the colonization of young chicks by *E. coli* O78: K80, *Salmonella enterica* serotype Enteritidis and *Clostridium perfringens* (La Ragione et al., 2001; La Ragione and Woodward, 2003). While the molecular mechanisms responsible for this effect remain poorly understood it is anticipated that the competitive exclusion of pathogens will result from one or more modes of action, including the production of antimicrobial agents, such as bacteriocins (Patterson and Burkholder, 2003; see also Chapter 15, this volume).

Although antimicrobial activity was detected for all the poultry *Bacillus* strains in our recent study, there were considerable differences in the spectrum and degree of inhibition observed (Figure 2). Several of the gut *Bacillus* isolates demonstrated an antagonistic activity toward a broad Gram-positive bacterial spectrum, which included food spoilage and important pathogenic bacteria, such *Bacillus* spp., *Enterococcus faecalis*, *Staphylococcus aureus*, *Clostridium perfringens* and *Listeria monocytogenes*. On the other hand, Gram-negative bacteria appeared to be less susceptible to the antagonistic activity of these compounds, and only one isolate of *B. subtilis* was partially inhibitory to *E. coli* O78:K80. The widespread production of antimicrobial compounds by the poultry faecal

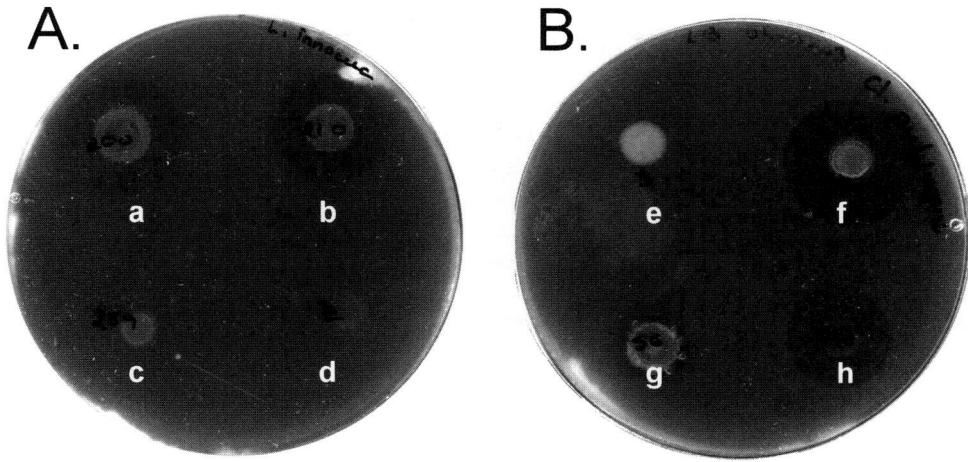

Figure 2. Representative examples of the antimicrobial activity displayed by some poultry faecal *Bacillus* isolates. Antimicrobial activity was screened by a colony overlay assay where producing strains (a, b, and f, *B. subtilis* isolates; c, *B. clausii*; d, *B. clausii* strain present in the probiotic Enterogermina; e, *B. megaterium*; g, *B. licheniformis*; h, *B. pumilus*) grown as spots for 24 hours on LB agar plates were overlayed with a specific indicator strain (A, *Listeria innocua*; B, *Clostridium perfringens*).

Bacillus isolates could be of importance in keeping a healthy gut microbial balance and to achieve the wanted probiotic protection against pathogens.

SPORULATION AND GERMINATION

All gut *Bacillus* isolates sporulated efficiently under laboratory conditions, with spore numbers in the order of 10^8-10^9 ml^{-1}. Noteworthy is the observation that heat treatment of sporulating cultures of *B. clausii* following completion of spore formation, resulted in an apparent increase in colony forming units. Similar observations have been made for other *Bacillus* strains (Hoa et al., 2000), where it was proposed to be associated with heat activation of spore germination. However, microscopic observation of DSM cultures of the *B. clausii* isolate revealed a considerable number of mature (phase bright) spores still inside the sporulating cells, which formed long chains (see above). Therefore, a more likely explanation is that heat treatment could separate the cells and/ or "burst" the mother cells. In any case, this would promote the release of spores, which would otherwise remain "aggregated", and thereby lead to erroneous counts.

We have also tested the ability of common laboratory germinants to induce the germination of spores derived from the poultry gut isolates. Compared to *B. subtilis* MB24 strain, all the gut isolates spores responded modestly to AGFK (3.3 mM L-asparagine, 5.6 mM glucose, 5.6 mM fructose. 10 mM KCl), and only for the *B. licheniformis* isolate, and one *B. subtilis* isolate, was the response to L- alanine comparable to the laboratory strain. That decreased rates of germination were observed for the natural *B. subtilis* isolates, strengthens the suggestion that the behavior of laboratory strains can differ significantly from natural *Bacillus* isolates. It is known that in *B. subtilis*, an alanine racemase is associated with the spore coat, presumably controlling the ratio of L-alanine to D-alanine, and thereby the extent of spore germination (Kanda-Nambu et al., 2000). Presumably, spores of the undomesticated *Bacillus* isolates have increased alanine racemase activity. Alternatively,

they may have an altered coat structure, which does not permit a small hydrophilic molecule such as L-alanine to interact with the spore and reach its receptor. The germination phenotype of *gerP* mutants in both *B. subtilis* and *B. cereus* can be overcome by genetically or biochemically eliminating most of the coat structure (Behravan et al., 2000). This suggests that the mutants have an altered coat, which does not interact properly with germinants (Behravan et al., 2000; see also Chapter 6). However, no gross morphological differences were seen in the coats of spores formed by the natural isolates of *B. subtilis*, when compared to those of the laboratory strain (Figure 3; see below). In any case, that spores of the undomesticated *B. subtilis* isolates fail to germinate efficiently in response to compounds that promptly trigger germination of a domesticated strain strongly suggests that other, as yet unidentified stimuli induce germination in the gut ecosystem. Further studies are needed to identify the natural or artificial agents that could induce germination of native *Bacillus* spores in the GI tract.

ULTRA-STRUCTURE AND SPORE COAT POLYPEPTIDE COMPOSITION

Bacillus spores are extremely resistant to external physical and chemical insults and survive exceptionally well in the environment (Driks, 1999; Henriques and Moran, 2000; Nicholson et al., 2000; see also Chapter 1). This is in part due to the structural organization of the spore, in which the spore core, carrying a copy of the genome, is surrounded by a thick cortex peptidoglycan layer, and further encased in a more or less complex protein coat. In spores of *B. subtilis*, as in most other spores, the coat is differentiated into three layers: a poorly defined undercoat closely apposed to the cortex, an intermediate lamellar inner coat, and an external dense outer coat (Driks, 1999; Henriques and Moran, 2000; Chapter 6, this volume).

Most of the variation seen in the ultrastructure of spores of different species is due to the presence or absence of extra layers, such as an exosporium surrounding, but loosely attached to the

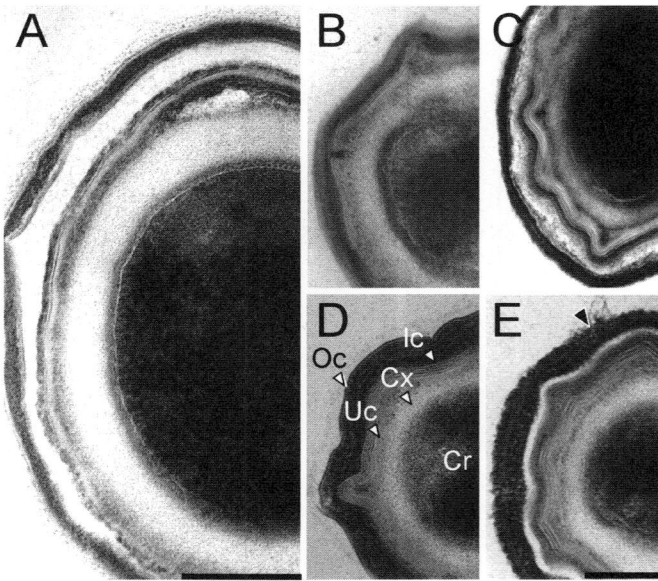

Figure 3. Thin section transmission electron micrographs of spores produced by selected gut *Bacillus* isolates, collected and processed from 24 h cultures in DSM. The following spore structures are indicated for a spore produced by a natural isolate of *B. subtilis* (panel D): Cr, spore core; Cx, spore cortex; Uc, Ic, and Oc, under, inner and outer spore coat. Other panels represent spores from isolates of the following species: A, *B. megaterium*; B, *B. licheniformis*; C, *B. pumilus*; E, *B. clausii*. Arrowhead in panel E indicates the hair-like structures visible with spores of the *B. clausii* isolate.

outer coat, as well as various appendages of poorly characterized nature and structure (Henriques and Moran, 2000). Spore appendages, for example, have been implicated in the well known ability of *B. cereus* spores to adhere to various surfaces (Hachisuka et al., 1984; Kozuka and Tochikubo, 1985).

In general, spores of the poultry gut isolates conformed to the overall pattern described above (Figure 3). Spores of all the *B. subtilis* isolates examined were virtually indistinguishable from spores of the laboratory strain MB24 (Henriques and Moran, 2000) (Figure 3, panel D, representative *B. subtilis* poultry isolate). Also, the structure of *B. megaterium* spores was similar to that of the "double" type spore coat described in the literature for this species (Aronson and Fitz-James, 1976) (Figure 3, panel A). However, in spores of *B. clausii*, the coat showed a relatively thin electron dense layer separating the inner coat from a more external lamellar layer (Figure 3, panel E). In addition, the spores were associated with an external electron dense layer with a porous appearance, of unknown origin, and also presented short hair-like structures protruding from the outermost coat layer (Figure 3, arrowhead in panel E). Interestingly, this pattern was very different from spores of the *B. clausii* probiotic Enterogermina (Green et al., 1999), for which the presence of an exosporium, but not of hair-like structures, has previously been described. None of the other spores examined had a structure that could be described as an exosporium or had any other type of spore appendages.

Analysis of the SDS-PAGE profiles of the polypeptides extracted from the coats of highly purified spore preparations showed that spores of *B. clausii*, *B. pumilus*, *B. subtilis*, *B.*

licheniformis and *B. megaterium* have clearly distinctive coat protein profiles (Figure 4). Note for example the three main components of less than 21 kDa that can be extracted from the coats of *B. clausii*. These results suggest that the polypeptide composition of the spore coat could within limits be a good fingerprint for a particular species. Intriguingly, the coat polypeptide composition of the *B. clausii* isolate and the *B. clausii* strain of the probiotic Enterogermina were remarkably similar (Figure 4, lanes 5 and 6), contrasting with the dissimilarity observed at the ultrastructural level. This suggests that a nearly identical collection of coat proteins can assume different types of structural organization. However, we note that not all the coat components are equally extractable and amenable to electrophoretic resolution, in part because of extensive cross-linking; also, even though the coat is mostly composed of protein, some coat proteins appear to be glycosylated (Chapter 6, this volume). Since the spore coat confers protection against harsh chemicals and peptidoglycan-breaking enzymes, and plays a decisive role in the response of spores to germinants (Henriques and Moran, 2000), it will be interesting to determine whether any of the features observed influence spore resistance, germination, persistence, or adhesion in the gut.

ACID AND BILE TOLERANCE

Gut bacteria need to be able to survive to the harsh conditions found in the GI tract. Spores should in theory be ideally suited for this, considering their extraordinary and well reported resistance properties (Driks, 1999; Henriques and Moran, 2000; Nicholson et al., 2000). Nevertheless, spores of some *Bacillus* probiotic strains were recently shown to be susceptible to bile salts, and to simulated gastric conditions (Duc et al., 2004), indicating that these properties are not universal among sporeformers.

Spores of all the poultry isolates tested, as well as those of *B. subtilis* MB24 were similarly resistant to both simulated gastric conditions and to bile salts. The same did not apply to the

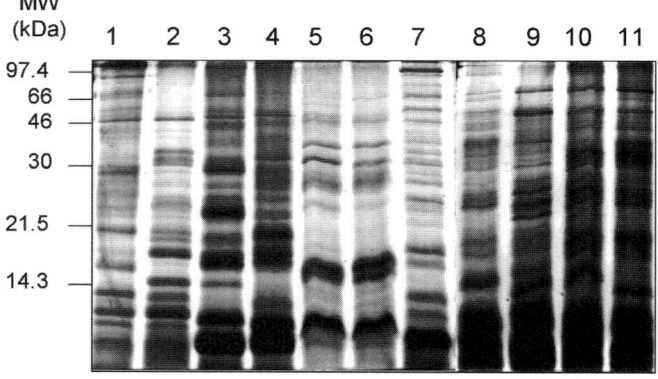

Figure 4. SDS-PAGE analysis of the coat polypeptide composition of spores of different *Bacillus* isolates. Lane 1, *B. megaterium* 899 (BGSC 7A1); lane 2, a *B. megaterium* natural isolate; lane 3, *B. licheniformis* 9945A (BGSC 5A2); lane 4, a *B. licheniformis* natural isolate; lane 5, *B. clausii* (Enterogermina); lane 6, a *B. clausii* natural isolate, lane 7, a *B. pumilus* natural isolate; lane 8, *B. subtilis* MB24; lanes 9, 10, and 11, different *B. subtilis* natural isolates. The position of molecular weight markers (in kDa), is shown on the left side of the panel.

vegetative cells, which were very susceptible to both conditions. These findings suggest that only spores of these isolates would survive transit through the GI tract. Their final outcome may then depend on whether or not an intestinal niche(s) exists where spores may persist and perhaps germinate. Recent findings from our laboratory (Costa, T., Barbosa, T.M., and A.O. Henriques, unpublished), suggest that the coat layers make an important contribution for the survival of spores to acid and bile salts, as chemically-decoated spores do not survive these conditions. The fact that resistance can be achieved with coats of different protein composition and slightly different structures, raises questions about the need for different degrees of coat complexity. Different studies have reported on the germination of *Bacillus* spores in the GI tract, including the jejunum and ileum (Mazza, 1994; Hoa et al., 2001; Casula and Cutting, 2002). Interestingly, *B. cereus* spores have been shown to survive the conditions of the GI tract and are capable of adhering to the gut epithelium (Andersson et al., 1998).

CONCLUDING REMARKS

There are several anatomically and physiologically different gut ecosystems, e.g. ruminants, mammals, birds, insects and fish. Nevertheless, all have a region that is characterized by a dense and diverse microbial population comprised of resident and transient microorganisms. It has been suggested that ~400 different species of bacteria are present in the human colon at a density of more than 10^{11} cells per gram of content (Wang et al., 1996). The large majority of these are obligate anaerobes, but aerobic and facultative organisms are also present. The nature and balance of the interactions between these organisms and the host is fundamental for their recognized beneficial effects in nutrition, physiology and protection of the host.

Bacillus spp. sporeformers are normally classified as aerobic bacteria, although several species are known to be capable of anaerobic growth (Drobniewski, 1993; Nakano and Zuber, 1998). This is of particular importance if we consider the potential for *Bacillus* spp. to colonize the GI tract, where species that are capable of sporulation have a natural advantage when it comes to surviving the adverse conditions found in this particular ecosystem. However in contrast to the anaerobic sporeformers *Clostridium* spp., which are common and numerically important members of the intestinal microbiota (Wang et al., 1996; Suau et al., 1999; Leser et al., 2002), microbial community-profiling studies frequently fail to detect *Bacillus* isolates in the gut. These observations may indicate that *Bacillus* represents only a small proportion of the resident gut bacteria or that their residence in the tract is only transient.

Despite these results some studies have reported the isolation of gut associated *Bacillus* strains. Earlier publications describe the isolation of *Bacillus*, particularly *B. cereus* from human faeces (Ghosh, 1978; Turnbull and Kramer, 1985; Macfarlane et al., 1986) and *Bacillus* spp. have also been isolated from the intestine of reared fish (Gonzalez et al., 1999). Species of the *B. cereus* group, such as *B. thuringiensis* and *B. cereus*, have also been suggested as natural inhabitants of invertebrate gut ecosystems. Here they have a symbiotic association with the gut of many different insects (Margulis et al., 1998; Nicholson,

2002; Jensen et al., 2003) but *B. thuringiensis* isolates can occasionally enter a pathogenic cycle in particular invertebrate or vertebrate hosts (Jensen et al., 2003). *B. thuringiensis* isolates were also found to be frequently associated with the faeces of different zoo-animals, and particularly in herbivores, suggesting that introduction to the gut could be through the ingestion of plant material (Lee et al., 2003).

Our recent study that isolated a large number of *Bacillus* strains from poultry faeces, suggests that a diverse collection of these organisms are indeed present in the poultry gut. While this study has yet to define the exact nature of this presence, the isolated sporeformers appear to be well equipped for both persistent and transient association with the gut ecosystem. In parallel work, aerobic sporeformers were also isolated from the faecal material of animals with quite different gut ecosystems, including sheep and cattle (our unpublished data).

Although *Bacillus* species are frequently found in food (Damgaard et al., 1996; Guinebretiere et al., 2001; Pepe et al., 2003), which can mediate their entry into the gut, with the use of *Bacillus* spores as human and animal probiotics, as well as agricultural biopesticides, the burden of food-borne spores that are reaching the human and animal GI ecosystem needs to be carefully monitored. In support of this, and as an example of the importance of monitoring, *B. thuringiensis* has been detected in feacal samples from workers of greenhouses that had used *B. thuringiensis*-pesticides (Jensen et al., 2002). In this study there was evidence for germination in the GI tract since only 10 % of the cells recovered from the faeces were resistant to heat treatment. However, no data was presented to indicate that *B. thuringiensis* was actually capable of colonizing the intestine, or germinating in the intestine before shedding.

Importantly, association of *Bacillus* with the GI tract is not always advantageous to the host. In this regard several species of *Bacillus* have been associated with GI tract disorders, such as food poisoning outbreaks. This is most commonly a consequence of the production of toxins (Salkinoja-Salonen et al., 1999; Mikkola et al., 2000; Agata et al., 2002), such as the emetic and the diarrheal toxin produced by *B. cereus* (Granum and Lund, 1997). Other species occasionally involved with food-poisoning episodes are *B. subtilis*, *B. pumilus*, *B. licheniformis*, *B. mycoides*, *B. thuringiensis*, *B. alvei*, *B. circulans*, *B. lentus* and *B. sphaericus* (Kramer and Gilbert, 1989; Jackson et al., 1995; Beattie and Williams, 1999). Another example is anthrax of the gastro intestinal tract caused by *B. anthracis* (Sirisanthana and Brown, 2002). It is suggested that the disease is related to the dose of spores ingested and the immune state of the host.

While these data suggest that *Bacillus* sp. can occupy almost any gut ecosystem, it remains a point of debate whether these bacteria are natural inhabitants of gut ecosystems or if their isolation results from ingestion with contaminated food. Nevertheless, even if transient, their frequent detection in faecal or GI tract samples suggests that their role in the intestinal microbial balance could well be underestimated and certainly, there is potential for exploitation. However, accurate identification of strains, and exhaustive characterization of possible harmful consequences will need to be performed. Nevertheless, as the number of studies in this area increases, and new molecular techniques are directed towards the identification and characterization of the complex gut microbial populations, we are sure to improve our understanding of the incidence and types of *Bacillus* that are to be found in different gut ecosystems and their physiological roles in these environments.

Recent studies have examined the phenotypic characteristics of some of the undomesticated *Bacillus* isolates, differing from or not exhibited by their domesticated counterparts (Branda et al., 2001; Kearns and Losick, 2003). These studies will be particularly informative in the model organism *B. subtilis*, which has been extensively studied in the laboratory. However, they will most likely be extended to other species, in particular those for which genome sequences are available, and may reveal new and exciting details of intricate biological processes such as spore development and spore germination.

ACKNOWLEDGEMENTS

Work in the laboratory at ITQB, Oeiras, Portugal, is supported by a grant from the European Union 5th Framework (QLK5-CT-2001-01729).

REFERENCES

Agata, N., Ohta, M., and Yokoyama, K. 2002. Production of *Bacillus cereus* emetic toxin (cereulide) in various foods. Int. J. Food Microbiol. *73*, 23-27.

Andersson, A., Granum, P.E., and Ronner, U. 1998. The adhesion of *Bacillus cereus* spores to epithelial cells might be an additional virulence mechanism. Int. J. Food Microbiol. *39*, 93-99.

Angert, E.R., Brooks, A.E., and Pace, N.R. 1996. Phylogenetic analysis of *Metabacterium polyspora*: clues to the evolutionary origin of daughter cell production in *Epulopiscium* species, the largest bacteria. J. Bacteriol. *178*, 1451-1456.

Angert, E.R., and Clements, K.D. 2004. Initiation of intracellular offspring in *Epulopiscium*. Mol. Microbiol. *51*, 827-835.

Angert, E.R., Clements, K.D., and Pace, N.R. 1993. The largest bacterium. Nature. *362*, 239-241.

Angert, E.R., and Losick, R. 1998. Propagation by sporulation in the guinea pig symbiont *Metanobacterium polyspora*. Proc. Natl. Acad. Sci. USA. *95*, 10218-10223.

Aronson, A.I., and Fitz-James, P. 1976. Structure and morphogenesis of the bacterial spore coat. Bacteriol. Rev. *40*, 360-402.

Beattie, S.H., and Williams, A.G. 1999. Detection of toxigenic strains of *Bacillus cereus* and other *Bacillus* spp. with an improved cytotoxicity assay. Lett. Appl. Microbiol. *28*, 221-225.

Behravan, J., Chirakkal, H., Masson, A., and Moir. A. 2000. Mutations in the *gerP* locus of *Bacillus subtilis* and *Bacillus anthracis* affect access of germinants to their targets in spores. J. Bacteriol. *182*, 1987-1994.

Bernhard, K., Schrempf, H., and Goebel, W. 1978. Bacteriocin and antibiotic resistance plasmids in *Bacillus cereus* and *Bacillus subtilis*. J. Bacteriol. *133*, 897-903.

Bozdogan, B., Galopin, S., and Leclercq, R. 2004. Characterization of a new *erm*-related macrolide resistance gene present in probiotic strains of *Bacillus clausii*. Appl. Environ. Microbiol. *70*, 280-284.

Branda, S.S., Gonzalez-Pastor, J.E., Ben-Yehuda, S., Losick, R. and Kolter, R. 2001. Fruiting body formation by *Bacillus subtilis*. Proc. Natl. Acad. Sci. U S A. *98*, 11621-11626.

Casula, G., and Cutting, S.M. 2002. *Bacillus* probiotics: spore germination in the gastrointestinal tract. Appl. Environ. Microbiol. *68*, 2344-2352.

Chase, D.G., and Erlandsen, S.L. 1976. Evidence for a complex life cycle and endospore formation in the attached, filamentous, segmented bacterium from murine ileum. J. Bacteriol. *127*, 572-583.

Damgaard, P.H., Larsen, H.D., Hansen, B.M., Bresciani, J., and Jorgensen, K. 1996. Enterotoxin-producing strains of *Bacillus thuringiensis* isolated from food. Lett. Appl. Microbiol. *23*, 146-150.

Donovan, W., Zheng, L.B., Sandman, K., and Losick, R. 1987. Genes encoding spore coat polypeptides from *Bacillus subtilis*. J. Mol. Biol. *196*, 1-10.

Driks, A. 1999. *Bacillus subtilis* spore coat. Microbiol. Mol. Biol. Rev. *63*, 1-20.

Drobniewski, F.A. 1993. *Bacillus cereus* and related species. Clin. Microbiol. Rev. *6*, 324-338.

Duc, L.H., Hong, H.A., Barbosa, T.M., Henriques, A.O., and Cutting, S.M. 2004. Characterisation of *Bacillus* probiotics available for human use. Appl. Environ. Microbiol. *70*,2161-2171.

Errington, J. 2003. Regulation of endospore formation in *Bacillus subtilis*. Nature Rev. Microbiol. *1*, 117-126.

Feinberg L., Jorgensen J., Haselton, A. Pitt A., Rudner R., and Margulis L. 1999. *Arthromitus* (*Bacillus cereus*) symbionts in the cockroach *Blaberus giganteus*: dietary influences on bacterial development and population density. Symbiosis. *27*, 109-123.

Ghosh, A.C. 1978. Prevalence of *Bacillus cereus* in the faeces of healthy adults. J. Hyg. (Lond.). *80*, 233-236.

Gonzalez, C.J., Lopez-Diaz, T.M., Garcia-Lopez, M.L., Prieto, M., and Otero, A. 1999. Bacterial microflora of wild brown trout (*Salmo trutta*), wild pike (*Esox lucius*), and aquacultured rainbow trout (*Oncorhynchus mykiss*). J. Food Prot. *62*, 1270-1277.

Granum, P.E., and Lund, T. 1997. *Bacillus cereus* and its food poisoning toxins. FEMS Microbiol. Lett. *157*, 223-228.

Green, D.H., Wakeley, P.R., Page, A., Barnes, A., Baccigalupi, L., Ricca, E., and Cutting, S.M. 1999. Characterization of two *Bacillus* probiotics. Appl. Environ. Microbiol. *65*, 4288-4291.

Guinebretiere, M.H., Berge, O., Normand, P., Morris, C., Carlin, F., and Nguyen-The, C. 2001. Identification of bacteria in pasteurized zucchini purees stored at different temperatures and comparison with those found in other pasteurized vegetable purees. Appl. Environ. Microbiol. *67*, 4520-4530.

Hachisuka, Y., Kozuka, S., and Tsujikawa, M. 1984. Exosporia and appendages of spores of *Bacillus* species. Microbiol. Immunol. *28*, 619-624.

Henriques, A.O., and Moran, C.P., Jr. 2000. Structure and assembly of the bacterial endospore coat. Methods. *20*, 95-110.

Hoa, N.T., Baccigalupi, L., Huxham, A., Smertenko, A., Van, P.H., Ammendola, S., Ricca, E., and Cutting, A.S. 2000. Characterization of *Bacillus* species used for oral bacteriotherapy and bacterioprophylaxis of gastrointestinal disorders. Appl. Environ. Microbiol. *66*, 5241-5247.

Hoa, T.T., Duc, L.H., Isticato, R., Baccigalupi, L., Ricca, E., Van, P.H., and Cutting, S.M. 2001. Fate and dissemination of *Bacillus* subtilis spores in a murine model. Appl. Environ. Microbiol. *67*, 3819-3823.

Hoffmann, T., Frankenberg, N., Marino, M., and Jahn, D. 1998 Ammonification in *Bacillus subtilis* utilizing dissimilatory nitrite reductase is dependent on *resDE*. J. Bacteriol. *180*, 186-189.

Hullo, M.F., Moszer, I., Danchin, A., and Martin-Verstraete, I. 2001. CotA of *Bacillus subtilis* is a copper-dependent laccase. J. Bacteriol. *183*, 5426-5430.

Jackson, S.J., Goodbrand, R.B., Ahmed, R., and Kasatiya, S. 1995. *Bacillus cereus* and *Bacillus thuringiensis* isolated in a gastrointestinal outbreak investigation. Lett. Appl. Microbiol. *21*, 103-105.

Jensen, G.B., Hansen, B.M., Eilenberg, J., and Mahillon, J. 2003. The hidden lifestyles of *Bacillus cereus* and relatives. Environ. Microbiol. *5*, 631-640.

Jensen, G.B., Larsen, P., Jacobsen, B.L., Madsen, B., Smidt, L., and Andrup, L. 2002. *Bacillus thuringiensis* in fecal samples from greenhouse workers after exposure to *B. thuringiensis*-based pesticides. Appl. Environ. Microbiol. *68*, 4900-4905.

Kanda-Nambu, K., Yasuda, Y., and Tochikubo, L. 2000. Isozymic nature of spore coat-associated alanine racemase of *Bacillus subtilis*. Amino Acids. *18*, 375-387.

Kearns, D.B., and Losick, R. 2003. Swarming motility in undomesticated *Bacillus subtilis*. Mol. Microbiol. *49*, 581-590.

Kozuka, S., and Tochikubo, K. 1985. Properties and origin of filamentous appendages on spores of *Bacillus cereus*. Microbiol. Immunol. *29*, 21-37.

Kramer, J.M., and Gilbert, R.J. 1989. *Bacillus cereus* and other *Bacillus* species. In: *Foodborne Bacterial Pathogens*. M.P. Doyle, ed. Marcel Dekker, New York and Basel. p. 21-70.

La Ragione, R.M., Casula, G., Cutting, S.M., and Woodward, M.J. 2001. *Bacillus subtilis* spores competitively exclude *Escherichia coli* O78,K80 in poultry. Vet. Microbiol. *79*, 133-142.

La Ragione, R.M., and Woodward, M.J. 2003 Competitive exclusion by *Bacillus subtilis* spores of *Salmonella enterica* serotype Enteritidis and *Clostridium perfringens* in young chickens. Vet. Microbiol. *94*, 245-256.

Lee, D.H., Shisa, N., Wasano, N., Ohgushi, A., and Ohba, M. 2003. Characterization of flagellar antigens and insecticidal activities of *Bacillus thuringiensis* populations in animal feces. Curr. Microbiol. *46*, 287-290.

Leser, T.D., Amenuvor, J.Z., Jensen, T.K., Lindecrona, R.H., Boye, M., and Moller, K. 2002. Culture-independent analysis of gut bacteria: the pig gastrointestinal tract microbiota revisited. Appl. Environ. Microbiol. *68,* 673-690.

Macfarlane, G.T., Cummings, J.H., and Allison, C. 1986. Protein degradation by human intestinal bacteria. J. Gen. Microbiol. *132,* 1647-1656.

Margulis, L., Jorgensen, J.Z., Dolan, S., Kolchinsky, R., Rainey, F.A., and Lo, S.C. 1998. The *Arthromitus* stage of *Bacillus cereus*: intestinal symbionts of animals. Proc. Natl. Acad. Sci. U S A. *95,* 1236-1241.

Martins, L.O., Soares, C.M., Pereira, M.M., Teixeira, M., Costa, T., Jones, G.H., and Henriques, A.O. 2002. Molecular and biochemical characterization of a highly stable bacterial laccase that occurs as a structural component of the *Bacillus subtilis* endospore coat. J. Biol. Chem. *277,* 18849-18859.

Mazza, P. 1994. The use of *Bacillus subtilis* as an antidiarrhoeal microorganism. Boll. Chim. Farm. *133,* 3-18.

Mikkola, R., Kolari, M., Andersson, M.A., Helin, J., and Salkinoja-Salonen, M.S. 2000. Toxic lactonic lipopeptide from food poisoning isolates of *Bacillus licheniformis*. Eur. J. Biochem. *267,* 4068-4074.

Monod, M., Denoya, C., and Dubnau, D. 1986. Sequence and properties of pIM13, a macrolide-lincosamide-streptogramin B resistance plasmid from *Bacillus subtilis*. J. Bacteriol. *167,* 138-147.

Nakano, M.M. and Zuber, P. 1998. Anaerobic growth of a "strict aerobe" (*Bacillus subtilis*). Annu. Rev. Microbiol. *52,* 165-190.

Nicholson, W.L. 2002. Roles of *Bacillus* endospores in the environment. Cell. Mol. Life Sci. *59,* 410-416.

Nicholson, W.L., Munakata, N., Horneck, G., Melosh, H.J., and Setlow, P. 2000. Resistance of *Bacillus* endospores to extreme terrestrial and extraterrestrial environments. Microbiol. Mol. Biol. Rev. *64,* 548-572.

Nicholson, W.L. and Setlow, P. 1990. Sporulation, Germination and Outgrowth. In: Molecular Biological Methods for *Bacillus*. C. R. Harwood and S.M. Cutting, ed. John Wiley &Sons Ltd., Chichester, United Kingdom. p. 391-450.

Okinaka, R.T., Cloud, K., Hampton, O., Hoffmaster, A.R., Hill, K.K., Keim, P., Koehler, T.M., Lamke, G., Kumano, S., Mahillon, J., Manter, D., Martinez, Y., Ricke, D., Svensson, R., and Jackson, P.J. 1999. Sequence and organization of pXO1, the large *Bacillus anthracis* plasmid harboring the anthrax toxin genes. J. Bacteriol. *181,* 6509-6515.

Palestrant, D., Holzknecht, Z.E., Collins, B.H., Parker, W., Miller, S.E., and Bollinger, R.R. 2004. Microbial biofilms in the gut: visualization by electron microscopy and by acridine orange staining. Ultrastruct. Pathol. *28,* 23-27.

Patterson, J.A. and Burkholder, K.M. 2003. Application of prebiotics and probiotics in poultry production. Poult. Sci. *82,* 627-631.

Pepe, O., Blaiotta, G., Moschetti, G., Greco, T., and Villani, F. 2003. Rope-producing strains of *Bacillus* spp. from wheat bread and strategy for their control by lactic acid bacteria. Appl. Environ. Microbiol. *69,* 2321-2329.

Probert, H.M., and Gibson, G.R. 2002. Bacterial biofilms in the human gastrointestinal tract. Curr. Issues Intest. Microbiol. *3,* 23-27.

Rowan, N.J., Caldow, G., Gemmell, C.G., and Hunter, I.S. 2003. Production of diarrheal enterotoxins and other potential virulence factors by veterinary isolates of *Bacillus* species associated with nongastrointestinal infections. Appl. Environ. Microbiol. *69,* 2372-2376.

Rowan, N.J., Deans, K., Anderson, J.G., Gemmell, C.G., Hunter, I.S., and Chaithong, T. 2001. Putative virulence factor expression by clinical and food isolates of *Bacillus* spp. after growth in reconstituted infant milk formulae. Appl. Environ. Microbiol. *67,* 3873-3881.

Salkinoja-Salonen, M.S., Vuorio, R., Andersson, M.A., Kampfer, P., Andersson, M.C., Honkanen-Buzalski, T., and Scoging, A.C. 1999. Toxigenic strains of *Bacillus licheniformis* related to food poisoning. Appl. Environ. Microbiol. *65,* 4637-4645.

Sirisanthana, T., and Brown, A.E. 2002. Anthrax of the gastrointestinal tract. Emerg. Infect. Dis. *8,* 649-651.

Smith, L.D. 1970. *Clostridium oceanicum*, sp. N., a spore-forming anaerobe isolated from marine sediments. J. Bacteriol. *103,* 811-813.

Sneath, P.H.A. 1986 Endospore-forming gram-positive rods and cocci. In: Bergey's Manual of Systematic Bacteriology. P.H.A. Sneath, N.S. Mair, M.E. Sharpe, and J.G. Holt, ed. Williams & Wilkins, Baltimore, Md. p. 1104-1207.

Suau, A., Bonnet, R., Sutren, M., Godon, J.J., Gibson, G.R., Collins, M.D., and Dore, J. 1999. Direct analysis of genes encoding 16S rRNA from complex communities reveals many novel molecular species within the human gut. Appl. Environ. Microbiol. *65,* 4799-4807.

Turnbull, P.C., and Kramer, J.M. 1985. Intestinal carriage of *Bacillus cereus*: faecal isolation studies in three population groups. J. Hyg. (Lond). *95,* 629-638.

Vaerewijck, M.J., De Vos, P., Lebbe, L., Scheldeman, P., Hoste, B., and Heyndrickx, M. 2001. Occurrence of *Bacillus sporothermodurans* and other aerobic spore-forming species in feed concentrate for dairy cattle. J. Appl. Microbiol. *91,* 1074-1084.

Wang, R.F., Cao, W.W., and Cerniglia, C.E. 1996. PCR detection and quantitation of predominant anaerobic bacteria in human and animal fecal samples. Appl. Environ. Microbiol. *62,* 1242-1247.

Chapter 17

Display of Molecules on the Spore Surface

Rachele Isticato, Giuseppina Cangiano, Maurilio De Felice and Ezio Ricca

SUMMARY

Surface display systems for the expression of heterologous molecules on the surface of viruses and cells has proven to be invaluable for a wide range of applications such as the development of bioadsorbents and biocatalysts, the identification of new antibiotics and antigens and the delivery of vaccines and drugs. The recent development of strategies to expose bioactive molecules on the surface of *Bacillus subtilis* spores, in which efficient display is coupled with spore robustness, has driven attention to the spore-based system as an attractive new tool to deliver antigens and enzymes of industrial or pharmaceutical relevance. In this chapter we first outline methods for presenting heterologous proteins on the surface of bacterial cells and then review the known examples of surface presentation on bacterial spores.

INTRODUCTION

Biological applications of surface display systems are numerous and include the development of bioadsorbents and biocatalysts, the identification of new antibiotics and antigens, and the delivery of vaccines and drugs. While certainly not complete, this list shows some of the potentially exciting and challenging areas that have recently attracted researchers' attention worldwide. Proteins able to bind metal ions or other pollutants as well as enzymes able to degrade polysaccharides when expressed in heterologous hosts can be used as bioadsorbents for heavy metal removal (Kotrba et al., 1999; Samuelson et al., 2000) or as whole-cell biocatalysts for detoxification of harmful organic contaminants from the environment (Dhillon et al., 1999), or for polysaccharide-degradation by the food and paper pulp industry. Peptides expressed in appropriate hosts are used to obtain combinatorial libraries, then used for i) the determination of epitope specificity of monoclonal antibodies, ii) the identification of interacting proteins and interaction sites (Georgiou et al., 1997) and iii) identification of new antibiotics (Murase et al., 2003; Kim et al., 2000; Huang *et a.,*. 2000). Proteins and peptides with high antigenic or pharmacological activity once expressed in appropriate hosts are used for the development of new vaccines and drugs (Liljeqvist et al., 1997; Lee et al., 2000, Chen and Schifferli, 2003).

Several approaches have been undertaken to develop efficient display systems expressing heterologous polypeptides on the surface of viruses (Rahman and Gopinathan, 2003), microbial (Georgiou et al., 1997; Stahl and Uhlen, 1997), mammalian (Whitehorn et al., 1995) and insect (Ernst et al., 1998) cells. All these systems share a common theme, targeting recombinant proteins to the cell surface by constructing gene fusions using sequences from membrane-anchoring domains of surface proteins (Lee et al., 2000).

In this chapter we will initially summarise the strategies so far developed to display heterologous proteins on the surface of bacterial cells and will then focus on the display of proteins on the surface of bacterial spores. The final part of the chapter will report on our recent studies on the use, as carrier proteins, of the CotB and CotC components of the *Bacillus subtilis* spore coat.

SURFACE DISPLAY COMPONENTS

Each surface display system involves at least two components: a carrier protein, anchored on the cell surface, and a heterologous passenger protein exposed outside the cell. Several characteristics of carrier proteins can affect the efficiency of surface display and have different effects on the stability and integrity of the host cell.

A successful carrier protein should meet the following requirements: i) it should have an efficient signal peptide or transporting signal, to allow the fusion protein to go through the membrane; ii) it should have a strong anchoring motif, to avoid detachment from the surface; iii) it should be resistant to the proteases present in the extracellular medium or in the periplasmic space.

The location of insertion, or fusion, of the heterologous protein into the carrier protein is another important factor, since it can influence the stability, the activity and post-translational modification of the fusion protein. Therefore, fusions at N-, C-terminus or interior of the carrier protein (sandwich fusions) are, in some cases, constructed with the same passenger to obtain an efficient display. Some characteristics of the passenger protein can also affect the translocation process and final surface display. The folding structure of the passenger protein such as the formation of disulfide bridges at the periplasmic side of the outer membrane (Maurer et al., 1997) or the presence of many charged or hydrophobic residues can affect translocation through the membrane (Nguyen et al., 1995). The choice of the host organism is also an essential step for efficient display. A good host should be compatible with proteins to be displayed and should be easy to manipulate and cultivate without cell lysis. As summarised below, both Gram-negative and Gram-positive bacteria have been successfully used with different advantages for several purposes.

DISPLAY OF HETEROLOGOUS POLYPEPTIDES IN GRAM-NEGATIVE BACTERIA

The display of proteins on the surface of Gram-negative bacteria is complex due to the nature of the cell envelope structure that consists of cytoplasmic membrane, periplasm and outer membrane. The recombinant polypeptide is exported from the

Table 1. Examples of surface display systems in Gram-negative bacteria.

Carrier protein	Passenger protein	Applications	References
N-terminal fusions			
Escherichia coli PAL (173aa)	Anti-atrazine antibody fragment (252aa)	Biosensor	Dhillon et al., 1999
Neisseria gonorrhoeae IgA1 protease (45 kDa)	Cholera toxin B subunit (CtxB) (13kDa)	Vaccines	Jose et al., 1995
E. coli adhesin (AIDA-I) (51,5kDa)	Cholera toxin B subunit (CtxB) (13kDa)	Vaccines	Jose et al., 2002
C-terminal fusions			
E. coli Lpp-OmpA (123aa)	Organophosphorus hydrolase (365aa)	Biodegradation	Richins et al., 1997
	Pytochelains (40aa)	Adsorption (bioaccumulation)	Bae et al., 2000
Pseudomonas syringae INP (36kDa)	Carboxymethylcellulase (CMCase) (33kDa)	Whole-cell biocatalyst	Jung et al., 1998
	Levansucrase (424aa)	Utilization of levan	Jung et al., 1998
	Epatitis B virus surface antigen (HbsAg) (168aa)	Vaccines, antibody production	Lee et al., 2000
Sandwich fusions			
E. coli OmpC (367aa)	Poly-His peptides (162aa)	Heavy metal removal	Xu and Lee 1999
E. coli LamB (446aa)	Human metallothionein (HMT) (66aa)	Heavy metal removal	Sousa et al., 1998
	Yeast metallothionein (YMT)	Vaccines, antibody	Martineau et al., 1991
	Epatitis B virus surface antigen (HbsAg) (168aa)	production	
E. coli FimH (30kDa)	Random Peptide library (33aa)	Screening of binding motif	Kjaergaard et al., 2001
E. coli FimA (180aa)	Cholera toxin B epitopes (CtxB)	Vaccines	Stentebjerg-Olesen et al., 1997
Caulobacter crescentus RsaA (1073aa)	Fragment from *Pseudomonas aeruginosa* K pilin (12aa)	Not indicated	Bingle et al., 1997

cytoplasm and, with the participation of the protein secretory apparatus of the cell, is targeted to the outer membrane and can go through it to be finally anchored to the external surface. Each step can be a potential problem for efficient display; insertion of aberrant proteins in the outer membrane can be toxic for the cell and poor exposure of surface-anchored heterologous polypeptides can result from steric effects caused by the lipopolysaccharide layer on the outer membrane that can prevent the interaction with antibodies and others ligands.

Targeting and anchoring mechanisms of carrier proteins vary among surface proteins and different approaches have been developed to overcome limitations of each system. As shown in Table 1, various surface proteins have been tested as carriers in Gram-negative bacteria. Most of these proteins, including the *E. coli* proteins LamB, OmpA, OmpC, OmpS, the lipoprotein TraT, fimbriae and flagellar proteins (FimH and PapA) (Georgiou et al., 1997) are located into the outer membrane.

Examples of the strategies used for the development of N-terminal, C-terminal and sandwich fusions, are found in the systems reported in Table 1. N-terminal fusions are generally preferred when the carrier protein presents an anchoring or translocation domain in its C-terminus. Peptidoglycan-associated lipoprotein (PAL) is a typical example of this kind of protein. It binds to the peptidoglycan with its C-terminal portion and to the outer membrane with its N-terminal cysteine modified by a lipid moiety, and it has been successfully used to expose, on the *E. coli* cell surface, an antibody fragment (scFv) that specifically binds to atrazine, a toxic molecule used as a herbicide that, due to its extended use, represents a serious environmental pollutant (Dhillon et al., 1999). Other examples of N-terminal fusions used as carriers are members of the immunoglobulin A (IgA) protease-like family, that contain an autotransporter structure at their C-terminus. The autotransporter proteins are generally

characterized by the feature that all the information required for the transport to the outer membrane and secretion through the cell envelope is contained in the protein itself. The autotransporter domain forms, after translocation through the cytoplasmic membrane, a barrel of 14 amphipatic β-sheets within the outer membrane. This porin-like structure promotes the translocation of N-terminally attached passenger proteins through the cell envelope of Gram-negative bacteria (Jose et al., 1995). The C-terminal domain of the *E. coli* adhesion protein (AIDA-I) is also recognized as a member of the autotransporter family and has been used to display dimeric bovine adrenodoxin (Adx) in a whole-cell steroid bioconversion system (Jose et al., 2002).

Several outer membrane proteins show targeting domains at the N-terminus and can be used in the construction of C-terminal fusions. Francisco et al. (1992) constructed a tripartite fusion consisting of the outer membrane localization domain of the major *E. coli* lipoprotein (Lpp) fused with the outer membrane protein A domain (OmpA), responsible for the transportation across the outer membrane of foreign proteins joined at the C-terminus. The Lpp-OmpA system has been the first tool to display full-length proteins on the surface of *E. coli* and it has been used to expose many heterologous proteins, including enzymes such as cellulases and esterases (Earhart *et.al.*, 2000).

Pseudomonas syringae ice nucleation protein (INP), which accelerates ice crystal formation in supercooled water (Green and Warren, 1985), is another example of a protein used to construct C-terminal fusions. Several viral antigens and enzymes have been successfully displayed by fusion to INP (Kwak et al., 1999; Jung et al., 1998). INP is an outer membrane protein that attaches to the bacterial cell surface via a glucosylphosphatidylinositol (GPI) anchor motif, widely used also in eukaryotes. However, unlike the GPI-anchor system of eukaryotes, the C-terminus is free and exposed on the cell surface, so that heterologous proteins fused to the C-terminus of INP can be displayed. INP-based cell surface display has many advantages compared to other Gram negative bacterial surface expression systems. It has a repeating domain with a catalytic role in the formation of ice crystals not essential for membrane anchoring, which can be used as a spacer between the heterologous protein and the cell surface. Moreover expression of INP-fusion proteins of up to 60 kDa does not disturb membrane structure nor bacterial growth and can be expressed on various Gram negative bacterial surfaces.

Sandwich fusions are, as shown in Table 1, the most common system for the surface display of heterologous proteins in Gram-negative bacteria. Proteins generally used as carriers in this kind of fusion do not have anchoring regions and the whole molecule is required for assembly. This class of molecules includes outer membrane proteins (OMP), subunits of extracellular appendages and S-layer proteins. OMPs contain a transmembrane domain composed of numerous antiparallel β-strand pairs connected by short loops on the periplasmic side and by long loops on the external side (Koebnik et al., 2000). The β-barrel structure anchors the protein to the outer membrane while the flexible external loops are less conservative and can tolerate a certain degree of modification in this region. These loops can be used as fusion sites for the display of heterologous proteins. The large majority of full-length proteins cannot be inserted into a surface loop without disruption of tertiary structure, and thus the size limit of tolerated peptides is normally ca. 70-80 amino acid residues (Sousa et al., 1998). However, the *E. coli* OmpC protein has been successfully used as partner in sandwich fusions with a 162 amino acids polypeptide (Xu and Lee, 1999).

Polymeric surface appendages, such as fimbriae or flagella, can be used as scaffolds for surface presentation. A large variety of fimbrial proteins have been tested, including the FimA and FimH proteins of type 1 fimbriae (Hedegaard and Klemm, 1989; Kjaergaard et al., 2001) and the FelA subunit of type P fimbriae (van Die et al., 1990). Fimbrial proteins are present in extremely high numbers on the cell surface, which makes them very interesting for display purposes. However, the principal structural proteins of various fimbriae can accommodate relatively small inserts (up to 30 amino acids) without altering organelle structure and display efficiency (Klemm and Schembri, 2000).

Flagella-based display has been obtained by fusing the passenger protein into a non essential region of flagellin without affecting the self-assembly features of the carrier (Westerlund-Wikstrom et al., 1997; Stentebjerg-Olesen et al., 1997).

The S-layer, present in some Gram-negative bacteria, can also be used to anchor heterologous proteins. In particular, the *Caulobacter crescentus* S-layer protein RsaA has been used as a carrier protein to display a peptide of 12 amino acids from *Pseudomonas aeruginosa* K (Bingle et al., 1997).

DISPLAY OF HETEROLOGOUS POLYPEPTIDES IN GRAM-POSITIVE BACTERIA

Although initial studies on surface engineering have been mainly focused on Gram-negative bacteria, several examples of display with Gram-positive bacteria systems have been reported (Table 2). Also, in Gram-positives several strategies have been developed to expose heterologous proteins on the cell surface by construction of N-terminal fusions, C-terminal fusions and "sandwich fusions". Gram-positives exhibit some cell surface properties which potentially make them more suitable for specific applications such as whole-cell adsorbents and biocatalysts. The surface receptors of Gram-positive bacteria seem more permissive for the insertion of extended sequences of heterologous proteins than the Gram-negative receptors (Fischetti et al., 1993) and the rigid structure of their cell wall makes it possible to use intact bacteria for separation/purification purposes without extensive cell lysis.

Anchored surface proteins must cross the thick cell wall to display their functional domains to the outer environment and this is facilitated by features shared by most surface anchored proteins. The N-terminal domains often contain binding or catalytic functions followed by a variable number of repeats that may or may not have a specific activity. Sometimes a stretch of proline-rich amino acids is found to precede a C-terminal anchoring tail consisting of approximately 35 amino acids, that includes a conserved LPXTG-motif followed by a stretch

Table 2. Examples of surface display systems in Gram-positive bacteria.			
Carrier protein	Passenger protein	Applications	References
N-terminal fusions			
Staphylococcal protein A (SpA) (472aa)	Variants of human respiratory sincitial virus (RSV) glycoprotein fragment	Live vaccines	Nguyen et al., 1995
	Cholera toxin B subunit (CtxB) (103aa)	Live vaccines	Liljeqvist et al., 1997
Bacillus subtilis CotB (275aa)	Tetanus toxin fragment C (TTF-C) (50Kda)	Protects against tetanus toxin	Isticato et al., 2001
C-terminal fusions			
Bacillus anthracis S-layer protein EA1 (209aa)	Tetanus toxin fragment C (TTF-C) (50Kda)	Protects against tetanus toxin	Mesnage et al., 1999
Streptococcus pyogenes M6 (262aa)	V3 domains of HIV-1 gp120	Vaccines	Oggioni et al., 1999
	Brucella abortus antigen L7/L12	Vaccines	Dieye et al., 2001
Bacillus subtilis CotB (275aa)	Tetanus toxin fragment C (TTF-C) (50Kda)	Protects against tetanus toxin	Isticato et al., 2001
Bacillus subtilis CotC (66aa)	*E. coli* Labile Toxin B subunit (eLTB) (459aa)	Adjuvant effect in vaccine formulations	Mauriello et al., 2004
	Tetanus toxin fragment C (TTF-C) (50Kda)	Protects against tetanus toxin	Mauriello et al., 2004
Sandwich fusions			
Bacillus subtilis CotB (275aa)	Tetanus toxin fragment C (TTF-C) (50Kda)	Protects against tetanus toxin	Isticato et al., 2001

of hydrophobic residues and a tail of six or seven mostly positive charged residues (Navarre and Schneewind, 1999). The mechanism of cell-wall sorting and anchoring involves a proteolytic cleavage between the threonine and glycine residues of the LPXTG-motif followed by the formation of a covalent linkage of the C-terminus to a free amino group of the peptide bridge in the peptidoglycan (Navarre and Schneewind, 1994; Ton-That et al., 2000). The enzyme responsible for this process in *Staphylococcus aureus* has been identified and called sortase (Mazmanian et al., 1999). In the N-terminal fusions construction, the principle for surface exposure of large polypeptides involves the fusion of the target gene between the N-terminal secretion signal and the C-terminal anchoring sequence of carrier proteins. One of the most used model systems of surface display in Gram-positive bacteria is the anchoring mechanism of *Staphylococcus aureus* protein A (SpA) (Schneewind et al., 1995) (Figure 1). This region has been used to create shuttle plasmids for surface display of proteins of various length (ranging from 15 to 397 amino acids) on recombinant *Staphylococcus xylosus, Staphylococcus carnosus* and *Lactococcus lactis* (Steidler et al., 1998).

Successful examples of surface expression of N-terminal fusions in Gram-positive bacteria are: i) the display of a malaria antigen and of the albumin-binding reporter protein on the cell surface of *Staphylococcus xylosus* and *Staphylococcus carnosus* (Hannsson et al., 1992; Samuelson et al., 1995); ii) the functional expression of single-chain antibodies on Staphylococci which suggested the use of whole-cells for diagnostic purposes (Gunneriusson et al., 1996), and iii) the display of chimeric proteins containing polyhystidyl peptides able to bind divalent metal ions and proposed for bioremediation of heavy metals (Samuelson et al., 2000).

The surface of group-A streptococci contain dimeric α-helical fibrillar proteins named M-proteins, having typical N-terminal leader sequences and C-terminal sorting signals (Navarre and Schneewind, 1999) and are useful for the construction of N-terminal fusions. The M6-protein of *Streptococcus pyogenes*, in fact, has been functionally expressed on the surface of *Streptococcus gordonii* (Pozzi et al., 1992); subsequently this method has been used for surface presentation of several different immunogens on recombinant *Streptococcus gordonii* (Oggioni et al., 1999) and also in *Lacotococcus lactis* (Dieye et al., 2001). This system has been used to develop a food-grade, live vaccine delivery system for heterologous antigens (Ribeiro et al., 2002).

Beside the LPXTG box, another anchoring motif, used for the construction of C-terminal fusions to display proteins on the outer surface of Gram-positive bacteria, is represented by the S-layer homology domain (SLH) found in several Gram-positives (Figure 1). These domains are present in single or multiples at the N-terminus of Gram-positive S-layer proteins and are composed of 70 amino acids arranged in two α-helices flanking a β-strand. Experimental data suggest that SLH domains mediate association to the cell wall and are covalently linked to the peptidoglycan layer. The SLH domain of the *Bacillus anthracis* S-layer protein, EA1, has been used to display Tetanus toxin fragment C in *B. anthracis* (Mesnage et al., 1999).

Another example of a C-terminal fusion comes from the use of the *B. subtilis* autolysin modifier protein CwbA as a carrier (Acheson et al., 1997). In this case a heterologous antigen, the highly antigenic domain of the *Yersinia pseudotuberculosis* invasin protein (Inv), was expressed on the surface of *B. subtilis* cells.

SPORE SURFACE DISPLAY

A strategy to engineer *Bacillus subtilis* spores to display heterologous antigens on the spore surface has been recently reported (Isticato et al., 2001). A spore-based display system provides several advantages with respect to systems described above, based on the use of bacterial cells. High stability and

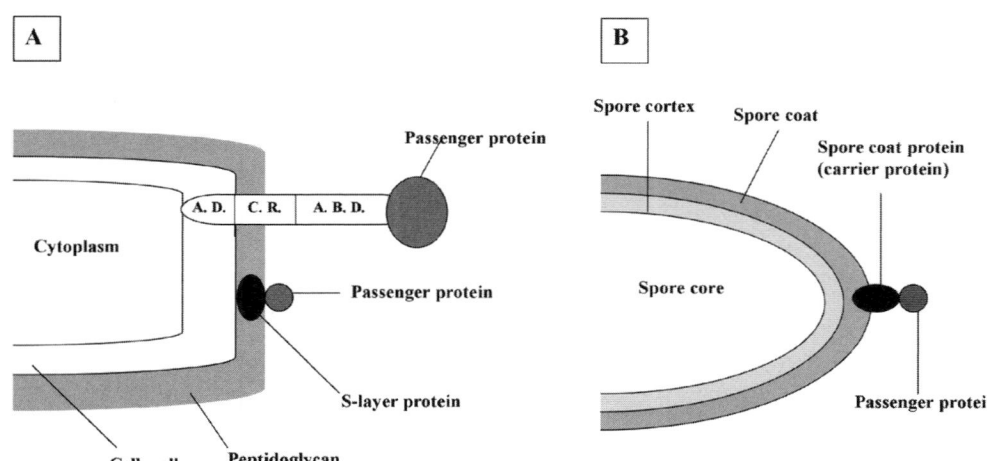

Figure 1. Cell-surface display systems in Gram-positive bacteria. (A) Cell-surface display system using staphylococcal protein A and S-layer proteins as examples of N-terminal fusion. A.D.: Anchoring Domain; C.R.: Charged Region; A.B.D.: Albumin Binding Domain. (B) Spore-surface display system using spore coat proteins.

A. D.	Anchoring Domain
C. R.	Charged Region
A. B. D.	Albumin Binding Domain

safety of the spore-display systems are expected because of the well-documented resistance (see Chapter 1) and safety of the bacterial spore (see Chapter 4 and 8).

As described in details in the previous paragraphs (see also Chapter 6), *Bacillus* spores are encased in a coat, a proteic structure whose rigidity and compactness immediately suggests the possibility of using its structural components as anchoring motifs for the expression of heterologous polypeptides on the spore surface. However, there is no information about the mechanisms of protein incorporation into the coat, the nature of structural components forming the most external part of the coat and whether there are anchoring motifs. To overcome the lack of this essential information initial attempts to expose heterologous proteins on the spore surface were focused on two coat components selected for its known surface location in the case of CotB (Isticato et al., 2001) or for its high relative abundance in the case of CotC (Mauriello et al., 2004). The observation that both these coat components were dispensable for the formation of an apparently normal spore as well as for its germination, was an additional positive reason to consider CotB and CotC as potentially suitable carrier candidates.

Two antigens were initially selected as model proteins to display on the spore surface: i) the non- toxic 459 amino acid C-terminal fragment of the tetanus toxin (TTFC), a well characterized and highly immunogenic (Helting and Zwisler, 1997) 51.8 kDa peptide, encoded by the *tetC* gene of *Clostridium tetani*; and ii) the 103 amino acid B subunit of the heat-labile toxin of enterotoxigenic strains of *Escherichia coli* (LTB), a 12 kDa peptide, encoded by the *eltB* gene (Douce et al., 1995).

CotB AS A CARRIER PROTEIN

Like other coat components, CotB has been associated to the outer coat layer on the base of genetic evidence (Zheng et al., 1988) and only recently an immunocytofluorimetric analysis performed on intact spores showed that CotB is accessible to CotB-specific antibodies and therefore that it is most probably exposed on the spore surface (Isticato et al., 2001).

The CotB structural gene, *cotB*, is under the dual transcriptional control of σ^K and the DNA-binding protein GerE. As a consequence, *cotB* is transcribed only in the mother cell compartment of the sporulating cell (Henriques and Moran, 2000). Once synthesised in the mother cell cytoplasm, CotB is assembled around the forming spore in a CotE, CotG and CotH dependent fashion (see Chapter 6). Therefore, CotB and the heterologous protein eventually fused to it, do not undergo a cell wall translocation step, typical of the other display systems discussed above.

CotB has a strongly hydrophilic C-terminal half formed by three 27 amino acid-repeats rich in serine, lysine and glutamine residues. Serine residues account for over 50% of the CotB C-terminal half. The lysine residues in the CotB repeats have been suggested to represent sites of intra- or inter-molecular cross-linking, by analogy to the connective tissue proteins collagen and elastin (Henriques and Moran, 2000; Kobayashi et al., 1998). The CotB protein has a deduced molecular mass of 46 kDa, but migrates on SDS-PAGE as a 66 kDa polypeptide. Recently the discrepancy between measured and deduced

molecular weight has been explained by showing that CotB is initially synthesized as a 46 kDa species, and converted into a 66 kDa dimer (Zilhao et al., 2004).

The strategy to obtain recombinant *B. subtilis* spores expressing CotB-TTFC or CotB-LTB on their surface was based on (i) use of the *cotB* gene and its promoter for the construction of translational fusions and on (ii) chromosomal integration of the *cotB-tetC* and *cotB-eltB* gene fusions into the coding sequence of the non-essential gene *amyE* (Cutting and Vander-Horn, 1990). Placing the fusion proteins under *cotB* transcriptional and translational signals ensured correct timing of expression during sporulation, while its chromosomal integration guaranteed the genetic stability of the construct. Due to the lack of information on CotB coat assembly and on the requirements for anchoring motifs, initial attempts were performed by positioning the passenger protein at the C-terminal, the N-terminal or in the middle of CotB (Figure 2A).

When TTFC and LTB were fused to the C-terminal end of CotB, the chimeric proteins failed to correctly assemble on the spore surface (Isticato and Ricca, unpublished). Such initial failures were attributed to a potential instability of the constructs, either at the DNA level or at the protein level. In order to bypass such problems TTFC and LTB were fused to the C-terminal end of a CotB form deleted of the three 27 amino acid-repeats, CotBD$_{105}$ (Figure 2). Different from the full length version, the CotBΔ_{105}-TTFC chimeric protein was correctly assembled and exposed on the spore surface (Isticato et al., 2001). A quantitative dot blot showed that each recombinant spore exposed an amount of CotBΔ_{105}-TTFC fusion protein equal to 0.00022 pg and made it possible to deduce that 1.5×10^3 chimeric molecules are present on the surface of each recombinant spore (Isticato et al., 2001).

Unlike CotBΔ_{105}-TTFC, CotBΔ_{105}-LTB was not properly assembled. The strain expressing this chimera showed reduced sporulation and germination efficiencies and its spores were not resistant to lysozyme. These observations, together with the SDS-PAGE analysis of the released coat proteins, suggested that the presence of CotBΔ_{105}-LTB strongly altered the spore coat layer. An *in-silico* analysis showed some homology between the chimeric product (in the fusion region) and LytF, a cell wall-associated endopeptidase produced by *B. subtilis* during vegetative growth, thus raising the possibility that the chimeric product could interfere with proper coat formation by degrading some coat components (Mauriello and Ricca, data not shown).

In addition to the C-terminal end fusion described above the model passenger protein TTFC has been fused also at the N-terminal and in the middle of CotB (Figure 2). In both cases the CotBΔ_{105} form of CotB was used to avoid the problems experienced with the C-terminal fusion (see above). Both the N-terminal and the sandwich fusions produced chimeric products that were properly assembled in the coat structure from both the qualitative and the quantitative point of view (Isticato et al., 2001). At least in the CotB case, it was then possible to conclude that where the passenger protein is posed does not affect display on the spore surface.

A

CotB as fusion partner

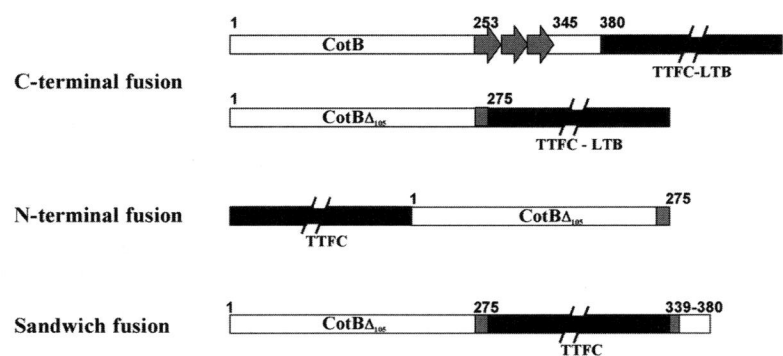

C-terminal fusion

N-terminal fusion

Sandwich fusion

B

CotC as fusion partner

C-terminal fusion

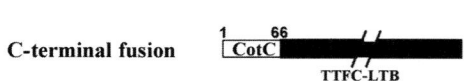

Figure 2. Spore-surface display system in *B.subtilis*. Black bars represent the two passenger proteins TTFC (459 amino acid) and LTB (103 amino acid). (A) Schematic representation of C-, N-terminal and sandwich fusions using CotB as carrier. Full lenght CotB (380 amino acid) and CotB$_{\Delta105}$ (275 amino acid) were used carrier proteins for the C-terminal fusions. The three 27-amino-acids repeats of the full lenght CotB are indicated by arrows. (B) Schematic representation of the two fusion proteins constructed using CotC as fusion partner.

CotC AS A CARRIER PROTEIN

CotC is a 12 kDa, alkali-soluble component of the *B. subtilis* spore coat, previously identified by reverse genetics (Donovan et al., 1987) and then associated to the outer coat layer based on genetic evidence (Zheng et al., 1988). CotC was initially considered as a carrier candidate for its relative abundance in the coat. Together with CotG and CotD, CotC represents about 50% of the total solubilized coat proteins. Such relatively high amounts could allow the assembly of a significant number of CotC-based chimeras on the coat, thus ensuring an efficient heterologous display. Expression of the *cotC* gene is under the control of the mother cell specific σ factor σ^K and of transcriptional regulators GerE and SpoIIID. As in the case of CotB, CotC is also transcribed in the mother cell and its assembly on the coat does not require membrane translocation. The primary product of the *cotC* gene is a 66 amino acid polypeptide extremely rich in tyrosine (30.3%) and lysine (28.8%) residues (Donovan et al., 1987; Kobayashi et al., 1994). However, it was recently shown that CotC is assembled into at least four distinct protein forms, ranging in size between 12 and 30 kDa (Isticato et al., 2004). Two of these, having molecular masses of 12 and 21 kDa and corresponding most likely to a monomeric and homodimeric form of CotC respectively, are assembled on the maturing spore immediately following their synthesis approximately eight hours after the onset of sporulation. The other two forms, 12.5 and 30 kDa, are probably the products of post-translational modifications of the two early forms, occurring directly on the coat surface during spore maturation (Isticato et al., 2004).

In the case of CotC only C-terminal fusions have so far been constructed (Table 2). Both CotC-TTFC and CotC-LTB gene fusions were obtained by cloning *tetC* or *eltB* in frame with the last *cotC* codon under the transcriptional and translational

control of the *cotC* promoter region. The gene fusion was then integrated into the *B. subtilis* chromosome at the *amyE* locus by double cross-over recombination (Mauriello et al., 2004). Both these two chimeric proteins were assembled on the coat of recombinant spores without major effect on the spore structure and/or function, since they appeared identical to wild type spores in terms of efficiency of sporulation and germination and resistance properties. Western blot, immunofluorescence microscopy and, for CotC-TTFC, cytofluorimetric analysis showed that both CotC-based chimeras were displayed on the surface of the recombinant spores. A quantitative determination of recombinant proteins exposed on *B. subtilis* spores revealed that ca. 9.7×10^2 and 2.7×10^3 molecules of CotC-TTFC and CotC-LTB, respectively, were extracted from each spore.

Although CotC appears more abundant than CotB within the coat, comparable amounts of heterologous proteins are exposed by the CotC-based and the CotB$_{\Delta105}$-based systems. This result was somewhat unexpected, since CotC appears to be much more abundant than CotB in the coat. A possible explanation comes from the recent finding that the C-terminal end of CotC is essential for the interaction of a CotC molecule with other CotC molecules and with other coat components (Isticato and Ricca, manuscript in preparation) and therefore indicate that the use of CotC as a carrier still needs to be optimised.

STABILITY OF SPORE-DISPLAYED PROTEINS

One of the main reasons to propose the use of the bacterial spore as a favourable display system is its well-documented stability. Spores can be simply stored at room temperature for a long time without reduction of their resistance and stability properties. This would be an extremely useful property for a variety of biotechnological applications. As an example, if the passenger protein is an antigen, the recombinant spore could

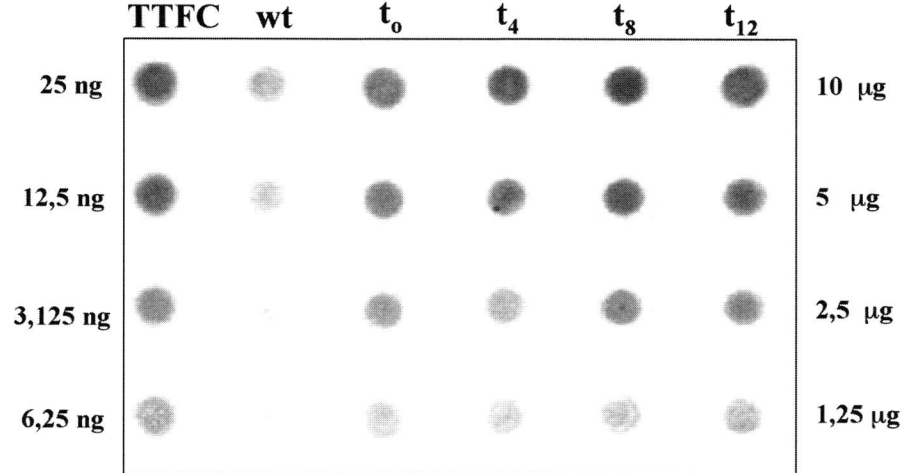

Figure 3. Dot blot analysis of proteins extracted from wild-type spores and from spores exposing the CotBD$_{105}$-TTFC chimera. Freshly purified spores expressing the CotBΔ$_{105}$-TTFC chimera (t0) or after 4, 8 or 12 weeks (t4, t8 and t12) of storage at room temperature were used to extract spore coat proteins and examined by dot blotting. Concentrations of purified TTFC (in nanograms) and of coat proteins (in micrograms) are indicated. Proteins were loaded and reacted with monoclonal anti-TTFC antibodies (Boeringher), then with phosphatase-conjugated secondary antibodies.

become an ideal heat-stable oral vaccine for use in developing countries, where heat-stability is of most concern due to poor distribution and storage.

However, while spore stability is well documented (see Chapters 1), stability of heterologous proteins exposed on the spore surface has been only recently investigated. Spores expressing CotBΔ$_{105}$-TTFC (see above) and parental spores were stored at -80°C, -20°C, +4°C and at room temperatures and assayed at different storage times up to 12 weeks. In all cases assayed the amount of heterologous protein present on the surface of recombinant spores appeared identical between freshly prepared spores and spores stored for up to 12 weeks (Figure 3). These results, indicating that heterologous proteins can be stably exposed on the surface of recombinant spores, confirm the spore-based system as a very promising display approach that could overcome some drawbacks of other systems and that could find applications in a variety of diverse biotechnological fields.

ACKNOWLEDGEMENTS
This work was supported by the European Union grant No. QLK5-CT-2001-01729 and by MIUR (Cofin 2002; FIRB 2002) grants to E.R.

REFERENCES
Acheson, D.,W., K., Sonenshein, A.,L., Leong, J.,M., Keusch, G., T. 1997 Heat-stable spore based vaccines: surface expression of Invasin-cell wall fusion proteins in *Bacillus subtilis*. In: F. Brown, D. Burton, P. Doherty, J. Mekalanos, E. Norrby (Eds) Vaccines 97. Cold Spring Harbor Laboratory Press N.Y., New York: 179-184.

Bae, W., Chen, W., Mulchandani, A., Mehra, R.,K. 2000. Enhanced bioaccumulation of heavy metals by bacterial cells displaying synthetic phytochelatins. Biotechnol Bioeng. *70*,518-524.

Bingle, W. H. 1997 Cell-surface display of a *Pseudomonas aeruginosa* strain K pilin peptide within the paracrystalline S-layer of *Caulobacter crescentus* Mol. Microbiol. *26*, 277-288.

Chen, H. and Schifferli D.M. 2003 Construction, characterization and immunogenicity of an attenuated *Salmonella enterica serovar typhimurium* pgtE vaccine expressing fimbriae with integrated viral epitopes from the spiC promoter. Infect. Immun. *71*, 4664-4673.

Cutting, S.M., and Vander-Horn, P.B. 1990. Genetic Analysis. In: Molecular Biological Methods for *Bacillus*. C.R. Harwood and S.M. Cutting, ed. John Wiley and Sons Ltd. p. 27-74.

Dhillon, J.,K., Drew, P.,D., Porter, A.,J.,R. 1999 Bacterial surface display of an anti-pollutant antibody fragment. Letters Appl. Microbiol. *28*, 350-354.

Dieye, Y., Usai, S., Clier, F., Gruss, A., Piard, J.,C. 2001 Design of a protein-targeting system for lactic acids bacteria. J. Bacteriol. *183*, 4157-4166.

Donovan, W., Zheng, L., Sandman, K., and Losick, R. 1987. Genes encoding spore coat polypeptides from *Bacillus subtilis*. J. Mol. Biol. *196*, 1-10. Chichester, England.

Douce, G., Turcotte, C., Cropley, I., Roberts, M., Pizza, M., Domenghini, M., Rappuoli, R., and Dougan, G. 1995. Mutants of *Escherichia coli* heat-labile toxin lacking ADP-ribosyltransferase activity act as nontoxic, mucosal adjuvants. Pro. Nat. Acad. of Sci. U.S.A. *92*, 1644-1648.

Earhart, C.,F. 2000 Use of Lpp-OmpA fusion vehicle for bacterial surface display. Methods Enzymol. *326*, 506-516.

Ernst, W., Grabherr, R., Wegner, D., Borth, N., Grassauer, A., Katinger, H. 1998 Baculovirus surface display: construction and screening of a eukaryotic epitope library. Nucl. Acids Res. *26*,1718-1723.

Fischetti, V.,A., Medaglini, D., Oggioni, M., Pozzi, G. 1993 Expression of foreign proteins on Gram-positive commensal bacteria for mucosal vaccine delivery. Curr. Opin. Biotechnol. *4*, 603-610.

Francisco, J.,A., Earhart, C.,F., Georgiou G. 1992 Transport and anchoring of beta-lactamase to the external surface of *Escherichia coli*. Proc. Natl. Acad. Sci. USA *89*, 2713-2717b.

Georgiou, G., Sthatopoulos, C., Daugherty, P.,S., Nayak, A.,R., Iverson, I.,R., Curtiss, R. 1997 Display of heterologous proteins on the surface of microorganisms: from the screening of combinatorial libraries to live recombinant vaccines. Nat. Biotechnol. *15*, 29-34.

Green, R.,L. and Warren, G.,J. 1985 Physical and functional repetition in a bacterial ice nucleation gene. Naure *317*, 645-648.

Gunneriusson, E., Samuelson, P., Uhlen, M., Nygren, P., Stahl, S. 1996 Surface display of a functional single-chain Fv antibody on Staphylococci. J. of Bacteriol. *178*,1341-1346.

Hannsson, M., Stahl, S., Nguyen, T.N., Bachi, T., Robert, A., Binz, H., Sjolander, A., Uhlen, M. 1992. Expression of recombinant proteins on the surface of the coagulase-negative bacterium *Staphylococcus xylosus*. J. Bacteriol *174*, 4239-4245.

Hedegaard, L., and Klemm, P. 1989 Type 1 fimbriae of *Escherichia coli* as carriers of heterologous antigenic sequences. Gene *85*, 115-124.

Helting, T.B., and Zwisler, O. 1977. Structure of tetanus toxin. I. Breakdown of the toxin molecule and discrimination between polypeptide fragments. J. Biol. Chem. *252*, 194–198.

Henriques, A.,O., and Moran, C.,P. 2000. Structure and assembly of the bacterial endospore coat. Methods *20*, 95–110.

Huang, W., Zhang, Z., Palzkill, T. 2000 Design of potent beta-lactamase inhibitors by phage display of beta-lactamase inhibitory protein. J Biol Chem. *275*, 14964-8.

Isticato, R., Cangiano, G., Tran, H.,T., Ciabattini, A., Medaglini, D., Oggioni, M.,R., De Felice, M., Pozzi, G., Ricca, E. 2001 Surface display of recombinant proteins on *Bacillus subtilis* spores. J. Bacteriol. *183*, 6294-6301.

Isticato, R., Esposito, G., Zilhão, R., Nolasco, S., Cangiano, G., De Felice, M., Henriques, A., O. and Ricca, E. 2004. Assembly of Multiple CotC Forms into the *Bacillus subtilis* Spore Coat. J. Bacteriol. *186*,1129-1135.

Lee, J.,S., Shin, K.,S., Pan, J.,G., Kim, C.,J. 2000 Surface-displayed viral antigens on *Salmonella* carrier vaccine. Nat. Biotechnol. *18*, 645-648.

Liljeqvist, S., Samulson, P., Hansson, M., Nguyen, T.,N., Binz, H., Stahl, S. 1997 Surface display of the cholera toxin B subunit on *Staphylococcus xylosus* and *Staphylococcus carnosus*. Appl. Environ. Microbiol. *63*, 2481-2488.

Jose, J., Jahnig F., Meyer T.,F. 1995 Common structural features of IgA1 protease-like outer membrane protein autotransporter. Mol. Microbiol. *18*, 378-380.

Jose, J., Bernhardt, R., Hannemann, F. 2002 Cellular surface display of dimeric Adx and whole cell P450-mediated steroid synthesis on *E. coli*. J. Biotechnol. *95*, 257-268.

Jung, H.,C., Lebeault J.,M., Pan, J.,G. 1998 Surface display of *Zymomonas mobilis* levansucrase by using the ice-nucleation protein of Pseudomonas syringae. Nat. Biotechnol. *16*, 576-580.

Kim, W.C., Rhee, H.I., Park, B.K., Suk, K.H., Cha, S.H. 2000 Isolation of peptide ligands that inhibit glutamate racemase activity from a random phage display library. J Biomol Screen. *5*, 435-40.

Kjaergaard, K., Schembri, M.,A., Klemm, P. 2001 Novel Zn(2+)-chelating peptides selected from a fimbria-displayed random peptide library. Appl. Environ. Microbiol. *67*, 5467-5473.

Klemm, P., and Schembri, M.,A. 2000 Fimbrial surface display systems in bacteria: from vaccines to random libraries. Microbiology 146 Pt *12*, 3025-3032.

Kobayashi, K., Kumazawa, Y., Miwa, K., and Yamanaka, S. 1994. e-(g-Glutamyl)lisine cross-links of spore coat proteins and transglutaminase activity in *Bacillus subtilis*. FEMS Microbiol. Lett. *144*, 157-160.

Kobayashi, K., Suzuki, S.I., Izawa, Y., Miwa, K., and Yamanaka, S. 1998. Transglutaminase in sporulating cells of *Bacillus subtilis*. J. Gen. Appl. Microbiol., *44*, 85-91.

Koebnik, R., Locher, K.,P., Van Gelder, P. 2000. Structure and function of bacterial outer membrane proteins: barrel in a nutshell. Mol. Microbiol. *37*, 239-253.

Kotrba, P., Doleckova, L., de Lorenzo V., Ruml, T. 1999 Enhanced bioaccumulation of heavy metal ions by bacterial cells due to surface display of short metal binding peptides. Appl. Environ. Microbiol. *65*, 1092-1098.

Kwak, Y.,D., Yoo, S.,K., Kim, E.,J. 1999 Cell surface Display of human immunodeficiency virus type 1 gp120 on *Escherichia coli* using ice nucleation protein. Clin. Diagn. Lab. Immunol. *6*, 499-503.

Martineau, P., Charbit, A, Leclerc, C., Werts, C., O'Callaghan, D., Hofnung, M. 1991. A genetic system to elicit and monitor antipeptide antibodies without peptide synthesis. Biotechnology *9*,170-172.

Maurer, J., Jose, J., Meyer, T.,F. 1997 Autodisplay: one-component system for efficient surface display and release of soluble recombinant proteins in *Escherichia coli*. J.Bacteriol. *179*, 794-804.

Mauriello, E., Duc, L.H., Isticato, R., Cangiano, G., Hong, H.A., De Felice, M., Ricca, E. and Cutting S.M. 2004 Display of heterologous Antigens on the *Bacillus subtilis* spore coat using CotC as a Fusion Partner. Vaccine *22*, 1177-1187.

Mazmanian, S.,K., Liu, G., Ton-That, H., Schneewind, O. 1999 *Staphylococcus aureus* sortase, an enzyme that anchors surface proteins to the cell wall. Science *285*, 760-763.

Mesnage, S., Weber-Levy M., Haustant, M., Mock, M., Fouet, A., 1999 Cell-surface exposed tetanus toxin fragment C produced by recombinant *Bacillus anthracis* protect against tetanus toxin. Infect. Immun. *67*, 4847-4850.

Murase, K., Morrison, K.L., Tam, P.Y., Stafford, R.L., Jurnak, F., Weiss, G.A. 2003 EF-Tu binding peptides identified, dissected, and affinity optimized by phage display. Chem Biol. *10*,161-8.

Navarre, W.,W., and Schneewind, O. 1994 Proteolytic cleavage and cell wall anchoring at the LPXTG motif of surface proteins in gram-positive bacteria. Mol. Microbiol. *14*, 115-121.

Navarre, W.,W., and Schneewind, O. 1999 Surface proteins of gram-positive bacteria and mechanisms of their targeting to the cell wall envelope. Microbiol. Mol. Biol. Rev. *63*, 174-229.

Nguyen, T.,N. Gourdon, M.,H., Hansson, M. Robert, A., Samuelson, P., Libon, C., Andreoni, C., Nygren, P.,A., Binz, H., Uhlen, M. 1995 Hydrophobicity engineering to facilitate surface display of heterologous gene products on *Staphylococcus xylosus*. J. Biotechnol. *42*, 207-219.

Oggioni, M.,R., Medaglioni, D., Maggi, T., Pozzi, G. 1999 Engineering the gram-positive cell surface for construction of bacterial vaccine vectors. Methods *19*, 163-173.

Pozzi, G., Contorni, M., Oggioni, M.,R., Manganelli, R., Tommasino, M., Cavalieri, F., Fischetti, V., A. Delivery and expression of a heterologous antigen on the surface of streptococci. Infect. Immun. *60*, 1902-1907.

Rahman, M.,M. and Gopinathan K.,P. 2003 Bombyx mori nucleopolyhedrovirus-based surface display system for recombinant proteins. J. Gen. Virol. *84*, 2023-2031.

Ribeiro, L.,A., Azevedo, V., Le Loyr, Y., Oliveira, S.,C., Dieye, Y., Piard, J.,C., Gruss, A., Langella, P. 2002 Production and targeting of the *Brucella abortus* antigen L7/L12 in *Lactococcus lactis*: a first step towards food-grade live vaccines against brucellosis. Appl. Environ. Microbiol. *68*, 910-916.

Richins, R.,D., Kaneva, I., Mulchandani, A., Chen, W. 1997. Biodegradation of organophosphorus pesticides by surface-expressed organophosphorus hydrolase. Nat Biotechnol. *15*,984-987.

Samuelson, P., Hansson, M., Ahlborg, N., Andréoni C., Gotz, F., Bachi, T., Nguyen, N., Binz, H., Uhlén, M., Stahl, S. 1995. Cell surface display of recombinant proteins on *Staphylococcus carnosus*. J. Bacteriol. *177*, 1470-1476.

Samuelson, P., Wernerus, H., Svedberg, M., Stahl, S. 2000 Staphylococcal surface display of metal-binding Polyhistidyl peptides. Appl. Environ. Microbiol. 66, 1243-1248.

Schneewind, O., Fowler, A., Faull, K.F. 1995 Structure of the cell wall anchor of surface proteins in *Staphylococcus aureus*. Science *268*, 103-106.

Sousa, C., Kotrba, P., Ruml, T., Cebolla, A., De Lorenzo V. 1998 Metalloadsorption by *E. coli* cells displayng yeast and mammalian methallothioneins anchored to the outer membrane protein LamB. J. Bacteriol. *180*, 2280-2284.

Stahl, S., and Uhlen, M. 1997 Bacterial surface display: trends and progress. Trends Biotechnol. *15*, 185-192.

Steidler, L., Viaene, J., Fiers, W., Remaut, E. 1998 Functional display of a heterologous protein on the surface of *Lactococcus lactis* by means of the cell wall anchor of *Staphylococcus aureus* protein A. Appl. Environ. Microbiol. *64*, 342-345.

Stentebjerg-Olesen, B., Pallesen, L., Jensen, L.,B., Christiansen, G., Klemm, P. 1997 Authentic display of a cholera toxin epitope by chimeric type 1 fimbriae: effects of insert position and host background. Microbiology *143*, 2027-2038.

Ton-That, H., Mazmanian, S.,K., Faull, K.,F., Schneewind, O. 2000 Anchoring of surface proteins to the cell wall of *Staphylococcus aureus*. Sortase catalyzed in vitro transpeptidation reaction using LPXTG peptide and NH(2)-Gly(3) substrates. J. Biol. Chem. *275*, 9876-9881.

van Die, I., van Oosterhout, J., van Megen, I., Bergmans, H., Hoekstra, W., Enger-Valk. B., Barteling, S., Mooi, F. 1990 Expression of foreign epitopes in P-fimbriae of *Escherichia coli*. Mol. Gen. Genet. *222*, 297-303.

Westerlund-Wikstrom, B., Tanskanen, J., Virkola, R., Hacker, J., Lindberg, M., Skurnik M., Korhonen T.,K. 1997 Functional expression of adhesive peptides as fusions to *Escherichia coli* flagellin. Protein eng. *10*, 1319-1326.

Whitehorn, E.,A., Tate, E., Yanofsky, S.,D., Kochersperger, L., Davis, A., Mortensen, R.,B., Yonkovich, S., Bell, K., Dower, W.J., Barrett, R.,W. 1995 A generic method for expression and use of "tagged" soluble versions of cell surface receptors. Biotechnology *13*, 1215-1219.

Xu, Z., and Lee, S.Y. 1999 Display of polyhistidine peptides on the *Escherichia coli* cell surface by using outer membrane protein C as an anchoring motif. Appl. Environ. Microbiol. *65*, 5142-5147.

Zheng, L., Donovan, W.P., Fitz-James, P.C., and Losick, R. 1988. Gene encoding a morphogenic protein required in the assembly of the outer coat of the *Bacillus subtilis* endospore. Genes Dev. *2*, 1047–1054.

Zilhao, R., Serrano, M., Isticato, R., Ricca, E., Moran Jr., C. ,P. and Henriques, A. ,O. 2004. Interactions among CotB, CotG, and CotH during assembly of the *Bacillus subtilis* spore coat. J. Bacteriol. *186*, 1110-1119.

Chapter 18

Spores as Oral Vaccines

Simon M. Cutting

SUMMARY

Recently, the first use of bacterial spores as oral vaccine vehicles has been reported showing that mice orally immunised with *Bacillus subtilis* spores expressing a tetanus antigen could be protected against lethal challenge with tetanus toxin. Unlike many 2nd Generation vaccine systems currently under development none offer the heat stability of bacterial spores or the flexibility for genetic manipulation. The current use of *Bacillus* spores as probiotics for both humans and animals may facilitate their eventual licensing as oral vaccines. Although this field is only now emerging initial proof of principle studies are outlined here as well as the potential advantages and disadvantages of the spore vaccine approach.

INTRODUCTION

Technology available today enables new vaccines to be rationally designed taking into account the specific nature of the disease and the preferred route of administration. Most pathogens enter the body at the mucosal surfaces (nasopharynx, lungs and gastrointestinal tract (GIT)) and it is here that local (mucosal) immunity is most important. The infectious agent first makes contact with and then colonises (or transverses) the mucosal surface to infect the host (*e. g.*, HIV, rotavirus, *Mycobacterium tuberculosis*). The primary components of mucosal immunity are the synthesis of secretory IgA (sIgA) at the mucosal surface and in the mucosal secretions as well as cellular immunity in the NALT (nasal associated lymphoid tissue) and the GALT (gut associated lymphoid tissue) (Medzhitov and Janeway, 1997). Traditional vaccination strategies relying on parenteral immunisation are unable to prevent the initial interaction of pathogen and host at the mucosal surface but rather resolve the resulting infection (Walker, 1994). For example, TB (tuberculosis) vaccines are administered by injection (systemically) yet must resolve an infection where the primary portal of entry is at the mucosa (the nasopharynx/lungs). A vaccine that enhances levels of local immunity in addition to providing systemic immunity would therefore be a hallmark of a successful vaccine.

Unfortunately, local immunity is not generated when an antigen is delivered systemically but this is not the case when an antigen is delivered orally or nasally. Improved vaccination strategies should ideally, then, use the oral or nasal route. Needle-less routes of administration are, of course, attractive for a number of other reasons and would simplify many of the steps in the vaccination process. However, protein antigens when delivered mucosally have been shown to be weakly immunogenic especially via the oral route (Van Ginkel, 2000) and there are a number of additional concerns with oral immunisation. First, is the antigen stable and protected from enzymatic degradation in the stomach? Second, is the antigen recognised as foreign or as a food (Czerkinsky et al., 1999)?

A final consideration with existing vaccination strategies is that almost all vaccines must be stored at 4°C and carry relatively short half-lives. Removal of the cold-chain using heat-stable vaccines would offer enormous advantages in developing countries where ineffective storage and transportation can effect the success of national immunisation programmes.

The combination of the known shortfalls of current vaccines has driven the research and development of better and improved vaccines, termed by the WHO as "2nd Generation Vaccines" (Green and Baker, 2002; Griffin, 2002; Shalaby, 1995). Ideally, such vaccines would be given by the oral (and less so, the nasal) route, be stable at room (ambient) temperature and provide long lasting immunity protecting the primary route of infection. Mucosal immunisation should induce secretory IgA antibodies directed against the specific pathogen as well as cellular responses, including responses from CD8[+] MHC restricted cytotoxic T-lymphocytes located in the mucosal epithelium and draining lymph nodes (LN) and well as induction of Th1 (T helper lymphocyte type 1) responses to provide an effective means for preventing infection.

A number of new vaccine systems are now under development, these include plants (Mason et al., 2002), non-living systems (e.g., liposomes (Han et al., 1997), ISCOMS (immune stimulating complexes) (Smith et al., 1998), PLG (poly-lactide-co-glycolide) polymers (Seong et al., 1999)) and bacteria (Drabner and Guzman, 2001; Green and Baker, 2002; Shata et al., 2000). The non-living systems systems require purification of a particular antigen that is then encapsulated or combined with the artificial packaging system prior to vaccination. As such, for large-scale, cost-effective, vaccine production these systems require careful consideration. Expression of heterologous antigens on, or in, a non-pathogenic bacterium could also provide a safe and effective method for delivering antigens and vaccination. By engineering the genome to faithfully reproduce the protective antigen or epitope, production is simplified and so long as appropriate quality control procedures are in place could enable cost-effective production within a developing country. Two broad groups of bacterial expression systems exist, those that colonise the GIT (e.g., *Salmonella* spp., *Streptococcus gordonii,* (Anderson et al., 1996; Medaglini et al., 2001)) and non-colonising bacteria such as *Lactococcus lactis* (Robinson et al., 1997). While these systems mentioned above are important none offer true heat-stability. Moreover, the technology required for the non-living systems is not really suitable for production

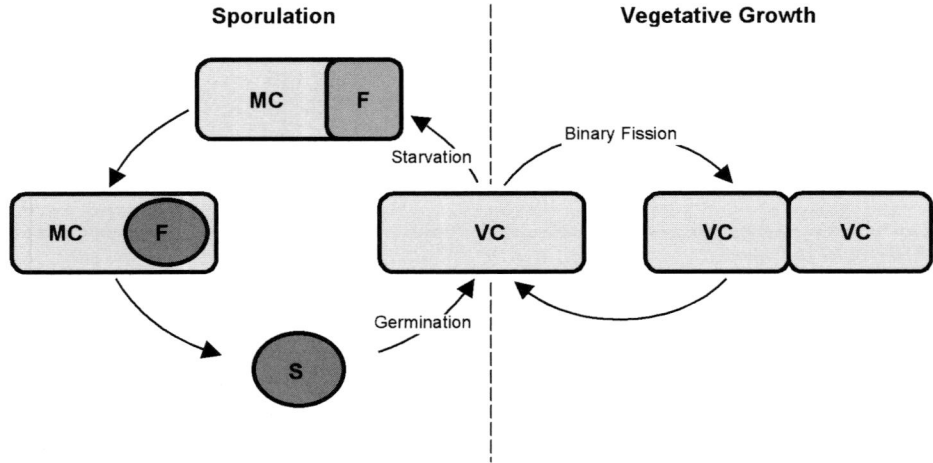

Figure 1. Unicellular differentiation of *B. subtilis*. Simplified diagram showing the two life cycles of bacterial endospores. Vegetative cells (VC) will grow by binary fission until starved. At this point individual cells commit to a program of irreversible differentiation to produce the bacterial spore (S). Sporulation takes approximately 8-10 hours to complete and is characterised by the asymmetric division of the cell into two chambers, the forespore (F) and the mother cell (MC). Both compartments carry a separate chromosome. The forespore is the germ-line cell and will develop into the spore which is released by lysis of the mother cell.

in developing countries while ethical and safety issues remain over the use of attenuated colonising bacterial strains.

THE BACTERIAL SPORE

Bacterial spores are unique in nature due to their extreme robustness. It is this feature (described in more detail in Chapter 1) that distinguishes bacterial spores (which are endospores) from fungal spores (exospores). As dormant life forms spores can survive indefinitely in a desiccated state and indeed, have been documented as surviving intact for 28 million years (Cano and Borucki, 1995)! The spore can resist temperatures as high as 90°C as well as exposure to noxious chemicals (Nicholson et

al., 2000). Most spore forming bacteria belong to two principal genera, *Bacillus* and *Clostridium* (NB. spore forming bacteria, belonging to other genera are known however). *Clostridia* spore-formers, unlike *Bacillus*, only differentiate under anaerobic conditions. For the purposes of this chapter only the use of *Bacillus* species will be discussed since the difficulties associated with producing *Clostridia* spores make it unlikely that this genus will be developed for use as vaccine vehicles.

Bacillus species produce a single spore (endospore) within the bacterial cell by a process of differentiation (Figure 1) requiring the coordinated action of hundreds of developmental genes (Errington, 1993). In *Bacillus subtilis*, which is the most

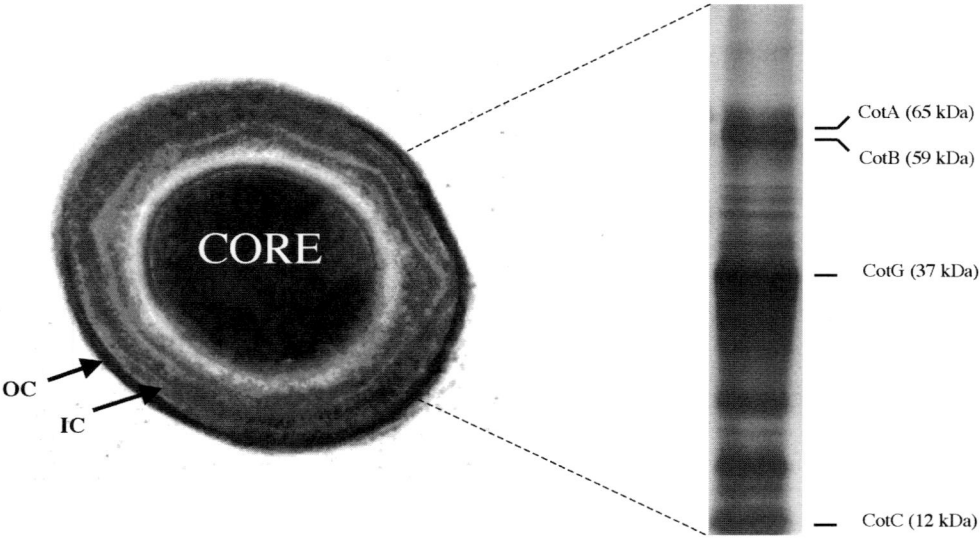

Figure 2. Spore structure. An electron micrograph of a mature spore of *B. subtilis* that has been released from the vegetative cell. The diameter of the spore is about 1μm in *B. subtilis* but can vary both in size and shape (spherical or ellipsoidal) between species. The outer coat (OC) is shown and this is often electron dense and lies above the laminated inner coat (IC). The coat layers protect the central core that carries the condensed chromosome. Shown to the right is a profile of the major spore coat proteins that are present in the spore coat layers. Proteins were extracted and fractionated by SDS-PAGE. The principal outer coat proteins with their molecular weights (note that some of the coat proteins, CotB and CotG run aberrantly and their observed molecular weights differ from the size predicted from the gene sequence). CotF (5 and 8 kDa) is not shown on this gel.

studied spore forming species, the morphology, biochemistry, physiology and genetics of sporulation are extremely well understood making *B. subtilis* a 'model' spore former (see (Errington, 1993; Piggot and Coote, 1976)). Together with *Escherichia coli* this microorganism offers unrivalled ease of genetic manipulation with simplified cloning tools readily available. Nutrient starvation serves as the trigger to initiate spore formation and after 8-10 hours the mature spore is released by lysis of the cell where it can remain dormant indefinitely in a desiccated state. Provision of appropriate nutrients will cause the spore to germinate and resume normal growth and reproduction (a process taking only minutes).

The spore is a desiccated life form and metabolically inactive. The single bacterial chromosome is condensed within the centre of the spore known as the spore core. Distinct layers of lipid membrane and modified peptidoglycan surround the core but the most important structure is the spore coat (Figure 2). This laminated proteinaceous shell provides the spore with resistance to organic solvents. In *B. subtilis* perhaps as many as 25 different coat proteins are present in three distinct coat layers but in other species there is evidence that the coat is less complex and may in some cases consist of only a few protein species (Aronson and Fitz-James, 1976; Driks, 1999; Henriques and Moran, 2000).

THE RATIONALE FOR DEVELOPING SPORES AS VACCINE VEHICLES

One of the principal reasons for using spores as oral vaccines is that they are currently being used as probiotics for both human and animals. Probiotics are live microbial preparations which, when taken orally, benefit the host (Fuller, 1989; Fuller, 1991). The use of *Bacillus* spores as probiotics is described elsewhere in this book (Chapters 11 and 13) and a summary of commercial products is listed in Appendix 1. Potential benefits are believed to include improved nutrition, growth, and prevention of various gastrointestinal disorders. In some countries, *Bacillus* probiotics

are taken as prophylactic agents, for example, to prevent childhood diarrhoea (Mazza, 1994). *B. subtilis* and *B. cereus* spores (in single doses of up to 10^9) are being used commercially as probiotics where they offer a number of advantages over the more common probiotics (*e. g.*, *Lactobacillus* species) in that they can be stored indefinitely in a desiccated form (Mazza, 1994). The prior and existing use of spores as probiotics may facilitate their eventual licensing as oral vaccines (conceptually comparable to plant vaccines). Of course, potential safety issues relating to the release of genetically modified organisms must be addressed and for veterinary use this aspect is particularly important where dormant recombinant spores would be released (via the faeces) directly into the environment. Regardless, compared to some of the other bacterial systems under development such as *Salmonella, Vibrio* and *Yersina* the use of *Bacillus* may prove to be more attractive.

Continued ingestion of large quantities of *Bacillus* spores raises the question of what happens to the spore in the GIT? While no evidence of *bona fide* colonisation has been found it is possible that the spore can interact with the GALT and in Chapter 7 a convincing case is made showing this. Rather than being transient passengers then, these "soil organisms" can germinate in the small intestine, and a proportion of them are able to cross the mucosal epithelium and enter the PPs where they appear to persist briefly within antigen presenting cells. This short-term persistence within the GALT may induce cellular responses as well as pro-inflammatory responses and this attribute is beneficial for new vaccines, particularly, vaccines to viral pathogens where CTL (cytotoxic T-lymphocyte), as opposed to humoral (antibody), responses are critical.

Spores are dormant bio-particles and, of course, can be stored indefinitely at ambient temperature. Potentially then, spore vaccines could be stored, long-term, as a capsule or tablet. For the traveller this feature would be attractive but perhaps this would be of most benefit to developing countries where the cost of storing and transporting vaccines using a cold chain is costly.

Table 1. Potential beneficial attributes of spore vaccines.		
Aspect	Potential advantages	Potential Disadvantages
Delivery route	Has been shown effective for oral delivery of an antigen and can survive the stomach barrier	Not known whether all antigens are immunogenic or stable when delivered orally. Oral tolerance has not been disqualified. Optimal and minimal dosing regimes have not yet been established.
Stability	Stable to 85-90°C depending on species, can be desiccated and stored indefinitely at ambient temperature as a powder or capsule	
Safety	Most *Bacillus* spore formers are non-pathogenic	
Production	Simplified production. Already being produced for use as probiotics.	Unlike other fermentation systems production of spores would require a separate fermentor because of the problems of decontamination
Genetic manipulation	Excellent genetic tools available for *B. subtilis*. Genome sequenced for a number of spore formers.	
Ethical	Current use of spores as human and animal probiotics in Europe, Asia and the US may facilitate licensing.	Potential problems with licensing of spores as veterinary vaccines due to the uncontrolled release of genetically modified dormant spores into the environment.
Other	1. Potential to express multiple antigens in spores. 2. Two routes for antigen expression available (spore and germinating spore) 3. Potential to express any biologically important protein	1. Stability of antigens when expressed on the spore coat can not be guaranteed and must established on a case by case basis. 2. No evidence for cellular/CTL responses has yet been shown

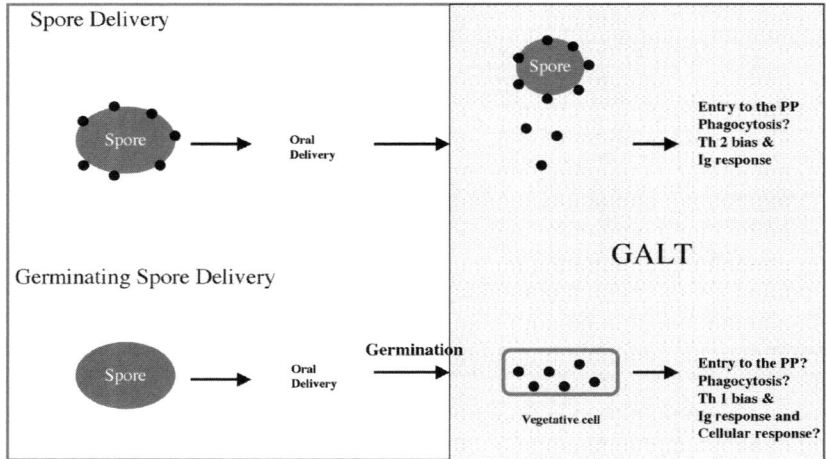

Figure 3. Antigen presentation strategies. The two routes for spore vaccination and antigen presentation are shown and are discussed in the text. Using the spore delivery approach the antigen gene is spliced to a spore coat gene enabling expression of a chimera (●) on the spore surface (Duc et al., 2003b). Following oral delivery the spore or spore integuements carrying the chimeric antigen interact with the GALT (stippled region) generating antibody responses. Using the germinating spore approach, the antigen is expressed using a strong *Bacillus* promoter and the spore enables safe delivery across the stomach barrier (Duc et al., 2003a). In the small intestine spores have been shown to germinate either in the lumen or in the PP following transit across the mucosal epithelium. Expression of the antigen in the vegetative cell produces antibody responses. Evidence in this laboratory shows that spores can germinate within antigen presenting cells and appear to induce expression of Th 1 cytokines suggesting potential cellular responses (Duc et al., 2004).

A final advantage of the spore vaccine approach is that there is an excellent technological platform supporting this genus. Although primarily supporting the use of *B. subtilis* extensive methodology (see Harwood and Cutting, 1990) has been established to genetically manipulate this organism and a large and vibrant research community uses this organism for pure and applied research. In sum, *Bacillus* spore formers can offer a number of advantages (listed in Table 1) for oral vaccine development and when weighed against those of comparable systems spores rank as an attractive system for future development.

USING THE SPORE FOR VACCINATION

The previous section has listed a number of attributes that make *Bacillus* spores particularly suitable candidates as vaccine vehicles. Unlike other bacterial vaccine systems, two, quite distinct, approaches are available for vaccination (Figure 3). In the first method an antigen or epitope is engineered onto the spore coat by fusing a spore coat (*cot*) gene with the antigen sequence. In the second approach, the antigen is expressed constitutively in the vegetative cell by fusion of the antigen gene to the transcriptional and translational sequences of a suitable *Bacillus* gene. Spores carrying this modified gene are used for oral delivery and then germinate in the GIT. The advantages and constraints of each of these two approaches are summarised in Table 1 but will be discussed in more detail.

Expression on the spore surface

At least five proteins are believed to exist on the outermost layer of the spore, CotA, CotB, CotC, CotD and CotG (Driks, 1999). The relative abundance of these proteins is poorly understood but ideally the most abundant protein would be chosen to facilitate the delivery of the highest possible dose of antigen.

A second consideration is whether the hybrid protein is stable on the spore coat. Can the protein fold properly and would it be subject to rapid proteolytic degradation in the stomach? Is the expressed antigen heat-stable on the spore surface? Is the chimeric coat protein still able to assemble onto the spore coat and is there a resultant defect in the spore structure affecting its robustness? Another question is whether the antigen is correctly presented to elicit immune responses? Specifically, are protective epitopes exposed or are they hidden within the spore coat? In fact, does an antigen need to be exposed at all, could it be contained within the spore? While many of these questions remain unanswered for one antigen the spore has been shown to serve as a good antigen delivery system. In a seminal study CotB was demonstrated as a suitable partner for antigen display in *B. subtilis* (Isticato et al., 2001 and see Chapter 17). The 43 kDa. CotB protein is present in the outer layers of the spore coat and was fused to the 459 amino acid C-terminal fragment of tetanus toxin fragment C (TTFC). Surface expression of the 82 kDa. chimeric CotB-TTFC protein was confirmed by Western blotting and FACS analysis. Dot blot quantification demonstrated that 9.75×10^{-5} pg of TTFC (representing 0.3% of extracted spore coat material) was present on each spore but that expression was dependent on the presence of an unmodified CotB protein also being present on the spore coat (Isticato et al., 2001). This was achieved by inserting the engineered CotB-TTFC gene at the *amyE* locus of *B. subtilis* using the pDG364 plasmid (Cutting and Vander-Horn, 1990). Many spore coat proteins have unusual amino acid sequences containing multiple repeats (e.g., CotB and CotG (Sacco et al., 1995; Zheng et al., 1988)) and abnormal numbers of specific amino acids such as serines or lysines (Driks, 1999). It is encouraging therefore that the spore coat is sufficiently malleable for antigen display and perhaps surprising that the spore coat can accommodate the stable presentation of the 47 kDa. TTFC moiety. A potential disadvantage of the spore coat approach is that the spore must be broken down to release its embedded antigens to the host immune system. Whether this arises from spore germination or breakage within an antigen presenting cell the outcome is the same, that is, to increase the dose more spores must be administered. To date too few studies have been conducted to address this but optimising dosing regimes must be established with each new spore vaccine using coat display. A further concern with the use of spores for vaccine delivery is for animal vaccines. For veterinary use recombinant spores would be excreted directly to the environment in the faces. Studies have shown that a large proportion of an orally administered dose of spores is excreted (Hoa et al., 2001) and since spores are dormant indefinitely this would lead to a steady accumulation of recombinant spores in the environment. In many countries with stringent rules regarding the disposal of genetically modified organisms this aspect would probably preclude the use of spores as veterinary vaccines unless the spore could be somehow incapacitated (eg, engineered to be germination-defective or replication-defective).

Proof of Principle for Oral Vaccination Using Tetanus as a Model Disease

Spores expressing CotB-TTFC were used to immunise mice by the oral route (Duc et al., 2003b). Serum IgG and faecal sIgA showed clear seroconversion to TTFC. The dosing schedule used three sets of three doses (1.67×10^{10}) over 5 weeks and was based on regimes optimised for oral immunisations (Challacombe, 1983; Robinson et al., 1997). The titres of TTFC-specific IgG after 33 days ($>10^3$) suggested that these were at protective levels and mice challenged with tetanus toxin corresponding to 10 LD_{50} were fully protected. Out of eight mice challenged with a 20 LD_{50} dose seven survived suggesting that this was the threshold for protection. A similar study was made using nasal immunisation with CotB-TTFC spores but with a lower dose and three immunisations (Duc et al., 2003b). Here, TTFC-specific IgG responses were lower but still showed seroconversion. These studies show that engineered spores expressing a heterologous antigen can be used for protective immunisation. Moreover, although mucosal responses are not important for protection against *Clostridium tetani* (a systemic pathogen) they are obviously important for mucosal pathogens. Further studies will be needed to optimise dosing regimes (less doses and less spores) but these seminal studies have opened the way for development against specific mucosal pathogens. Although these studies are encouraging and demonstrate humoral responses there is no clear evidence yet that indicates cellular responses. However, spores have been shown to disseminate to the GALT and are found in the PP and MLN. The small size of the spore (1 µm) would allow it to be taken up by M cells and transported to the PP where it could interact with antigen presenting cells. Initial studies in this laboratory have shown that spores can germinate and persist for a short period of time within intestinal macrophages as well as elicit Th 1 cytokines *in vivo* such as IFN-γ (Duc et al., 2004).

The Use of the Germinating Spore for Vaccination

The potential of an alternative route for antigen display and vaccination has recently been demonstrated (Duc et al., 2003a). β-galactosidase from *Escherichia coli* was used as a model antigen and fused to the *rrnO* promoter of *B. subtilis*. *rrnO* encodes a ribosomal rRNA and is expressed in vegetative cells. The recombinant *PrrnO-lacZ* gene was placed at the *amyE* locus (in *trans* to an intact *rrnO* gene) using the pDG364 vector. Dot blot quantification showed that the *PrrnO* promoter lead to extremely high levels of β-galactosidase expression amounting to 3.14% of total cellular protein. Spores carrying the *PrrnO-lacZ* gene were administered to mice by the oral route and generated significant β-galactosidase specific IgG titres (Duc et al., 2003a). These serum responses could only occur if the spore had germinated in the GIT lumen or in the GALT (by dissemination across the mucosal epithelium). Indeed, germination-defective spores carrying *PrrnO-lacZ* failed to generate systemic responses. These studies showed then that antigens can be delivered using the spore simply as a vehicle to provide safe passage across the stomach barrier and perhaps this route offers a more sophisticated route for vaccination since protection of the encoded antigen can be assured. Compared to the spore coat display method using the germinating spore could facilitate multiple rounds of cell replication and antigen expression since the germinating spore has been shown to facilitate limited growth within the GIT (Hoa et al., 2001). Ultimately, the germinating spore approach may enable less doses to be used for vaccination but a number of potential concerns exist and have yet to be addressed. Firstly, what proportion of spores germinate and can this be controlled? Presumably, germination must be influenced by the nutrient composition of the small intestine and to ensure a precise dose parameters must be established to ensure controlled germination. Another concern is whether persistence of germinated spores and prolonged expression of the antigen could lead to oral tolerance.

ETHICAL AND SAFETY ISSUES RELATING TO THE USE OF SPORES

The use of *B. subtilis* as a vaccine delivery system is attractive yet it is important to consider potential safety and ethical issues relating to this organism although this is addressed in much greater detail in Chapters 4 and 8. In the *Bacillus* genus only two species are known pathogens (*Bacillus anthracis* and *B. cereus*). *B. subtilis* is considered non-pathogenic but has very occasionally been associated with opportunistic infections in immune compromised patients (de Boer and Diderichsen, 1991). Development of spore formers as vaccines must, however, use well characterised species. The extensive and increasing use of *Bacillus* species as probiotics may facilitate ethical approval of spores as vaccines and in this regard the public interest is similar to that shown in plant vaccines.

CONCLUSIONS

Spores have been shown to be effective vehicles for oral immunisation against tetanus and the uniqueness of this approach compared to most other 2nd Generation Vaccine systems is the heat stability offered by spores. There seems then, a strong argument to further exploit spores for vaccination both for human and for animal use. Of course, so far, spores have been shown effective as vaccines but the ability to express a foreign protein on the spore coat, or indeed, in the germinating spore, would allow the delivery of other molecules of biomedical or industrial importance (e.g., cytokines or enzymes). To date, evidence supports the use of spores as oral vaccines where humoral responses are important for protection. This will include most bacterial diseases but for viruses convincing evidence of cellular (CTL-cytotoxic T-lymphocyte) responses is still needed. Similarly, for some mucosal pathogens where localised local immunity is required (e.g. TB) optimisation of nasal delivery must be addressed. The versatility of the spore, though, means that it should be possible to express more than one antigen on the spore surface as well as co-expression of an adjuvant such as CTB (cholera toxin subunit B). In sum, the spore offers unique attributes for vaccine development which could ultimately lead to improved public health both in developed and developing countries.

REFERENCES

Anderson, R., Dougan, G., and Roberts, M. (1996). Delivery of the Pertactin/ P.69 polypeptide of *Bordetella pertussis* using an attenuated *Salmonella typhimurium* vaccine strain: expression levels and immune response. Vaccine *14*, 1384-1390.

Aronson, A. I., and Fitz-James, P. (1976). Structure and morphogenesis of the bacterial spore coat. Bacteriological Reviews *40*, 360-402.

Cano, R. J., and Borucki, M. (1995). Revival and identification of bacterial spores in 25- to 40-million-year-old Dominican amber. Science *268*, 1060-1064.

Challacombe, S. J. (1983). Salivary antibodies and systemic tolerance in mice after oral immunisation with bacterial antigens. Ann N Y Acad Sci *409*, 177-192.

Cutting, S. M., and Vander-Horn, P. B. (1990). Genetic Analysis. In Molecular Biological Methods for *Bacillus*, C. R. Harwood, and S. M. Cutting, eds. (Chichester, England, John Wiley & Sons Ltd.), pp. 27-74.

Czerkinsky, C., Anjuere, F., McGhee, J. R., George-Chandy, A., Holmgren, J., Kieny, M. P., Fujiyashi, K., Mestecky, J. F., Pierrefite-Carle, V., Rask, C., and Sun, J. B. (1999). Mucosal immunity and tolerance: relevance to vaccine development. Immunol Rev *170*, 197-222.

de Boer, A. S., and Diderichsen, B. (1991). On the safety of *Bacillus subtilis* and *B. amyloliquefaciens*: a review. Appl Microbiol Biotechnol *36*, 1-4.

Drabner, B., and Guzman, C. A. (2001). Elicitation of predictable immune responses by using live bacterial vectors. Biomol Eng *17*, 75-82.

Driks, A. (1999). *Bacillus subtilis* spore coat. Microbiol Mol Biol Rev *63*, 1-20.

Duc, L. H., Hong, H. A., and Cutting, S. M. (2003a). Germination of the spore in the gastrointestinal tract provides a novel route for heterologous antigen presentation. Vaccine *21*, 4215-4224.

Duc, L. H., Hong, H. A., Fairweather, N., Ricca, E., and Cutting, S. M. (2003b). Bacterial spores as vaccine vehicles. Infect Immun *71*, 2810-2818.

Duc, L. H., Hong, H. A., Uyen, N. Q., and Cutting, S. M. (2004). Immunogenicity and intracellular fate of *B. subtilis* spores. Vaccine *22*, 1873-1885.

Errington, J. (1993). *Bacillus subtilis* sporulation: regulation of gene expression and control of morphogenesis. Microbiol Rev *57*, 1-33.

Fuller, R. (1989). Probiotics in man and animals. J Appl Bacteriol *66*, 365-378.

Fuller, R. (1991). Probiotics in human medicine. Gut *32*, 439-442.

Green, B. A., and Baker, S. M. (2002). Recent advances and novel strategies in vaccine development. Curr Opin Microbiol *5*, 483-488.

Griffin, J. F. T. (2002). A strategic approach to vaccine development: animal models, monitoring vaccine efficacy, formulation and delivery. Advanced Drug Delivery Reviews *54*, 851-861.

Han, M., Watarai, S., Kobayashi, K., and Yasuda, T. (1997). Application of liposomes for development of oral vaccines: study of in vitro stability of liposomes and antibody response to antigen associated with liposomes after oral immunization. J Vet Med Sci *59*, 1109-1114.

Harwood, C. R., and Cutting, S. M. (1990). Molecular Biological Methods for *Bacillus*. (Chichester, England., John Wiley & Sons Ltd.).

Henriques, A. O., and Moran, C. P. (2000). Structure and assembly of the bacterial endospore coat. Methods, a Companion to Methods in Enzymology *20*, 95-110.

Hoa, T. T., Duc, L. H., Isticato, R., Baccigalupi, L., Ricca, E., Van, P. H., and Cutting, S. M. (2001). Fate and dissemination of *Bacillus subtilis* spores in a murine model. Appl Env Microbiol *67*, 3819-3823.

Isticato, R., Cangiano, G., Tran, H. T., Ciabattini, A., Medaglini, D., Oggioni, M. R., De Felice, M., Pozzi, G., and Ricca, E. (2001). Surface display of recombinant proteins on *Bacillus subtilis* spores. J Bacteriol *183*, 6294-6301.

Mason, H. S., Warzecha, H., Mor, T., and Arntzen, C. J. (2002). Edible plant vaccines: applications for prophylactic and therapeutic molecular medicine. Trends Mol Med *8*, 324-329.

Mazza, P. (1994). The use of *Bacillus subtilis* as an antidiarrhoeal microorganism. Boll Chim Farm *133*, 3-18.

Medaglini, D., Ciabattini, A., Spinosa, M. R., Maggi, T., Marcotte, H., Oggioni, M. R., and Pozzi, G. (2001). Immunization with recombinant *Streptococcus gordonii* expressing tetanus toxin fragment C confers protection from lethal challenge in mice. Vaccine *19*, 1931-1939.

Medzhitov, R., and Janeway, C. (1997). Innate immunity: the virtues of a nonclonal system of recognition. Cell *91*, 295-298.

Nicholson, W. J., Munakata, N., Horneck, G., Melosh, H. J., and Setlow, P. (2000). Resistance of *Bacillus* endospores to extreme terrestrial and extraterrestrial environments. Microbiol Mol Biol Rev *64*, 548-572.

Piggot, P. J., and Coote, J. G. (1976). Genetic aspects of bacterial endospore formation. Bacteriological Reviews *40*, 908-962.

Robinson, K., Chamberlain, L. M., Schofield, K. M., Wells, J. M., and Le Page, R. W. F. (1997). Oral vaccination of mice against tetanus with recombinant *Lactococcus lactis*. Nat Biotechnol *15*, 653-657.

Sacco, M., Ricca, E., Losick, R., and Cutting, S. (1995). An additional GerE-controlled gene encoding an abundant spore coat protein from *Bacillus subtilis*. J Bacteriol *177*, 372-377.

Seong, S. Y., Cho, N. H., Kwon, I. C., and Jeong, S. Y. (1999). Protective immunity of microsphere-based mucosal vaccines against lethal intranasal challenge with *Streptococcus pneumoniae*. Infect Immun *67*, 3587-3592.

Shalaby, W. S. W. (1995). Development of oral vaccines to stimulate mucosal and systemic immunity: barriers and novel strategies. Clin Immunol Immunopathol *74*, 127-134.

Shata, M. T., Stevceva, L., Agwale, S., Lewis, G. K., and Hone, D. M. (2000). Recent advances with recombinant bacterial vaccine vectors. Mol Med Today *6*, 66-71.

Smith, R. E., Donachie, A. M., and Mowat, A. M. (1998). Immune stimulating complexes as mucosal vaccines. Immunol Cell Biol *76*, 263-269.

Van Ginkel, F. W. (2000). Vaccines for mucosal immunity to combat emerging infectious diseases. Emerg Infect Dis *6*, 123-132.

Walker, R. I. (1994). New strategies for using mucosal vaccination to acheive more effective immunisation. Vaccine *12*, 387-400.

Zheng, L., Donovan, W. P., Fitz-James, P. C., and Losick, R. (1988). Gene encoding a morphogenic protein required in the assembly of the outer coat of the *Bacillus subtilis* endospore. Genes Dev *2*, 1047-1054.

Chapter 19

Recombinant Clostridial Spores in Tumour Therapy

Oliver J Pennington, Lieve Van Mellaert, Jan Theys,
Sofie Barbé, Philippe Lambin, Jozef Anné, and Nigel P Minton

SUMMARY

Effective tumour therapies are reliant on the selective delivery of anti-cancer agents at high therapeutic doses. In recent years, gene therapy using viral vectors as the delivery vehicle has received considerable attention. However, such vectors exhibit a lack of tumour specificity, promote poor levels of transgene expression and are inefficiently distributed throughout the tumour mass. In contrast, the ability of intravenously injected clostridial spores to infiltrate and thence selectively germinate in the hypoxic regions of solid tumours appears a totally natural phenomenon. The organism requires no fundamental alterations, and its ability to target solid tumours is exquisitely specific. As a consequence, clostridial spores have now been used to deliver various therapeutic proteins to the tumours of selected animal models. In the case of prodrug converting enzymes, delivery has been shown, in certain models, to result in complete tumour regression following the systemic administration of prodrug.

INTRODUCTION

The genus *Clostridium* is one of the largest prokaryotic genera (Minton and Clarke, 1989). Characterised as anaerobic, Gram-positive rods they are unified by their ability to form spores. As a grouping, they have achieved greatest prominence as a consequence of their more notorious representatives, *Clostridium botulinum, Clostridium tetani, Clostridium perfringens* and *Clostridium difficile*. In this context, the spore production represents one of, if not the most important, virulence factors. Thus the capacity of *C. botulinum* spores to survive extreme adversity has meant that this organism has been the principal target in food processing for almost a century. Similarly, within the healthcare setting, the ability of the spores of *C. difficile* to survive in the hospital and nursing home environment, makes this particular clostridia one of the more important causes of nosocomial infection, and one of the most difficult examples of microbial contamination to eradicate.

Paradoxically, it is the very ability to form spores that presents the genus with perhaps its greatest potential to benefit mankind, through the exploitation of clostridial spores as a delivery system for treating cancer. This is because intravenously administered spores localise to, and thence exclusively germinate in, the hypoxic centres of solid tumours. The net result is the establishment of an actively growing population of vegetative cells, specifically restricted to the anaerobic environment within the tumour mass. This unique feature provides the opportunity to selectively deliver therapeutic agents to solid tumours, by endowing the clostridial species used with appropriate genes capable of directing the production of the desired therapeutic.

CLOSTRIDIAL TUMOUR ONCOLYSIS

The notion of using bacteria to treat tumours has been around for nearly 140 years. The first documented case of a cancer patient being purposefully injected with live bacteria as a form of cancer treatment was undertaken by the German physician W. Busch in 1868 (*cited* Hall, 1988). The patient, who had an inoperable sarcoma, was infected with *Streptococcus pyogenes* by placing her into a bed previously occupied by a patient with "erysipelas" (a *S. pyogenes* infection). Within a week the primary tumour had decreased by 50%, and the lymph nodes in the neck had also shrunk in size. Unfortunately the patient died some 9 days later. It provided, however, the first indication that tumours can be reduced by bacterial infection. Treatment with *S. pyogenes* was also performed by William B. Coley at New York Hospital, and Friedrich Fehleisen in Germany around 1880–1890 who achieved some degree of tumour regression and patient survival (Pawelek et al., 2003), although the studies were anecdotal and difficult to reproduce. Subsequent animal studies indicated that the results were in part attributed to the infection and in part to the stimulation of the host immune response.

The concept of using bacteria as tumour vectors has, however, been most vigorously pursued using *Clostridium* species. As early as 1813, Vautier reported that cancer patients who suffered gas gangrene (*C. perfringens*), were apparently cured of their cancer following the clostridial infection (*cited* in Hall, 1988). The potential of deliberately using clostridia was explored in 1935, when Connell (1935) used sterile filtrates of *C. histolyticum* for the treatment of advanced cancers. The clinical improvements that were observed were attributed to the production of proteolytic enzymes that preferentially degraded cancerous tissue without affecting normal tissue. These observations subsequently led to several experiments in which such proteolytic enzymes were designed to be produced *in situ* by deliberately infecting normal tissue with clostridial spores. Thus, Parker et al. (1947) demonstrated that intratumoural injection of spores of *C. histolyticum* into the transplanted sarcomas of mice resulted in germination of the clostridia and a noticeable lysis of the tumour tissue. These studies gave the first indication of the potential to use clostridial spores to induce the lysis of tumours.

Following on from Parkers' work, Malmgren and Flanigan (1955) further demonstrated the extreme selectivity of clostridia for the hypoxic/necrotic areas of tumours. Systemic administration of *C. tetani* into mice bearing a variety of tumours resulted in the mice succumbing to tetanus within 48 hours. The healthy control mice, that were not transplanted with a tumour, remained unaffected, demonstrating no signs of disease over the entire time of the study. This observation showed that spore germination and production of the tetanus toxin occurred exclusively in the anaerobic/necrotic cores of tumours.

Möse and co-workers (Möse and Möse, 1959; Möse, 1960) were the first to consider using non-pathogenic clostridial spores as a means of treating cancer via clostridial-induced lysis. For this purpose they isolated *C. butyricum* M55, a strain they later termed *C. oncolyticum* but which has now been reclassified as *C. sporogenes* ATCC 13732. The intravenous injection of M55 spores into mice transplanted with solid Ehrlich tumours resulted in the softening of tumours followed by the spontaneous discharge of a "brownish liquid necrotic mass". Of the few animals that survived, most had recurrence of the tumour at the same site, indicating the presence of an outer viable rim of cells from which tumour regrowth could occur. The results were confirmed with studies which used other rodent tumour models and non-pathogenic *Clostridium* species (Möse and Möse, 1964; Gericke and Engelbart, 1964; Thiele et al, 1964).

The data obtained in these early studies were sufficient for the initiation of clinical studies. These were undertaken using *C. oncolyticum* M55 by Möse and Möse, following the demonstrated that this organism lacked human pathogenicity by administering a spore suspension to themselves (*cited* Carey et al., 1967). Initial trials were performed with 36 patients, who received a dose of $10^6 - 10^9$ spores either intratumourally or intravenously. The injections were tolerated well, and were followed by a high-grade fever which decreased to a low-grade fever for $1 - 3$ days. Oncolysis occurred in the largest of the tumours, with no visible signs of oncolysis in the smaller tumours, metastases, and the surrounding tissues, some $5 - 8$ days later. Trials were also performed in the USA (*cited* Carey et al., 1967) with five patients with advanced neoplastic disease. Oncolysis occurred in three cases only in the largest of tumours, and in one case a transient clinical benefit was attributed to clostridial-induced oncolysis. Several clinical trials have assessed the use of clostridia for the treatment of inoperable malignant brain tumours (Kretschmer, 1972; Heppner and Möse, 1978; Heppner et al., 1983). Oncolysis of the glioblastomas did occur, but complications arose before the completion of oncolysis and the tumours had to be removed via conventional surgery. The clinical trials showed that most patients can tolerate *Clostridium* well, although the presence of an outer viable rim of cells hampered progress.

Experiments with wild-type clostridial species clearly demonstrated that spore treatment is remarkably well tolerated and that vegetative growth frequently leads to the destruction of large parts of the tumour ('oncolysis'). Invariably, however, an outer viable rim remains from which tumour regrowth occurs. This has led to the concept of combining drug treatment with spore administration to enhance the already observed therapeutic effect. In early studies, alkylating agents of the ethyleneimino type (such as tetramin, E-39, trenimon and mitomycin C) were shown to produce positive results when the dose and timing

of drug/spore administration was adjusted to optimum levels (Thiele et al., 1964). The concept of enhancing the therapeutic properties of the clostridial cells used through genotypic changes was first suggested by Schlechte and Elbe (1988). These workers attempted to make a recombinant strain of *C. butyricum* M55 that produced Colicin E3, an *E. coli* bacteriocin suggested to possess canceriostatic properties. At this time, the methodology required to generate such a recombinant strain was not available, and the evidence presented to support the creation of such a strain was not convincing. It has only been with the advances made in the late 80s and early 90s in clostridial gene transfer systems (Minton et al., 1993; Mauchline et al., 1999) that the desired changes could be reproducibly made. As these gene systems were developed for non-pathogenic, saccharolytic strains, the initial experiments have been undertaken with the clostridial species *C. acetobutylicum* and *C. beijerinckii*.

USE OF SACCHAROLYTIC STRAINS

As a prelude to the enhancement of the therapeutic properties of clostridial strains through recombinant approaches, the ability of the various saccharolytic clostridial species to colonise tumours has been re-assessed. Lemmon et al. (1997) systemically applied 10^8 spores of *C. beijerinckii* NCIMB 8052 to mice bearing mammary EMT6 tumours and showed that vegetative Gram-positive clostridial rods were only present in hypoxic/necrotic regions of the tumour. Histological samples of tissues such as brain, heart, kidney, lung and spleen indicated the complete absence of spores by Gram stain. In another study (Lambin et al., 1998), the colonisation ability of four different saccharolytic clostridia: *C. beijerinckii* ATCC 17778, *C. limosum* DSM 1400, *C. acetobutylicum* ATCC 824 and NI-4082 (now reclassified as *C. saccharoperbutylacetonicum*), were compared to *C. oncolyticum*. It was found that a spore titre of at least 10^7 was required for colonisation of rhabdomyosarcomas implanted in WAG/Rij rats, and that *C. acetobutylicum* colonised better than the other three saccharolytic strains and obtained a similar population level to *C. oncolyticum*. Heat treatment of the tumours indicated that the numbers of cfu's obtained there (up to 10^9 per gram of tissue after $4 - 5$ days) were in fact due to vegetative cells, where as the 10^4 to 10^6 cfu's in normal tissues were due to spores as heat treatment had no effect on colony counts. This result was confirmed with Gram staining and histochemical staining of the tumour and normal tissues. No clostridial spores were present in urine samples taken at 4 and 8 days after spore injection.

Having established that saccharolytic clostridial strains are able to colonise tumours, efforts have focussed on the introduction of genes encoding anti-cancer agents. To date, the introduction of two classes of recombinant therapeutic protein have been explored: (i) proteins that are antitumourogenic in their own right such as toxins and cytokines, and; (ii) enzymes which turnover a prodrug to a toxic drug, so called prodrug-converting enzymes (PCEs).

Tumour necrosis factor (TNFα)

TNFα acts as a therapeutic agent in several ways. These include selective destruction of the neovasculature leading to thrombosis and necrosis of tumours, stimulation of T-cell immunity, and direct cytotoxicity to tumour cells, mainly via apoptosis (Fiers, 1991; Larrick and Wright, 1990; Laster et al., 1988; Zheng et al., 1995). Furthermore, enhancement of the anti-tumour effect of TNFα has been demonstrated when combined with irradiation (Hallahan et al., 1995; Sersa et al., 1988). The downside to its use, however, is the systemic toxicity that ensues from direct administration. Its delivery via clostridia therefore provides an opportunity for the localised production, thereby minimising any toxic side effects.

In the investigations of Theys et al. (1999), mouse TNF α (mTNFα) was placed under the control of the *eglA* promoter and *eglA* signal sequence for secretion of TNFα into the tumour mass. Biologically active mTNFα was detected by evaluating spectrophotometrically the cytotoxicity of mTNFα towards WEH164 clone13 cells via the MTT assay. The presence of mTNFα in both the culture supernatants and cell lysates was further demonstrated by Western blot. The cleaved, processed form of mTNFα was detectable in culture supernatants, with both the precursor and processed forms of mTNFα detectable in cell lysates. Lysates and supernatants were taken at various times throughout bacterial growth and added to the highly TNFα sensitive WEHI164 clone 13 cells. Up to $10^{5.5}$ U TNFα per ml of lysate and approximately 10^4 U TNFα per ml of supernatant were detectable. However, it was noted that the TNFα activity in the supernatant decreased after 20 hours, unlike the TNFα activity in the cell lysates which was maximal at mid-log growth phase and maintained that level for at least 20 hours. This was attributed to the acidification of the media during growth influencing TNFα stability as the phenomenon was less pronounced in pH buffered media.

Prodrug-converting enzymes (PCEs)

Greatest attention has focussed on the use of prodrug converting enzymes (PCEs). These represent the crucial component of Directed Enzyme Prodrug Therapy (DEPT). In DEPT strategies the anti-cancer drug is introduced into the bloodstream as harmless "prodrug". This is subsequently converted into the active drug by an "enzyme" that is specifically targeted ("directed") to tumour cells, prior to injection of the prodrug. Specific targeting of the enzyme ensures that high therapeutic doses of the drug are exclusively achieved within the vicinity of the tumour, and not elsewhere in the body. As each enzyme molecule is able to catalyse the generation of large quantities of therapeutic drug, not every tumour cell needs to be specially targeted, resulting in the so-called 'bystander effect'.

In order for enzyme/prodrug therapy to be effective, both the enzyme and prodrug should meet certain requirements. The enzyme should be either of a non-human origin or a human protein that is absent or expressed only at a low concentration in normal tissue (Rigg and Sikora, 1997; Rainov et al., 1998). The protein must achieve a sufficient level of expression in the tumours and have high catalytic activity (Niculescu-Duvaz et al., 1998). The prodrug should be a good substrate for the expressed enzyme and not be activated by exogenous enzymes present in normal tissue. The prodrug must be able to pass through the tumour cell membrane in order for it to take its effect on the

tumour cell. It is also preferential to have as high as possible cytotoxicity differential between the prodrug and toxic drug. Finally it is beneficial if the drug is highly diffusible or is actively taken up by nonexpressing cancer cells in order to elicit a "bystander killing effect" (Niculescu-Duvaz et al., 1998).

The first DEPT strategy was proposed by Bagshawe (1987). Here, tumour delivery is mediated by the fusion of the PCE to a monoclonal antibody raised against a tumour-specific antigen, *ie.*, Antibody-Directed Enzyme Prodrug Therapy (ADEPT). Whilst such an approach remains under investigation (Francis et al., 2002; Spooner et al., 2003), ADEPT suffers from a number of fundamental drawbacks. Tumours exhibit great heterogeneity with regard to the type of antigen present, necessitating the generation of a multitude of different reagents for different forms of cancer. Moreover, in some instances tumours do not possess an enriched antigen, and so may not be targeted by this approach. This has led, in part, to a greater emphasis in recent years on strategies which seek to deliver the gene (Gene-Directed Enzyme Prodrug Therapy, GDEPT) encoding the PCE, most often through the use of a viral vector (Green et al., 2003; Okabe et al., 2003), in so-called Viral-Directed Enzyme Prodrug Therapy (VDEPT). However, GDEPT/VDEPT approaches also exhibit fundamental deficiencies, most notably a lack of tumour specificity, poor levels of transgene expression and the inefficient distribution of the vector throughout the tumour mass (Xu and McLeod, 2001).

The suggested (Lemmon et al., 1994) delivery of PCEs through the use of clostridial spores (Clostridial-Directed Enzyme Prodrug Therapy, CDEPT) would overcome many of these problems. The strategy does not rely on the presence or absence of a particular antigen. Rather, it achieves its specificity through the existence of the hypoxic environment present in the centre core of the majority of solid tumours (Figure 1). Two PCEs have been explored to date, based on the bacterial enzymes cytosine deaminase (CD) and nitroreductase (NTR). Both enzymes essentially have no human equivalent.

Nitroreductase (NTR)

The first prodrug-converting enzyme to be cloned and expressed in clostridia was the *E. coli* B *nfnB* gene encoding nitroreductase (Michael et al., 1995), which was introduced into *C. beijerinckii* NCIMB 8052 (Minton et al., 1995). NTR reduces the prodrug CB1954 (5-(aziridin-1-yl)-2,4-dinitrobenzamide) into either the 2-hydroxylamino, or 4-hydroxylamino products (Knox et al., 1988a), and it is the 4-hydroxylamino product (5-(aziridin-1-yl)-4-hydroxylamino-2-nitrobenzamide) that is particularly toxic because it leads to the formation of interstrand DNA cross-links, which are poorly repaired by the cell (Knox et al., 1998b). On a dose by dose basis, the 4-hydroxylamine species is 10^4- to 10^5-fold more cytotoxic than the CB 1954 progenitor.

The *nfnB* gene was efficiently expressed in *C. beijerinckii* NCIMB 8052, where recombinant NTR was estimated to represent 8% of the cell soluble protein. Following the intravenous injection of the recombinant spores into mice bearing EMT6 tumours, tumour lysates were shown, by Western blots, to contain the *E. coli*-derived enzyme (Minton et al., 1995). Subsequently Lemmon et al. (1997) confirmed

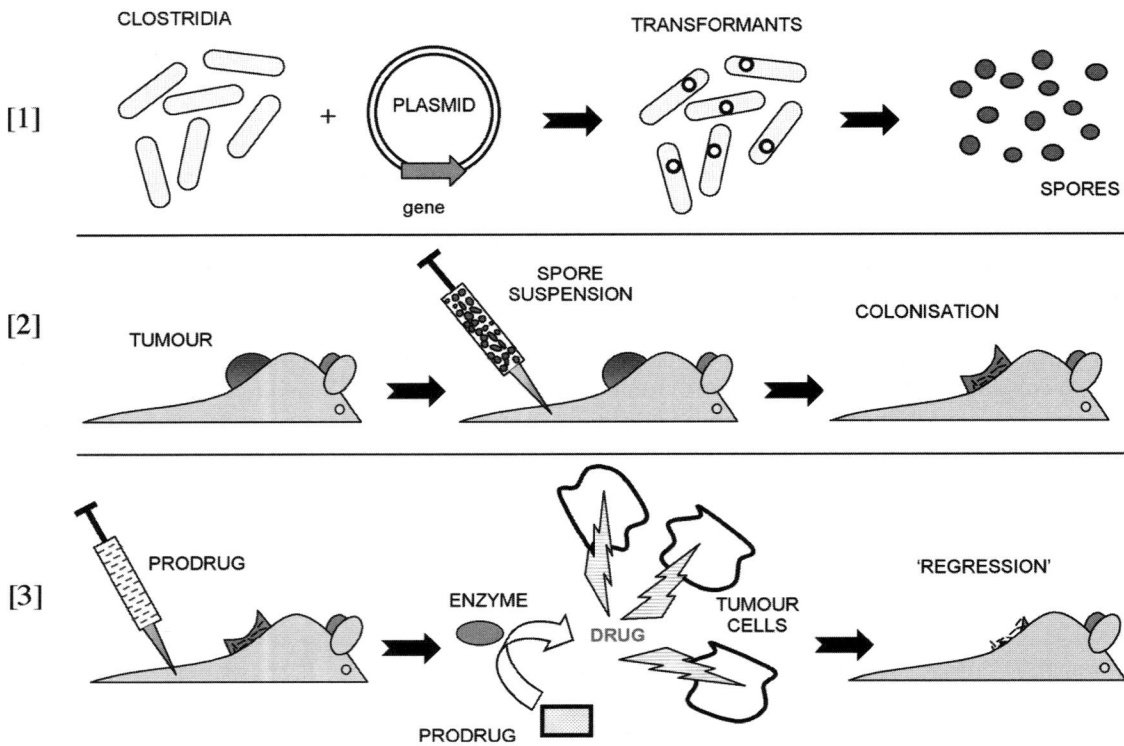

Figure 1. Clostridial-Directed Enzyme Prodrug Therapy (CDEPT).
Panel 1. The gene encoding a prodrug converting enzyme (*eg.,* cytosine deaminase) is inserted into a clostridial expression plasmid and the recombinant vector generated transformed into the clostridial host. Spore stocks are then prepared from the resultant recombinant cells.
Panel 2. A tumour bearing animal is then injected intravenously with a spore suspension (approx. 10^8 spores in 0.1ml saline). Spores are dispersed throughout the body, but only those that encounter the hypoxic environment of the tumour are able to germinate and multiply as vegetative cells. Colonisation and the number of vegetative cells in a tumour maybe increased by the administration of a vascular targeting agent shortly (4 h) after spore administration.
Panel 3. Following effective colonisation (1-2 weeks), a prodrug is administered systemically (*eg.,* 5-FC). This is converted into the active drug (*eg.,* 5-FU) within the tumour by recombinant prodrug converting enzyme (*eg.,* cytosine deaminase) being produced by the vegetative clostridial cells. Repeated administrations of prodrug are possible.

the result obtained with *C. beijerinckii* and measured a 22-fold increase in killing in EMT6 tumours treated with CB1954. However, despite the fact that recombinant NTR could be detected in tumours, enzyme activity could not be detected in tumour tissue.

Cytosine deaminase (CD)

The other type of prodrug converting enzyme to be expressed in clostridia is cytosine deaminase (CD). This enzyme converts the non-toxic prodrug 5-fluorocytosine (5-FC) into the cytotoxic drug 5-fluorouracil (5-FU). 5-FU acts as an anticancer agent as it is further metabolised into 5-fluorouridine-5'-triphosphate and 5-fluoro-2'-deoxyuridine 5'-monophophate, which inhibit DNA and RNA synthesis (Polak et al., 1976). This enzyme is of particular interest in DEPT strategies as the difference in toxicity between prodrug (5-FC) and drug (5-FU) is large (10^4). Moreover, both 5-FU and 5-FC are currently approved for clinical applications in the treatment of breast and gastrointestinal cancers. Fox et al. (1996) were the first to report expression of the *E. coli* CD gene (*codA*) in *C. beijerinckii*. Supernatants taken from the genetically modified *Clostridium* culture increased the sensitivity of murine EMT6 carcinoma cells to 5-FC 500-fold, a level comparable to other studies where the *codA* gene is transfected into mammalian cells.

More recently, *C. acetobutylicum* was engineered to secrete cytosine deaminase specifically at the site of the tumour by fusion of the *closI* signal sequence to the 5' end of *codA* (Theys et al., 2001a). Western blot analysis revealed the 52-kDa CD protein in both culture filtrates and cell lysates from early-logarithmic growth phase samples. *In vitro* and *in vivo* tests ascertained a maximal enzyme activity of 700 pmol 5-FC converted to 5-FU/ml in recombinant bacterial supernatants during early-log phase, a level which would give a sufficient conversion efficiency (3%) *in vivo* of 5-FC to 5-FU to be therapeutically viable. The level of cytosine deaminase activity decreased after early-log phase possibly due to denaturation of the enzyme from acidification of the media or by proteolytic breakdown from extracellular proteases produced by the bacterial host.

USE OF PROTEOLYTIC STRAINS

The lack of tumourigenic effects seen following administration of prodrug to animals colonised with engineered saccharolytic clostridia may have been due to a number of factors, including segregational instability and insufficient production of the recombinant PCE gene. The latter could most simply be accounted for by the attainment of relatively low cell densities in the colonised tumour, compared to, for instance *C. sporogenes* M55. Indeed, it was apparent that the bacterial cell numbers in tumours colonised by *C. beijerinckii* NCIMB 8052 was two orders of magnitude lower than comparable tumours colonised by strain M55 (Liu et al., 2002).

These experiments suggest that a strain such as *C. sporogenes* M55 may represent a more effect delivery vehicle for anti-cancer agents. As this strain initially proved recalcitrant to gene transfer, another *C. sporogenes* (NCIMB 10696) was identified into which plasmid could be transformed. Accordingly, an expression plasmid (pMTL540CD) carrying *codA* was successfully introduced into strain NCIMB 10696 and its effectiveness assessed (Liu et al., 2002). Cell extracts taken from recombinant *C. sporogenes* increased the cytotoxicity of 5-FC to SCCVII cells by a factor of 10^4. Immunoblot analysis of CD and detection of enzyme activity *in vitro* were performed both 7 and 14 days post-intravenous injection of 10^8 spores of recombinant *C. sporogenes* into SCCVII tumour-bearing mice. CD was detected by immunoblot analysis in both the 7 and 14 day tumour samples. Also, extracts from tumours from mice injected with recombinant spores increased the cytotoxicity of 5-FC by a factor of 10^3 with no increase observed for mice injected with wild-type spores. The exquisite selectivity of the process was demonstrated by showing (via immunoblots) that CD was confined to the tumour and could not be found in any of the other tissues screened (brain, heart, lung, liver, kidney and spleen).

To further confirm that *C. sporogenes* was producing sufficient CD to convert 5-FC to 5-FU at a clinically relevant rate, colonised mice were injected with either 5-FC or 5-FU at maximally tolerated doses. Control groups of animals received either recombinant spores alone, or saline as a negative control. Injection of spores alone caused a small amount of tumour lysis as seen in other experiments, whereas a combination of 5-FC and recombinant spores produced a greater growth delay than that given by the maximally tolerated dose of 5-FU. However, in both cases, the tumours became mildly resistant to 5-FU after the first week of injections.

The benefits of a more aggressive coloniser have additionally been shown in a study that utilised a non-toxinogenic, proteolytic strain of *C. novyi* (Dang et al., 2001). These authors assessed a variety of anaerobic bacteria for their capacity to grow extensively and uniquely in the anoxic zones of transplanted tumours. Of the 26 bacterial species tested, 12 species were of the non-pathogenic anaerobe *Bifidobacterium*, 2 species were *Lactobacillus* and 8 were members of the genus *Clostridium*. Intravenous injection of *Bifidobacterium* species resulted in tight clusters of colonies uniquely in the tumour, rather than dispersed throughout the necrotic regions as preferred, whereas the Clostridia, in particular *C. novyi* and *C. sordellii* exhibited extensive spreading throughout the poorly vascularised area of the tumour mass when injected intratumourally. Toxicity of the *Clostridium* species is a barrier when injected intravenously, (injection of up to 10^8 spores led to the death of all tumour bearing mice within 16-18 hours, due to the release of potent lethal bacterial toxins). To overcome this toxicity, the lethal toxin gene was eliminated from *C. novyi* by heat shock, the toxin gene being located within a phage episome. The result was confirmed by PCR and led to a new strain, *C. novyi*-NT (*C.*

sordellii was not chosen due to the presence of two homologous toxin genes). Intravenous injection of spores of *C. novyi*-NT led to germination of the spores in the tumour, and resulted in greatly expanded areas of necrosis, without the toxicity side-effects observed earlier. However, as in earlier studies, a viable rim of tumour cells still existed at the periphery of the tumour.

COMBINATION THERAPIES

At an early stage in the evolution of clostridial-based strategies, it was recognised that procedures that enhance tumour hypoxia might increase oncolysis. Thus, Dietzel et al. (1978) looked at increasing hypoxia in the tumours by raising the temperature within the tumour using microwaves (High-Frequency Hyperthermia, H-FH). When this was combined with irradiation in repeated cycles, a larger reduction in tumour weight in animal tumour models was noticed (Gericke et al., 1979). Increasing the degree of hypoxia through restriction of the oxygen level in the respiratory air supplied to animals has also been investigated. A reduction of between 11 and 12% in the air supplied to animals carrying both Ehrlich and Hardy Passey melanoma dramatically improved the extent of tumour lysis, resulting in complete tumour eradication in 30% of cases (Möse, 1979). Most recently, a more effective route to increasing tumour hypoxia has been demonstrated, using vascular targeting agents.

Vascular targeting agents

It had been observed that the efficiency with which tumours are colonised following spore inoculation was directly proportional to tumour volume (Theys et al., 2001b), with a tumour size of approximately 3 cm being necessary to guarantee colonisation. This size constraint is directly linked to the degree of hypoxia/ necrosis present, *ie.*, the larger the tumour the greater the extent of hypoxia/ necrosis within the centre. This led to the hypothesis that the use of angiogenesis inhibitors and/or vascular targeting agents would increase the level of hypoxia/necrosis and thereby lead to more effective colonisation by clostridia. Moreover, it would allow smaller, more oxic, tumours to be colonised.

This hypothesis was tested in WAG/Rij rats bearing rhabdomyosarcomas using Combretastatin A4-phosphate (CombreAp, OXIGENE, Lund, Sweden), an example of a new class of tumour vasculature targeting agent. CombreAp selectively attacks and destroys tumour-specific blood vessels formed by angiogenesis (Dark et al., 1997) leaving normal vasculature unharmed, due to the morphologically and functionally abnormal tumour blood vessels (Denekamp, 1993). Systemic administration of CombreAp was shown to result in severe tumour vascular shutdown some 3 to 6 hours after administration, leading to obvious necrosis within 1 to 3 days (Landuyt et al., 2000). Moreover, if administration of CombreAp was preceded 4 hours earlier by an intravenous dose of clostridial spores then small tumours (< 1 cm^3) were colonised. Equivalent sized tumours in the control animals that did not receive any CombreAp were not colonised (Theys et al., 2001b), indicating a strong relationship between the level of hypoxia/necrosis and the likelihood of tumour colonisation.

The therapeutic benefit of combining vascular targeting agents with enzyme prodrug therapy using engineered proteolytic clostridia has now been demonstrated (Liu et al., 2003). For these studies the recombinant plasmids carrying *nfnB* or *codA* that had previously been introduced into saccharolytic clostridia (Fox et al., 1996; Lemmon et al., 1997) were transformed into *C. sporogenes* NCIMB 10696. The resultant clones were then used in a therapy experiment in which vascular targeting agent DMXAA (5,6 dimethylxantheone-4-acetic acid) was administered some 4 hours after spore injection. Administration of DMXAA was found to elicit a 4-fold increase in the extent of tumour colonisation, and complete tumour regression was achieved with both subcutaneously transplanted murine SCCVII and human HT29 carcinomas in C3H and nude mice, respectively. These experiments also evaluated the benefit of using more soluble derivatives of the prodrug CB1954. One of the drugs tested (SN24927) was shown to cause more effective tumour regression in combination with clostridial cells overproducing NTR.

The benefits of vascular targeting agents were also shown in the studies of Dang et al. (2001) using *C. novyi*-NT. In this case the vascular targeting agent employed was dolastatin-10 (D10), which was also used in combination with the DNA damaging agent mitomycin C (MMC). Treatment with spores, D10, and MMC resulted in dramatic effects on the large subcutaneous tumours of the colorectal cancer cell line HCT116 in nude mice. Twenty four hours post-administration of the spores the tumour mass swelled and became edematous. D10 was then given intravenously and followed by MMC 24 hours later. Six hours after D10 treatment a zone of haemorrhagic necrosis could be seen. This zone increased in size over 24 hours and often completely enveloped the tumour. The necrotic masses then shrunk over the following 2 – 4 weeks. The control mice, which received only D10 and MMC, did not display a reduction in tumour mass. The downside to this combination bacteriolytic therapy (COBALT), is the association of significant toxicity: 15–45% of animals died depending on tumour size, when *C. novyi*-NT spores were combined with chemotherapy. A possible explanation could be tumour lysis syndrome, a phenomenon seen in other cases where large tumours are rapidly destroyed by antineoplastic agents.

Temporal regulation of gene expression

A further avenue that is under investigation is to position the gene encoding the therapeutic agent under the control of a radiation-inducible promoter. This would place the production of the desired therapeutic protein, in combination with appropriate doses of ionising irradiation, under both temporal and spatial control. In bacteria, DNA damage is restored by a number of systems, the most important of which is the SOS repair mechanism (Miller et al 1990). The *recA* gene is a central component of this system. Using both Northern blot analysis and promoter systems, the expression of the *recA* gene of *C. acetobutylicum* has been shown to be induced by 30% following irradiation at clinically relevant doses of 2 Gy (Nuyts et al., 2001a; Nuyts et al., 2001b). The feasibility of using the *recA* promoter to regulate therapeutic protein production was tested through its fusion to a TNFα gene, modified to include *eglA* signal sequence (Nuyts et al., 2001c),. Secretion of mTNFα

was shown to increase by 44% increase 3.5 hours after a single irradiation dose of 2 Gy. As radiotherapy is usually given in smaller fractionated doses, a second dose of 2 Gy was given 3 hours after the first. This lead to a 1.33- to 1.36-fold increase in TNFα production. The level of induction achieved correlates well with the increase seen after a single dose indicating that the *recA* promoter can be reactivated by a second dose of radiotherapy (Nuyts et al., 2001c).

Radioinducibility is mediated by binding of DinR (equivalent to LexA in *E.coli*) to operator sequences in the *recA* promoter, termed Cheo boxes. In the original experiments, expression from the *recA* promoter was not completely repressed, as TNFα was produced even in the absence of irradiation. To exert greater regulatory control, a second Cheo box was subsequently incorporated into the *recA* promoter (Nuyts et al., 2001d). This resulted in an increase in mTNFα secretion from 44% for the wild-type promoter to 412% for the promoter with an extra Cheo box after a single irradiation dose of 2 Gy. Moreover, these authors were able to show that the constitutive promoter of the endo-β-1,4-glucanase promoter gene (*eglA*) could be converted to an ionising irradiation responsive promoter through the introduction of a Cheo box. Thus, the presence of a Cheo box resulted in a 242% increase in mTNFα secretion. The increase in mTNFα secretion, from irradiated and non-irradiated control cultures, was confirmed by RT-PCR to be caused by enhanced transcription rather than any other factors.

Radiation-inducible promoters show promise for use in CDEPT strategies as the therapeutic protein is only induced in response to irradiation. Therefore, radiation may be employed as a 'molecular switch', activating transcription/expression of the therapeutic gene only in tumour tissues rather than in other non-tumoural hypoxic tissues, such as abscesses or infracted tissues.

OTHER BACTERIAL DELIVERY SYSTEMS

Whilst clostridia have been under the greatest scrutiny, other bacterial-based tumour delivery systems are under investigation, including the facultative anaerobe *Salmonella typhimurium* and the obligate anaerobe *Bifidobacterium*. Growth of *S. typhimurium* within tumours is mediated by severe disablements, that both reduces toxicity (by alteration of the lipopolysaccharide) and imposes a requirement for adenine. The latter is presumed not to be readily available in normal healthy tissues, but is enriched in the tumour environment as a consequence of cellular breakdown. This attenuation allows the organism to accumulate and persist in tumours and exert an inherent anti-tumour efficacy in their own right. However, whilst this efficacy has been observed in animal models (Pawelek et al., 1997; Low et al., 1999), it was not seen in human trials (Toso et al., 2002). As with clostridia, the use of *S. typhimurium* has been combined with DEPT approaches, most notably with *codA*, where appropriately manipulated strains have been shown, in combination with 5-FC, to cause significant tumour suppression in rodent models (Zheng et al., 2000). This particular system has been taken into a Phase I clinical trial (Cunningham and Nemunaitis, 2001), but the final outcome has yet to be published.

The obligate anaerobe *Bifidobacterium longum* has also been shown to selectively colonise solid tumours (Yazawa et al., 2000). Compared to clostridia, the population densities achieved in tumours are significantly less, *eg.,* $<10^3$ cfu / g tumour tissue compared to 10^9 cfu/ g in the case of *C. sporogenes*. Given the results obtained with saccharolytic clostridial species, these low numbers might be predicted to be insufficient for the production of effective levels of a recombinant therapeutic protein. Surprisingly, however, recent studies have shown that a strain engineered to produce CD caused significant regression of DMBA-induced mammary tumours in rats following administration of 5-FC (Fujimori et al., 2002). Similarly, *Bifidobacterium adolescentis* has been used to deliver an endostatin gene to solid tumours in mice, where a strong inhibition of angiogenesis stemmed tumour growth (Li et al., 2003).

CONCLUSIONS

It is now some 10 years since the CDEPT concept was first proposed (Lemmon et al., 1994) as a means of treating solid tumours. In the intervening years a number of key steps have been taken towards proof of principle. The necessary techniques and vector systems have been developed that allow various clostridial species to be engineered to produce desired therapeutic proteins. However, it is only in the last few years that the most significant progress has been made, with the development of strains and combination therapies to promote more effective colonisation of tumours. The use of vascular targeting agents, in particular ensure reproducible colonisation, even of small tumours, and allow the attainment of higher numbers of bacterial cells, and as a consequence higher concentrations of therapeutic enzymes.

Whilst *Clostridium* is a fastidious anaerobe, the therapeutic formulation (the spore suspension) is oxygen insensitive. Unlike other anaerobes, such as *Bifidobacterium*, it is therefore easily stored, handled and administered by systemic injection. The clostridial treatment appears to be well tolerated, and does not appear to elicit an immune response (Theys et al., 2001b). In contrast to classical gene therapy approaches, the therapeutic gene carried by *Clostridium* is not incorporated into the genome of the tumour cell, but remains in the bacterial cell. Furthermore, clostridia, unlike viral vectors, may be removed at any point through the use of an appropriate antibiotic, *eg.,* metranidizole (Theys et al., 2001b).

Clearly, there are still a number of additional refinements that have to be put in place before this approach can move into a clinical setting. Whilst proteolytic strains such as *C. sporogenes* and *C. novyi* are more effective in tumour colonisation, and achieve higher population densities, the more aggressive properties responsible may cause toxicity (Dang et al., 2001). An alternative may yet be to use saccharolytic strains, in combination with vascular targeting agents to improve the effectiveness of colonisation. The data with *B. longum* has shown that high bacterial cell numbers are not necessarily required (Fujimori et al., 2002), and the higher levels of therapeutic protein necessary could be achieved through the use of more effective transcription and translation signals. Regardless of the host employed, for clinical evaluation, strains need to be generated which do not carry bacterial antibiotic resistance markers. This may be most simply achieved through the integration of the therapeutic genes concerned into the genome.

In addition to optimising the production of recombinant protein, the strategy may also benefit from maximising the catalytic activity of the enzyme utilised, through the use of the most effective enzyme and prodrug. The benefit of a more soluble NTR prodrug has already been demonstrated (Liu et al., 2003). The use of more effective enzymes is also likely to prove fruitful (Anlezark et al., 2002). The deliberate secretion of the therapeutic protein may extend the area of drug generation away from the immediate vicinity of the foci of colonisation within the tumour. However, secreted enzymes are prone to proteolysis. This may present a particular problem with proteolytic clostridial strains. Moreover, a wider distribution of enzyme may give rise to a specific immune response directed against the protein. Immune responses in ADEPT prevent repeat therapies. On the other hand, recent studies have shown that the immunogenicity of specific enzymes can be rationally attenuated (Spencer et al., 2002).

Despite the above issues, CDEPT shows considerable promise as a novel therapy for treating solid tumours. Whilst concerns over the concept of deliberately 'infecting' patients with a live clostridial species may surface, it is to be anticipated that clinical evaluations should allay any such fears, allowing the full potential of CDEPT to be realised.

ACKNOWLEDGEMENT
The author wishes to thank Nicola Minion for typing this manuscript, Ben Minton for artwork, and the financial support of the European Union (QLK3-2001-01737).

REFERENCES
Anlezark, G.M., Melton, R.G., Sherwood, R.F., Coles, B., Friedlos, F, and Knox, R.J. (1992). The bioactivation of 5-(aziridin-1-y1)-2,4-dinitrobenzamide (CB1954). I. Purification and properties of a nitroreductase enzyme from *Escherichia coli*: a potential enzyme for antibody-directed enzyme prodrug therapy (ADEPT). Biochem. Pharmacol. *44*, 2289-2295.

Anlezark, G.M., Vaughan, T., Stubbs, S., Michael, N.P., Murdoch, H., Sims, M.A., Wigley, S., Fashola-Stone, E., and Minton, N.P. (2002). The *Bacillus amyloliquefaciens* orthologue of *B. subtilis ywrO* encodes a nitroreductase enzyme which activates the prodrug CB 1954. Microbiol. *148*, 297-306.

Bagshawe, K.D. (1987). Antibody directed enzymes revive anti-cancer prodrug concept. Br. J. Cancer. *56*, 531-532.

Carey, R.W., Holland, J.F., Whang, H.Y., Neter, E., and Bryant, B. (1967). Clostridial Oncolysis in Man. Eur. J. Cancer. *3*, 37-46.

Connell, H.C. (1935). The study and treatment of cancer by proteolytic enzymes. A preliminary report. Can. Med. Ass. J. *33*, 364-370.

Cunningham, C., and Nemunaitis, J. (2001). A phase I trial of genetically modified *Salmonella typhimurium* expressing cytosine deaminase (TAPET-CD, VNP20029) administered by intratumoral injection in combination with 5-fluorocytosine for patients with advanced or metastatic cancer. Protocol no: CL-017. Hum. Gene. Ther. *12*, 1594-1596.

Dang, L.H., Bettegowda, C., Huso, D.L., Kinzler, K.W., and Vogelstein, B. (2001). Combination bacteriolytic therapy for the treatment of experimental tumors. Proc. Natl. Acad. Sci. U S A. *98*, 15155-15160.

Dargatz H., Diefenthal T., Witte V., Reipen G., and von Wettstein D. (1993). The heterodimeric protease clostripain from *Clostridium histolyticum* is encoded by a single gene. Mol. Gen. Genet. *240*, 140-145.

Dark, G. G., Hill, S. A., Prise, V. E., Tozer, G. M., Pettit, G.R., and Chaplin, D. J. (1997). Combretastatin A-4, an agent that displays potent and selective toxicity toward tumour vasculature. Cancer Res. *57*, 1829-1834.

Denekamp, J. (1993). Review article: angiogenesis, neovascular proliferation and vascular pathophysiology as targets for cancer therapy. Br. J. Radiol. *66*, 181-196.

Dietzel, F., Gericke, D., Schumacher, L., and Linhart, G. (1978). Combination of radiotherapy, microwave-hyperthermia and clostridial oncolysis on experimental mouse tumours. In Cancer Therapy by Hyperthermia and Radiation, C. Streffer, ed. (Baltimore-Munich: Essen. Urban and Schwarzenberg), pp. 689-694.

Fiers, W. (1991). Tumour necrosis factor. Characterisation at the molecular, cellular, and *in vivo* level. FEBS Lett. *285*, 199-212.

Fox, M.E., Lemmon, M.J., Mauchline, M.L., Davis, T.O., Giaccia, A.J., Minton, N.P., and Brown, J.M. (1996). Anaerobic bacteria as a delivery system for cancer gene therapy: *in vitro* activation of 5-fluorocytosine by genetically engineered clostridia. Gene Ther. *3*, 173-178.

Francis, R.J., Sharma, S.K., Springer, C., Green, A.J., Hope-Stone, L.D., Sena, L., Martin, J., Adamson, K.L., Robbins, A., Gumbrell, L., O'Malley, D., Tsiompanou, E., Shahbakhti, H., Webley, S., Hochhauser, D., Hilson, A.J., Blakey, D., and Begent, R.H. (2002). A phase I trial of antibody directed enzyme prodrug therapy (ADEPT) in patients with advanced colorectal carcinoma or other CEA producing tumours. Br. J. Cancer. *87*, 600-607.

Fujimori, M., Amano, J., and Taniguchi, S. (2002). The genus *Bifidobacterium* for cancer gene therapy. Curr. Opin. Drug. Discov. Devel. *5*, 200-203.

Gericke, D., Dietzel, F., König, W., Rüster, L., and Schumacher, L. (1979). Further progress with oncolysis due to apathogenic clostridia. Zentralbl. Bakt. 1. Abt. Orig. A. *243*, 102-112.

Gericke, D. and Engelbart, K. (1964). Oncolysis by clostridia. II. Experiments of a tumor spectrum with a variety of clostridia in combination with heavy metal. Cancer Res. *24*, 217-221.

Green, N.K., McNeish, I.A., Doshi, R., Searle, P.F., Kerr, D.J., and Young, L.S. (2003). Immune enhancement of nitroreductase-induced cytotoxicity: studies using a bicistronic adenovirus vector. Int J Cancer. *104*, 104-112.

Hall, S. S. (1988). A commotion in the blood. (New York Press. Henry Holt and Company).

Hallahan, D. E., Mauceri, H. J., Seung, L. P., Dunphy, E. J., Wayne, J. D., Hanna, N. D., Toledano, A., Hellman, S., Kufe, D. W., and Weichselbaum, R. R. (1995). Spatial and temporal control of gene therapy using ionizing radiation. Nature Med. *1*, 786-791.

Heppner, F. Möse, J.R, Ascher, P.W., and Walter, G. (1983). Oncolysis of malignant gliomas of the brain. Proceedings of the 13th International Congress of Chemotherapy. 38-44.

Heppner, F., and Möse, J.R. (1978). The liquefaction (Oncolysis) of malignant gliomas by a non pathogenic *Clostridium*. Acta. Neurochir. *42*, 123-125.

Knox, R.J., Friedlos, F., Jarman, M., and Roberts, J.J. (1988a). A new cytotoxic DNA interstrand crosslinking agent, 5-(aziridin-1-yl)-4-hydroxylamino-2-nitrobenzamide, is formed from 5-(aziridin-1-yl)-2,4-dinitrobenzamide (CB 1954) by a nitroreductase enzyme in Walker carcinoma cells. Biochem. Pharmacol. *37*, 4661-4669.

Knox, R.J., Boland, M.P., Friedlos, F., Coles, B., Southan, C., and Roberts, J.J. (1988b). The nitroreductase in Walker cells that activates 5-(aziridin-1-yl)-4-hydroxylamino-2-nitrobenzamide is a form of NAD(P)H dehydrogenase (quinone) (EC 1.6.99.2). Biochem. Pharmacol. *37*, 4671-4677.

Kretschmer, H. (1972). Treatment of malignant brain tumours by *Clostridium butyricum* M55. Proceedings of the VII th International Congress of Chemotherapy. 721-723.

Lambin, P., Theys, J., Landuyt, W., Rijken, P., van der Kogel, A., van der Schueren, E., Hodgkiss, R., Fowler, J., Nuyts, S., de Bruijn, E., van Mellaert, L., and Anné, J. (1998). Colonisation of the *Clostridium* in the body is restricted to hypoxic and necrotic areas of tumours. Anaerobe. *4*, 183-188.

Landuyt, W., Verdoes, O., Darius, D. O., Drijkoningen, M., Nuyts, S., Theys, J., Stockx, L., Wynendaele, W., Fowler, J. F., Maleux,G., Van den Bogaert, W., Anné, J., van Oosterom, A., and Lambin, P. (2000). Vascular targeting of solid tumours: a major 'inverse' volume-response relationship following combretastatin A-4 phosphate treatment of rat rhabdomyosarcomas. Eur. J. Cancer *36*, 1833-1843.

Larrick JW, Wright SC. (1990) Cytotoxic mechanism of tumor necrosis factor-alpha. FASEB J. *4*, 3215-23.

Laster, S. M., Wood, J. G., and Goodding, L. R. (1988). Tumour necrosis factor can induce apoptosis and necrotic forms of cell lysis. J. Immunol. *141*, 2629-2634.

Lemmon, M.J., Elwell, J.H., Brehm, J.K., Mauchline, M.L., Michael, P.M., Minton, N.P., Giaccia, A.J. and Brown, J.M. (1994) Anaerobic bacteria as a gene delivery system to tumours. Proc. Am. Assoc. Cancer Res. *35*, 374.

Lemmon, M.J., van Zijl, P., Fox, M.E., Mauchline, M.L., Giaccia, A.J., Minton, N.P., and Brown, J.M. (1997). Anaerobic bacteria as a gene delivery system that is controlled by the tumor microenvironment. Gene Ther. *4*, 791-796.

Li, X., Fu, G.F., Fan, Y.R., Liu, W.H., Liu, X.J., Wang, J.J., and Xu, G.X. (2003). *Bifidobacterium adolescentis* as a delivery system of endostatin for cancer gene therapy: Selective inhibitor of angiogenesis and hypoxic tumor growth. Cancer Gene Ther. *10*, 105-111.

Liu, S. C., Dorie, M. J., Giaccia, A. J., Patterson, A., Denny, W., Wilson, W., and Brown, J. M. (2003). Tumour-specific enzyme prodrug gene therapy using genetically engineered *C. sporogenes* as a gene delivery system combined with vascular targeting agents. Abstract of the 4th International Meeting on the Molecular Genetics and Pathogenesis of Clostridia, p. 23, Woods Hole, MA, USA, April 2003.

Liu, S-C., Minton, N. P., Giaccia, A. J., and Brown, J. M. (2002). Anticancer efficacy of systemically delivered anaerobic bacteria as gene therapy vectors targeting tumour hypoxia/necrosis. Gene Ther. *9*, 291-296.

Low K.B., Ittensohn M., Le T., Platt J., Sodi S., Amoss M., Ash O., Carmichael E., Chakraborty A., Fischer. J ., Lin. S.L.0., Luo. X., Miller. S.I., Zheng. L., King. I., Pawelek, J.M., and Bermudes, D. (1999). Lipid A mutant *Salmonella* with suppressed virulence and TNFα induction retain tumor-targeting *in vivo*. Nat. Biotechnol. *17*, 37-41.

Malmgren, R.A., and Flanigan, C.C. (1955). Localization of the vegetative form of *Clostridium tetani* in mouse tumours following intravenous spore administration. Cancer Res. *15*, 473-478.

Mauchline, M.L., Davis, T.O., and Minton, N.P. (1999). Genetics of Clostridia. In Manual of Industrial Microbiology and Biotechnology, 2nd Edition, A.L. Demain, J.E. Davies, eds. (Washington DC, USA: ASM Press), pp. 475-490.

Michael, N.P., Brehm, J.K., Anlezark, G.M., and Minton, N.P. (1995). Physical characterisation of the *Escherichia coli* B gene encoding nitroreductase and its over expression in *Escherichia coli* K-12. FEMS Microbiol. Letts. *124*, 195-202.

Miller, R. V., and Kokjohn, T. A. (1990). General microbiology of recA: environmental and evolutionary significance. Annu. Rev. Microbiol. *44*, 365-394.

Minton, N.P., and Clarke, D.J. (1989). Clostridia Biotechnology Handbooks, Vol. 3. (New York, USA: Plenum).

Minton, N. P., Swinfield, T. J., Brehm, J. K., Whelan, S. M., and Oultram, J. D. (1993). Vectors for use in *Clostridium acetobutylicum* In Genetics and Molecular Biology of Anaerobic Bacteria, M. Sebald, ed. (New York, USA: Springer-Verlag), pp. 120-140.

Minton, N.P., Mauchline, M.L., Lemmon, M.J., Brehm, J.K., Fox, M, Michael, N.P., Giaccia, A.J., and Brown, J.M. (1995). Chemotherapeutic tumour targeting using clostridial spores. FEMS Microbiol. Rev. *17*, 357-364.

Möse, J.R. (1979). Versuche zur verbesserung der onkolyse mit dem clostridienstamm. Zbl. Bakt. Hyg., I. Abt. Orig. A *244*, 541-545.

Möse, J.R. (1960). Zur Beeinflussbarkeit verschiedener tiertumoren durch einen apathogenen Clostridienstamm. Z. Krebsforsch *63*, 447-455.

Möse, J.R., and Möse G. (1964). Oncolysis by clostridia. I. Activity of *Clostridium butyricum* (M-55) and other nonpathogenic clostridia against the Ehrlich carcinoma. Cancer Res. *24*, 212-216.

Möse, J.R., and Möse, G. (1959). Onkolyseversuche mit apathogen, anaerogen sporenbildern am Erhlich-Tumor der maus. Z. Krebsforsch *63*, 63-74.

Niculescu-Duvaz, I., Spooner, R., Marais, R., and Springer, C. J. (1998). Gene directed enzyme prodrug therapy. Bioconjug. Chem. *9*, 4-22.

Nuyts, S., Theys, J., Landuyt, W., van Mellaert, L., Lambin, P., and Anné, J. (2001a). Increasing specificity of anti-tumour therapy: cytotoxic protein delivery by non-pathogenic clostridia under regulation of radio-induced promoters. Anticancer Res. *21*, 857-861.

Nuyts, S., van Mellaert, L., Theys, J., Landuyt, W., Lambin, P., and Anné, J. (2001b). The use of radiation-induced bacterial promoters in anaerobic conditions: a means to control gene expression in clostridium-mediated therapy for cancer. Radiat. Res. *155*, 716-723.

Nuyts, S., Van Mellaert, L., Theys, J., Landuyt, W., Bosmans, E., Anné, J., and Lambin, P. (2001c). Radio-reponsive *recA* promoter significantly increases TNFα production in recombinant clostridia after 2 Gy irradiation. Gene Ther. *8*, 1197-1201.

Nuyts, S., van Mellaert, L., Barbé, S., Lammertyn, E., Theys, J., Landuyt, W., Bosmans, E., Lambin, P., and Anné, J. (2001d). Insertion or deletion of the Cheo Box modifies radiation inducibility of *Clostridium* promoters. Appl. Environ. Microbiol. *67*, 4464-4470.

Okabe S, Arai T, Yamashita H., and Sugihara K. (2003). Adenovirus-mediated prodrug-enzyme therapy for CEA-producing colorectal cancer cells. J. Cancer Res. Clin. Oncol. *129,* 367-373.

Parker, R.C., Plumber, H.C., Siebenmann, C.O., and Chapman, M.G. (1947). Effect of histolyticus infection and toxin on transplantable mouse tumours. Proc. Soc. Exp. Biol. Med. *66,* 461-465.

Pawelek, J. M., Low, K. B., and Bermudes, D. (2003). Bacteria as tumour-targeting vectors. The Lancet Oncology. *4,* 548–556.

Polak, A, Eschenhof, A., Fernex, M., and Scholer, H.J. (1976). Metabolic studies with 5-fluorocytosine-6-^{14}C in mouse, rat, rabbit, dog and man. Chemotherapy *22,* 137-153.

Rainov, N. G., Dobberstein, K. U., Sena-Esteves, M., Herrlinger, U., Kramm, C. M., Philpot, R. M., Hilton, J., Chiocca, E. A., and Breakefield, X. O. (1998). New prodrug activation therapy for cancer using cytochrome p450 4B1 and 2-aminoanthracene/4-Ipomeanol. Human Gene Ther. *9,* 1261-1273.

Rigg, A., and Sikora, K. (1997). Genetic prodrug activation therapy. Mol. Med. Today *3,* 359–366.

Schlechte, H., and Elbe, B. (1988). Recombinant plasmid DNA variation of *Clostridium oncolyticum* – model experiments of cancerostatic gene transfer. Zbl. Bakt. Hyg. A. *268,* 347-356.

Sersa, G., Willingham, V., and Milas, L. (1988). Anti-tumor effects of tumor necrosis factor alone or combined with radiotherapy. Int. J. Cancer *42,* 129-134.

Spencer, D.I.R., Robson, L., Purdy, D., Whitelegg, N.R., Michael, N.P., Bhatia, J., Sharma, S.K., Rees, A.R., Minton, N.P., Begent, R.H.J., and Chester, K.A. (2002). A strategy for mapping and neutralizing conformational immunogenic sites on protein therapeutics. Proteomics *3,* 271-279.

Spooner, R.A., Friedlos, F., Maycroft, K., Stribbling, S.M., Roussel, J., Brueggen, J., Stolz, B., O'Reilly, T., Wood, J., Matter, A., Marais, R., and Springer, C.J. (2003). A novel vascular endothelial growth factor-directed therapy that selectively activates cytotoxic prodrugs. Br. J. Cancer. *88,* 1622-1630.

Theys, J., Nuyts, S., Landuyt, W., Van Mellaert, L., Dillen, C., Böhringer, M., Dürre, P., Lambin, P., and Anné, J. (1999). Stable *Escherichia coli – Clostridium acetobutylicum* shuttle vector for secretion of murine tumour necrosis factor alpha. Appl. Environ. Microbiol. *65,* 4295-4300.

Theys, J., Landuyt, W., Nuyts, S., Van Mellaert, L., de Bruijn, E., Lambin, P., and Anné, J. (2001a). Specific targeting of cytosine deaminase to solid tumors by engineered *Clostridium acetobutylicum*. Cancer Gene Ther. *8,* 294-297.

Theys, J., Landuyt, W., Nuyts, S., Van Mellaert, L., Bosmans, E., Rijnders, A., Van Den Bogaert, W., van Oosterom, A., Anné, J., and Lambin P. (2001b). Improvement of *Clostridium* tumour targeting vectors evaluated in rat rhabdomyosarcomas. FEMS Immunol. Med. Microbiol. *30,* 37-41.

Thiele, E.H., Arison, R.N., and Boxer, G.E. (1964). Oncolysis by clostridia. III. Effects of clostridia and chemotherapeutic agents on rodent tumors. Cancer Res. *24,* 222-233.

Toso, J.F., Gill, V.J., Hwu, P., Marincola, F.M., Restifo, N.P., Schwartzentruber, D.J., Sherry, R.M., Topalian, S.L., Yang, J.C., Stock, F., Freezer, L.J., Morton, K.E., Seipp, C., Haworth, L., Mavroukakis, S., White, D., MacDonald, S., Mao, J., Sznol, M., and Rosenberg, S.A. (2002). Phase I study of the intravenous administration of attenuated *Salmonella typhimurium* to patients with metastatic melanoma. J. Clin. Oncol. *20,* 142-152.

Xu, G., and McLeod, H.L. (2001). Strategies for enzyme/prodrug cancer therapy. Clin. Cancer. Res. *7,* 3314-3324.

Yazawa, K., Fujimori, M., Amano, J., Kano, Y., and Taniguchi, S. (2000). *Bifidobacterium longum* as a delivery system for cancer gene therapy: selective localization and growth in hypoxic tumors. Cancer Gene Ther. *7,* 269-274.

Zheng, L. X., Fisher, G., Miller, R. E., Peschon, J., Lynch, D. H., and Lenardo, M. J. (1995). Induction of apoptosis in mature T cells by tumour necrosis factor. Nature *377,* 348–351.

Zheng, L.M., Luo, X., Feng, M., Li, Z., Le, T., Ittensohn, M., Trailsmith, M., Bermudes, D., Lin, S.L., and King, I.C. (2000). Tumor amplified protein expression therapy: *Salmonella* as a tumor-selective protein delivery vector. Oncol. Res. *12,* 127-135.

Appendix I

Commercial Probiotic Products Containing *Bacillus* Spores

Simon M. Cutting

Table 1. Commercial Probiotic Products Containing *Bacillus* Spores[1].

Product	Target	Manufacturer	Comments	References
AlCare	Swine	Alpharma Inc.,Melbourne, Australia www.alpharma.com.au/ alcare.htm	*B. licheniformis* at 10^9-10^{10} spores/kg	
Bactisubtil®	Human	1) Aventis Pharma , Mem Martins, Portugal, www.aventispharma.pt 2) Marion Merrell Down Laboratories, Levallois-Perret, France.	Capsule carrying 1 X 10^9 spores of *Bacillus cereus* strain IP 5832	(Ciffo, 1984; Green et al., 1999; Hoa et al., 2000; Mazza, 1994; Osipova et al., 1998; Senesi et al., 2001; Smirnov et al., 1994; Sorokulova et al., 1997b) Chapter 8 and 11
BaoZyme-Aqua	Aquaculture -shrimps	Sino-Aqua Corp., Kaohsiung, Taiwan www.sino-aqua.com	*B. subtilis* strains Wu-S and Wu-T at 10^8 CFU/g, product also contains *Lactobacillus* and *Saccharomyces* spp.	
Bibactyl	Human	Tendiphar Corporation, Ho Chi Minh City, Vietnam	Sachet (1g) carrying 10^7-10^8 spores of *B. subtilis*.	
BioGrow®	Poultry and pigs	Provita Eurotech Ltd., Omagh, Northern Ireland, UK. www.provita.co.uk	Listed as containing spores of *B. licheniformis* (1.6 X 10^9 CFU/g) and *B. subtilis* (1.6 X 10^9 CFU/g).	Chapter 13
BioPlus 2B®	a) Piglets[2] b) Sows, pigs, Chickens, turkeys for fattening[3]	Christian Hansen, Hoersholm, Denmark www.chbiosystems.com	Mixture (1/1) of *B. licheniformis* (DSM 5749) and *B. subtilis* (DSM 5750) at 1.6 X 10^9 CFU/g of each bacterium	Chapter 13, 15 and Appendix II
Biosporin®	Human	1) Biofarm, Dniepropetrovsk, Ukraine 2) Garars, Russia.	Biosporin is a mixture of two strains of living antagonistic bacteria *B. subtilis 2335* (sometimes referred to as *B. subtilis 3*) and *B. licheniformis 2336* (ratio is 3:1). Originally isolated from animal fodder. There are a number of versions of this product produced in different countries	(Bilev, 2002; Furzikova et al., 1999; Furzikova et al., 2000; Osipova et al., 1998; Pinchuk et al., 2001; Smirnov et al., 1994; Sorokulova, 1996; Sorokulova, 1997; Sorokulova, 1998; Sorokulova et al., 1997a; Sorokulova et al., 1997b) Chapter 11 and 15
Biosubtyl	Human	Biophar Company, Da lat, Vietnam	Sachet (1 g) carrying 10^6-10^7 of *B. cereus* spores mixed with tapioca. Product labelled as *B. subtilis*. The strain is closely related by 16S rRNA analysis to IP 5832 used in Bactisubtil®. (Hoa et al., 2000)	(Hoa et al., 2000) Chapter 8 and 11
Biosubtyl	Human	Biophar Company, Nha Trang, Vietnam	Sachet (1 g) carrying 10^6-10^7 of *B. pumilus* spores mixed with tapioca. Product labelled as *B. subtilis*. (Green et al., 1999; Hoa et al., 2000)	(Green et al., 1999; Hoa et al., 2000) Chapter 8 and 11
Bispan®	Human	Binex Co. Ltd, Busan, S. Korea www.bi-nex.com	Tablet carrying spores (1.7 X 10^7) of *B. polyfermenticus*	(Lee et al., 2001) Chapter 15
Domuvar	Human	BioProgress SpA, Anagni, Italy www.giofil.it	Vial carrying 1 X 10^9 spores of *Bacillus clausii* in suspension, labelled as carrying *B. subtilis*. No longer marketed.	(Hoa et al., 2000)
Enterogermina®	Human	Sanofi Winthrop SpA, Milan, Italy www.automedicazione.it	Vial carrying 1 X 10^7 spores of *B. clausii* in suspension. At least four different strains of *B. clausii* present and product originally labelled as carrying *B. subtilis*.	(Ciffo, 1984; Green et al., 1999; Hoa et al., 2000; Mazza, 1994; Mazza et al., 1992; Senesi et al., 2001) Chapter 8, 11 and 15
Flora-Balance	Human	Flora-Balance, Montana, USA www.flora-balance.com	Capsules labelled as carrying *B. laterosporus* BOD[4] but containing *Brevobacillus laterosporus* BOD	Chapter 15

Lactipan Plus	Human	Istituto Biochimico Italiano SpA, Milan, Italy	Capsule carrying spores of *Bacillus subtilis* labelled as carrying 2 X 10^9 spores of *Lactobacillus sporogenes*[4]	(Hoa et al., 2000)
Lactopure	Human Animal feed	Pharmed Medicare, Bangalore, India www.pharmedmedicare.com	Labelled as *Lactobacillus sporogenes*[4] but contains *B. coagulans*	Chapter 15
Lactospore	Human	Sabinsa Corp., Piscataway, NJ, USA www.sabinsa.com	Labelled as carrying *Lactobacillus sporogenes*[4] but contains *B. coagulans* 6-15 billion/g	Chapter 15
Medilac	Human	Hanmi Pharmaceutical Co. Ltd., Beijing, China www.hanmi.co.kr	*B. subtilis* strain R)179 (at 10^8/g) in combination with *Enterococcus faecium*	
Nature's First Food	Human	Nature's First Law, San Diego, CA, USA www.rawfood.com	42 species listed as probiotics including: *B. subtilis, B. polymyxa, B. pumilus* and *B. laterosporus*.	
Primal Defense™	Humans	Garden of Life® www.thehomeostasisprotocol.com/ mall/Primal-defense/article3.htm	14 bacterial components including *B. subtilis* and *B. licheniformis*	See Chapter 15
Promarine®	Aqaculture -shrimps	Sino-Aqua company, Kaohsiung, Taiwan www.sino-aqua.com	Carries 4 strains of *B. subtilis*	See Chapter 15
Subtyl	Human	Mekophar, Pharmaceutical Factory No. 24, Ho Chi Minh City, Vietnam	Capsule carrying 10^6-10^7 spores of a *B. cereus* species termed *B. cereus var* vietnami (Hoa et al., 2000). Product labelled as carrying *B. subtilis*.	(Hoa et al., 2000) Chapter 8
Toyocerin®	Chickens, laying hens, calves, cattle for fattening[3]	Asahi Vet S.A., Tokyo (Head Off.), Japan www.asahi-kasei.co.jp	*B. cereus var toyoi* (NCIMB-40112/CNCMI-1012) at a minimun concentration of 1 X 10^{10} CFU/g.	(Baum et al., 2002; De Cupere et al., 1992; Jadamus et al., 2002; Jadamus et al., 2001; Kyriakis et al., 1999) Chapter 13 and Appendix II
WC Cultuur	Biological cleaning fluid	Noble BioProducts BV, Oldenzaal, The Netherlands www.noblebio.com	Contains spores of multiple *Bacillus* species at 6.76 X 10^7/ml.	

[1]this list is probably not complete and new products are being introduced or updated continously
[2]authorised by the EU for unlimited use
[3]authorised by the EU on a provisional basis
[4]not officially recognised as a *Bacillus* species (www.bacterio.cict.fr)

REFERENCES

Baum, B., Liebler-Tenorio, E. M., Enss, M. L., Pohlenz, J. F., and Breves, G. (2002). *Saccharomyces boulardii* and *Bacillus cereus var. Toyoi* influence the morphology and the mucins of the intestine of pigs. Z Gastroenterol *40*, 277-284.

Bilev, A. E. (2002). Comparative evaluation of probiotic activity in respect to *in vitro* pneumotropic bacteria and pharmacodynamics of biosporin-strain producers in patients with chronic obstructive pulmonary diseases. Voen Med Zh *323*, 54-57.

Ciffo, F. (1984). Determination of the spectrum of antibiotic resistance of the *Bacillus subtilis* strains of Enterogermina. Chemioterapia *3*, 45-52.

De Cupere, F., Deprez, P., Demeulenaere, D., and Muylle, E. (1992). Evaluation of the effect of 3 probiotics on experimental *Escherichia coli* enterotoxaemia in weaned piglets. Zentralbl Veterinarmed B *39*, 277-284.

Furzikova, T. M., Sergeichuk, M. G., Sorokulova, I. B., and Smirnov, V. V. (1999). The effect of the cultivation conditions on the properties of bacilli comprising the basis of probiotics. Mikrobiol Zh *61*, 19-27.

Furzikova, T. M., Sorokulova, I. B., Serhiichuk, M. H., Sichkar, S. V., and Smirnov, V. V. (2000). The effect of antibiotic preparations and their combinations with probiotics on the intestinal microflora of mice. Mikrobiol Zh *62*, 26-35.

Green, D. H., Wakeley, P. R., Page, A., Barnes, A., Baccigalupi, L., Ricca, E., and Cutting, S. M. (1999). Characterization of two *Bacillus* probiotics. App Env Microbiol *65*, 4288-4291.

Hoa, N. T., Baccigalupi, L., Huxham, A., Smertenko, A., Van, P. H., Ammendola, S., Ricca, E., and Cutting, A. S. (2000). Characterization of *Bacillus* species used for oral bacteriotherapy and bacterioprophylaxis of gastrointestinal disorders. Appl Env Microbiol *66*, 5241-5247.

Jadamus, A., Vahjen, W., Schafer, K., and Simon, O. (2002). Influence of the probiotic strain *Bacillus cereus var. toyoi* on the development of enterobacterial growth and on selected parameters of bacterial metabolism in digesta samples of piglets. J Anim Physiol Anim Nutr (Berl) *86*, 42-54.

Jadamus, A., Vahjen, W., and Simon, O. (2001). Growth behaviour of a spore forming probiotic strain in the gastrointestinal tract of broiler chicken and piglets. Arch Tierernahr *54*, 1-17.

Kyriakis, S. C., Tsiloyiannis, V. K., Vlemmas, J., Sarris, K., Tsinas, A. C., Alexopoulos, C., and Jansegers, L. (1999). The effect of probiotic LSP 122 on the control of post-weaning diarrhoea syndrome of piglets. Res Vet Sci *67*, 223-228.

Mazza, P. (1994). The use of *Bacillus subtilis* as an antidiarrhoeal microorganism. Boll Chim Farm *133*, 3-18.

Mazza, P., Zani, F., and Martelli, P. (1992). Studies on the antibiotic resistance of *Bacillus subtilis* strains used in oral bacteriotherapy. Boll Chim Farm *131*, 401-408.

Osipova, I. G., Sorokulova, I. B., Tereshkina, N. V., and Grigor'eva, L. V. (1998). Safety of bacteria of the genus *Bacillus*, forming the base of some probiotics. Zh Mikrobiol Epidemiol Immunobiol, 68-70.

Pinchuk, I. V., Bressollier, P., Verneuil, B., Fenet, B., Sorokulova, I. B., Megraud, F., and Urdaci, M. C. (2001). *In vitro* anti-*Helicobacter pylori* activity of the probiotic strain *Bacillus subtilis* 3 is due to secretion of antibiotics. Antimicrob Agents Chemother *45*, 3156-3161.

Senesi, S., Celandroni, F., Tavanti, A., and Ghelardi, E. (2001). Molecular characterization and identification of *Bacillus clausii* strains marketed for use in oral bacteriotherapy. Appl Environ Microbiol *67*, 834-839.

Smirnov, V. V., Rudenko, A. V., Samgorodskaia, N. V., Sorokulova, I. B., Reznik, S. R., and Sergeichuk, T. M. (1994). Susceptibility to antimicrobial drugs of strains of bacilli used as a basis for various probiotics. Antibiot Khimioter *39*, 23-28.

Sorokulova, I. B. (1996). Outlook for using bacteria of the genus *Bacillus* for the design of new biopreparations. Antibiot Khimioter *41*, 13-15.

Sorokulova, I. B. (1997). A comparative study of the biological properties of Biosporin and other commercial *Bacillus*-based preparations. Mikrobiol Zh *59*, 43-49.

Sorokulova, I. B. (1998). Effect of probiotics from bacilli on macrophage functional activity. Antibiot Khimioter *43*, 20-23.

Sorokulova, I. B., Beliavskaia, V. A., Masycheva, V. A., and Smirnov, V. V. (1997a). Recombinant probiotics: problems and prospects of their use for medicine and veterinary practice. Vestn Ross Akad Med Nauk, 46-49.

Sorokulova, I. B., Kirik, D. L., and Pinchuk, I. I. (1997b). Probiotics against Campylobacter Pathogens. J Travel Med *4*, 167-170.

Appendix II

EU Regulations on Bacillary Probiotics for Animal Feeds

Elinor McCartney

SUMMARY

The use of live micro-organisms (probiotics) in animal feeds has come under legislative pressure in the EU during the last decade, resulting in increased costs of obtaining marketing authorisation. For example, current estimates to license a probiotic for broilers and piglets, two common target animal categories, are around €1,400,000, a considerable cost for an industry where sales volumes and margins are significantly lower than for feed or pharmaceutical products. Of particular concern to regulators has been the risk of transference of antibiotic resistance in probiotic bacteria to other bacteria in the environment and the food chain. Another perceived risk concerning the use of *Bacillus* species as animal probiotics, linked to the robustness and longevity of bacillary spores, is the potential for toxin production and enteropathogenicity in some strains. The intense scrutiny of animal probiotics in relation to possible transfer of antibiotic resistance is in marked contrast to regulatory attitudes to live bacteria used in human pharmaceuticals or in food or beverages. The recently established EFSA (European Food Safety Authority) is charged, among other tasks, with harmonising food and feed law in the EU. This may result in tougher legislation for live micro-organisms used in human food, beverages and as pharmaceuticals.

THE HISTORY OF ANIMAL FEED ADDITIVE LEGISLATION IN THE EU, 1970-2004

Influence of European History and Culture

Understanding current EU feed additive legislation is simpler if viewed from a historical perspective, taking into account several "cultures" surrounding the human pharmaceutical, veterinary, animal feed, and human nutrition/food environments, as well as broader "*STEEPLE*" factors (Social, Technological, Economic, Ethical, Political, Legal and Ecological).

In the first half of the twentieth century, hunger was widespread in Europe, but in the decades following the 2nd World War European agriculture responded well to the challenge and by the 1960s Western Europe enjoyed an abundant food supply at affordable prices. Indeed, some would say that EU agriculture has killed itself with its own success, as the cost of maintaining beef and milk 'mountains' has consumed taxpayers' funds in the latter decades of the twentieth century. Table 1 illustrates some major changes in focus concerning animal production during the 20th century. Today, the EU is wealthy in global terms, and Europeans can afford to be particular about what they eat. Quality has replaced quantity as the production imperative.

Pharmaceutical Legacy To Feed Additives

Moving onto a separate historical track, the human pharmaceutical industry also took off in the post-war era in Europe. Although penicillin was discovered in the 1920s, it was only in the decades after the Second World War that many new, effective antibiotics were discovered or invented. Bacteria, of course, have always had to co-exist with antimicrobials and developed antibiotic resistance as a parallel phenomenon (SCAN 2003, http://europa.eu.int/comm/food/fs/sc/scan/out108_en.pdf). Indeed, the increase of antibiotic resistance resulting from human activity has been a key driver of EU feed additive legislation today, as discussed later.

If antibiotics were the success story of the twentieth century, then thalidomide was one of the tragedies (http://www.thalidomide.ca/wit.html). Thalidomide is the primary reason that current guidelines on teratogenicity testing of veterinary and human pharmaceuticals as well as food additives and chemically-based feed additives specify using at least two mammalian species, including one non-rodent species. Interestingly, thalidomide was never marketed in the USA because its adverse effects were discovered while the drug was still in the registration process at the FDA (Food and Drug Administration). Hence, thalidomide is one reason why regulatory authorities tend to adopt a very cautious approach to innovation – everything is toxic until proven otherwise.

Table 1. Food and feed in Europe during the 20th Century.	
DECADE/S	FOCUS
1900's to 1950s	Cycle of wars, depression, hunger and recovery
1950s	Start of intensive animal production and industrial feed production (e.g. broiler industry)
1960s	Further advances in animal nutrition, breeding and production, leading to affordable food in abundance
1970s	Least cost formulations for animal feeds – competitive focus on reducing costs and producing more
1980s	Focus on food flavour and savour – quality replaces quantity as the priority
1990s	Food safety, animal welfare and environmental considerations come to the fore
2000s	Consumer health and consumer perceptions drive the food chain, from "Farm to Fork"
The cost of food is no longer a key issue in Europe	

Table 2. Structure of a 70/524 feed additive dossier.					
Rapporteur's Report to the EU Commission and Member States					
SECTION I	SECTION II	SECTION III	SECTION IV	SECTION V	SECTION VI
Summary	Quality	Efficacy	Safety	Monograph	Identification Note
	Expert Reports on Quality, Efficacy and Safety				
Text	Text	Text	Text	Text	Text
Annexes					
Original reports (and original data), statistical analyses, statistical meta-analysis					
Bibliography, scientific references and marginal data					

Pharmaceutical Triad: Safety, Quality and Efficacy

The veterinary pharmaceutical industry was born from its human pharmaceutical parents. Indeed many veterinary drugs in the 20th Century, especially antibiotics, were simply adapted formulations of human drugs. Other products derived from the veterinary pharmaceutical industry in the 1960s and 1970s were the classic feed additives, antimicrobial growth promoters and coccidiostats (AGPCs). Not surprisingly, the initial data package requirements for these feed additive products were similar or identical to human drugs, involving "*GLP*" (Good Laboratory Practice) toxicity studies in laboratory animals and appropriate "*GMP*" (Good Manufacturing Practice) manufacturing standards and quality control procedures.

It is worth remembering three words in relation to traditional human or veterinary pharmaceutical regulations – "*safety*", "*quality*" and "*efficacy*". There is some divergence between data packages for human pharmaceuticals and veterinary pharmaceuticals or feed additives, especially in relation to food-producing animals, where the safety package often requires tissue residue and environmental data. In addition, products of food animals (e.g. meat, milk and eggs) can be evaluated for quality, appearance and flavour, which are often important with respect to feed additive efficacy. Most important, as a logical consequence of their common origins in human medicaments, the pharmaceutical principles and regulatory "culture" of safety, quality and efficacy have tended to apply to both veterinary pharmaceuticals and feed additives.

CURRENT EU LEGISLATION GOVERNING FEED ADDITIVES (70/524), AND CHANGES UNDER EFSA

In 1970 the EU published Council Directive 70/524/EEC, the first feed additive Directive. Its main aims were to harmonise and centralise Community legislation on the registration and use of feed additives, and of course the pharmaceutical triad of safety, quality and efficacy was applied, in keeping with the pharmaceutical origins of the classic feed additives (AGPCs). At the date of writing (October 2003), 70/524 has undergone numerous modifications and 5 major amendments, the last major refurbishment in 1996 (Council Directive 96/51/EC). All current EU legislation in force and under preparation is published on the relevant EU web page (http://europa.eu.int-eur-lex).

Currently, the obtention and maintenance of EU approval for a feed additive requires the investment of considerable resources – time, money and skill. The applicant chooses an EU Member State as "*Rapporteur*", and submits to the *Rapporteur* a product dossier comprising data on the safety, quality and

efficacy of the feed additive in question (Table 2). The specific data requirements for each type of feed additive are listed in detailed guidelines, published and updated from time to time by the EU. Until the 1980's, the majority of EU feed additives were chemically-based products, and these are governed by guidelines not dissimilar to those regulating veterinary pharmaceuticals (Commission Directive 2001/79/EC, Part I).

When the Member State *Rapporteur* is satisfied with the dossier, copies are sent to the EU Commission in Brussels, to all other EU Member States, and eventually to SCAN (Scientific Committee on Animal Nutrition), a group of independent scientists appointed by the EU Commission to give opinions on feed additives and related matters. Member States may also use their own in-house or independent scientific experts and committees to evaluate a feed additive dossier. Member States, their experts and SCAN may ask questions to the applicant before issuing an opinion on the dossier. Many such questions require the investment of additional time and money to generate data that can provide answers. All questions must be answered to everyone's satisfaction before the dossier can proceed through the registration process. In theory, a feed additive could pass through this system in just over 2 years (Table 3), but in practice 5-10 years are the current rates of passage, since the clock stops every time a question is raised and, apart from the time required to generate new data to address each question, the whole system is slowed down by bureaucratic delays and limited resources.

Once all safety and quality questions have been answered satisfactorily, and initial, limited evidence of efficacy is accepted, SCAN issues an opinion. If this SCAN opinion is

Table 3. Timescale for EU Feed Additive Approval, Council Directive 96/51/EC.	
Process	Theoretical timescale
Validation of dossier with *Rapporteur*	~365 days
Submit to EU Commission and Member States	60 days for format approval (Member States)
	30 days to place on STAN agenda
	320 days for decision approval
Total	775 days (>2 years)
Clock stops	When STAN* asks question/s
	When SCAN** asks question/s
	Bureaucratic delays
*STAN = Standing Committee on Animal Nutrition (Member States), now part of the Standing Committee on the Food Chain and Animal Health **SCAN = Scientific Committee on Animal Nutrition (Independent Experts), now known as FeedAp, and forming part of the European Food Safety Authority Scientific Committee and Panels	

favourable, The Standing Committee on Animal Nutrition (STAN) votes, usually supporting SCAN's opinion, and the additive is approved provisionally for 4 years. Efficacy is then further assessed under EU conditions and for definitive approval, a feed additive product must produce at least 3 high quality (near-"*GLP*" standards) for each animal category and claim statistically significant improvements in some aspect of animal feed, nutrition, performance, production, product quality, animal welfare, or environmental effects. Veterinary claims must be avoided, since these would classify the product as a veterinary pharmaceutical, governed by different legislation. These more detailed efficacy studies are again evaluated by EU Member States (STAN) and SCAN, questions asked or clarification requested, a SCAN opinion issued and a STAN vote taken, before a feed additive may attain definitive and permanent EU approval. The STAN vote must pass by a qualified majority, consisting of 2/3 of EU Member States and 2/3 of the vote (Table 4).

A new Feed Additive Regulation (FAR, Regulation (EC) N° 1831/2003) will apply from late 2004, under the European Food Safety Authority (EFSA). One of the objectives of EFSA and the new FAR is to separate risk assessment from risk management. The new system will dispense with Member State *Rapporteurs* and provisional approvals. Feed additive dossiers will be evaluated by permanent EFSA staff, assisted by EFSA's Scientific Committee and Panels (e.g. FeedAp). There will be a strong focus on safety (risk assessment). Once EFSA issues an opinion, the EU Commission will be responsible for risk management via the Comitology procedure, which means that, as happens currently, a qualified majority of Member States will be required for EU approval of a feed additive. One important difference from the current system is that EFSA may issue a positive opinion on a feed additive, but the Comitology procedure, which allows consideration of non-scientific criteria (e.g. societal, economic, ethical), may result in the feed additive being prohibited in the EU.

Table 4. EU member states, population and N° of votes.		
Country	Population	N° votes
Germany	83 M	10
UK	60 M	10
France	58 M	10
Italy	56 M	10
Spain	40 M	8
The Netherlands	16 M	5
Greece	11 M	5
Belgium	10 M	5
Portugal	10 M	5
Austria	8 M	4
Sweden	8 M	4
Finland	7 M	3
Denmark	5 M	3
Eire	3 M	3
Luxembourg	1 M	2
EU 15	376 M	87

SPECIFIC REQUIREMENTS AND COSTS FOR EU APPROVAL OF A PROBIOTIC FEED ADDITIVE

During the 1980's and 1990s two new types of product became popular in animal feeds: feed enzymes and probiotics. From around the same time, there was increasing consumer concern over the safety of the food chain, stimulated by several food/feed/pharmaceutical scares:

- Increase in antibiotic resistance, due to heavy use of antibiotics in various spheres (human and veterinary medicine, antimicrobial growth promoters in animal feeds).
- Illegal uses of diethyl stilboestrol and clenbuterol in food-producing animals.
- Food poisoning, especially due to *Salmonella* spp. or *E. coli 0157* strains.
- Bovine Spongioform Encephalopathy (BSE).
- GMO (Genetically-Modified Organisms) Foods and Crops.
- Dioxin contaminants in animal feeds and animal products.

During the 1990s, the EU decided to ban antibiotic growth promoters (AGPs) and encourage the development of alternative products and management systems. Probiotics seemed an attractive option for animal feeds, with a long history of safe use in human foods and beverages, namely yogurts and fermented lactic beverages, products with "healthy" and "natural" connotations for EU consumers. Hence, initial requirements for obtention of EU approval for a feed probiotic were considerably lighter than for a classic AGPC, with no requirement for chronic toxicity testing, residue or environmental studies (Commission Directive 94/40). However, there was a parallel and increasing concern in the EU surrounding the prevalence of antibiotic resistance as a result of the use and abuse of antibiotics in human and animal medicine and as feed additives. This concern over antibiotic resistance and its potential transference and spread throughout the food chain has led to the imposition of additional safety requirements for probiotics, namely that they should not be capable of the production of antimicrobial substances relevant to the use of antibiotics in humans or animals, nor should they contribute further to the reservoir of antibiotic resistance genes already present in the gut flora of animals and the environment (SCAN 2001, http://europa.eu.int/comm/food/fs/sc/scan/out68_en.pdf).

Whilst the differences in data requirements for registering a new coccidiostat or a new animal probiotic in the EU are still considerable – EU approval of a coccidiostat may cost an estimated €20 million - the cost of obtaining definitive EU approval for a probiotic is still substantial and estimated at around €1.4 million, a considerable cost for an industry where sales volumes and margins are significantly lower than for feed or pharmaceutical products (Table 5). The most recent, specific data requirements for an EU animal probiotic registration application were published on the SCAN web page in late 2001, and are applied by Member State experts and SCAN when assessing probiotic feed additive dossiers submitted for

Item	Description	Estimate
Section I Summary	Translate summary text to French and German Expert reports on Quality, Efficacy, Safety and Statistics	€10,000 €40,000
Section II Quality	Required data (e.g. strain identity; genetic stability; antibiotic production and antibiotic resistance; stability of product, product in premixes, during feed pelleting, and storage; dusting and homogeneity data; development of quality control procedures during product manufacture; validation of analytical methods)	€250,000
Section III Efficacy	Required data (e.g. efficacy in broilers and weaned piglets under EU conditions to near "GLP" standards)	€500,000
Section IV Safety	Required data (e.g. *in vitro* and laboratory animal safety studies, tolerance tests in weaned piglets and broilers)	€300,000
Staff Costs	Registration staff costs (e.g. time spent designing and co-ordinating studies, visiting trial sites, meeting the *Rapporteur*, EU Commission and Member States, writing the dossier)	€250,000
Contingency		€50,000
Total	€1,400,000	

Table 5. Cost of definitive EU approval of a bacillary probiotic in broilers and weaned piglets.

evaluation in the EU (SCAN 2001, http://europa.eu.int/comm/food/fs/sc/scan/out68_en.pdf).

ADVANTAGES AND DISADVANTAGE OF *BACILLUS* SPECIES AS AN ANIMAL PROBIOTIC FEED ADDITIVE

Bacillus spp. have a distinct advantage over other feed additive probiotics, due to their dormant spore state. Spores are resistant to a range of environmental stressors, notably heat. Most animal feeds are produced as pellets, which are subjected to moisture, heat and friction during feed processing and pelleting. The majority of bacteria in yoghurts or lactic drinks, which may be considered the human equivalents of animal feed probiotics, and especially *Lactobacillus* and *Bifidus* spp., would not survive the temperatures involved in producing feed pellets. Bacillary probiotics, presented as viable spores, readily survive packaging, transport, storage and feed processing, under a variety of conditions that would be damaging to less resistant micro-organisms, resulting in considerable loss of viable counts in final feed. However, the known robustness and longevity of the bacillary spore has attracted special regulatory focus, for example in relation to potential toxin production. *Bacillus cereus*, *Bacillus subtilis* and B*acillus licheniformis*, commonly used as human or animal probiotics, have all been described as opportunistic pathogens and SCAN originally considered that toxin production among *Bacillus* spp. is more widespread than previously thought (SCAN 2000, http://europa.eu.int/comm/food/fs/sc/scan/out41_en.pdf), although later or updated SCAN opinions have recognised the possibility that unfiltered, low molecular weight material in culture supernatants may confound *in vitro* assays designed to detect cytotoxins (SCAN 2001, http://europa.eu.int/comm/food/fs/sc/scan/out73_en.pdf).

THE REGULATORY FATE OF BACILLARY PROBIOTIC FEED ADDITIVES IN THE EU

The twin concerns of potential toxin production and antibiotic resistance have seriously hindered attempts to obtain EU approvals for probiotics in animal feeds. In fact, only 2 animal probiotics have been successful in achieving definitive EU approval, Bioplus 2B® (Christian Hansen, Denmark) and Toyocerin® (Toyo Jozo, Japan) both of these based on *Bacillus*

spp. These 2 products were approved in the EU only after demonstrating that resistance to antibiotics of their strains was not transferable, and that the strains were not toxin producers. A number of other bacillary probiotics have been rejected by SCAN on safety grounds related to potential toxin production and/or potential transference of antibiotic resistance, including several *Bacillus* spp (Table 6).

Of particular interest is SCAN's comment on Neoferm BS-10®: "*The 2 strains of bacteria which are the active components of the feed additive Neoferm BS-10® derive from a pharmaceutical preparation registered for human use. Because the pharmaceutical preparation was designed for use as an adjunct to antibiotic therapy, its constituent strains were deliberately selected to demonstrate resistance to a number of clinically important antibiotics,............all important in human and veterinary medicine. While the selection of antibiotic-resistant strains was appropriate for short-term therapeutic use in humans, the same justification cannot be applied to a product designed for long-term use as a feed additive. Animals are thought to act as a reservoir of bacteria resistant to antibiotics and their transfer to humans via the food chain or other routes may be a contributory factor in the developing resistance to antibiotics in clinical use............Thus, because of the risk of dissemination of genes that confer resistance to clinically important antibiotics, SCAN consider that the use of Neoferm BS 10® as a feed additive would be unsafe.*" (SCAN 1999, http://europa.eu.int/comm/food/fs/sc/scan/out28_en.pdf).

QUALIFIED PRESUMPTION OF SAFETY (QPS) AND CONTRAST WITH PROBIOTICS USED IN HUMANS

As indicated by their 1999 opinion on Neoferm BS-10®, SCAN considered that long-term use of bacillary probiotics in animals would constitute a risk to human health, especially if the probiotic strains had the potential to transfer antibiotic resistance through the food chain. In 2001, SCAN issued an opinion on the criteria for assessing the safety of micro-organisms resistant to antibiotics of human clinical and veterinary importance. Whilst this opinion had been requested of SCAN by the EU Commission specifically in relation to micro-organisms used as feed additives, SCAN took the opportunity to highlight an obvious inconsistency between feed and food additive legislation in the EU: "*Implementation conclusions above*

Table 6. EU regulatory fate of bacillary probiotics for animal feeds after scan opinions[1].

Product	Marketed by	*Bacillus* spp.	Resistant to	Toxin Production	SCAN Opinion	EU Status
Bioplus 2B®	Christian Hansen Biosystems	*Bacillus licheniformis* *Bacillus subtilis*	Flavomycin Zinc bacitracin Clindamycin	None	Positive	Approved
Toyocerin®	Asahi Vet	*Bacillus cereus var. toyoi*	Tetracyclines Chloramphenicol Sulphonamides	None	Positive	Approved
Alcare™	Alpharma	*Bacillus licheniformis*	Sulphonamdides Erythromycin Bacitracin	None	Negative	Not approved
Esporafeed Plus®	Norel	*Bacillus cereus*	Tetracyclines	Insufficient data	Negative	Not approved
Neoferm BS-10®	?	*2 strains of Bacillus clausii*	β-lactams Macrolides Lincosamides Rifampicin	Not mentioned	Negative	Not approved
Paciflor®	Intervet	*Bacillus cereus*	Not mentioned	Enterotoxicity	Negative	Not approved

[1]Note: SCAN opinions on the above animal feed probiotics are listed in full in the reference section, and can be found on the SCAN web site: http://europa.eu.int/comm./food/fs/sc/scan/index_en.html.

are consistent with the position adopted by the Commission on the need to take action to preserve the value of antibiotics in human and veterinary medicine. They should lead to the exclusion of any microbial feed additive containing one or more bacteria carrying resistance genes capable of being transferred to other bacteria. They are, however, far more stringent than those currently applied to live micro-organisms used in foods and consumed directly by humans. This seems to SCAN contrary to the stated desires of the Commission and iniquitous for producers of animal products. Accordingly, SCAN recommends that a consistency of approach should be adopted for all microbial products entering the food chain." (SCAN 2001, revised in 2003, http://europa.eu.int/comm/food/fs/sc/scan/out108_en.html).

In addition, other researchers had already highlighted potential dangers from the use of human bacillary probiotics. For example, Green et al. (1999) and Hoa et al. (2000) studied several bacillary probiotics marketed for human use in Europe and elsewhere, and found 6 of 7 products examined to be incorrectly identified on the product label. These researchers were also concerned about the high levels of antibiotic resistance, commenting that at least 2 probiotic strains had been subject to a history of mutagenesis in the process of creating antibiotic resistance (Hoa et al., 2000).

In 2002, SCAN issued a position paper on the safety assessment and regulatory aspects of micro-organisms in feed and food applications and proposed a Qualified Presumption of Safety (QPS) system to evaluate certain safety aspects of all micro-organisms entering the food chain, notably the presence of transmissible antibiotic resistance or known virulence factors (SCAN 2002, http://europa.eu.int/comm/food/fs/sc/scan/out85_en.pdf). This QPS proposal has been expanded and clarified recently by a Working Group consisting of members of SCAN, SCF (Scientific Committee on Food) and SCP (Scientific Committee on Plants) (SCAN 2003, http://europa.eu.int/comm/food/fs/sc/out178_en.pdf). In this second paper, the Working Group make specific reference to

Bacillus spp. and potential toxigenicity, and re-emphasise the tenet that all live bacteria entering the food chain, whether via animal feed or directly consumed by humans should be free of any acquired resistance to antibiotics of importance in clinical and veterinary medicine. Two bacterial examples are discussed in more detail:

- Dairy *Lactobacilli spp.* – QPS status considered, provided that evidence indicates the absence of acquired antibiotic resistance.
- *B. subtilis* and related *Bacillus* spp. – QPS status considered, provided that: the strain is free of any acquired resistance to antibiotics of importance in human and veterinary medicine; there is absence of a capacity to produce antibiotics with structural similarities to those of importance in human and veterinary medicine likely to encourage development of resistance; and there is provision of PCR-based evidence of the absence of a toxigenic potential and evidence of absence of effects in cytotoxicity assays.

THE EUROPEAN FOOD SAFETY AUTHORITY (EFSA) AND THE PRECAUTIONARY PRINCIPLE

The recently established EFSA is charged, among other tasks, with harmonising food and feed law in the EU (Regulation (EC) 178/2002). Under EFSA, risk assessments should be undertaken in an independent, objective and transparent manner, based on available scientific information and data. Scientific Committees such as SCAN have been traditionally responsible for risk assessments in the EU, so in fact there is continuity in this aspect. Another EU principle, which will also continue under EFSA is the precautionary principle, which should be used in a uniform way throughout the European Community. The precautionary principle has never been fully defined in EU law, especially in relation to human or animal health, but suffice it to say that the EU reserves the right to manage risks based on a risk assessment in which potential dangers arising from a phenomenon, product or process have been identified, but scientific evaluation does not allow the risk to be determined

```
                    ┌──────────────────────┐
                    │ Screen for resistance* │
                    └──────────┬───────────┘
                               │         ┌─────────────────────┐
                               │         │ No resistance detected │
                               │         └─────────────────────┘
                    ┌──────────┴───────┐
                    │    Resistance    │
                    └─────────┬────────┘
              ┌───────────────┴──────────────┐
              │  For each resistance identified │
              └───────────────┬──────────────┘
                    ┌──────────┴──────────┐
                    │ Test for transferability │
                    └──────────┬──────────┘
    ┌─────────────┐            │
    │ Transferable │            ├──────────────────────────────┐
    └─────────────┘            │   No transfer of genes detected │
                               └──────────────────────────────┘
              ┌────────────────────────────────────────────────────┐
              │ Examine for the presence of exogenous, known resistance genes │
              └────────────────────────────────────────────────────┘
        ┌──────────────────────────────┐    ┌──────────────────────────┐
        │ Presence of a known resistance gene │    │ No known resistance gene present │
        └──────────────────────────────┘    └──────────────────────────┘
                                             ┌──────────────────────────┐
                                             │  Provision of evidence     │
                                             │  that resistance is        │
                                             │  intrinsic or mutational   │
                                             └──────────────────────────┘
              ┌─────────────────────┐    ┌──────────────────────┐
              │ Not convincing evidence │    │  Convincing evidence   │
              └─────────────────────┘    └──────────────────────┘
    ┌──────────────┐                          ┌──────────────────────┐
    │ Not acceptable │                          │      Acceptable        │
    └──────────────┘                          └──────────────────────┘
```

Figure 1. SCAN Flow Chart to Assess Safety of Microroganisms Resistant to Antibiotics of Clinical and Veterinary Importance.

with sufficient certainty. Such a risk management exercise may include any action from a "watching brief" or a recommendation to a legally binding measure.

THE FUTURE OF LIVE *BACILLUS* SPECIES IN PHARMACEUTICALS, FOODS AND FEEDS IN THE EU

Bearing in mind the precautionary principle, the growing unease among scientists concerning the use of human probiotics that are resistant to antibiotics, and taking into account EFSA's mission to harmonise feed/food legislation, the next few years could herald new regulatory requirements for live bacteria used in human foods and pharmaceuticals. For *Bacillus* spp. used as adjuncts to human pharmaceuticals, future EU legislation may parallel current EU requirements for animal feed probiotics, summarised as a flow chart in the relevant SCAN opinion (Figure 1, reproduced from SCAN 2003, http://europa.eu.int/comm/food/fs/sc/scan/out108_en.pdf).

In addition, there is encouraging data from the field of animal probiotics, which suggests that probiotics, despite *in vitro* sensitivity to antibiotics, may exhibit *in vivo* compatibility. Hence, there may be no reason to include antibiotic resistance in probiotics intended for use as an adjunct to antibiotic therapy in humans. The active Bacillary strain of Toyocerin®, approved in the EU as a feed additive probiotic in sows, piglets and other target animal categories, is sensitive *in vitro* to many commonly used veterinary antibiotics, yet may be used in feeds

containing such antibiotics without any significant reduction in viable counts of the probiotic strain, and in addition, with no reduction in its bioregulatory capacity (Shimura et al., 1979). Shimura and his colleagues fed 8 groups of piglets for a two-week study period as follows:

Group 1 – Negative Control (no antibiotic or Toyocerin®)
Group 2 – Positive Control (Toyocerin® only)
Group 3 – Toyocerin® + Kitasamycin

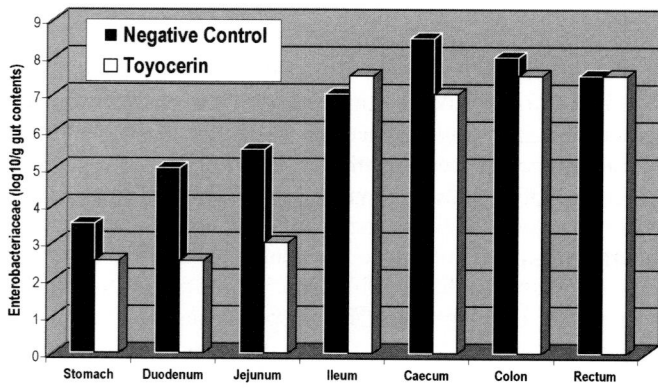

Figure 2. Toyocerin® Reduces Enterobacteriaceae in the Piglet Upper Gut. Animals dosed with 10^6 spores/g. Data taken from week 2. Note: animals were not fed antibiotics.

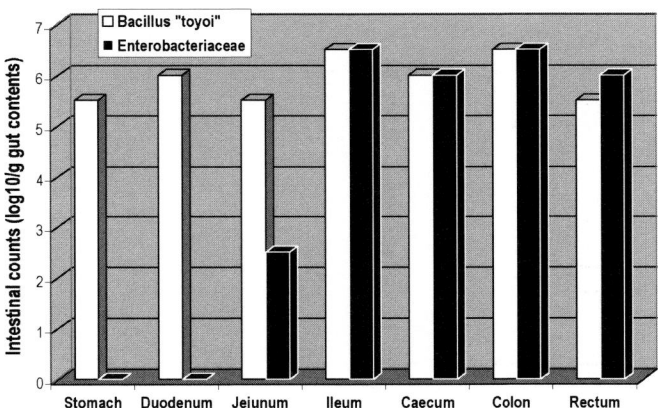

Figure 3. Toyocerin® is compatible with Kitasamycin *in vivo*. Animals dosed with 10^6 spores/g. Feed also contained the macrolide Kitasamycin at 100 ppm. Data taken from week 2.

Group 4 – Toyocerin® + Colistin
Group 5 – Toyocerin® + Carbadox
Group 6 – Toyocerin® + Kitasamycin + Colistin
Group 7 – Toyocerin® + Thiopeptin + Colistin
Group 8 – Toyocerin® + Enramycin + Colistin.

Samples of gut contents were taken at Time 0, Week 1 and Week 2, and cultured for the Toyocerin® probiotic strain and for *Enterobacteriaceae*. Negative Control piglets had significant numbers of *Enterobacteriaceae* throughout the gut ($\sim10^3$–10^6/ g in upper gut; $\sim10^7$-10^9/g in lower gut) and throughout the study, but no Toyocerin® probiotic strain was isolated. Positive Control piglets had high numbers of the Toyocerin® strain throughout the gut at Week 1 and Week 2, with reduced *Enterobacteriaceae* at the same time points, especially in the upper gut. Piglets fed antibiotics combined with Toyocerin® exhibited similar results to the Positive Controls. This research was published in Japanese, hence selected data are reproduced in Figures 2 and 3.

The reason for this unexpected *in vivo* compatibility between probiotics and antibiotics is not known, but may be related to the dosage of probiotic applied in the feed, and to the fact that a bacillary probiotic incorporated into animal feed provides constant seeding of the gastrointestinal tract with viable spores that germinate and multiply to a limited extent, sufficient to exert a probiotic effect. Such observations from the world of animal probiotics may stimulate scientists to research and develop human probiotics that may be used in conjunction with antibiotic therapies, without the necessity, real or perceived, to select probiotic strains for multiple antibiotic resistance.

REFERENCES

Commission Directive 94/40/EC of 22 July 1994 amending Council Directive 87/153/EEC fixing guidelines for the assessment of additives in animal nutrition. http://europa.eu.int-eur-lex

Commission Directive 2001/79//EC of 17 September 2001 amending Council Directive 87/153/EEC fixing guidelines for the assessment of additives in animal nutrition. http://europa.eu.int-eur-lex

Council Directive 70/524/EEC of 23 November 1970 concerning additives in feedingstuffs. http://europa.eu.int-eur-lex

Council Directive 96/51/EC of 23 July 1996 amending Directive 70/524/EEC concerning additives in feeding stuffs. http://europa.eu.int-eur-lex

Green, D.H., Wakeley, P.R., Page, A., Barnes, A., Baccigalupi, L., Ricca, E., and Cutting, S.M. (1999) Characterisation of two *Bacillus* probiotics. Appl. Environ. Microbiol. 65, 4288-4291.

Hoa, N.T., Baccigalupi, L., Huxham, A., Smertenko, A., Van, P.H., Ammendola, S., Ricca, E., and Cutting, S.M. (2000) Characterisation of *Bacillus* species used for oral bacteriotherapy and bacterioprophylaxis of gastrointestinal disorders. Appl. Environ. Microbiol. 66, 5241-5247.

Regulation (EC) Nº 178/2002 of the European Parliament and of the Council of 28 January 2002 laying down the general principles and requirements of food law, establishing the European Food Safety Authority and laying down procedures in matters of food safety. http://europa.eu.int-eur-lex

Regulation (EC) Nº 1831/2003 of the European Parliament and of the Council of 22 September 2003 on additives for use in animal nutrition. http://europa.eu.int-eur-lex

SCAN (1999) Assessment by the Scientific Committee on Animal Nutrition (SCAN) of a micro-organisms product: Esporafeed Plus®. http://europa.eu.int/comm/food/fs/sc/scan/out39_en.html

SCAN (1999) Assessment by the Scientific Committee on Animal Nutrition (SCAN) of a micro-organisms product: Neoferm BS-10®. http://europa.eu.int/comm/food/fs/sc/scan/out28_en.html

SCAN (1999) Assessment by the Scientific Committee on Animal Nutrition of the safety of product Paciflor® for use as a feed additive. http://europa.eu.int/comm/food/fs/sc/scan/out62_en.html

SCAN (2000) Opinion of the Scientific Committee on Animal Nutrition on the safety of use of *Bacillus* species in animal nutrition. http://europa.eu.int/comm/food/fs/sc/scan/out41_en.html

SCAN (2000) Report of the Scientific Committee on Animal Nutrition on Product Bioplus 2B® for use as feed additive. http://europa.eu.int/comm/food/fs/sc/scan/out49_en.html

SCAN (2001) Commentary by the Scientific Committee on Animal Nutrition on data relating to toxin production submitted by Intervet International GmbH on another product (Toyocerin® produced by Asahi Vet S.A.). http://europa.eu.int/comm/food/fs/sc/scan/out73_en.html

SCAN (2001) Guidelines for the assessment of additives in feedingstuffs. Part II: enzymes and micro-organisms. http://europa.eu.int/comm/food/fs/sc/scan/out68_en.html

SCAN (2001) Report of the Scientific Committee on Animal Nutrition on Product Toyocerin® for use as feed additive. http://europa.eu.int/comm/food/fs/sc/scan/out72_en.html

SCAN (2002) Opinion of the Scientific Committee on Animal Nutrition on the use of *Bacillus licheniformis* NCTC 13123 in feedingstuffs for pigs (Product AlCare™). http://europa.eu.int/comm/food/fs/sc/scan/out79_en.html

SCAN (2002) Position paper of the Scientific Committee on Animal Nutrition on safety assessment and regulatory aspects of micro-organisms in feed and food applications. http://europa.eu.int/comm/food/fs/sc/scan/out85_en.html

SCAN (2003) On a generic approach to the safety assessment of micro-organisms used in feed/food and feed/food production. http://europa.eu.int/comm/food/fs/sc/scan/out178_en.html

SCAN (2003) Opinion of the Scientific Committee on Animal Nutrition on the criteria for assessing the safety of micro-organisms resistant to antibiotics of human clinical and veterinary importance. http://europa.eu.int/comm/food/fs/sc/scan/out108_en.html

Shimura, T., Nogami, M., Hattori, Y., Yamane, Y., Masagaki, H., Murofushi, M. and Kozasa, M. (1979) Stability in the intestine of piglets and effects on the porcine intestinal bacterial flora of Toyocerin® (feed additive containing viable *Bacillus toyoi* spores). Journal of Veterinary Medicine (Japan). 343-347.

Appendix III

The Fungicidal Activity of Spore-Forming Bacteria

Rudolf R. Azizbekyan

SUMMARY

Today, especially in less developed countries, food deficits are being observed. One of the primary reasons behind the global food deficit are fungal diseases of commercially important plants which amount to annual losses to agriculture of between 20 to 30%. Current methods for control of plant diseases is by the use of chemical pesticides, which are harmful to non target organisms and pollute the environment. For this reason, in the last decade increasing attention has been given to biological methods of plant protection. One major attribute of *Bacillus* species is their ability to synthesize biologically active compounds, including insecticidal toxins and antifungal antibiotics. *B. thuringiensis*–based insecticides have been used for more than forty years for the control of the harmful insects (e.g., *Lepidoptera, Coleoptera* and *Diptera*). The advantages of the large-scale application of biological insecticides is the relative simplicity and cost-effectiveness of their manufacture. *Bacillus*-based fungicides have not yet received widespread attention nor application. Nevertheless, an increasing number of *Bacillus* strains with anti-fungal activity are being reported. In most cases, *Bacillus* species synthesize a number of fungicidal factors that enhance their antagonistic effect on plant pathogens and have been shown to reduce the probability of infection during storage and some examples are given in this chapter. Biological fungicides do not pollute the environment and produce no measurable effect on the microenvironment. In the near future we predict an increased interest in *Bacillus*-based biofungicides and their application in plant protection.

INTRODUCTION

Fungal infections of plants are major factors that reduce levels of agricultural production. Some fungi not only cause losses but also produce mycotoxins, that accumulate in the kernels of grain. Modern practices for plant disease control are based on chemical pesticides, but their use can also lead to unwanted problems. Biological control of fungal diseases offers an attractive alternative to synthetic chemical pesticides. Antagonistic microorganisms or compounds produced by these organisms can be used as biological agents for plant protection. Included in this group are spore forming bacteria (*Bacillus* spp. including *B. subtilis, B. megaterium, B. pumilus, B. amyloliquefaciens, B. cereus, B. mycoides, B. licheniformis, B. anthracis, B. thuringiensis, B. mojavensis, B. brevis, B. laterosporus*) that have been shown to exhibit antifungal activity.

Fungicidal properties of *Bacillus* species can be associated with different compounds. For example, the closely related organisms, *B. cereus, B. mycoides, B. anthracis* and *B.*

thuringiensis produce zwittermicin-A (a linear aminopolyol molecule). *Bacillus* spp. also produce a large number of peptide, antifungal, antibiotics, which differ in their structure and mechanism of biosynthesis. Most antifungal peptides produced by *Bacillus* species have a molecular weight of less than 2000 Da and are synthesized non-ribosomally. These antibiotics are organized in a linear, cyclic or branched cyclic structure. The cyclic lipopeptides produced by *B. subtilis* belong to different families (iturin, surfactin and plipastatin-fengycin) based on their structural relationships. Antifungal activity of the cyclic decapeptide antibiotics tyrocidine family (gramicidin S and loloatins) from *Brevibacillus brevis* and *Brevibacillus laterosporus* respectively, have also been described.

Antifungal antibiotics differ in their modes of action and the efficacy of antibiotics can be different towards fungal spores and mycelia. The three principle modes of action of the antifungal lytic antibiotics are:

1) formation of pores on membranes,
2) disruption of membranes, and
3) synthesis of peptides that penetrate through membranes and bind with organic molecules.

Biofungicides may be applied to different plant parts for treatment but, in general, biofungicides are being used for the treatment of seeds and for post–harvest treatment of fruits. Biological control of disease in the root and leaf zones is less effective though. In this chapter we examine the variety of antifungal antibiotics produced by *Bacillus* species. In addition we present data showing our results of using *Bacillus spp.* for the control of fungal diseases during storage of potato tubers and wheat under field conditions in Russia.

Many fungal pathogens reduce the quality and quantity of plant production and they can be grouped according to the location of the plant they infect or in which they produce their spores. Fungi can attack plant vegetative organs such as the leaf, stem, and sometimes also the root. Some fungi (*Fusarium, Aspergillus*) not only cause crop losses but also produce mycotoxins, which can subsequently contaminate flour. Mycotoxins can decrease food potential by 25-30%. For example, infection of seed by *Fusarium graminearum* reduces seed germination, seedling vigor, plant emergence, and adversely affects grain quality and mycotoxin contamination of corn remains a serious economic problem (Cleveland et al., 2003).

Economic damage due to agricultural pests and phytopathogenic fungi is estimated to mount to billions of dollars annually and various strategies have been used in attempting to control pests. One strategy is the use of chemical pesticides with a broad range of activity although there are a number of disadvantages to using such chemical pesticides. First, due to their broad spectrum of activity, chemical pesticides may adversely affect non-target organisms such as beneficial insects, birds, and mammals. Second, chemical pesticides are frequently toxic to animals and humans. Third, targeted pests often develop resistance when repeatedly exposed to such substances. Current practice for plant disease control based on chemical pesticides is therefore highly controversial. The widespread use of chemical pesticides has led to the pollution of water sources and agricultural production areas. To minimize these negative consequences of chemical pesticides for human health and the environment, the biological control of plant pests is a rationale alternative. Substitution of chemical pesticides by environmentally friendly biological pesticides that are highly specific will preserve biological diversity and help to maintain an ecologically clean environment. Therefore, there is now an urgent interest in developing effective alternative pesticides. This review will describe our current knowledge concerning antifungal activity of spore forming bacteria and metabolites produced by these microorganisms.

THE BIOLOGICAL PESTICIDES

Biological control offers an attractive alternative to the use of synthetic chemical pesticides. The use of the terms "biological control" ("biocontrol") or "biological pesticide" ("biopesticides") means a decrease in or the elimination of phytopatogenic fungi by microorganisms or substances produced by these bacteria. At present many microorganisms exhibit antagonistic effects on phytopatogenic fungi and either the living microorganisms, or their naturally produced compounds, can be used as biological pesticides for plant protection.

The antagonistic effect of microorganisms to fungi can result from a number of factors including the antibiosis and competition for space and nutrients between microbe and fungus. Compared to chemical pesticides, biopesticides possess some distinct advantages. First, they are usually less harmful than chemical pesticides. Second, biopesticides generally affect only the target pest and closely related organisms. Third, they are more biodegradable, thus avoiding potential pollution problems. Fourth, in most cases, biopesticides are less phytotoxic. Fifth, biopesticides are less expensive to develop than synthetic chemical pesticides. Finally, the use of biopesticides as a component of Integrated Pest Management programs can help to decrease the use of chemical-based pesticides.

At the same time we must recognize potential disadvantages of biopesticides. In some cases they are incompatible with the use of chemical based pesticides. In most cases biopesticides act slowly and suppress the growth of pest organisms rather than eliminate them. Biopesticides have a relatively critical application time and have limited lifespan once applied. Some efforts to apply live microorganisms for plant protection have been limited by their narrow range of activity, or by the instability of these organisms under field conditions involving relatively high temperatures, desiccation after spraying and inactivation by ultraviolet sunlight. In most cases biocontrol products are also more expensive to use.

Microorganisms with potential antifungal activity can be identified using a number of bioassay methods. These include those based on agar diffusion and using liquid-media assays and fungal spores or mycelia as the test inoculate.

BACILLUS AS AN ANTIFUNGAL AGENT

Presently certain spore-forming bacteria (*Bacillus* spp. includes *B. subtilis*, *B. megaterium*, *B. pumilus*, *B. amyloliquefaciens*, *B. cereus*, *B. mycoides*, *B. licheniformis*, *B. anthracis*, *B. thuringiensis*, *B. mojavensis*, *B. brevis*, *B. laterosporus*) have been shown to exhibit antifungal activity. The application of spore-forming bacteria and/or its metabolites, as agents in the protection of some agricultural plants, is promising. *Bacillus*-based fungicides can be used during the vegetative phase of plant growth, for seed and soil treatment and post-harvest protection of crops. *Bacillus* and its antifungal substances offer some advantages such as rapid killing of pathogenic fungi and suppression of resistant mutants. They are also straightforward to produce and to use, and do not need any special equipment for application and storage.

Bacillus spp. are known to produce a variety of antifungal antibiotics and metabolites. Many of these antifungal substances have been identified. Antibiotic production by the *Bacillus* strains has been found to depend on the components of the culture media, fermentation conditions, pH etc. (Leifert et al., 1995; Milner et al., 1995; Akpa et al., 2001). In some cases *Bacillus* produce several antifungal factors simultaneously (Katz and Demain, 1977; Feignier et al., 1995; Sandrin et al., 1990; Razafindralambo et al., 1993; Hiraoka et al., 1992; Tsuge et al., 1996; Schneider et al., 1999; Hiradate et al., 2002; Roongsawang et al., 2002). At present the use of different strains of *Bacillus* spp. as biocontrol agents of fungal plant pathogens have been described.

Antifungal activity of some *B. cereus*, *B. mycoides*, *B. anthracis*, *B. thuringiensis* strains is determined by production of two antibiotics- zwittermicin-A and kanosamine (Smith et al., 1993; Osburn et al., 1995; Milner et al., 1999 a; Stohl et al., 1999 a,b). Zwittermicin A is a water-soluble, acid-stable, linear aminopolyol molecule with a molecular weight of 396 Da (He et al., 1994; Stab et al., 1994). A second fungicidal metabolite produced by some strains of this group of bacteria have been identified as 3-amino-3-deoxy-D-glucose, also known as kanosamine (Milner et al., 1996 a). *Bacillus* spp. also produce a large number of peptide antifungal antibiotics which differ in their structure and mechanism of biosynthesis. Some bacillary antibiotics are gene-encoded and synthesized ribosomally, another group of antibiotics are synthesized non-ribosomally (Zuber *et al.*,1993; Huang et al., 1993; Kleinkauf and von Döhren. 1997; Duitman, et al., 1999; Lin et al., 1999; Steller et al., 1999).

Most antifungal peptides secreted by *Bacillus* have a molecular weight of less than 2000 Da. Antibiotics produced non-ribosomally are composed of 2 to 20 amino acids organized in a linear, cyclic or branched cyclic structure. There is a known group of cyclic lipopeptides produced by *B. subtilis* including iturin (Peypoux et al., 1978; Besson and Michel, 1987; Arrendale et al., 1988; Gueldner et al., 1988), bacillomycin (Besson et al., 1977; Peypoux et al., 1980; Peypoux et al.,1981; Mhammedi et al., 1982; Peypoux et al., 1984; Peypoux et al., 1985; Eshita et al., 1995; Moyne et al., 2001,b), bacillopeptin (Kajimura et al., 1995), mycosubtlin (Peypoux et al., 1986), surfactin (Arima et al., 1968; Kluge et al., 1988), fengycin (Vanittanakom et al., 1986), plipastatin (Tsuge et al., 1996), agrastatins (Heins et al., 2001), some of which demonstrate antifungal activity.

The *Bacillus* antifungal lipopeptides are members of a particular antibiotic class formed by the iturin (Maget-Dana and Peypoux, 1994), surfactin (Peypoux et al., 1999), and plipastatin-fengycin families (Umezawa, et al., 1986; Vanittanakom et al., 1986). These lipopeptides belong to different families based on their structural relationships. The general structure of these heteropeptides is a peptide cycle directly related to a fatty acid chain. Peptide antibiotics of the iturin family consist of a cyclic octapeptide with seven alpha-amino acids and one beta-amino acid with an aliphatic side chain. The molecular weight of this family of antibiotics ranges from 1030 to 1100 Da. Lipopeptide antibiotics fengycin and plipastatin consist of ten amino acids and a lipid attached to the N-terminal end of the molecule. They differ from iturin and surfactin by the presence of unusual amino acids such as ornithine and allo-threonine. Agrastatin structure is similar to plipastatin with the substitution of the some amino acid residues. The molecular weight of agrastatin A was determined as 1448 Da (Heins et al., 2001).

Surfactin is also a cyclic lipopeptide containing seven residues of D- and L-amino acids and one residue of a C-14 or C-15 β-hydroxy fatty acid linked with a heptapeptide by a lactone ring. An amino acid sequence of surfactin is completely different from the peptides of the iturin group (Kluge et al., 1988)**.** Most strains of *B. subtilis* produce antibiotics of the iturin family. The members of the iturin group consist of one β-amino fatty acid and 7 α-amino acids while the members of the surfactin and plipastatin-fengycin groups are composed of one β-hydroxy fatty acid and 7 and 10 α-amino acids, respectively. The presence of the β-amino fatty acid is the most important characteristic of the iturin A group and distinguishes this group from the other two groups. The length of the fatty acid chains can vary from C-14 to C-17 for iturins, from C-14 to C-18 for fengycins and from C-13 to C-16 for surfactins. It is of interest to note that lipopeptides modify the surface hydrophobicity of host bacteria (Ahimou et al., 2000).

The strain *B. licheniformis* M-4 produces a 3400 Da hydrophilic peptide fungicin M4 with narrow-spectrum antifungal activity which is composed of 34 amino acid residues of seven types. Fungicin M-4 is resistant to proteinase K, peptidases, trypsin, lipase and other enzyme (Lebbadi et al., 1994).

A hexadepsipeptide fusaricidin A containing a 15-guanidino-3-hydroxypentadecanoic acid as a side chain has been isolated from *B. polymyxa*. Fusaricidin A is active against fungi (*F. oxysporum*) and Gram-positive bacteria (Kajimura and Kaneda, 1996). Antifungal metabolites produced by *B. pumilus* inhibited mycelial growth of many species of *Aspergillus, Penicillium and Fusarium* (Munimbazi and Bullerman, 1998).

The cyclic decapeptide antibiotics in the tyrocidine family gramicidin S and loloatins from *Brevibacillus brevis and Brevibacillus laterosporus* (formerly *Bacillus brevis* Nagano *Bacillus laterosporus* (Shido et al., 1996) respectively and have been found to show fungicidal activity (Murray et al., 1986; Smirnova et al., 1993; Gerard et al., 1999; O'Donell, 1997). The loloatins and tyrocidines compose a family of antibiotic cyclic decapeptides containing four aromatic amino acid residues.

Most of the antifungal metabolites secreted into the surrounding medium have been shown to be resistant to heat, hydrolysis by various proteases, nucleases and other enzymes (Lebbadi et al., 1994; Nair et al., 2002). The cyclic lipopeptides of the Bacillomycin D family a1 and a2 isolated from *B. subtilis* have molecular weights of 1045 and 1059 Da respectively. Neither lipopeptides showed a decrease in their antifungal activity after being boiled or treated with lipase (Moyne et al., 2001a).

The antifungal metabolite AB1 from *B. mojavensis* has been characterized as a thermostable, protease-, phospholipase C- and alkali-resistant substance that is secreted into the surrounding medium (Nair et al., 2002). An antifungal metabolite isolated from *B. pumilus* (with no identified chemical structure) was shown to be heat-stable and resistant to hydrolysis by various proteases, peptidases and other enzymes (Munimbazi and Bullerman, 1998). Four lipopeptides have been isolated from *B. thuringiensis* ssp. *kurstaki*. These lipopeptides, kurstakins, have the same amino acid sequence, but different fatty acids (Hathout et al., 2003).

MODE OF ACTION OF ANTIFUNGAL AGENTS

Antifungal antibiotics differ in their mode of action. It is important to note that the efficacy of antifungal antibiotics can differ towards fungal spores and mycelia. Purified aminopolyol antibiotic zwittermicin A reversibly reduced elongation of germ tubes derived from cysts of fungi and kanosamine, caused swelling of the germ tube and delayed cyst germination (Silo-Suh et al., 1994; Silo-Suh et al., 1998; Shang et al., 1999). Antifungal peptide antibiotics are classified by their mode of antifungal action. There are three principal modes of fungicidal activity of lytic peptides. First is the formation of pores on membranes as a result of the aggregation of peptides and then leakage of vital ions from within the cell. Second is the disruption of membranes. Third is the penetration of the peptides through membranes and binding with some organic molecule(s).

Antibiotics of the iturin group affected fungal membrane surface tension, which caused pore formation and created *trans* membrane ion-conducting channels and the leakage of K^+ and other ions (Besson et al., 1984; Maget-Dana et al., 1985; Latoud et al., 1986; Thimon et al., 1992). Iturins were found to act upon the sterols present in the cytoplasmic membrane of the organism as well (Quentin et al., 1982). The lipopeptide bacillomycin L from *B. subtilis* belongs to the iturin family of antifungal agents and acts with a strict sterol-phospholipid dependence on biological membranes (Volpon et al., 1999). Unfortunately, bacillomycin L and iturin A are hemolytic, which may reduce their potential use as antifungal agents (Latoud et al., 1986). Iturin A and bacillomycin F cause the lysis of fungal conidiospores. The iturin-like compound produced by *B. subtilis* YM 10-20 increases the permeability of fungal spores and blocks germination of fungal spores (Chitarra et al., 2003).

Gramicidin forms cation-specific channels in membranes. Interestingly gramicidin S showed sporicidal action to conidia of *Botrytis cinerea* and is less inhibitory towards mycelial growth (Edwards and Seddon, 2001). As a result of the co-production of multiple antifungal agents *Bacillus* can reveal other modes of fungicidal action. Conidial germination of *Botrytis cinerea* has been inhibited by gramicidin S of *B. brevis*. The second antagonistic effect of *B. brevis* was determined by the production of a biosurfactant which reduced leaf wetness and inhibited the development of the pathogen (McHugh and Seddon, 2001).

APPLICATION OF BIOFUNGICIDES

While various biocontrol agents for control of pathogenic fungi are known, presently the field application of biofungicides is limited and they occupy a small part of the crop protection market. The biofungicide may be applied to the plant in different suitable forms, such as a spray, powder, granule, or as a liquid suspension. Some strains of *Bacillus* simultaneously produce a number of lipopeptides, among them surfactin. In these cases the use of supernatants and whole broth cultures results in high antifungal efficacy.

Fungal pathogens can be grouped according to the plant parts they mainly infect or where they sporulate. The biofungicide may be applied to the leaf, the stalk, the root (soil) and the seeds. In general, biofungicides are being used for the treatment of seeds or post–harvest treatment of crops (Klich et al., 1991; Klich et al., 1994; Leifert et al., 1997). The biological control of diseases commonly infecting plants in the root zone (rhizosphere) and the leaf zone (phylloplane) is less effective. In field conditions, most of biofungicides have been used for the control of root fungi (soil treatment) (Handelsman et al., 1990) but not for the plant foliar treatment (Asaka and Shoda, 1996).

The use of biofungicides for pathogen control in the leaf zone is similar to the use of chemical fungicides. The application of biofungicides for root pathogen control has some peculiarities. To control root fungi living microorganisms (spores) are used. The general mode of antagonistic action when using living microorganisms is colonization of the plant root system and competition with plant pathogens. For effective control these microorganisms have to possess some specific characteristics. First, spores of antagonistic microorganisms must be able to germinate rapidly for effective colonization of the root and competition for space and /or nutrients with specific microbial communities that occupy this niche. Second, in field conditions, the living microorganisms (spores) have to survive long-term, be resistant to ultraviolet light, and able to tolerate changes in temperature and humidity. Third, the living microorganisms (spores) must be able to reproduce in field conditions under fluctuating conditions. Hence, successful application of antagonistic bacteria introduced into soil for plant protection requires that a sufficiently high number of bacteria survive and continue antifungal activity for a long time. The possible

Table 1. Sensitivity of fungi to spore-forming isolates[1].

Fungi	Tested bacterial strains*	
	Brevibacillus laterosporus # 1	*Bacillus subtilis # 11*
Alternaria tenuis	+	+
Aspergillus niger	-	-
Fusarium nivalae	+	+
Fusarium solani	+	+
Fusarium graminearum	-	+
Phomopsis helianthi	+	-
Phytophthora infestans	+	+
Rhizoctonia solani	+	+
Sclerotinia sclerotiorum	+	+

[1]Zoospores of fungi (about 5×10^4) were spread onto nutrient agar. A cultural broth of antagonistic bacteria (50 µl) was introduced into 8-mm wells cut into the agar. The plates incubated at room temperature, and zones of inhibition were determined. Data reported are representative of 3-4 independent experiments.
* "+" means the sensitivity of fungi to tested strain; "-" means the resistance of fungi to tested strain.

solution to these problems is to use a mixture of microorganisms possessing antifungal activity. In this case, the bacterial "cocktail" will have a better chance for effective colonization and competition, survival and propagation. However, the large-scale production of this type of biofungicide (with separate fermentation of each bacterial strains and subsequent mixing) will increase the cost of application.

RESULTS FROM THIS LABORATORY

In the course of our screening program for new biological control agents we have collected spore forming microorganisms that were isolated from different samples: soil, water, insects and leaves. We have conducted the laboratory studies and small-scale field trials of several naturally occurring bacterial isolates. Our biocontrol studies were mainly focused on the protection of potato and wheat. Since the application of most live *Bacillus* or its metabolites has been limited by their narrow range of antifungal activity, we tried to isolate the bacterial antagonistic strains effective against specific kinds of phytopathogenic fungi. In this case we hoped to use these types of bacteria for the control of both the rhizosphere and the phylloplane infecting fungi.

POTATO PROTECTION

The most harmful fungi to potato are *Phytophthora infestans*, *Fusarium solani*, *Rhizoctonia solani* and *Alternaria tenuis*. By using the agar diffusion bioassay a few bacterial strains were preliminarily identified for their ability to suppress potato

Table 2. Treatment of potato tubers and planting[1].

Variant of tubers treatment	Infected tubers, %	Harvest of healthy tubers	
		Tons per ha	%
Untreated	27.9	21.8	100.0
B. laterosporus #1 -based formulation	20.6	26.8	119.3

[1]10 kg tubers were treated with 1 ml (5×10^{10} spores per ml) - *B. laterosporus #1* -based formulation dispersed into 30 ml water before planting. Each assay was conducted in duplicate in three independent experiments.

Table 3. The Effect of *B. laterosporus # 1* -based formulation on potato tubers cultivar "Zhukovsky" at storage[1].

Variant of treatment	Storage under controlled conditions, days			
	120		180	
	Infected tubers, %	Waste, %	Infected tubers, %	Waste, %
Untreated	16.0	10.2	16.0	10.2
B. laterosporus –based formulation, 80 g per ton /3L	8.5	5.5	8.5	5.5

[1]10 kg tubers were treated with 1 ml ($5x10^{10}$ spores per ml) of a *B. laterosporus #1* -based formulation dispersed into 30 ml water and then stored at 4° C at a defined humidity. Each assay was conducted in duplicate in six independent experiments during 3 growing season.

diseases caused by *Phytophthora infestans* and *Alternaria tenuis*. The isolates with fungicidal activity belonged to *Bacillus* (Azizbekyan et al., 2001; Azizbekyan et al., 2002). To broaden the range of use of these strains we evaluated their fungicidal potential. The zoospores or conidia of fungi were spread on potato agar and a culture broth of strains was poured into wells cut in the agar. The isolates had different spectrums of fungicidal activity. Two strains had the highest activity, strain #1, which belonged to *B. laterosporus,* and strain #11, which belonged to *B. subtilis* (Table 1).

In an assay for inhibition of germination, tubers were treated by mixing with *Rhizoctonia solani* spores (10^5) and *B. laterosporus* #1 spores (10^7) and then planted. Treatments were scored by a qualitative comparison with the germination of tubers that were treated with *Rhizoctonia solani* spores (10^5) only. Treatment with *B. laterosporus* #1 spores kept the tubers healthy and with active germination. Untreated tubers were visibly affected by the fungus and exhibited delayed germination. Fungicidal factors of *B. laterosporus* #1 were associated with the cell fraction. To determine the mechanism of fungicidal action we tested this culture on *Fusarium solani*. Treatment with *B. laterosporus* #1 delayed *Fusarium solani* conidia germination and germ tube growth rate was lower than in the control.

The strain *B. laterosporus # 1* was evaluated for its effect on the harvest of several cultivars of potato over five seasons at three Moscow regional field sites where phytopathogenic fungi–*Phytophthora infestans, Rhizoctonia solani, Fusarium solani, Alternaria tenuis* - were present naturally. Potato harvests treated with *B. laterosporus* #1 were significantly greater than the harvest of the untreated potato (22.8 and 21.9 tons per ha, respectively) and the percentage of infected tubers was less (9.8 and 12.0 %, respectively). This data suggests that fungal infection was a factor in reducing the harvest of untreated potato (Azizbekyan et al., 2001).

Hereinafter, for potato protection a *B. laterosporus* #1-based proprietary formulation was developed, which consists of *B. laterosporus* #1 spores and components that provide adhesion and spreading of biological fungicide. Potato tubers were treated with spore-based formulations prior to sowing. The treatments significantly improved the potato harvest in all growing seasons (Table 2).

To estimate the potential use of the biofungicide for control of fungi during post harvest storage, the potato tubers were treated by different concentrations of this formulation and stored in controlled conditions. The best results were obtained by using 80 g of biofungicide in 3l of water per ton of tubers (Table 3).

The fungicidal component was extracted from the culture pellet and purified using reversed-phase column chromatography and reversed-phase HPLC. According to Nuclear Magnetic Resonance and Mass Spectrometry data the isolated compound is a novel cyclic decapeptide. This peptide (of the tyrocidine family) is not secreted into the culture liquid and accumulates in cells and/or spores. However, since tyrocidines show activity against only Gram-positive bacteria we conclude that this new antibiotic appears to be the first member of a new cyclic decapeptide antibiotic family, active simultaneously against Gram-positive and Gram-negative bacteria. This new peptide antibiotic has a broad range of antifungal activity. During germination of *B. lateroposrus* #1 spores the peptide is released and inhibits phytopatogenic fungi. We believe that the biocontrol activity of *B. lateroposrus* #1 can be attributed mainly to production of this peptide, although some other compounds can show fungicidal effect too.

WHEAT PROTECTION

Wheat is obviously one of the most important crops in the world and fungal diseases of wheat are believed to decrease the global yield by as much as 30-50% annually. Fungal pathogens can infect different parts of wheat (head, stalk, seed). The seed-infecting fungal pathogens are primarily *Fusarium* sp. and *Aspergillus*, which both cause disease and produce mycotoxins, which contaminate flour. The fungal disease known as wheat scab (also known as 'head blight') is an increasing danger to

Table 4. Mode of fungicidal activity of *B. subtilis* #11 (24 hours of contact)[1].

Fungi	Germ tube growth, μm		Mode of activity (by light microscopy)
	Untreated	*B. subtilis #11* treated	
Fusarium graminearum	29.7	00.0	Lysis of conidia
Fusarium solani	67.0	00.0	Protoplasting of conidia
Phytophthora infestans	36.3	00.0	Zoospores don't germinated

[1]For the germ tube elongation assay 10^2 fungal spores were mixed with an equal volume of *B. subtilis* cultural broth in microtiter plates wells. After 24 h the wells were examined microscopically for spore germination. Treatments were scored by comparison with an untreated control.

Table 5. Additional winter wheat yield by use *B. subtilis*–based fungicide[1].

Variant of treatment	HARVEST, kg per ha			SUM	HARVEST	
	I field	II field	III field		kg per ha	%
Untreated	4280	5140	5590	15010	5000	100.0
B. subtilis#11–based formulation	4830	5990	6105	16925	5641	112.8

[1]Cultivar "Bezostaya 1" of winter wheat was selected for field trials. Experiments were conducted at field sites in the North Caucasus region. Untreated seeds and seeds treated by a *B. subtilis*-based liquid formulation (the cultural broth 1:50 as a fungicide factor was used) were planted with standard equipment. Each assay was conduct in duplicate in three separate fields.

sustainable wheat production in the world. The wheat scab pathogen is *Fusarium graminearum* and is a common problem in Europe, Asia, South Africa, and the Midwestern and Eastern America. In Russia about 5 million ha. are located in the scab risk area making wheat protection against scab an important national problem. Scab often causes reduction in weight and lowers the grade of the wheat. Low-grade wheat must be blended with higher-grade wheat; which results in higher costs. Therefore, scabby wheat often has to be used only as animal feed and to date there is no efficient means of scab control.

Biological control with antagonistic microorganisms has been shown to successfully suppress some fungal diseases of wheat. However, *Fusarium graminearum* Shwabe displays high resistance to the antagonistic microorganisms. During our screening program we found only three bacterial strains with high levels of wheat scab suppression. The isolate *B. subtilis* #11 inhibited the growth of *Fusarium graminearum* when fungi were spread on potato agar and an inoculate of the strain (grown in culture broth) was introduced into agar wells. Strain #11 exhibited a wide spectrum of antifungal activity. Both culture broth with cells and cell-free filtrates were active against various fungi. The cell-free filtrates may be used immediately, lyophilized, refrigerated or frozen for future use. The supernatant may be diluted, the minimal inhibiting concentration of the fungicide-containing supernatant for *Fusarium* spp. was at least 1:200.

To determine the mode of fungicidal activity we performed a microscopic study. Several modes of antagonistic effect of *Bacillus* spp. on fungi have been identified. Treatment by *Bacillus subtilis* #11 led to complete inhibition of germ tube growth and total lysis of conidia of *F. graminearum*, cell wall lysis of *F. solani* and complete inhibition of *Phytophthora infestans* zoospores germination (Table 4).

The fungicide-containing supernatant or whole broth culture can be formulated as an aqueous suspension. This preparation has been used for the treatment of seed and vegetative plants at different growth stages of the wheat. Antagonistic efficacy in controlling fungi has been confirmed in field studies. The harvest of the wheat treated with the antagonistic bacteria during growing stages was significantly higher (10-15%) than the harvest of the untreated wheat (Table 5).

The prevalence of fungi infection on wheat had been reduced 2-fold The results of post harvest seed monitoring confirmed that contamination of seed by phytopathogenic fungi was decreased by 3-4 times (Table 6). The antifungal compound from a liquid culture of *B. subtilis* #11 has been isolated. This compound was soluble in methanol and ethanol, and was insoluble in water, ether and acetone. Antifungal activity was not lost after treatment with heat (80⁰ C- 30 min), trypsin, chymotrypsin and pronase E proteolytic enzymes (10 μg per ml, 37⁰ C, 1 hour).

In conclusion, our data confirm that the use *of B. laterosporus*-based fungicides leads to an increased potato yield and a 2-fold decrease in the number of infected tubers under storage.

The treatment of wheat in the vegetative phase of growth by *B. subtilis*-based formulations has enabled an improved (up to 15-20%) harvest of wheat. Finally, the new *Bacillus* fungicides isolated in this laboratory have potential as effective agents for protection of plants to fungal pathogens.

REFERENCES

Ahimou, F., Jacques, P., and Deleu, M. 2000. Surfactin and iturin A on *Bacillus subtilis* surface hydrophobicity. Enzyme Microb. Technol. *27*, 749-754.

Akpa, E., Jacques, P., Wathelet, B., Paquot, M., Fuchs, R., Budzikiewicz, H., and Thonart, P. 2001. Influence of culture conditions on lipopeptide production by *Bacillus subtilis*. Appl. Biochem. Biotechnol. 91-*93*, 551-61.

Table 6. Determination of infected seeds and prevalence of Fusarium infected seed[1].

Variant of plant treatment	Harvest, %		Seed infected by fungi, %	Number of tested plants	*Fusarium* infected seed among infected seed,%				
	kg per ha	%			Repeated trials			Sum	Average
					I	II	III		
Untreated	4800	100.0	32	1091	27.7	57.2	65.1	150.0	50.0
B. subtilis #11-based formulation	5730	119.3	8	956	27.0	28.4	35.1	90.5	30.2

[1]Cultivar "Bezostaya 1" of winter wheat was selected for field trials. Experiments were conducted at field sites in the North Caucasus region. Untreated seeds and seeds treated by a *B. subtilis*-based liquid formulation (the cultural broth 1:80 as a fungicide factor was used) were planted with standard equipment. Number of infected seeds (seed monitoring) was scored after wheat was harvested. Among infected seeds *Fusarium* infected seed was determined.

Arima, K., Kakinuma, A., and Tamura, G. 1968. Surfactin, a crystalline peptide-lipid surfactant produced by *Bacillus subtilis*: isolation, characterization and its inhibition of fibrin clot formation. Biochim. Biophys. Res. Commun. *31*, 488-494.

Arrendale, R.F., Gueldner, R.C., Chortyk, O.T., and Crumley, F.G. 1988. Characterization of the amino acids of antifungal peptides from *B. subtilis* by cold on-column injection capillary GC/MS. J. Microbiol. Methods. *8*, 249-257.

Asaka, O., and Shoda, M. 1996. Biocontrol of *Rhizoctonia solani* damping-off of tomato with *Bacillus subtilis* RB14. Appl. Environ. Microbiol. *62*, 4081-4085.

Azizbekyan, R., Kuzin, A., Nikolaenko, M., Smirnova, T., and Shamshina, T. 2001. Biological control of plant fungal diseases. In Biological control of Fungal and Bacterial Plant Pathogens. IOBC wprs Bulletin,v.*24*, 93-95.

Azizbekyan, R., Kuzin, A., Kuznetsova, N., Nikolaenko, M., Smirnova, T., and Shamshina,T. 2002. Antagonistic effect of spore forming bacteria on phytopatogenic fungi. Abstracts 1-st International Congress "Biotechnology - state of the art and prospects of development" 136-137.

Besson, F. and Michel, G. 1987. Isolation and characterization of new iturins: iturin D and iturin E. J. Antibiotics. *40*, 437-442.

Besson, F., Peypoux, F., Michel, G., and Delcambe, L. 1977. The structure of bacillomycin L, an antibiotic from *Bacillus subtilis*. Eur. J. Biochem. 1;*77*, 61-67.

Besson, F., Peypoux, M., Quentin, J., and Michel, G. 1984. Action of antifungal peptolipids from *Bacillus subtilis* on the cell membrane of *Saccharomyces cerevisiae*. J. Antibiot. *37*, 172-177.

Chitarra, G.S., Breeuwer, P., Nout M.J.R., van Aelst, A.C., Rombouts, F.M., and Abee, T. 2003. An antifungal compound produced by *Bacillus subtilis* YM 10-20 inhibits germination of *Penicillium roqueforti* conidiospores. J. Appl. Microbiol. *94*, 159-168.

Cleveland, T.E, Dowd, P.F., Desjardins, A.E., Bhatnagar, D., Cotty, P.J. 2003. United States Department of Agriculture-Agricultural Research Service research on pre-harvest prevention of mycotoxins and mycotoxigenic fungi in US crops. Pest Manag Sci. *59*, 629-42.

Duitman, E. H., Hamoen, L.W., Rembold, M., Venema, G., Seitz, H., Saenger,W., Bernhard, F., Reinhardt, R., Schmidt, M., Ullrich, C., Stein, T., Leenders, F., and Vater, J. 1999. The mycosubtilin synthetase of *Bacillus subtilis* ATCC6633, a multifunctional hybrid between a peptide synthetase, an amino transferase, and a fatty acid synthase. Proc. Natl. Acad. Sci. USA. *96*, 13294-13299.

Feignier, C., Besson, F., and Michel, G. 1995. Studies on lipopeptide biosynthesis by *Bacillus subtilis*: isolation and characterization of iturin-, surfactin+ mutants. FEMS Microbiol. Lett. *15*, 127, 11-15.

Gerard, J.M., Haden, P., Kelly, M.T., and Andersen, R.J.1999. Loloatins A-D, cyclic decapeptide antibiotics produced in culture by a tropical marine bacterium. J. Nat. Prod. *62*, 80-85.

Gueldner, R.C., Reuilly, C.C., Pusey, P.L., Costello, E.C., Arrendale, R.F., Cox, R.H., Himmelsbach, D.S., Crumley, F.G., and Cutler, H. 1988. Isolation and identification of iturins as antifungal peptides in the biological control of peach brown rot with *Bacillus subtilis*. J. Agricultural and Food Chemistry. *36*, 366-370.

Handelsman, J., Raffel, S., Mester, E.H., Wunderlich, L., and Grau, C.R.1990. Biological control of damping-off of alfalfa seedlings with *Bacillus cereus* UW85. Appl. Environ. Microbiol. *56*, 713-718.

Hathout, Y., Ho, Y-P., Ryzhov, V., Demirev, P., and Fenselay, C. 2000. Kurstakins: a new class of lipopeptides isolated from *Bacillus thuringiensis*. J. Nat. Prod. *63*, 1492-1496.

He, H., Silo-Suh, L. A., Clardy, J., and Handelsman, J. 1994. Zwittermicin A, an antifungal and plant protection agent from *Bacillus cereus*. Tetrahedron Lett. *35*, 2499-2502.

Heins S.D., Manker D.C., Jimenez D.R., McCoy, R.J., Marrone, P.G., and Orjala J.E.2001. Strain of *bacillus* for controlling plant diseases and corn rootworm. US Patent, No. 6239103.

Hiradate, S., Yoshida, S., Sugie, H., Yada, H., and Fujii, Y. 2002. Mulberry anthracnose antagonists (iturins) produced by *Bacillus amyloliquefaciens* RC-2. Photochemistry. *61*, 693-698.

Hiraoka, H., Asaka, O., Ano, T., and Shoda, M. 1992. Characterization of *Bacillus subtilis* RB14, coproducer of peptide antibiotics iturin A and surfactin. J. Gen. Appl. Microbiol. *38*, 635-640.

Huang, C. C., Ano, T., and Shoda, M. 1993. Nucleotide sequence and characteristics of the gene, *lpa-14*, responsible for biosynthesis of the lipopeptide antibiotics iturin A and surfactin from *Bacillus subtilis* RB14. J. Ferment. Bioeng. *76*, 445-450.

Edwards, S.G., and Seddon, B. 2001. Mode of antagonism of *Brevibacillus brevis* against *Botrytis cinerea in vitro*. J. Appl.Micrbiol, *9*, 652-659.

Eshita, S.M., Roberto, N.H., Beale, J.M., Mayima, B.M. and Workman, R.F. 1995. Antibiotic, Bacillomycin Lc, a new antibiotic of the iturin group: isolations, structures, and antifungal activities of the congeners. J. Antibiot. (Tokyo). *48*, 1240-1247.

Isogai, I., Takayama, S., Murakoshi, S., and Suzuki, A. 1982. Structures of β-amino acids in antibiotics iturin A. Tetrahedron Lett. *23*, 3065-3068

Katz, E., and Demain, A.L.1977. The peptide antibiotics of *Bacillus*: chemistry, biogenesis, and possible functions. Bacteriol. Rev. 41, 449-474.

Kajimura, Y., Sugiyama, M., and Kaneda, M.1995. Bacillopeptins, new cyclic lipopeptide antibiotics from *Bacillus subtilis* FR-2. J. Antibiot. (Tokyo). *48*, 1095-1103.

Kajimura, Y., and Kaneda, M. 1996. Fusaricidin A, a new depsipeptide antibiotic produced by *Bacillus polymyxa* KT-8. Taxonomy, fermentation, isolation, structure elucidation and biological activity. J. Antibiot.(Tokyo). *49*, 129-135.

Kleinkauf, H., von Dohren, H. 1997. Applications of peptide synthetases in the synthesis of peptide analogues. Acta Biochim. Pol.*44*, 839-847.

Klich, M.A., Arthur, K.S., Lax, A.R., and Bland, J.M. 1994. Iturin A: a potential new fungicide for stored grains. Mycopathologia. *127*, 123-127.

Klich, M.A., Lax, A.R., and Bland, J.M.1991. Inhibition of some mycotoxigenic fungi by iturin A, a peptidolipid produced by *Bacillus subtilis*. Mycopathologia. *116*, 77-80.

Kluge, B., Vater, J., Salnikow, J. and Eckart, K. 1988. Studies on the biosynthesis of surfactin, a lipopeptide antibiotic from *Bacillus subtilis* ATCC 21332. FEBS Lett. *11*, 107-110.

Latoud, C., Peypoux, F., Michel, G., Genet, R., and Morgat, J. L. 1986. Interactions of antibiotics of the iturin group with human erythrocytes. Biochim. Biophys. Acta. *856*, 526-535.

Lebbadi, M., Galvez, A., Maqueda, M., Martinez-Bueno, M., and Valdivia, E. 1994. Fungicin M4, a narrow spectrum peptide antibiotic from *Bacillus licheniformis* M-4. J. Appl. Bacteriol. *77*, 49-53.

Leifert, C., Li, H., Chidburee, S., Hampson, S., Workman, S., Sigee, D., Epton, H.A., and Harbour, A. 1995. Antibiotic production and biocontrol activity by *Bacillus subtilis* CL27 and *Bacillus pumilus* CL35. J. Appl. Bacteriol. *78*, 97-108.

Leifert, C., Epton, H., and Sigee, D. 1997. Antibiotics for biological control of post harvest diseases U.S. Pat. No. 5597565.

Lin, T.S., Chen, C.L., Chang, L.K., Tschen, J.S., and Liu, S.T. 1999. Functional and transcriptional analyses of a fengycin synthetase gene, *fenC*, from *Bacillus subtilis*. J. Bacteriol. *181*, 5060-5067.

Maget-Dana, R., Peypoux, F.1994. Iurins, a special class of pore-forming lipopeptides: biological and physicochemical properties. Toxicology. *87*, 151-174.

Maget-Dana, R., Ptak, M., Peypoux, F., and Michel, G.1985. Pore-forming properties of iturin A, a lipopeptide antibiotic. Biochim. Biophys. Acta. *815*, 405-409.

McHugh,R., and Seddon, B.2001. Mode of action of *Brevibacillus brevis*: biocontrol and biorational control. In Biological control of Fungal and Bacterial Plant Pathogens. IOBC wprs Bulletin,v.*24*, 17-20.

Mhammedi, A., Peypoux, F., Besson, F., and Michel, G. 1982. Bacillomycin F, a new antibiotic of iturin group: isolation and characterization. J. Antibiot. (Tokyo). *35*, 306-311.

Milner, J.L., Raffel, S.J., Lethbridge, B.J., Handelsman, J. 1995. Culture conditions that influence accumulation of zwittermicin A by *Bacillus cereus* UW85. Appl. Microbiol. Biotechnol. *43*, 685-91.

Milner, J.L., Silo-Suh, L., Lee, J.C., He, H., Clardy, J., and Handelsman, J. 1996,a. Production of kanosamine by *Bacillus cereus* UW85. Appl. Environ. Microbiol. *62*, 3061-3065.

Milner, J.L., Stohl, E.A., and Handelsman, J. 1996,b. Zwittermicin A resistance gene from *Bacillus cereus*. J. Bacteriol. *178*, 4266-4272.

Moyne, A.L., Cleveland, T.E., and Tuzun, S. 2001,a. Small peptides with antipathogenic activity, treated plants and methods for treating same. US Patent, No.6183736.

Moyne, A.L., Shelby, R., Cleveland, T.E., and Tuzun, S. 2001,b. Bacillomycin D: an iturin with antifungal activity against *Aspergillus flavus*. J. Appl. Microbiol. *90*, 622-629.

Munimbazi, C., and Bullerman, L.B. 1998. Isolation and partial characterization of antifungal metabolites of *Bacillus pumilus*. J. Appl. Microbiol. *84*, 959-968.

Murray, T., Leighton, F.C., Seddon, B. 1986. Inhibition of fungal spore germination by gramicidin S and its potential use as a biocontrol against fungal plant pathogens. Letters Appl. Microbiol. *3*, 5-7.

Nair, J.R., Singh,G., and Sekar, V. 2002. Isolation and characterization of a novel *Bacillus* strain from coffee phyllosphere showing antifungal activity. J. Appl. Microbiol. *93*, 772-780.

O'Donell, B. 1997. Treatment of soil and plants with a composition containing *Bacillus laterosporus* . US Patent No. 5702701.

Osburn, R.M., Milner, J.L., Oplinger, E.S, Smith,R.S., and Handelsman, J. 1995. Effect of *Bacillus cereus* UW85 on the yield of soybean at two field sites in Wisconsin. Plant Dis. *79*, 551-556.

Peypoux, F., Besson, F., Michel, G., Lenzen, C., Dierickx, L., and Delcambe, L. 1980. Characterization of a new antibiotic of iturin group: bacillomycin D. J. Antibiot. (Tokyo). *33*, 1146-1149.

Peypoux, F., Besson, F., Michel, G., and Delcambe, L. 1981. Structure of bacillomycin D, a new antibiotic of the iturin group. Eur. J. Biochem. *118*, 323-327

Peypoux, F., Bonmatin, J. M., and Wallach, J. 1999. Recent trends in the biochemistry of surfactin. Appl. Microbiol. Biotechnol. *51*, 553-563.

Peypoux, F., Guinand, M., Michel, G., Delcambe, L., Das, B.C., and Lederer, E. 1978. Structure of iturin A, peptidolipid antibiotic from *Bacillus subtilis*. Biochemistry.*17*, 3992-3996.

Peypoux, F., Marion, D., Maget-Dana, R., Ptak, M., Das, B.C., and Michel, G. 1985. Structure of bacillomycin F, a new peptidolipid antibiotic of the iturin group. Eur. J. Biochem. *153*, 335-340.

Peypoux, F., Pommier, M.T., Das, B.C., Besson, F., Delcambe, L., and Michel, G. 1984. Structures of bacillomycin D and bacillomycin L peptidolipid antibiotics from *Bacillus subtilis*. J.Antibiot. *37*, 1600-1604.

Peypoux, F., Pommier, M.T., Marion, D., Ptak, M., Das, B.C., and Michel, G. 1986. Revised structure of mycosubtilin, a peptidolipid antibiotic from *Bacillus subtilis*. J. Antibiot. *39*, 636-641.

Quentin, M.J., Besson, F., Peypoux, F., and Michel, G. 1982. Action of peptidolipidic antibiotics of the iturin group on erythrocytes. Effect of some lipids on hemolysis. Biochim. Biophys.Acta.*684*, 207-211.

Razafindralambo, H., Paquot, M., Hbid, C., Jacques, P., Destain, J., and Thonart, P.. 1993. Purification of antifungal lipopeptides by reversed-phase high-performance liquid chromatography. J Chromatogr. *639*, 81-85.

Razafindralambo, H., Popineau, Y., Deleu, M., Hbid, C., Jacques, P., Thonart, P., and Paquot, M. 1998. Foaming properties of lipopeptides produced by *Bacillus subtilis*: effect of lipid and peptide structural attributes. J. Agric. Food Chem. *46*, 911-916.

Roongsawang, N., Thaniyavarn, J., Thaniyavarn, S., Kameyama, T., Haruki, M., Imanaka,T., Morikawa, M., and Kanaya, S. 2002. Isolation and characterization of a halotolerant *Bacillus subtilis* BBK-1 which produces three kinds of lipopeptides: bacillomycin L, plipastatin, and surfactin. Extremophiles. *6*, 499-506.

Sandrin, C., Peypoux, F., and Michel, G. 1990. Coproduction of surfactin and iturin A, lipopeptides with surfactant and antifungal properties, by *Bacillus* subtilis. Biotechnol. Appl. Biochem. *12*, 370-375.

Shang, H., Chen, J., Handelsman, J., and Goodman, R.M. 1999. Behavior of *Pythium torulosum* zoospores during their interaction with tobacco roots and *Bacillus cereus*. Curr Microbiol. *38*, 199-204.

Shido, O., Takagi, H., Kadowaki, K., and Komagata, K.1996. Proposal of two new genera, *Brevibacillus* gen nov and *Aneurinibacillus* gen nov. International J. Systemic Bacteriol. *46*, 939-946.

Silo-Suh, L.A, Lethbridge, B.J., Raffel, S.J., He, H., Clardy, J., and Handelsman, J. 1994. Biological activities of two fungistatic antibiotics produced by *Bacillus cereus* UW85. Appl. Environ. Microbiol. *60*, 2023-2030.

Silo-Suh, L.A., Stabb, E.V.S., Raffel, S.J., and Handelsman, J. 1998. Target range of zwittermicin A, an aminopolyol antibiotic from *Bacillus cereus*. Curr. Microbiol. *37*, 6-11.

Smirnova, T.A., Shamshina, T.N., Konstantinova, G.E., Ganushkina L.A., Kuznetsova, N.I., Minenkova, I.B., Nikolaenko, M.A., and Azizbekyan, R.R. 1993. Strain of *Bacillus laterosporus* with plural biological activity. Biotechnologiya (Russia).*1*, 11-16.

Smith, K.P., Havey, M.J., and Handelsman, J.1993. Suppression of cottony leak of cucumber with *Bacillus cereus* strain UW85. Plant Dis. *77*, 139-142.

Schneider, J., Taraz, K., Budzikiewicz, H., Deleu, M., Thonart, P., and Jacques, P. 1999. The structure of two fengycins from *Bacillus subtilis* S499. Z. Naturforsch. *54*, 859-865.

Stabb, E.V., Jacobson, L.M., and Handelsman J. 1994. Zwittermicin A – producing strains of *Bacillus cereus* from diverse soils. Appl. Environ. Microbiol. *60*, 4404 -4412.

Steller, S., Vollenbroich, D., Leenders, F., Stein, T., Conrad, B., Hofemeister, J., Jacques, P., Thonart, P., and Vater, J. 1999. Structural and functional organization of the fengycin synthetase multienzyme system from *Bacillus subtilis* b213 and A1/3.Chem Biol. *6*, 31-41.

Stohl, E.A., Brady, S.F., Clardy, J., and Handelsman, J. 1999,a. ZmaR, a novel and widespread antibiotic resistance determinant that acetylates zwittermicin A. J. Bacteriol. *181*, 5455-5460.

Stohl, E.A., Milner, J.L., and Handelsman, J. 1999,b. Zwittermicin A biosynthetic cluster. Gene. *237*, 403-411.

Thimon, L., Peypoux, F., Maget-Dana, R., and Michel,G. 1992. Surface-active properties of antifungal lipopeptides produced by *Bacillus subtilis*. J. Am. Oil Chem. Soc. *69*, 92-93.

Tsuge, K., Ano, T., and Shoda, M. 1996. Isolation of a gene essential for biosynthesis of the lipopeptide antibiotics plipastatin B1 and surfactin in *Bacillus subtilis* YB8. Arch. Microbiol. *165*, 243-251.

Umezawa, H., Aoyagi, T., Nishikiori, T., Okuyama, A., Yamagishi, Y., Hamada, M., and Takeuchi, T. 1986. Plipastatins: new inhibitors of phospholipase A$_2$, produced by *Bacillus cereus* BMG202-fF67. I. Taxonomy, production, isolation and preliminary characterization. J. Antibiot. *39*, 737-744.

Vanittanakom, N., Loeffler, W., Koch, U., and Jung, G. 1986.Fengycin- a novel antifungal lipopeptide antibiotic produced by *Bacillus subtilis* F-29-3. J. Antibiot. (Tokyo). *9*, 888-901.

Volpon, L., Besson, F., and Lancelin, J.,M. 1999. NMR structure of active and inactive forms of the sterol-dependent antifungal antibiotic bacillomycin L. *Eur J. Biochem. 264*, 200-210.

Zuber, P., Nakano, M.M., and Mahariel, M.A. 1993. Peptide antibiotics. In: *Bacillus subtilis* and Other Gram-Positive Bacteria. A.L. Sonenshein, ed. American Society of Microbiology, Washington, D.C. p. 897-916.

Index